D1161512

RHEOLOGY

Volume 3: Applications

RHEOLOGY

Volume 3: Applications

Edited by

Giovanni Astarita
Giuseppe Marrucci
Luigi Nicolais

University of Naples
Naples, Italy

PLENUM PRESS · NEW YORK AND LONDON

Library of Congress Cataloging in Publication Data

International Congress on Rheology, 8th, Naples, 1980.
  Rheology.

  Proceedings of the International Congress on Rheology; 8th, 1980)
  Includes indexes.
  1. Rheology – Congresses. 2. Polymers and polymerization – Congresses.
3. Fluid dyanmics – Congresses. 4. Suspensions (Chemistry) – Congresses. I.
Astarita, Giovanni. II. Marrucci, G. III. Nicolais, Luigi. IV. Title. V. Series:
International Congress on Rheology. Proceedings; 8th, 1980.
  QC189.I52 8th, 1980 [QCl89.5.A1] 531'.11s [531'.11]
  ISBN 0-306-40467-2 (v. 3)   80-16929

Proceedings of the Eighth International Congress of Rheology, held in Naples, Italy,
September 1–5, 1980, published in three parts of which this is Volume 3.

© 1980 Plenum Press, New York
A Division of Plenum Publishing Corporation
227 West 17th Street, New York, N.Y. 10011

Printed in the United States of America

VIII INTERNATIONAL CONGRESS ON RHEOLOGY

Naples, September 1–5, 1980

HONORARY COMMITTEE

PRESIDENT

Prof. J. Kubat, *President, International Committee on Rheology*

MEMBERS

Dr. G. Ajroldi, *Past President, Italian Society of Rheology*
Prof. U. L. Businaro, *Director of Research, FIAT*
Dr. E. Cernia, *Director, Assoreni*
Prof. C. Ciliberto, *Vice-President, C.N.R.*
Prof. G. Cuomo, *Rector, University of Naples*
Dr. A. Del Piero, *Director, Tourism Bureau, Town of Naples*
Dr. D. Deuringer, *Director RAI, Radio-Television Network, Naples*
Prof. F. Gasparini, *Dean, Engineering School, University of Naples*
Porf. L. Malatesta, *President, Chemistry Committee, C.N.R.*
Prof. L. Massimilla, *Past Dean, Engineering School, University of Naples*
Prof. A. B. Metzner, *Fletcher Brown Professor, University of Delaware*
Prof. N. Polese, *President, University Social Services, Naples*
Prof. M. Silvestri, *President, Technical Committee, C.N.R.*
Prof. N. W. Tschoegl, *Secretary, International Committee on Rheology*
Sen. M. Valenzi, *Mayor, Town of Naples*
Prof. A. Valvassori, *Director, Istituto Donegani*

ORGANIZING COMMITTEE

Prof. G. Astarita, *President*
Prof. G. Marrucci
Prof. L. Nicolais, *Secretary*

ACKNOWLEDGMENTS

Support from the following Institutions in gratefully acknowledged:

Alitalia, Linee Aeree Italiane, Rappresentanza di Napoli

Assoreni

Azienda Autonoma di Soggiorno, Cura e Turismo di Napoli

Azienda Autonoma di Soggiorno, Cura e Turismo di Sorrento

Centro Ricerche FIAT, S.p.A.

Comitato per la Chimica del Consiglio Nazionale delle Ricerche

Comitato Tecnologico del Consiglio Nazionale delle Ricerche

Istituto Donegani S.p.A.

Opera Universitaria, Napoli

RAI, Radiotelevisione Italiana, Sede Regionale per la Campania

Società Italiana di Reologia

U.S. Air Force

Università di Napoli

CONTENTS OF THE VOLUMES

VOLUME 1: PRINCIPLES

Invited Lectures (IL)
Theory (TH)

VOLUME 2: FLUIDS

Fluid Dynamics (FD)
Rheometry (RH)
Polymer Solutions (PS)
Polymer Melts (ML)
Suspensions (SS)

VOLUME 3: APPLICATIONS

Polymer Processing (PC)
Rubber (RB)
Polymer Solids (SD)
Biorheology (BR)
Miscellaneous (MS)
Late Papers (LP)

PREFACE

At the VIIth International Congress on Rheology, which was held in Goteborg in 1976, Proceedings were for the first time printed in advance and distributed to all participants at the time of the Congress. Although of course we Italians would be foolish to even try to emulate our Swedish friends as far as efficiency of organization is concerned, we decided at the very beginning that, as far as the Proceedings were concerned, the VIIIth International Congress on Rheology in Naples would follow the standards of timeliness set by the Swedish Society of Rheology. This book is the result we have obtained. We wish to acknowledge the cooperation of Plenum Press in producing it within the very tight time schedule available.

Every four years, the International Congress on Rheology represents the focal point where all rheologists meet, and the state of the art is brought up to date for everybody interested; the Proceedings represent the written record of these milestones of scientific progress in rheology. We have tried to make use of the traditions of having invited lectures, and of leaving to the organizing committee the freedom to choose the lecturers as they see fit, in order to collect a group of invited lectures which gives as broad as possible a landscape of the state of the art in every relevant area of rheology. The seventeen invited lectures are collected in the first volume of the proceedings. We wish to express our thanks, for agreeing to prepare these lectures on subjects suggested by ourselves, and for the effort to do so in the scholarly and elegant way that the reader will appreciate, to all the invited lectures: R.B.Bird, D.V.Boger, B.D.Coleman, J.M.Dealy, P.De Gennes, C.D.Denson, H.Janeschitz-Kriegl, A.Y.Malkin, R.A. Mashelkar, S.Onogi, C.J.S.Petrie, R.F.Schwarzl, J.Silberberg, K.Te Nijenhuis, C.A.Truesdell, K.Walters, K.Wichterle.

As for the organization of the Congress itself, at the time
of writing it is still in the future, and we can only hope that it
will work out smoothly. If it does, a great deal of merit will be
due to the people who have agreed to act as Chairmen of the
individual sessions, and we wish to acknowledge here their help:
J.J.Benbow, B.Bernstein, H.C.Booij, B.Caswell, Y.Chen, M.Crochet,
P.K.Currie, M.M.Denn, A.T.Di Benedetto, H.Giesekus, J.C.Halpin,
A.Hoffmann, Y.Ivanov, L.P.B.Janssen, T.E.R.Jones, W.M.Jones,
H.Kambe, J.L.Kardos, E.A.Kearsley, J.Klein, K.Kirschke, S.L.Koh,
J.Kubat, R.F.Landel, R.L.Laurence, G.L.Leal, C.Marco, J.Meissner,
B.Mena, A.B.Metzner, S.Middleman, Y.F.Missirlis, S.L.Passman,
S.T.T.Peng, J.R.A.Pearson, R.S.Porter, P.Quemada, A.Ram, C.K.Rha,
W.R.Schowalter, J.C.Seferis, C.L.Sieglaff, S.S.Sternstein, R.I.
Tanner, N.Tschoegl, J.Vlachopoulos, J.L.White, C.Wolff, L.J.Zapas.

The contributed papers have been grouped in eleven subject
areas: theory; fluid dynamics; rheometry; polymer solutions;
polymer melts; suspensions; polymer processing; rubber; polymeric
solids; biorheology; miscellaneous. Of these, the first one (theory)
has been included in the first volume together with the invited
lectures; the next five, which all deal with fluid-like materials,
have been included in the second volume, and the last five have
benn included in the third volume. Categorizations such as these
invariably have a degree of arbitrariness, and borderline cases
where a paper could equally well have been included in two different
categories do exist; we hope the subject index is detailed enough
to guide the reader to any paper which may be placed in a category
unexpected from the reader's viewpoint.

Rheology is not synonymous with Polymer Science, yet sometimes
it almost seems to be: papers dealing with polymeric materials
represent the great majority of the content of this book. Regret-
ting that not enough work is being done on the rheology of non-
polymeric materials is an exercise in futility; yet this does
seem an appropriate time for reiterating this often repeated
consideration.

We would like to have a long list of people whose help in
organizing the Congress we would need to acknowledge here.
Unfortunately, there are no entries to such a list, with the
exception of young coworkers and students who have helped before
the time of writing, and will help after it. To these we extend
our sincere and warmest thanks; their unselfishness is further
confirmed by our inability to report their names. With this

exception, we have organized the technical part of the Congress
singlehandedly, and we state this not because we are proud of it,
but only as a partial excuse for any mishaps that may, and
unfortunately will, take place.

We regret that only the abstract of some papers appear in the
Proceedings. The mail service being what it is, some papers did
not reach us in time for inclusion in the Proceedings; others
reached us in time, but were not prepared in the recommended form.
Also, some abstracts reached us so late that there was no time
left for preparation of the final paper.

At the very end of the third volume, we have collected whatever information (title, abstract, or complete paper) we could on
contributed papers the very existence of which became known to
us after we had prepared the Table of Contents, Author Index and
Subject Index. Again, we apologize for this.

Finally, we want to express our most sincere wishes of
success to whoever will be in charge of organizing the IXth
International Congress in 1984. Based on our own experience, and
in view of the Orwellian overtones of the date, we cannot avoid
being pleased at the thought that, whoever it is, it will not
be us.

Naples, 1st March 1980   Gianni Astarita
                                                    Giuseppe Marrucci
                                                    Luigi Nicolais

CONTENTS

VOLUME 3 - APPLICATIONS

NOTE: Papers identified by the ° sign were not received in time for inclusion in this book, and only the abstract is included.

Preface

POLYMER PROCESSING

POLYMER PROCESSING

NON-ISOTHERMAL FLOWS OF VISCOELASTIC FLUIDS

R. K. Gupta and A. B. Metzner

University of Delaware
Newark, Delaware 19711

INTRODUCTION

During the last two decades, considerable progress has been made in developing stress-deformation rate relationships for viscoelastic liquids: the ground rules pertaining to an admissible equation of state were set forth by Coleman and Noll (1961) and Coleman (1964). According to their theory, the extra stress in a material is assumed to be a functional of the history of the deformation process that the material has experienced as well as of the history of the temperature. Unfortunately, almost all existing theories ignore temperature -- treating it as a parameter and not as a variable. In such analyses, once a constitutive equation has been derived for isothermal conditions allowance is made for varying temperature conditions by incorporating within the equation the experimentally-observed time-temperature superposition, or other equivalent empirical observations. While this ensures the correct behavior at different fixed temperature levels, it need not give the correct results for the transition process between two temperatures, and should not if the temperature history is important.

In this work the simplest model of polymer molecules in solution, namely, the elastic dumbbell model, is employed. A key assumption, that of a constant mean square end-to-end distance at equilibrium, is relaxed, since this variable is known experimentally to be temperature-dependent (Flory, 1953). With this modification, dramatically new results, in accord with experimental observations, are predicted for the idealized experiment of stretching of a polymer sample at a constant stretch rate but under non-isothermal conditions.

3

DEVELOPMENT OF THE EQUATION

If one uses the equivalent chain model, each isolated polymer molecule can be divided into a number of statistical segments whose distribution function for the orientation and length is a Gaussian one (Treloar, 1975). Further, for the molecule as a whole the mean-square end-to-end distance at equilibrium, $<L^2>$, is independent of temperature, and this leads to the prediction (see Bueche, 1962 or Schultz, 1974) that the polymer may be idealized as a spring whose spring constant increases linearly with the absolute temperature. Thus,

$$F = -2kT\beta R \tag{1}$$

where F is the spring force, k is Boltzmann's constant, T is temperature, $\beta = 3/(2<L^2>)$ and R is the end-to-end vector.

When a polymer molecule is dissolved in a solvent, however, the equilibrium end-to-end distance increases, often quite drastically, as the temperature of the solution increases (Flory, 1953). Hence, $<L^2>$ must be taken to depend on temperature; this makes β variable and a decreasing function of temperature.

The position of one bead of the dumbbell relative to the other can be described by means of a distribution function $\phi(R)$ such that $\phi(R)dR$ represents the probability that the second bead will be found at a distance R to $R + dR$ away from the first bead. If one neglects the inertia of the beads, a force balance on each bead yields (Bird et al., 1971):

$$\zeta(L \cdot R - \dot{R}) - kT\partial(\ln\phi)/\partial R + F = 0 \tag{2}$$

In the above, ζ is the drag coefficient of the solvent on the molecule and L the velocity gradient. The first term arises due to the drag while the second term is due to Brownian motion. Since the beads are neither created nor destroyed, conservation of beads leads to:

$$D\phi/Dt = -\text{div}(\dot{R}\phi) \tag{3}$$

Also, the polymer contribution to the stress tensor is given by (Bird et al., 1971):

$$P = 2ckT\beta<RR> - ckT \; 1 \tag{4}$$

where c is the polymer concentration. The first term on the right hand side results from the spring in the dumbbell while the second term is the contribution of the momentum of the beads.

Manipulating Eqs. 2-4 can be shown to yield (Gupta, 1980):

$$P + \theta \delta P/\delta t = 2ckT\theta D + \theta PD(\ln T\beta)/Dt + \theta ckT \, D(\ln\beta)/Dt \; 1 \qquad (5)$$

where $\quad \theta = \zeta/(4kT\beta)$

and $\quad \delta P/\delta t = DP/Dt - L\cdot P - P\cdot L^T$

If the temperature is constant, the last two terms of Eq. 5 are identically zero, and one obtains the upper convected Maxwell model (White and Metzner, 1963). One can, therefore, identify the coefficient of $D = (L+L^T)/2$ as 2μ, in which μ denotes the viscosity function.

One knows that β is normally a decreasing function of temperature; a reasonable form of the temperature dependence of β is:

$$\beta = \nu T^{-(B+1)} \qquad (6)$$

Any other functional form would yield essentially the same result. With the help of (6) one may rewrite (5) as:

$$P[1+\theta B \, D(\ell nT)/Dt] + \theta \delta P/\delta t = \mu\{2D - (B+1)[D(\ell nT)/Dt]1\} \qquad (7)$$

Equation (7) is the final result of this analysis: it is the usual and useful contravariant Maxwell model generalized to account for the transient non-isothermal effects frequently encountered in polymer processing operations. Two new terms $[\theta BD(\ln T)/Dt]P$ and $\mu(B+1)[D(\ln T)/Dt]1$, are found to arise. The major consequence of the parameter B in Eq. (7) is to stiffen the material during cooling and soften it while heating provided that the imposed temperature changes are rapid enough. In particular, during cooling, the stress generated due to deformation is predicted to be greater than that obtained with the use of the time-temperature superposition principle.

For polymer solutions the single new parameter B can be determined entirely from a knowledge of the temperature dependence of the solvent viscosity and the solution viscosity (Gupta, 1980). Thus, at least in principle, no new adjustable parameters arise. For melts, however, B becomes an empirical quantity whose value is determined from non-isothermal experimental data after the temperature dependence of μ and θ has been defined from isothermal measurements over a range of temperatures.

APPLICATION OF THE EQUATION TO NON-ISOTHERMAL UNIAXIAL EXTENSION EXPERIMENTS

We considered the following six cases with the stretch rate d_{11} being constant at a level of 0.0528 sec.^{-1} in all instances:

a) $R' = 0$, temperature = 120°C (isothermal)

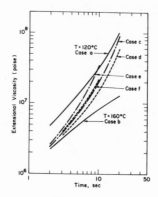

Fig. 1.   Illustrative calculations of stress development, expressed
          as the instantaneous extensional viscosity, for uniaxial
          extension at a constant stretch rate of 0.053 sec^{-1}.

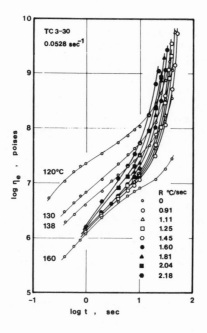

Fig. 2   Stress development (as above).   Reproduced with permission
         from Matsumoto and Bogue (1977).

    b) $R' = 0$, temperature = 160°C (isothermal)
    c) $R' = -2$°C/sec, initial temperature = 160°C, B = 5
    d) $R' = -2$°C/sec, initial temperature = 160°C, B = 0
    e) $R' = -4$°C/sec, initial temperature = 160°C, B = 5
    f) $R' = -4$°C/sec, initial temperature = 160°C, B = 0

Herein R' denotes the cooling rate (a negative quantity). Specific values of d_{11} and R' were chosen to conform to published experimental results kindly made available to us in detail by the authors (Matsumoto and Bogue, 1977).

    Figure 1 shows the predictions for the isothermal extensional viscosity (cases a and b) along with the non-isothermal predictions (cases c-f). It is seen that the curve for $R' = -2$°C/sec (case c) approaches the 120°C curve but does not cut it at t = 20 seconds, the time needed for the sample to cool down to 120°C from 160°C. The curve for $R' = -4$°C/sec (case e), however, actually <u>intersects</u> the 120°C curve <u>before</u> the ten seconds required for the sample to cool to 120°C, showing that as the cooling rate increases the unusual thermal effects become more pronounced. The predictions of the two remaining cases are lower than those of the corresponding cases with a non-zero value of B. Also, these will always be bounded by the isothermal predictions at 120°C and 160°C if the sample temperature lies between these two limits.

    It needs to be emphasized that the prediction of a higher stress in a sample A deformed in an identical manner to another sample B, but which is always kept hotter than B, is highly unusual. To the best of our knowledge, no prior constitutive equations are capable of making a similar prediction at arbitrary temperature levels. However, as the glass transition temperature of a melt is approached, the analysis of Matsumoto and Bogue (1977) does predict similar thermal effects due to shifts in Tg with cooling rate. Their study has recently been extended by Carey, Wust and Bogue (1980).

COMPARISON WITH EXPERIMENTAL DATA

    In a recent publication, Matsumoto and Bogue (1977) reported the results of experiments in which polystyrene samples were stretched at a constant stretch rate of 0.0528 sec^{-1} and cooled simultaneously at rates of up to 2.38°C/sec. Their experimental results are reproduced in Fig. 2. An examination of the temperatures at which the 2.18°C/sec curve intersects the various isothermal curves shows that at the points of intersection the temperature in the non-isothermal sample is always <u>higher</u> than that in the corresponding isothermal sample. A similar, though less pronounced, trend is observed for the lower cooling rates also. Additionally, recent data for the 2-dimensional stretching of polystyrene films (Gupta, 1980) show similar trends.

IN CONCLUSION, the influence of the thermal history upon rheological properties may be predicted using the flexible dumbbell molecular model, in which these effects arise naturally from changes in molecular conformation with temperature. The effects are of sufficient magnitude to influence polymer processing operations; during cooling they appear to stabilize the process and vice versa.

REFERENCES

Agrawal, P. K., 1977, "Extensional Rheology of Molten Polymers," M.Ch.E. Thesis, University of Delaware.

Agrawal, P. K., Lee, W. K., Lorntson, J. M., Richardson, C. I., Wissbrun, K. F., and Metzner, A.B., 1977, "Rheological behavior of molten polymers in shearing and in extensional flows," Trans. Soc. Rheol., 21:355.

Bird, R. B., Warner, H. R., and Evans, D.C., 1971, "Kinetic theory and rheology of dumbbell suspensions with Brownian motion," Adv. Polym. Sci., 8:1.

Bueche, F., 1962, "Physical Properties of Polymers," Wiley-Interscience, New York.

Carey, D. A., Wust, C. J., Jr., and Bogue, D. C., 1980, "Studies in non-isothermal rheology: behavior near the glass transition temperature and in the oriented glassy state," J. Appl. Poly. Sci. (in press).

Coleman, B. D., 1964, "Thermodynamics of materials with memory," Arch. Ration. Mech. Anal., 17:1.

Coleman, B. D., and Noll, W., 1961, "Foundations of linear viscoelasticity," Rev. Mod. Phys., 33:239.

Flory, P. J., 1953, "Principles of Polymer Chemistry," Cornell University Press, Ithaca.

Gupta, R. K., 1980, "A New Non-isothermal Rheological Constitutive Equation and its Application to Industrial Film Blowing Processes," Ph.D. Thesis, University of Delaware.

Matsumoto, T., and Bogue, D. C., 1977, "Non-isothermal rheological response during elongational flow," Trans. Soc. Rheol.,21:453.

Schultz, J. M., 1974, "Polymer Materials Science," Prentice Hall, Englewood Cliffs.

Treloar, L. R. G., 1975, "Physics of Rubber Elasticity," 3rd edition, Clarendon Press, Oxford.

White, J. L., and Metzner, A. B., 1963, "Development of constitutive equations for polymeric melts and solutions," J. Appl. Polym. Sci., 7:1867.

NOTE ADDED AFTER COMPLETION OF THE ABOVE. Our comments re. earlier work are incomplete. An ex. review is that of Pearson and McIntire (1979, JNNFM 6, 81); esp. sig. are papers by Crochet and Naghdi (1978, J. Rheol. 22, 73 and earlier). Although Eq. 5 has been given before β (Eq. 6) was taken as temp.-indep., thus losing the principal prediction of the present analysis.

SENSITIVITY OF THE STABILITY OF ISOTHERMAL MELT SPINNING TO

RHEOLOGICAL CONSTITUTIVE ASSUMPTIONS

Jing-Chung Chang and Morton M. Denn

University of Delaware
Newark, Delaware 19711 U.S.A.

INTRODUCTION

An instability in polymer melt spinning known as <u>draw resonance</u> is characterized by the onset of sustained oscillations in tension and drawn-filament diameter at a critical draw ratio (takeup velocity/initial velocity). The oscillations have a well-defined period and amplitude. Theory and experiment are reviewed by Petrie and Denn (1976) and Denn (1980). The most general theory is that of Fisher and Denn (1976,1977), which is based on a single-relaxation-time Maxwell model with a deformation-dependent viscosity. Broadly, the critical draw ratio is predicted to decrease as the viscosity function becomes more shear thinning, and to increase with increasing elasticity (increasing relaxation time or, equivalently, decreasing modulus.) Absolute stabilization is predicted above a critical value of a dimensionless group $\lambda v_0/L$, where v_0 is the initial velocity, L is the length of the melt zone, and λ is the relaxation time evaluated at v_0/L. Cooling is usually stabilizing.

All analyses of extensional flows based on single-relaxation-time models seem to require a relaxation time that is larger than that measured in shear in order to match steady-state data. This is because the relaxation spectrum is averaged differently in shearing and extensional flows (Denn and Marrucci, 1977; Denn, 1977; Petrie, 1979). Given this uncertainty in the relaxation time, the Fisher-Denn theory is consistent with most experimental data for isothermal systems, although a throughput effect noted by Chang and Denn (1979) on a polyacrylamide-corn syrup solution and a die flow effect observed by Matsumoto and Bogue (1978) on polypropylene are not explainable within this framework. The

9

former (and perhaps the latter as well) may be associated with differences in the boundary conditions. The theory appears to be in general agreement with most non-isothermal experiments, but data here are quite limited. A destabilizing effect of cooling observed by Matsumoto and Bogue (1978) in the spinning of polystyrene is not explainable within the context of the Fisher-Denn theory. Apparent contradictions of the theory reported by Han and Apte (1979) cannot be evaluated, since their experiments do not deal with the onset of instability. In any event, the fluctuations shown by Han and Apte do not have the characteristic periodic behavior of draw resonance and may be associated with other system transients (c.f. Petrie and Denn, 1976), perhaps amplification of fluctuations in cross-flow air.

An important question in a theoretical analysis of this type is the sensitivity of the result to specific constitutive hypotheses; this is the question of structural sensitivity. We provide a partial answer by considering a general rheological constitutive equation that includes the Maxwell fluid as a limiting case. Only isothermal spinning is considered, so hypotheses related to heat transfer mechanisms are not examined.

CONSTITUTIVE EQUATION

The constitutive equation used was developed by Phan-Thien (1978; Phan-Thien and Tanner, 1977). The general form for the extra stress $\underset{\sim}{\tau}$ in terms of the deformation gradient $\underset{\sim}{D}$ is

$$\underset{\sim}{\tau} = \sum_{k=1}^{N} \underset{\sim}{\tau}_k \tag{1}$$

$$\lambda_k \left\{ \frac{\partial \underset{\sim}{\tau}_k}{\partial t} + \underset{\sim}{v} \cdot \nabla \underset{\sim}{\tau}_k - \underset{\sim}{L} \cdot \underset{\sim}{\tau}_k - \underset{\sim}{\tau}_k \cdot \underset{\sim}{L}^T \right\} + \exp(\frac{p}{G_k} \, \mathrm{tr} \underset{\sim}{\tau}_k) \underset{\sim}{\tau}_k = 2\lambda_k G_k \underset{\sim}{D} \tag{2}$$

$$\underset{\sim}{L} = \nabla \underset{\sim}{v} - q\underset{\sim}{D}, \qquad \underset{\sim}{D} = \tfrac{1}{2}\left[\nabla \underset{\sim}{v} + (\nabla \underset{\sim}{v})^T \right] \tag{3}$$

The Maxwell fluid corresponds to $N = 1$, $p = q = 0$. Stresses are bounded at large extension rates for $p > 0$, while $q > 0$ leads to viscous shear thinning and a finite second normal stress difference. The relaxation times λ_k and moduli G_k are obtained from linear viscoelasticity measurements, and q is obtained from the viscosity function. p must be measured in an extensional experiment and is currently available only for low density polyethylene, where the model is fit with $p = 0.015$.

STABILITY ANALYSIS

The linear stability analysis follows that of Fisher and Denn

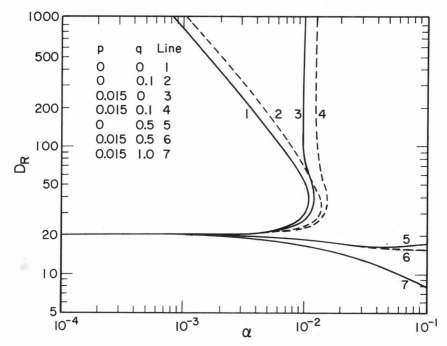

Fig. 1. Critical draw ratio, N = 1, as a function of α for
various p and q.

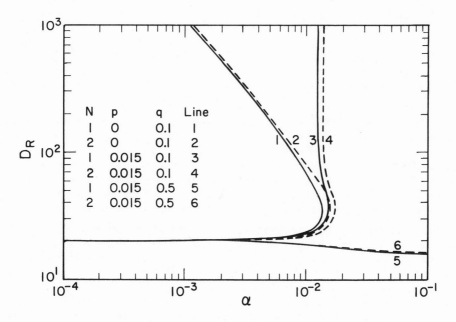

Fig. 2. Critical draw ratio. For N = 2, $\lambda_{max}/\lambda_{min}$ = 2, $G_1 = G_2$.

Table 1.   Critical Draw Ratios for the Phan-Thien Model,
           p = 0.015, N = 1, and a Power-law Fluid Giving
           Approximately the Same Steady-state Velocity Profile.

q	α	D_R	Equivalent n	Power law D_R
0.5	0.02	17	0.90	18
1.0	0.01	13.5	0.86	15.8
1.0	0.06	8.0	0.65	10.1
1.0	0.10	7.0	0.60	8.0

(1976). Steady state equations are as given by Phan-Thien (1978),
except that there is a misprint in his Eq. (60). (Replace ξP_i with
$(2 - \xi)P_i$.) Detailed perturbation equations are given by Chang
(1980). The critical draw ratio is expressed in all cases as a
function of α, where $\alpha = \lambda_{max} v_0/L$; λ_{max} is the maximum zero-shear
(linear viscoelastic) relaxation time.

RESULTS

     The effects of the parameters p and q are shown in Fig. 1 for
N = 1. The overall qualitative behavior of the Maxwell model is
retained: decreased stability from shear thinning (q > 0) and
stability at all draw ratios beyond a critical value of $\alpha("\alpha_m")$.
The initial decrease in the D_R vs. α curve is the result of a
transition from a zero-shear to deformation-dependent viscosity,
while the subsequent upturn is from the stabilizing effect of
elasticity.

     The Fisher-Denn theory uses a power-law viscosity function, so
direct comparison is not possible. For $\alpha < \alpha_m$ the critical draw
ratio from the Fisher-Denn theory is close to that for an inelastic
power-law fluid. We have fit the steady-state velocity profiles
at several points with a power-law model and compared the critical
draw ratios; results are shown in Table 1. Clearly, the two
approaches give comparable results in the shear-thinning region.

     The effect of two relaxation times is shown in Fig. 2 for
$G_1 = G_2$ (a "box spectrum") and $\lambda_{max}/\lambda_{min} = 2$; results are essentially the same for $\lambda_1 G_1 = \lambda_2 G_2$ (a "wedge spectrum"). The stability
behavior is the same for N = 1 and N = 2 as long as α is based on
λ_{max} and not on a mean relaxation time, although there are quantitative differences near α_m. Direct comparison with experimental
data for α near α_m would require the use of a spectrum appropriate
to the material and N much greater than two. Phan-Thien (1978)

used N = 8 for his comparisons with steady spinning data on a polystyrene, for example.

The only important qualitative difference between these results
and those of Fisher and Denn is the absence of a second stable
region at high draw ratios for p > 0 except at most in a finite
range of $\alpha < \alpha_m$. This contrasts with the Maxwell fluid, where the
unbounded stress growth leads to a second stable region at large
draw ratios for all $0 < \alpha < \alpha_m$.

ACKNOWLEDGMENT

This work was supported by the U.S. National Science
Foundation under Grant Eng. 76-15880.

REFERENCES

Chang, J. C., 1980, Ph.D. Dissertation, Univ. Delaware, Newark.
Chang, J. C. and Denn, M. M., 1979, J. Non-Newtonian Fluid Mech.,
    5:369.
Denn, M. M., 1977, in Rivlin, R., ed., "The Mechanics of Visco-
    elastic Fluids," AMD-22, A.S.M.E., N.Y.
Denn, M. M., 1980, Ann. Rev. Fluid Mech., 12:365.
Denn, M. M., and Marrucci, G., 1977, J. Non-Newtonian Fluid Mech.,
    2:159.
Fisher, R. J., and Denn. M. M., 1976, AIChE J., 22:236.
Fisher, R. J., and Denn. M. M., 1977, AIChE J., 23:23.
Han, C. D., and Apte, S. M., 1979, J. Appl. Poly. Sci., 24:61.
Matsumoto, T., and Bogue, D. C., 1978, Poly. Eng. Sci. 18:564.
Petrie, C. J. S., 1979, "Elongational Flows," Pitman, London.
Petrie, C. J. S., and Denn, M. M., 1976, AIChE J., 22:209.
Phan-Thien, N., 1978, J. Rheol., 22:259.
Phan-Thien, N., and Tanner, R. I., 1977, J. Non-Newtonian Fluid
    Mech., 2:353.

are then fitted to the three-parameter equation proposed by Mieras and van Rijn[14]:

$$\eta = \frac{\eta_o}{\left[1 + \left(\frac{\sigma_t}{\sigma_c}\right)^n\right]^{1.7}} \qquad (1)$$

in which η is the viscosity, η_o the zero-shear viscosity, σ_t the shear stress, σ_c a so-called elasticity parameter and n an index between 0 and 2. The parameters η_o and σ_c are related to the molecular structure of polypropylene, η_o being a measure of the weight average molecular weight $\overline{M}_w$ and σ_c being a measure of the width of the molecular weight distribution in terms of the ratio $R = \overline{M}_z/\overline{M}_w$. σ_c increases with decreasing values of R. The values obtained for η_o and σ_c are included in Table 1.

In addition to this rheological characterisation, the polypropylenes were analysed by Gel Permeation Chromatography (GPC) to provide the actual molecular parameters. The values obtained for $\overline{M}_w$, $Q = \overline{M}_w/\overline{M}_n$ and $R = \overline{M}_z/\overline{M}_w$ are also listed in Table 1.

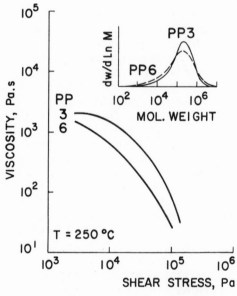

Fig. 1. Flow curves of two polypropylenes being different in width of the molecular weight distribution. GPC chromatograms of the polypropylenes are also inserted.

As an example Fig. 1 shows the shear viscosity functions obtained at 250 °C together with the GPC chromatograms for two of the samples.

Apparatus and Procedure

The experimental set-up used consists of a single-screw laboratory extruder, a deflection head fitted with a capillary die and a Rheotens apparatus as take-up system. Stretching took place under non-isothermal conditions as the extrudates were allowed to cool under ambient conditions. In all experiments the polymer strand was still in the molten state when reaching the take-up roll. By using this set-up it was also possible to measure the tensile forces applied to the melt.

The polypropylenes were extruded at a constant mass flow rate of 10 g/min through a 3 mm

capillary having an l/d ratio of 10. The melt temperature in the die
was 220 °C (±1 °C) for all tests. In some experiments the mass flow
rate and the l/d ratio of the die were varied. The draw-down distance was varied between 5 and 28 cm. Stretching was carried out at
constant take-up speeds. The experiments were started at low speed,
which was then increased in steps into the region where draw resonance occurs. At each take-up speed the maximum and minimum values
of the extrudate diameter were measured at the position of the take-
up roll by means of a microscope. In the case of draw resonance, the
diameter data were collected when a steady-state pulsation of the
diameter was attained. The period of resonance was determined by
measuring the elapsed time for 10 cycles at the same position. In all
experiments extrudate swell was small in the presence of stretching
and swell differences were less than 10 %. Therefore, draw ratios
were based on extrusion velocity.

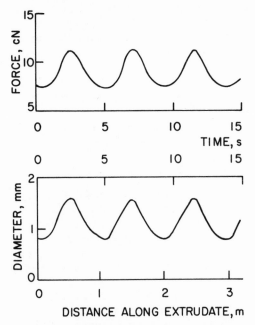

Fig. 2. Typical profiles of force
and diameter during draw
resonance of polypropylene
melts.

RESULTS AND DISCUSSION

Characteristics of Draw Resonance

Fig. 2 shows typical
plots of the diameter and of
the corresponding take-up
force versus time obtained
during draw resonance of polypropylene melts. Both force and
diameter vary in a periodic
manner with a well defined
amplitude and frequency. Moreover, it appears that both
oscillations are nearly in-
phase and that the waveform of
resonance is skewed, which are
also characteristic features of
the resonance phenomenon. Most
of these observations support
the view that the mechanism of
draw resonance is a mass accumulation process initiated by
local thinning of the extrudate
near the capillary die[15].

Effects of Molecular Structure

The effects of molecular structure upon draw resonance behaviour
were studied for a fixed set of processing conditions, viz. mass flow
rate = 10 g/min, l/d ratio capillary = 10, draw-down distance = 15 cm
and non-isothermal stretching.

Molecular weight. Fig. 3 shows the resonance behaviour of a
series of five polypropylenes differing widely in molecular weight
but similar in distribution width. The data are plotted as the ratio
of the maximum to minimum diameter of the extrudate d_{max}/d_{min} as a
function of draw ratio v/v_o. The draw ratio used is based on extrusion velocity, i.e. it is defined as the ratio of the take-up velocity v to the average velocity in the die v_o. It follows from Fig. 3
that the onset of resonance considerably shifts to higher draw ratios
with decreasing molecular weight. Similar molecular weight effects
have been reported by other investigators[1,8].

Molecular weight distribution. Fig. 4 shows the effect of molecular weight distribution width for a pair of polypropylenes of the
same average molecular weight. It appears that the polymer with the
narrow molecular weight distribution can be stretched to higher draw
ratios.

Comparison with theory. From Figs. 3 and 4 it follows that for
all samples the critical draw ratio is well below 20, being the
theoretical value for Newtonian liquids under isothermal conditions.
Values lower than 20 have also been observed in other experimental
studies on polypropylene[8,9,11,16] and according to the Fisher and
Denn analysis for viscoelastic liquids this can be attributed to the
non-Newtonian shear behaviour of the samples.

Effects of Processing Variables

Draw-down distance. Fig. 5 shows the effect of draw-down distance upon the resonance behaviour of three polypropylene grades
differing in molecular weight but with a similar (rather narrow)
molecular weight distribution. It appears that the onset of resonance is not much influenced by varying the draw-down distance but

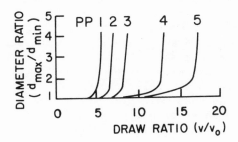

Fig. 3. Effect of molecular weight
upon draw resonance for
polypropylenes of equal
distribution width.

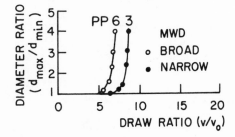

Fig. 4. Effect of molecular weight
distribution upon draw
resonance for a pair of
polypropylenes of the
same average molecular
weight.

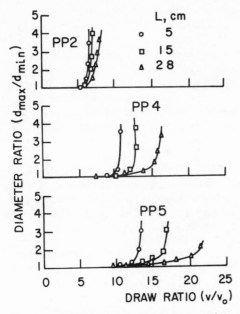

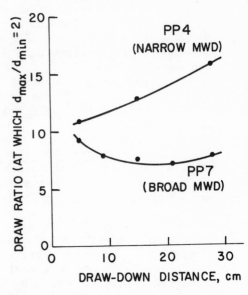

Fig. 5.   Effect of draw-down dis-
          tance upon draw resonance
          at three molecular weight
          levels.

Fig. 6.   Effect of draw-down dis-
          tance upon resonance
          behaviour for polypropyl-
          enes of different mole-
          cular weight distribution.

that, when arrived in the resonance region, the oscillation is damped
with increasing distance especially for the lower molecular weight
materials. Similar experimental observations were made by Christensen[1]
and Lamb[16] on polypropylene and by Ishihara and Kase[5] on polyethylene
terephthalate. Physically, the proposed mechanism[15] is that thin
regions in the extrudate cool down more rapidly and so attain a
higher viscosity; this, in turn, reduces the rate of extension of the
thinner portions and thus suppresses the occurrence of draw resonance.

A different picture is observed for polypropylenes having a broad
molecular weight distribution. At large draw-down distances the stabilising effect of cooling is still present, but at small distances the
critical draw ratio is found to increase with decreasing distance.
This behaviour is illustrated in Fig. 6 for polypropylene PP 7 and
compared with that of PP 4 having the same molecular weight but a
much narrower distribution. Similar observations were reported by
Ishihara and Kase[5] for a polyethylene terephthalate melt.

According to Shah and Pearson's work on Newtonian liquids the
stabilising effect of cooling appears to be determined by the quantity

$$k \ominus St\, e^{-St} \tag{2}$$

in which k is a temperature-viscosity coefficient, Θ the initial melt
temperature minus ambient temperature and St is the dimensionless
Stanton number. The stability of the process can be enhanced by
increasing the value of $k \Theta St e^{-St}$ i.e. either by an increase in k -
which determines the rate of increase in viscosity with decreasing
temperature - or by an increase in Θ or St (only at low St values!)
which determine the rate of cooling. The Stanton number itself is
given by equation (3)

$$St = 1.67 \times 10^{-4} \frac{L}{\rho^{1/3} c_p G_o^{2/3} A_o^{1/6}} \tag{3}$$

(from Kase's work[4]; all data in c.g.s. units)

where L is the draw-down distance, ρ the melt density, c_p the specific
heat, G_o the mass flow rate and A_o the initial extrudate cross-sectional area. The experiments reported in this work were all carried out
in the low St region. Calculation of the Stanton number leads to a
value of 6.5×10^{-3} for a mass flow rate of 10 g/min and a draw-down
distance of 5 cm. The enhanced stability observed when the draw-down
distance is increased i.e. when St is increased is, therefore, in
agreement with theory although in our experiments the critical draw
ratio itself remains practically unchanged.

The stabilising effect at small draw-down distances as observed
for sample PP 7 is also predicted theoretically by Fisher and Denn[6]
and according to this theory it should be ascribed to the elasticity
of the melt.

Mass flow rate. The effect of mass flow rate was studied for
sample PP 2 at a draw-down distance of 5 cm, for which distance the
stabilising effect of cooling is considered to be small. It was found
that the use of lower mass flow rates allows higher critical draw
ratios to be obtained.

Die geometry. Of the geometrical variables only the effect of
length-to-diameter (l/d) ratio of the capillary was investigated. It
was found that long dies give rise to higher critical draw ratios
than short dies. Very striking l/d effects were also observed in
other studies[10,11]. In such studies the importance of the die swell
region is stressed, but the observed l/d effects are not yet clearly
understood.

Summarising it can be said that our results confirm the general
statement of Han and Kim[10] that factors which lower the orientation
of the melt (low mass flow rates, long dies) make the stretching
process more stable.

Periodicity of draw resonance. In examining the available experimental data it was found that the periodicity of the resonance phenomenon depends on the processing variables rather than on the molecular characteristics of the polypropylene samples. The period of resonance t_r, for instance, is determined by the extrusion velocity v_o, the draw ratio v/v_o and the draw-down distance L

$$t_r = f (v_o, v/v_o, L) \qquad (4)$$

When the periodicity is expressed in terms of the wavelength of the extrudate taken up, λ_r, it only depends on the draw ratio and the draw-down distance.

$$\lambda_r = v\, t_r = f (v/v_o, L) \qquad (5)$$

Fig. 7 shows that λ_r slightly increases with increasing draw ratio and is proportional to the draw-down distance. The left-hand part of Fig. 7 also shows that the wavelength is not affected by the molecular structure of the polymer (cf. e.g. PP 2 and PP 7) and by varying the extrusion velocity v_o by a factor of 5 (open symbols). These results strongly suggest that draw resonance is essentially a dynamic and not a viscoelastic process.

In Kase's theoretical work[4] on the isothermal melt spinning of Newtonian liquids the following equation was derived for the resonance wavelength

$$\lambda_r = \frac{v/v_o}{\ln (v/v_o)}\, g\, (v/v_o)\, L \qquad (6)$$

in which g (v/v_o) is a function of draw ratio. According to this equation λ_r depends only on draw ratio and draw-down distance which is in line with the experimental observations of this work. Less perfect quantitative agreement, however, was found between the theoretically predicted wavelengths using Equation (6) and the experimental values as follows from Fig. 7. The much higher experimental values are most probably caused by the strong non-Newtonian character of the polypropylene melts studied.

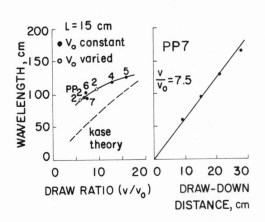

Fig. 7.   Resonance wavelength as a function of draw ratio and draw-down distance.

Knowledge of the relation between the periodicity of resonance and the processing variables can be of use in recognising the draw resonance phenomenon in practice.

CONCLUSIONS

It has been shown that the draw resonance behaviour of poly-propylene melts depends on the molecular structure of the polymer but also heavily on the processing variables. It is found that the critical draw ratio for the onset of resonance increases with decreasing molecular weight and with decreasing width of its dis-tribution. Of the processing variables, the length of the stretching zone in particular plays an important role in stabilising the stretching operation. Long stretching zones can be stabilised by cooling and short stretching zones can be stabilised by melt elas-ticity, particularly in the case of polypropylenes having a broad distribution. In addition, it is found that the use of low mass flow rates and long dies leads to enhanced stability. As regards the periodicity of resonance it is found that the resonance wave-length is a function only of draw ratio and draw-down distance and independent of molecular structure. Most of the observations can be explained in terms of existing theories.

Knowledge of the relationships found is of vital importance for our ability to remedy draw resonance defects in practice. It should be stressed, however, that the relationships are in no way universal, so that its application to conditions far outside the range of variables investigated might not be justified.

REFERENCES

1. R.E. Christensen, "Extrusion Coating of Polypropylene", S.P.E.
    Journal, 18, 751 (1962).
2. J.R.A. Pearson and M.A. Matovich, "Spinning a Molten Threadline",
    Ind. Eng. Chem. Fundam., 8, 606 (1969).
3. Y.T. Shah and J.R.A. Pearson, "On the Stability of Nonisothermal
    Fiber Spinning", Ind. Eng. Chem. Fundam., 11, 145 (1972).
4. S. Kase, "Studies on Melt Spinning. IV. On the Stability of Melt
    Spinning", J. Appl. Pol. Sci., 18, 3279 (1974).
5. H. Ishihara and S. Kase, "Studies on Melt Spinning. VI. Simula-
    tion of Draw Resonance Using Newtonian and Power Law Viscosities",
    J. Appl. Pol. Sci., 20, 169 (1976).
6. R.J. Fisher and M.M. Denn, "A Theory of Isothermal Melt Spinning
    and Draw Resonance", AIChE J., 22, 236 (1976).
7. R.J. Fisher and M.M. Denn, "Mechanics of Nonisothermal Polymer
    Melt Spinning", AIChE J., 23, 23 (1977).
8. A. Bergonzoni and A.J. DiCresce, "The Phenomenon of Draw
    Resonance in Polymeric Melts", Pol. Eng. Sci., 6, 45 (1966).
9. C.B. Weinberger, G.F. Cruz-Saenz and G.J. Donnelly, "Onset of
    Draw Resonance During Isothermal Melt Spinning: A Comparison
    Between Measurements and Predictions", AIChE J., 22, 441 (1976).
10. C.D. Han and Y.W. Kim, "Studies on Melt Spinning. VI. The Effect
    of Deformation History on Elongational Viscosity, Spinnability,
    and Thread Instability", J. Appl. Pol. Sci., 20, 1555 (1976).
11. T. Matsumoto and D.C. Bogue, "Draw Resonance Involving Rheolo-
    gical Transitions", Pol. Eng. Sci., 18, 564 (1978).
12. C.D. Han and S.M. Apte, "Studies on Melt Spinning. VIII. The
    Effects of Molecular Structure and Cooling Conditions on the
    Severity of Draw Resonance", J. Appl. Pol. Sci., 24, 61 (1979).
13. C.J.S. Petrie and M.M. Denn, "Instabilities in Polymer
    Processing", AIChE J., 22, 209 (1976).
14. H.J.M.A. Mieras and C.F.H. van Rijn, "Influence of Molecular
    Weight Distribution on the Elasticity and Processing Properties
    of Polypropylene Melts", J. Appl. Pol. Sci., 13, 309 (1969).
15. L.L. Blyler and C. Gieniewski, "Melt Spinning and Draw Resonance
    Studies on a Poly (α-methylstyrene/silicone) Block Copolymer",
    Pol. Eng. Sci. in press.
16. P. Lamb, "Analysis of Fabrication Processes", SCI Monograph
    No. 26, 296 (1967).

QUANTITATIVE INVESTIGATIONS OF ORIENTATION DEVELOPMENT IN VITRYFYING
DEFORMING POLYMER MELTS WITH APPLICATION TO PROCESSING

J.L. White

University of Tennessee, Knoxville, Tennessee, U.S.A.

(Abstract)

The state of orientation and morphology in polymer products
formed from the melt has long been known to depend upon kinematics
and cooling conditions immediately prior to solidification. The
authors and their coworkers have carried out extensive investigations
of vitrification and crystallization of deforming polymer melts
during the past several years. Here we discuss investigations of
vitrifying polymers.

Studies of vitrifying atactic polystyrene (PS) and polyethylene
terephthalate (PET) indicate that orientation (or birefringence)
levels in the melt immediately prior to vitrification are simply
"frozen in". As these melts obey the Rheo-Optical Law relating
birefringence to the difference in principal stresses, birefringence
distributions are determined by the stress field. This has been
verified for melt spun PS and PET filaments as well as for PS
vitrified in shear flow and blown tubular film.

The polarizability tensor may be used to define uniaxial and
biaxial orientation factors which can represent the second moment
of the orientation distribution in filaments and films. The Rheo-
Optical Law represents a linear relationship between these
orientation factors and applied stresses.

Birefringence in vitrified polymers can also be caused by
thermal stresses induced by spatial variations in specific volume.

This contribution to birefringence is especially important when the magnitude of the room temperature stress-optical constant approaches that in the melt. In PS and PET this apparently not an important contribution to birefringence though it may be with other polymers.

The birefringence'-stress at vitrification hypothesis' can be used to predict birefringence or orientation distributions in objects formed by molding processes such as injection or blow molding. However fluid mechanical, heat transfer and stress relaxation models must be incorporated.

Orientation has a major effect on the mechanical properties of many glassy plastics such as PS. Most importantly it results in a "brittle-ductile" transition.

QUANTITATIVE INVESTIGATIONS OF CRYSTALLIZATION KINETICS AND
CRYSTALLINE MORPHOLOGY DEVELOPMENT IN SOLIDIFYING POLYMER MELTS
WITH APPLICATION TO PROCESSING

J.L. White, J.E.Spruiell

University of Tennessee, Knoxville, Tennessee, U.S.A.

(Abstract)

We have investigated i, the crystallization kinetics and
ii, development of crystalline morphology occurring in deforming
polymer melts during solidification with special reference to
polymer processing.

Induction times to crystallization in polyolefin melts in shear
and elongation flow and interpreted in terms of nucleation ratios.
It is found that induction time data for both types of flow may be
correlated in terms of the difference in principal stresses. This
reflects through consideration of the Rheo-Optical Law a relationship to molecular orientation in the melt prior to crystallization.
It is possible to develop a theory of nucleation which reflects
this behavior and depends on the stress field during flow prior to
solidification.

Extensive studies of melt spinning of polyolefin and polyamide
fibers over a range of conditions (extrusion temperature, take-up
velocity, extrusion rate, molecular weight) indicates that the
crystalline orientation and morphology is primarily controlled by
the stresses in the spinline at the point of solidification.
Orientation levels are substantially high than that existing in the
melt prior to crystallization. However, large variations in quench
rate can induce changes in crystalline form or even the formation
of crystallinity. Preliminary studies indicate that this is also
the case for fiber melt spun using air jects as in spun bonding

processes.

Investigations of the products of tubular film extrusion of polyethylene using pole figure analysis will also be discussed. These indicate orientation bias in the products to a much greater extent than expected from the Rheo-Optical Law.

RHEOLOGY IN CALENDERING OF THERMOPLASTICS

John Vlachopoulos

Department of Chemical Engineering
McMaster University
Hamilton, Ontario, Canada L8S 4L7

INTRODUCTION

Calendering is a process involving a pair of corotating heated rolls which is fed with a thermoplastic melt for the production of sheet or film. Nearly all models proposed in the literature for this process are based on the simplification of the general equations for the conservation of mass, momentum and energy by making use of the lubrication approximation. Actually, most recent theoretical analyses are extensions of Gaskell's model [1]. Various aspects of the theory can be found in the textbooks of McKelvey [2], Torner [3], Middleman [4], and Tadmor and Gogos [5]. Recent developments in calendering theory include the works of Vlachopoulos et al [6,7,8], Dobbels and Mewis [9], Agassant et al [10,11], Dimitrijew and Sporjagin [12] and Woskressenski et al [13].

Despite the relatively large number of publications on calendering, the various models have remained without any experimental verification. Bergen and Scott's pressure distribution data [14] were obtained by calendering a plasticized resin of unknown rheological behavior. These data have been often used as a test of the hydrodynamic theory [2,7]. It must be noted, however, that pressure distribution data without a rheological law for the material may offer some information as to the shape of the curve, pressure versus distance, but constitute a poor test of the predictive power of any given model. Also, Unkrüer's pressure distribution measurements [15] in calendering of a rigid PVC resin were obtained without the required rheological data.

29

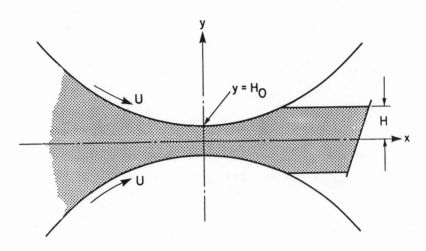

Fig. 1 Calendar gap geometry

THE HYDRODYNAMIC MODEL WITHOUT SLIP

The general equations for the conservation of mass and momentum can be simplified by using the lubrication approximation [2]. For the rectangular coordinate system of Figure 1 we have

$$Q = 2 \int_{o}^{h} u \, dy \tag{1}$$

$$\frac{\partial P}{\partial x} = \frac{\partial}{\partial y} (\tau_{xy}) \tag{2}$$

where the shear stress is approximated by the power law expression

$$\tau_{xy} = m \left(\frac{\partial u}{\partial y}\right)^{n} \tag{3}$$

It is easy to show that if u(h)=U (no slip) the pressure gradient takes the following form (4):

$$\frac{\partial P}{\partial x^{*}} = -m \left(\frac{2n+1}{n}\right)^{n} \left(\frac{U}{H_{o}}\right)^{n} \sqrt{\frac{2R}{H_{o}}} \frac{(\lambda^{2} - x^{*2}) |\lambda^{2} - x^{*2}|^{n-1}}{(1+x^{*2})^{2n+1}} \tag{4}$$

where x^{*} is a dimensionless distance $x^{*} = x/\sqrt{2RH_{o}}$. The pressure distribution is obtained by integrating the above equation and

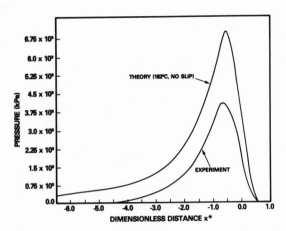

Fig. 2 Comparison of experimental pressure distribution with
        predictions of the model without slip

using the boundary condition $P(\lambda) = 0$, with the result

$$P = m(\frac{2n+1}{n})^n (\frac{U}{H_o})^n \sqrt{\frac{2R}{H_o}} \int_{x^*}^{\lambda} \frac{|\lambda^2 - x^{*2}|^{n-1} (\lambda^2 - x^{*2})}{(1+x^{*2})^{2n+1}} dx^* \qquad (5)$$

where λ is a dimensionless flow parameter defined by

$$\lambda^2 = \frac{Q}{2UH_o} - 1 \qquad (6)$$

A comparison between theory and experiments is shown in Fig. 2.
The experimental results were kindly provided to the author by
J.C. Chauffoureaux [16] of Solvay & Cie, S.A. The discrepancies
are larger than anticipated and it was assumed that they were
mainly due to slip at the wall.

THE HYDRODYNAMIC MODEL WITH SLIP

    Chauffoureaux [17] developed a method for measuring wall
slip. Measurements for the PVC resin which was used to produce
the data of Fig. 2, gave a slip velocity which could be
approximated by the expression

$$U_s = \frac{1}{\beta} \tau_w^\alpha \qquad (7)$$

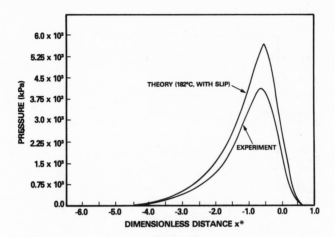

Fig. 3 Comparison of experimental pressure distribution with predictions of the model with slip

Vlachopoulos and Hrymak [18] developed a model which takes into account slip at the wall.

The analysis begins with the simplified momentum equation

$$\frac{\partial P}{\partial x} = \frac{\partial}{\partial y} (\tau_{xy}) \tag{8}$$

where

$$\tau_{xy} = m \left(\frac{\partial u}{\partial y}\right)^n \tag{9}$$

The boundary conditions for calendering with slip at the roll surface are

at $y = 0$ $\frac{\partial u}{\partial y} = 0$

$$\tag{10}$$

at $y = h$ $u = U - U_o - \frac{1}{\beta} \tau_w^\alpha$ (here $U_o = 0$)

Upon integrating the momentum equation and after applying the boundary conditions we obtain the following expression for the velocity profile in the region where the pressure gradient is positive

$$u = \frac{n}{n+1} \left(\frac{1}{m}\frac{\partial P}{\partial x}\right)^{1/n} (y^{n+1/n} - h^{n+1/n}) + U - U_o - \frac{1}{\beta}\left(\frac{\partial P}{\partial x}\right)^{\alpha} h^{\alpha} \qquad (11)$$

and another expression for the region of negative pressure gradient

$$u = -\frac{n}{n+1} \left(-\frac{1}{m}\frac{\partial P}{\partial x}\right)^{1/n} (y^{n+1/n} - h^{n+1/n}) + U - U_o - \frac{1}{\beta}\left(-\frac{\partial P}{\partial x}\right)^{\alpha} h^{\alpha} \qquad (12)$$

Either of the equations for the velocity profile may then be integrated to give the volumetric flow rate

$$Q = 2\int_o^h u\,dy \qquad (13)$$

The result may be written as

$$\frac{Q}{2h} = U - U_o - \frac{1}{\beta}\left(\frac{\partial P}{\partial x}\right)^{\alpha} h^{\alpha} - \frac{n}{(2n+1)} \left(\frac{1}{m}\frac{\partial P}{\partial x}\right)^{1/n} h^{n+1/n} \qquad (14)$$

A comparison between theory and experiment is given in Fig. 3.

VISCOUS DISSIPATION EFFECTS

Kiparissides and Vlachopoulos [8] developed a finite difference technique to solve the nonisothermal problem with the lubrication approximation. The simplified conservation equations

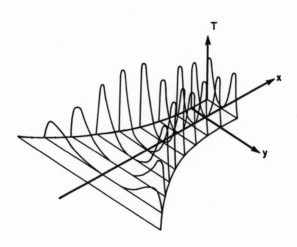

Fig. 4 Development of temperature profiles

are

$$Q = 2 \int_{o}^{h} u\,dy \tag{15}$$

$$\frac{\partial P}{\partial x} = \frac{\partial}{\partial y}\,(\tau_{xy}) \tag{16}$$

$$\rho\,C_p\,u\,\frac{\partial T}{\partial x} = k\,\frac{\partial^2 T}{\partial y^2} + \tau_{xy}\,\left(\frac{\partial u}{\partial y}\right) \tag{17}$$

The development of the temperature profile for a power-law fluid is schematically shown in Fig. 4.   It is interesting to note the two maxima near the roll surfaces where viscous dissipation is large.   The numerical calculations [8] further showed that it is possible to achieve relatively high local temperatures which might have a detrimental effect on temperature sensivite materials.

VISCOELASTIC EFFECTS

     A simple order of magnitude approximation shows that the normal stresses may account only for 1 or 2% of total pressure developed in the calender gap [4,18].   Elongational flow develops in the entry region where recirculation patterns appear.   It is believed that instabilities may originate in this region, propagate downstream and appear as distortions on the surface of the calendered sheet.   In the entry to the calender gap the polymer melt flows from a nearly stationary region to a rapidly accelerating flow field.   Consequently fluid relaxation times might be of importance.   A study of viscoelastic effects is currently underway by the author.

EFFECTS OF FLOW GEOMETRY

Calendering may be symmetric or asymmetric depending on the relative size of roll diameters and relative roll speeds.   These matters have been dealt with by Kiparissides and Vlachopoulos [7] and by Takserman-Krozer et al [19].   An additional complication in the analysis arises because of the fact that flow in the calender gap is nearly always in three dimensions.   The melt sheet as it flows in the direction of the gap it also spreads in the direction of roll axis.   For the PVC data shown earlier, the sheet at entry had a width of 22.5 cm and the exit 37.0 cm.   The author is not aware of any publications relating to the three dimensional problem.

REFERENCES

[1] R.E. Gaskell, J. Appl. Mech. (Trans. (ASME), 17, 334 (1950).

[2] J.M. McKelvey, "Polymer Processing", John Wiley, New York (1962).

[3] R.V. Torner, "Grundprozesse der Verarbeitung von Polymeren", VEB Deutscher Verlag fuer Grundstoffindustrie, Leipzig, G.D.R. (1974).

[4] S. Middleman, "Fundamentals of Polymer Processing", McGraw-Hill, New York (1977).

[5] Z. Tadmor and C.G. Gogos, "Principles of Polymer Processing", John Wiley, New York (1979).

[6] G. Ehrmann and J. Vlachopoulos, Rheol. Acta, 14, 761 (1975).

[7] C. Kiparissides and J. Vlachopoulos, Polym. Eng. Sci., 16, 712 (1976).

[8] C. Kiparissides and J. Vlachopoulos, Polym. Eng. Sci., 18, 210 (1978).

[9] F. Dobbels and J. Mewis, AIChE J., 23, 224 (1977).

[10] J.F. Agassant and P. Avenas, J. Macromol. Sci.-Phys., 14, 345 (1977).

[11] J.L. Bourgeois and J.F. Agassant, J. Macromol. Sci.-Phys., 14, 367 (1977).

[12] J.G. Dimitrijew and E.A. Sporjagin, Plast. Kautsch., 24, 484 (1977).

[13] A.M. Woskressenski, W.N. Krassowski, L.K. Sewastjanow, C. Kohlert and E.O. Reher, Plast. Kautsch., 26, 92 (1979).

[14] J.T. Bergen and G.W. Scott, J. Appl. Mech. (Trans. ASME), 18, 101 (1951).

[15] W. Unkrüer, "Beitrag Zur Ermittlung des Druckverlaufes und der Fliessvorgaenge im Walzspalt bei der Kalanderverabeitung von PVA Hart zu Folien", Doctoral Thesis IKV, Techn. Hoch. Aachen (1970).

[16] J.C. Chauffoureaux, private communication, Solvay & Cie., S.A., Brussels, Belgium.

[17] J.C. Chauffoureaux, C. Dehennau and J. Van Rijckevorsel, J. Rheol., 21, 1 (1979).

[18] J. Vlachopoulos and A.N. Hrymak, "Calendering of PVC: Theory and Experiments", Pol. Eng. Sci. (in press).

[19] R. Takserman-Krozer, G. Schenkel, and G. Ehrmann, Rheol. Acta, 14, 1066 (1975).

FLOW IN INJECTION MOULDS

D.P. Isherwood, J.G. Williams and Y.T. Yap

Department of Mechanical Engineering
Imperial College of Science & Technology
London SW7 2BX, England.

INTRODUCTION

In all but the simplest injection moulding configurations, there
will occur in the mould filling process a situation in which two, or
more, distinct melt streams will interact with each other.   This
may occur because of flow splitting by an insert within the mould
and the subsequent recombination of the separated flows, or because
of the collision of flow fronts originating from two, or more,
individual gates.   (This may be considered as flow splitting of melt
before entering the mould.)   At any event, two separate flow streams
moving in different directions will merge with each other, setting up
an interface, which will be retained in the final component to a
greater of lesser degree, dependent on melt properties and processing
conditions.   This interface, which is frozen into the moulding, has
become known as a weld, or knit, line and its effect on the overall
strength of a plastic component is becoming the source of much
interest, particularly as there is relatively little information in
the published literature.

Since the origin of the weld line is a disruption caused by the
collision of at least two flow streams, it has been seen as a
potential source of weakness.   Previous studies have revealed that
this is, indeed, the case [1,2,3].   The objective of this investigation is to explore the weld line effect, with a view to determining how processing parameters contribute to the strength of components containing weld lines.

EXPERIMENT

The design of mould used in this study is illustrated in
Figure 1.   The size of the mould cavity is 140 mm × 140 mm × 3 m,
which is nearly the maximum shot volume of the injection moulding
machine, a Bipel 70/17 single reciprocating-screw machine with a
75 tonnes mould clamping force.   The design of the obstacle, which
is removable, and the mould layout in general was adopted in order
that a minimum disruption is imposed on the flow stream; the object
being to divide, separate and then recombine the melt with little
disorder so that extraneous effects do not cloud the comparison
between the strength of components with and without the weld line
(i.e. with and without the obstacle).   The particular obstacle shape
chosen is also very similar to the mandrel supporting spiders used
in pipe extrusion, so some results may be applicable to the extrusion
process.   A force transducer was positioned at the exact centre of
the mould cavity, enabling a measurement of cavity pressure to be
made.   A pressure transducer was located in the nozzle of the
injection moulder and a displacement transducer was used to measure
injection rates.   Melt temperature was determined by thermocouples.
The mould temperature could be held constant over the range 15°C to
65°C using a R.A. Stephens temperature controller.

The materials chosen for this investigation were a low density
polyethylene (ICI grade WNC71) and a polystyrene (ICI grade 2CL).
These materials were selected, not particularly for the rheological
properties of their melts, but rather for their distinct fracture
behaviour; polyethylene being ductile and polystyrene brittle.

Specimens were moulded under conditions of varying melt
temperature, mould temperature and injection speed, respectively.
Whilst each of these parameters was altered, all others were held
constant.   After moulding, the specimens were cut normal to the
principal flow direction from each component.   These specimens are
designated A and B, the latter being nearer the obstacle or gate.
The polyethylene specimens were subjected to monotonic tensile and
reversed fatigue tension tests, whilst the polystyrene specimens
were subjected to bending and notched impact tests.   The choice of
tests was determined by the relative stiffness and ductility of the
specimens of each material.   Some of the polyethylene mouldings
were also viewed under polarised light in order to examine the
degree of molecular orientation.

In order to assess the effect of the obstacle in the flow of
polymer in the mould, a series of 'short shot' mouldings for each
material was produced.   The flow front pattern obtained for
polyethylene is illustrated in Figure 1.   The amount of drag
produced in polystyrene was rather less than for polyethylene,
reflecting the higher viscosity of the latter at moulding temperatures.

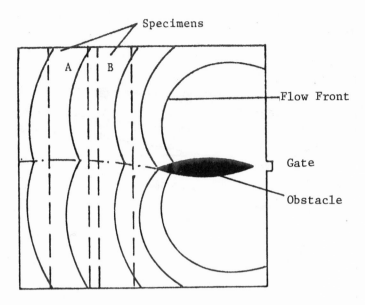

Figure 1 :-Mould layout, showing flow fronts
and specimen location.

RESULTS

Space precludes the quantitative presentation of all but the
most striking effects here.   Some broad conclusions may, however,
be drawn.   The clearest of these is that the effect of the obstacle,
and hence the weld line on the flexural and impact strength of
polystyrene is quite small.   The reduction in flexural strength is
on average 5% and impact strength 10%.   Further, the effect of
variation of processing conditions is also small as is the
difference in relative strength with distance from the obstacle.

For polyethylene, the tensile strength is reduced by an average
of 9% when the weld line is present.   The percentage elongation to
fracture is reduced by an average of 32% and the fatigue life by
90%.   In all cases, the performance of the specimen cut nearer to
the obstacle is superior to that of the specimen located further
from the obstacle, but it should be noted that neither specimen was
cut adjacent to the obstacle (see Figure 1) so that the immediate
effects of flow disruption will not be apparent in the specimens.
The effect of mould temperature was inconclusive;   a great deal of
scatter and no discernible trends were evident, so no conclusions
regarding the effect of the temperature of the mould can be drawn.

The effects of the variation of the other processing conditions
(i.e. melt temperature and injection speed) were much clearer and,
as is implied from the comparison of performance of specimens with
and without weld lines, the effect of such variations is greater for

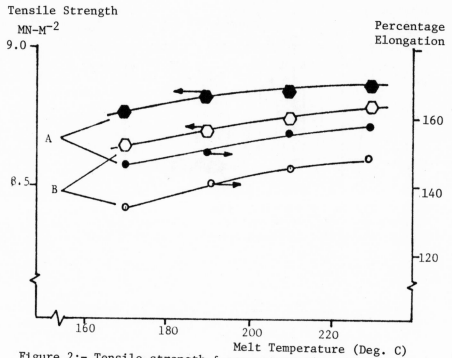

Figure 2:- Tensile strength & percentage elongation
against melt temperature.

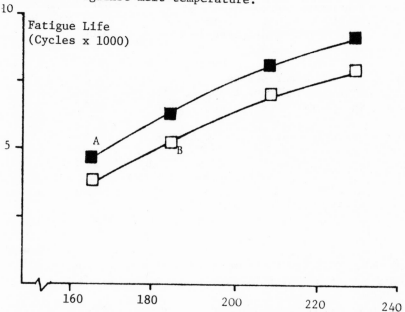

Figure 3:- Fatigue life against melt temperature.

fatigue than for percentage elongation which, in turn, is greater than for tensile strength. The effects of melt temperature are illustrated in Figures 2 and 3, and a consistent increase, slight in tension but significant in fatigue, of performance with temperature is demonstrated. The upper limit of the temperature range was determined by poor surface finish and moulding quality. The effect of injection speed is demonstrated in Figure 4. The tensile strength results are omitted; they show similar but weaker trends than the other test results. There is a clear maximum in performance at an injection rate of about 80 cm^3/sec in all cases.

The birefringence examination revealed that at an injection rate of 12 cm^3/sec, there is a relatively broad wake (about half the maximum width of the obstacle) downstream of the obstacle, whereas at 60 cm^3/sec the wake reduces to a thin distinct line.

DISCUSSION

The flow front pattern shown in Figure 1 clearly demonstrates the drag effect due to the obstacle, which, although it decays, is retained downstream of the obstacle. As stated previously, the

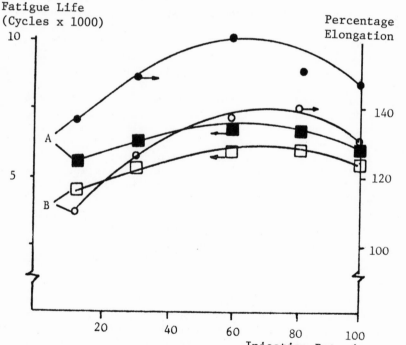

Figure 5:- Fatigue life & percentage elongation against injection rate.

behaviour is less marked for polystyrene than for polyethylene
because of its lower viscosity at moulding temperatures.   This, in
part, explains why the effect of the weld line is less pronounced
for polystyrene, since the alteration to the flow by introducing an
insert is less significant than for polyethylene.   A further
explanation for the difference in weld line effects is the fact that,
whilst polystyrene is amorphous, polyethylene is crystalline.
Hence, the orientation induced into the moulding along the weld line
will tend to be frozen into the specimen of polyethylene, so that
the weakening effect is retained in the final cooled component.
The relatively low viscosity of polystyrene also contributes to
explain why components cut in different downstream locations have
similar strengths in polystyrene compounds;   since the local
disruption due to the obstacle decays more rapidly.   Bearing in
mind that no specimens were cut close to the obstacle, where the
flow is most disorganised, the reasons for the greater strength of
the polyethylene sample cut nearer the obstacle seems to be
attributable to the earlier recombination of flows in this situation.
In this situation, the temperature at recombination is higher than
further downstream and so the diffusion process relieving induced
effects is enhanced.

The fact that increasing melt temperature improves the
performance of final components is clearly demonstrated in Figures
2 and 3.   Higher temperatures of melt not only decrease viscosity,
but also allow for higher diffusion rates and more time for relaxing
and relieving processes to take place.   Unfortunately, this also
implies longer process cycle times and a conflict between output
rates and product quality is revealed.   This conflict is not
apparent in the same way in the effect of increasing injection
speeds.   Indeed, up to injection rates of about 80 cm^3/sec,
strength is improved but it decreases as the rate is further
increased.   These opposing mechanisms seem to be governing the weld
line strength as injection rate is changed.   Higher injection
speeds mean that recombination takes place at higher temperatures,
but they also imply a greater degree of orientation along the weld
line which becomes intense and local at very high rates.   The
former mechanism, improving strength, dominates at low rates,
whilst the latter, reducing strength, becomes more significant at
high rates.   It may reasonably be expected that the relative
strengths of components cut at different locations may be altered
by such a change in mechanism but Figure 4 shows that this is not
the case.

A further outcome of this study is to reveal that whilst a
simple static tension test does not reveal strong effects, dynamic
fatigue testing does.   Clearly, the weld line is more susceptible
to reversed rather than monotonic loading.

CONCLUSION

This prototype study has revealed that the interface formed by recombining melt streams represents a source of weakness in a finished component. The effect, however, is slight in polystyrene, which is brittle, amorphous and has relatively low viscosity at moulding temperatures. Crystalline, relatively viscous poly-ethylene demonstrates a greater weakening effect due to the presence of a weld line, but even in this material the reduction of tensile strength is small; the reduction in percentage elongation to fracture is greater and the weakening effect in fatigue is strong. This ranking in terms of sensitivity to the weld line is also evident in the response to the variation of processing variables. Mould temperature effects are inconclusive, but increasing melt temperature increases strength whilst increasing injection rate reveals a maximum strength at about 80 cm^3/sec for the configuration adopted. Explanations for the behaviour have been offered in terms of material properties.

REFERENCES

[1] Hagerman, E.M., "Weld-Line Fracture in Moulded Parts", S.P.E., October 1973.

[2] Hagerman, E.M., "Weld-Line Fracture in ABS Polymers", S.P.E. Tech. Papers, 19, 1973.

[3] Hobbs, S.Y., "Some Observations on the Morphology and Fracture Characteristics of Knit Lines", S.P.E., September 1974.

CHARACTERIZATION OF EXTRUSION DIES BY THE STRESS DISTRIBUTION AT THE DIE EXIT AND THE FREE RECOVERY OF EXTRUDATE ELEMENTS

E. Fischer and H.H. Winter

Institut für Kunststofftechnologie
Universität Stuttgart
Böblinger Str. 70, 7000 Stuttgart 1

1. INTRODUCTION

The properties of polymer extrudates are mostly determined by the flow in the extrusion die and possibly by the flow history and temperature history upstream in the processing equipment. The analysis of these effects is quite involved since the flow in the extrusion die is inhomogeneous, i.e. the strain and temperature history is different for a polymer element of a layer near the wall and a polymer element flowing more in the middle of the flow channel (Fig. 1).

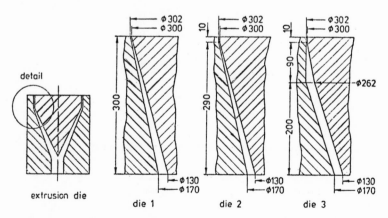

Fig. 1: Sketch of 3 annular die geometries to be investigated. For comparison, a fourth geometry will be used: An annulus of cross section (300 mm to 302 mm) which is constant throughout ("die 4") .

2. RHEOLOGICAL CONSTITUTIVE EQUATION

The rheological behaviour of the flowing molten polymer will be described by the rubberlike liquid equation of Lodge[1] as modified by Bernstein, Kearsley and Zapas[2], and by Kaye[3]:

$$\underline{\underline{\sigma}}(t) = -p \, \underline{\underline{1}} + \int_{-\infty}^{t} m(t,t') \cdot \underline{\underline{C}}^{-1}(t,t') \, dt' \tag{1}$$

where $\underline{\underline{\sigma}}(t)$ is the stress at time t, p the isotropic pressure contribution, $\underline{\underline{1}}$ the unit tensor, $m(t,t')$ the memory functional, and $\underline{\underline{C}}^{-1}(t',t)$ the relative Finger strain tensor between time t' and t. — The material behaviour is contained in the memory functional $m(t',t)$ which not only depends on the time difference t-t' (as the rubberlike liquid does) but also on the time dependent invariants of the strain tensor $\underline{\underline{C}}^{-1}$. — A factorized discrete memory

$$m(t,t') = h(I,II) \sum_{i=1}^{N} G_i \cdot \exp\left[-(t-t')/\tau_i\right] \tag{2}$$

with h = relaxation damping function; I,II = first and second invariants of $\underline{\underline{C}}^{-1}$; N = number of relaxation times; G_i = relaxation moduli of linear viscoelasticity; τ_i = relaxation times of linear viscoelasticity, tested in a most comprehensive test serie, was found to describe all phenomena in shear and uniaxial extension of a low density polymer sample (called "melt I") 4-6 . The damping function used was

$$h = \sum_{j} f_j \cdot \exp\left\{-n_j \left(\alpha I + (1-\alpha) II\right)^{1/2}\right\} \tag{3}$$

where f_j, n_j and α were material parameters. — In most applications, the strain invariants I, II are decreasing functions of time difference t-t'. Counter examples are recovery experiments (extrudate swell, i.e.). In these cases the minimum value of h

$$h^*(t,t') = \min_{t=t'}^{t=t} \left[h(t,t')\right] \tag{4}$$

has to be used in the memory functional (irreversibility assumption of Wagner[8]). — The temperature dependence of the rheological model can be described by the temperature dependence of relaxation times τ_i. The relaxation moduli G_i and damping function seem to be independent of temperature. — The components of Finger tensor are

$$\underline{\underline{C}}^{-1} - \begin{pmatrix} (1+\gamma^2)/\delta_x^2 & -\gamma/(\delta_x \delta_y) & 0 \\ -\gamma/(\delta_x \delta_y) & 1/\delta_y^2 & 0 \\ 0 & 0 & 1/\delta_z^2 \end{pmatrix} \tag{5}$$

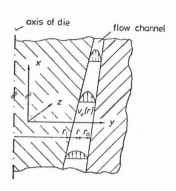

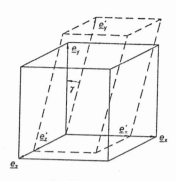

x = flow direction
y = radial direction
z = circumferential direction

Fig. 2: Coordinate system
x,r for annular flow

Fig 3: Deformation t'→t
of a material element

where δ_x, δ_y, δ_z are dilatations in the (carthesian) directions[7]. The shear angle is given by $\gamma(t') = \int_t^{t'} \dot{\gamma}\, dt'$ (Fig.2,3).
Shear and extension are chosen in such a manner that
shear direction and its normal to the shear plane are
principal directions of the extensional flow component.
The shear plane (x-z-plane) changes its area due to the
extensional component.

3. CALCULATION OF FLOW HISTORY IN EXTRUSION DIES

Integral constitutive equations are most suitable
for calculating the stress in a flow of known kinematics.
The kinematics in extrusion dies, however, can be calculated by means of very well established numerical solution procedures assuming locally steady shear flow[9].
The kinematics then will be used to track the fluid
elements along their paths in the extrusion die. The
deformation- (and temperature-)history along each path
determines the state of the polymer at the exit of the
die. - The simplifying assumptions for calculating the
kinematics in annular die flow are the following:
a) constant temperature throughout the flow-channel;
b) no pressure dependence of the rheological properties;
c) constant density;
d) steady shear flow in each cross section.
The coordiante system of the following analysis is
shown in Fig. 2. Annular flow will be described in cylindrical coordinates (x,r) while the strain history
of a fluid element is most easily described in carthesian coordinates (x,y,z). - The assumption of steady

shear flow reduces the x-component of equation of motion to

$$\frac{d\rho}{dx} = \frac{1}{r} \frac{d}{dr} \left(r \cdot \tau_{rx} \right) \tag{6}$$

where the shear stress τ_{rx} is determined by the local
shear rate

$$\dot{\gamma} = \frac{d\, v_x}{d\, r} \tag{7}$$

and the rheological constitutive equation (eq.(1) -
applied to steady shear flow). - The volume flow rate
remains constant with x, since the density is assumed
to be constant throughout the flow. The average velocity

$$\overline{v} = \frac{\dot{V}}{\pi \left(r_a^2 - r_i^2 \right)} = \frac{2}{r_a^2 - r_i^2} \int_{r_i}^{r_a} v_x \cdot y \cdot dy \tag{8}$$

then adjusts to the change in cross-sectional area along
the die. - The radial position of a fluid element is
determined by its value of the stream function, which
strays constant along the annulus. The stream function

$$\psi (r) = \left(\int_{r_i}^{r} v_x (r) \cdot r \cdot dr \right) \bigg/ \left(\int_{r_i}^{r_a} v_x (r) \cdot r \cdot dr \right) \tag{9}$$

is used for integration at constant ψ. Integration along
the stream leads to an evaluation of shear strain $\gamma(t', \psi)$.

4. STRESS RATIO $N_1/2\,\tau_{12}$ AT DIE EXIT

The stress at the exit of extrusion die is a measure of the memorized flow history in the channel. We
therefore calculate the normal stress distribution
$N_1(\psi)$ and the shear stress distribution $\tau_{12}(\psi)$ from
the kinematics as shown above. The ratio of these stresses is found to be the recoverable strain[1]

$$\gamma_r = N_1 / (2 \cdot \tau_{12}) \tag{10}$$

when the flow is steady shear flow. This result, theoretically obtained by Lodge for steady shear flow at
small shear rates, seems to be applicable to high shear
rates too. This has been confirmed experimentally by
Laun[10]. - Flow in extrusion dies cannot be considered
to be steady shear flow, even if the kinematics are
locally very similar. In Fig. 4 the stress ratio as
calculated for the three extrusion dies is compared
with the corresponding recoverable strain distribution
(stress ratio at the exit of "die 4" which is an annulus
of constant cross-section throughout). The difference
is very pronounced: The largest stress ratios are found
in "die 1" and "die 3" in the middle of the flow channel. The stress ratio in "die 2" is much smaller than

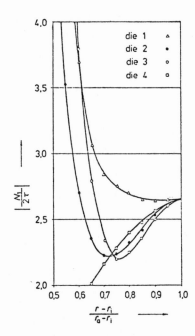
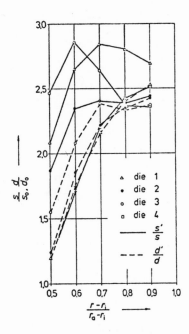

Fig. 4: Stress ratio at exit of extrusion dies of Fig. 1. Rheological data of Laun 5 with ṁ=210 kg/h, T=180 °C.

Fig. 5: Recovery of fluid elements after 100 s.

in the annulus of constant cross section ("die 4"). Tapered sections obviously lead to a significant increase of the stress ratio (due to extensional flow component) while shear flow reduces the stress ratio to a smaller value. - For all four dies close to the wall the stress ratios approach the same value, since the flow approaches steady shear flow there. You can find, however, only a small fraction of the extruded polymer which originates from the layer near the walls. The largest part of the polymer melt passes the annulus in the middle area, where the differences between the four dies are very pronounced. Fig. 4 shows the influence of small changes in die geometry to the stress distribution.

5. FREE RECOVERY AT DIE EXIT

Recoverable strain can experimentally confirmed by measurement of extrudate swell. The polymer stream increases its cross-sectional area when leaving the extrusion die. This area-change depends on time and is related to the strain history in the die. Extrudate swell, however, is an integral measure which smoothes out inhomogeneities within each cross section. It cannot be distinguished between the recoverable strain of an element near the surface of the extrudate and an element from the middle axis. We therefore propose a method describing details of the effects influenced by inhomogeneous strain history across the extrusion die. - It is assumed that the extrudate at the die exit is instantaneously chopped into <u>small elements of homogeneous strain history</u>. The isothermal free recovery of these elements is simulated by a computer study: Each polymer element in the extrusion die undergoes shear and biaxial extension. At time t' = 0 the stress is made isotropic. We then look for the instantaneous and the delayed recovery at times t > 0. - The equation of the stress is used as

$$\underline{\underline{\sigma}}(t) = \underline{\underline{0}} = -\rho \, \underline{\underline{1}} + \int_{-\infty}^{0} [m \, \underline{\underline{C}}^{-1}]_D \, dt' + \int_{0}^{t} [m \, \underline{\underline{C}}^{-1}]_R \, dt' \quad (11)$$

where $[m\underline{\underline{C}}^{-1}]_D$ = strain history of polymer elements along stream line in extrusion die; $[m\underline{\underline{C}}^{-1}]_R$ = strain history during free recovery at constant temperature. - The system of equations (11) was solved numerically. Results of the calculations are shown in Fig. 5. Close to the wall the recovery approaches the values corresponding to steady shear flow. Large differences are found in the middle of the annuli. They are similar to the distribution of $N_1 / (2 \, \tau_{12})$.

REFERENCES

1. Lodge, A.S., Trans.Faraday Soc. 52, 120 (1956).

2. Bernstein, B., E.A. Kearsley, and L.J. Zapas, Trans. Soc.Rheol. 7, 391 (1963).

3. Kaye, A., College of Aeronautics, Note 134, Cranfield, 1962.

4. Wagner, M.H., Rheol. Acta 15, 136 (1976).

5. Laun, H.M., Rheol. Acta 17, 1 (1978).

6. Wagner, M.H., H.M. Laun, Rheol. Acta 17, 138 (1978).

7. Wolff, R., students thesis, unpublished, Univ. of Stuttgart, IKT (1979).

8. Wagner, M.H., S.E. Stephenson, Rheology 23(4),489(1979)

9. Winter, H.H., Adv.Heat Transfer 13, 205 (1977).

10. Laun, H.M., Elast.Deformationsanteile b.d.Dehnung u. Scherung v.Kunststoffschmelzen. Jahrestg.d.deutsch. rheol.Gesellschaft, Aachen, 1979.

INFLUENCE OF WALL SLIP IN EXTRUSION

Günter Mennig

Deutsches Kunststoff-Institut
Schloßgartenstraße 6 R
6100   Darmstadt, Germany

INTRODUCTION

It is known that some polymer melts like rigid PVC[1,2] show wall slipping properties. Although it may be immediately concluded that it leads to different velocity profiles, there is little information available on the influence of wall slip on processing in extrusion and quality of the end product.

Apart from an investigation[3] on the different melting mechanism in a single screw extruder there are only few recent papers on this subject. Worth et al.[4,5] experimented with a wall slipping model fluid in the metering zone. They also calculated the effect of wall slip on throughput and power consumption in case of a one-dimensional isothermal Newtonian situation with simplified assumptions in regard to slip velocities. A more sophisticated study was done by Schlegel[6], who calculated down channel and transverse flow for a Bingham plastic. The results of experiments with ceramics in a screw pump (no compression ratio) were in good accordance with the calculated ones for small pressures.

From an earlier work presented by the author[7], it was shown that the slip coefficient z (i.e. ratio of characteristic lengths of slip flow volume to shear flow volume) is influenced by temperature and pressure. In Fig. 1 results which were obtained by a particular evaluation method[8] are shown for a rigid S-PVC. The pronounced decrease of slip with increasing temperature and decreas-

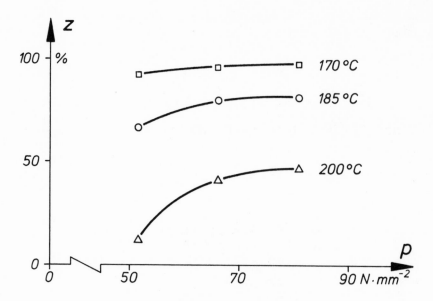

Fig. 1. Influence of temperature and pressure on slip
coefficient z for rigid S-PVC

ing pressure must have its effects in extrusion. However, since this is true for the metering zone only it
cannot be predicted that the usually obtained integral
values e.g. throughput or screw torque will differ considerably for wall sticking and wall slipping conditions. But it may be expected that via velocity profile
there should be an effect on residence time.

EXPERIMENTAL

    The experiments were carried out with a single screw
laboratory scale extrusion unit of 20 mm dia. The barrel
had provision to press small quantities of differently
coloured melt into the screw channel during operation.
The choosen screw (compression ratio 2.5 : 1) exhibited
a very "stiff" processing performance, i.e. within certain limits the throughput was not dependent on mass
temperature or pressure but on screw speed only (horizontal screw characteristics in a throughput-pressure-dia-
gram). The wall slipping S-PVC was the previously
used[2,7] with lubricants added to make it suitable for
extrusion. The wall sticking material was a commercial

LDPE. Dies of 2, 3 and 4 mm dia but same length and the
following barrel temperature program (BTP) were choosen:
> BTP 1 :   150, 155, 160 °C
> BTP 2 :   170, 175, 180 °C

(Increasing temperatures from hopper to screw tip).

As a matter of routine, throughput Q, pressure at the
die entrance p, temperature of out coming strand T and
screw torque M were recorded as a function of screw
speed n. Amongst these values both p and M showed an unusual (and unexpected) decrease with increasing n in
case of PVC for BTP 1 whereas for BTP 2 the curves showed the characteristics of PE for all conditions. Q-p-
plots drawn from these results ressemble the findings
for extra-high molecular weight PE which also slips at
the wall[9].

Minimum residence times t obtained by measuring the
time between pressing differently coloured material into
the screw channel and its first appearance at the out-
coming strand are given in Fig. 2 for PE and Fig. 3 for
PVC (t_3, t_2, t_1 for middle of feeding zone, transition
zone and metering zone). It can be clearly seen that in
case of wall sticking PE t is not affected by different
processing conditions but is a function of Q only. But
for PVC there is a pronounced difference in t for different barrel temperatures.

THEORETICAL CONSIDERATIONS

Following the classical simplified method of describing the flow in a screw channel as the isothermal flow
between parallel plates[10,11,12] Q can be written as

$$Q = b(uh/2 + \Delta ph^3/12\mu)$$ (1)

(b = channel width, u = velocity of upper plate, h =
distance between plates, Δp = pressure difference (=p),
μ = viscosity).

Since the first term (drag flow) is dependent on
n only, it will not change with varied processing conditions or material properties. If moreover the total
throughput   - as in the present case -   is also dependent on n only, the second term must remain constant
for n = const, irrespective of changes in Δp or μ .

In case of non-Newtonian fluids, this is true for
constant flow index of power law[13], i.e. within a cer-

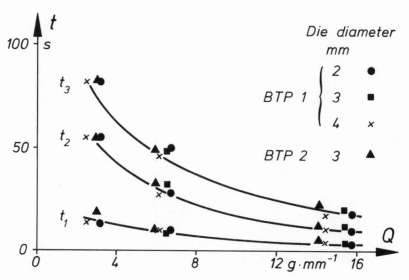

Fig. 2. Minimum residence time t in extrusion of PE

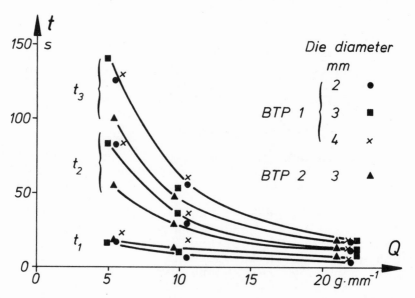

Fig. 3. Minimum residence time t in extrusion of rigid
        S-PVC

tain temperature range[14]. Therefore it may be fairly well assumed for practical cases that if Q = f(n) only, there should be little influence of processing conditions on velocity profile and subsequently on residence time (particularly near the walls). Any result as shown in Fig. 3 must be the expression of different boundary conditions e.g. wall slip.

Assuming slip velocities at both the walls the velocity distribution for the one-dimensional, isothermal Newtonian case is given by[15]:

$$v_x = uy/h + (y^2-hy)\Delta p/2\mu + (v_{su}-v_{sl})y/h + v_{sl} \qquad (2)$$

with v_{su} and v_{sl} = slip velocities at upper and lower plate.

Calculated results are shown in Fig. 4 for different boundary conditions[15].

Extending eq. (2) by the ratio of drag flow to pressure flow Ø

$$2Q = bh(u(1 - \emptyset) - v_{su} - v_{sl}) \qquad (3)$$

shows clearly that the influence of slip on output is dependent on the magnitude of Ø (for given processing conditions). This is valid for both negative and positive pressure gradients along the channel.

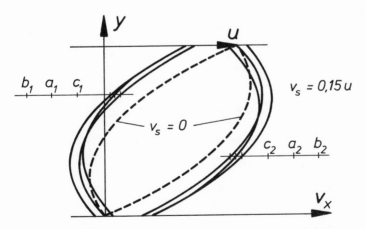

Fig. 4. Velocity profile for different slipping (a,b,c) and non-slipping boundary conditions and positive (1) and negative (2) pressure gradient

CONCLUSIONS

Wall slip may be detected from negative slopes of die characteristics in a throughput-pressure plot. These curves cannot be extrapolated to the origin.

Slip leads to differences in velocity profile. However, the velocity of the material at one of the walls may be zero or very small for certain boundary conditions. Subsequently, residence times may be very long.

The influence of slip on velocity and throughput and thus on quality and efficiency is affected by the ratio between drag flow and pressure flow.

REFERENCES

1. H. Offermann, PhD Thesis, Univ. Aachen (1972)
2. G. Mennig, Rheol. Acta 15:199 (1976)
3. K. P. Klenk, Plastverarbeiter 21:537 (1970)
4. R. A. Worth and H. A. A. Helmy, Plast. Rubber: Process. 2:3 (1977)
5. R. A. Worth, J. Parnaby and H. A. A. Helmy, Polym. Eng. Sci. 17:257 (1977)
6. D. Schlegel, Chem. Ing. Tech. MS 546/77; Synopsis: Chem. Ing. Tech. 49:985 (1977)
7. G. Mennig, J. Macromol. Sci.-Phys. B14:231 (1977)
8. R. Hegler, Project work, Univ. Darmstadt (1978)
9. D. Boes, Kunststoffe 60:294 (1970)
10. E. C. Bernhardt (Ed.), "Processing of Thermoplastic Materials", Reinhold, New York (1959)
11. J. M. McKelvey, "Polymer Processing", John Wiley, New York (1962)
12. G. Schenkel, "Kunststoff-Extrudertechnik", Carl Hanser, München (1963)
13. B. S. Glyde and W. A. Holmes-Walker, Int. Plast. Eng. (1962), ref. in: Kunststofftechnik Kt 73.12-1 (1973)
14. G. Mennig and G. Ehrmann, Kunststofftechnik 8:159 (1969)
15. G. Mennig, Rheol. Acta, in press

ACKNOWLEDGMENT

The author gratefully acknowledges the financial support provided by the Arbeitsgemeinschaft Industrieller Forschungsvereinigungen e. V. for the pursuit of this investigation.

AN INELASTIC APPROACH TO EXTRUDATE SWELLING

R.I. Tanner

Department of Mechanical Engineering
University of Sydney
Sydney, N.S.W. 2006

INTRODUCTION

Some time ago we began to apply finite element computer methods to extrusion problems and jet flow problems. In this paper we survey the progress that has been made by ourselves and others in the solution of these problems using computer methods. We also put forward a mechanism of swelling that became evident from the computer evidence. The object of our research is to understand the losses at inlet and exit, the shape of the extrudate (or the jet) and ultimately the stability of the flow in terms of basic parameters. In the present paper, we will concentrate on the exit problem only; some progress has been made with the inlet problem but it is a simpler problem than the exit one and we expect that when we understand the latter, then the former can be easily attacked. Although we are talking about extrusion here, it is believed that the methods and some of the results have relevance to other flow problems such as melt spinning, film-blowing and wire coating.

To begin with, let us concentrate on the simplest problem of a long (plane slit or axisymmetric circular) die. This will ensure that the flow is a fully developed viscometric flow far upstream of the exit plane. Let us now consider the swelling ratio χ. This ratio is defined (see Fig. 1) as

$$\chi = R_f/R_o \tag{1}$$

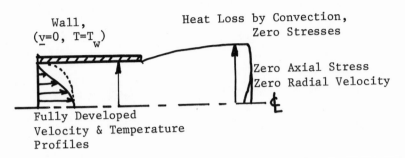

Fig. 1. Definition Sketch for Extrusion Problem.

The swelling ratio is a function of several parameters. These are:

(1) Geometry – R_o is the only length parameter.

(2) Flow Kinematics – only $\bar{w}$, the average entry speed, is needed.

(3) Gravity (g) and surface tension (σ). For most polymer melts, ignoring these factors is a good assumption.

(4) Fluid properties. Clearly we need density (ρ) and viscosity (η) for an isothermal incompressible Newtonian fluid (we only consider incompressible fluids). For other fluids, especially viscoelastic fluids, other parameters are needed, and these will be introduced as required.

(5) If heat transfer is important, then the wall temperature (T_w), the surrounding temperature (T_∞), a surface heat transfer coefficient (h), the product of density and specific heat (ρc) and the conductivity (k) will need to be specified.

(6) Other factors such as slip at the wall can also be considered.

FACTORS INFLUENCING EXTRUDATE SWELL

We can study the influence of the above factors by dimensional methods; once the relevant dimensionless quantities have been selected, then computer solutions can be made to explore the range of parameters required.

Newtonian Fluids

In this case we have (ignoring gravity)

$$\chi = \chi(R_o, \bar{w}, \rho, \eta, \sigma) \tag{2}$$

Here R_b is the tube radius and the other parameters are defined above.

Formation of the dimensionless groups shows that

$$\chi = \chi(2\rho\bar{w}R_o/\eta, \sigma/\eta\bar{w}) \tag{3}$$

The influence of the Reynolds number (Re = $2\rho wR_o/\eta$) and the surface tension number ($\sigma/\eta\bar{w}$) on χ have been thoroughly investigated;[1] neither produces much change in jet diameter. In all cases, we used a finite element program that has been previously described.[2]

From the rheological point of view, the Reynolds number is usually negligible, and in the following we shall ignore it; we shall also ignore surface tension effects which are also small. In brief, inertia tends to make jets smaller, and surface tension tends to pull the jet shape back towards a cylinder of radius R_o. When we ignore ρ and σ in our list of parameters in (2), then no dimensionless group can be formed from R_o,$\bar{w}$ and η, and we have simply that χ is a constant χ_o. For the circular tube extrusion, we find[2] χ_o = 1.13, while for a plane sheet extrusion,[3] χ_o = 1.19. These are to be regarded as base cases for the rest of the work.

Inelastic Non-Newtonian Fluids

The next level of complexity is to permit the viscosity to vary with shear rate. To describe a realistic viscosity-shear rate curve will usually involve several parameters as the changeover from Newtonian behaviour at low shear rates ($\dot{\gamma}$) to a power-law behaviour at higher shear rates. The most important features can be investigated by using the simple power-law rule which gives (in simple shearing)

$$\eta = K|\dot{\gamma}|^{(n-1)} \tag{4}$$

where K is a consistency parameter and n is the flow index. (In a general flow, we replace $|\dot{\gamma}|$ by $\sqrt{I_2}$, where I_2 is the second invariant of the rate-of-deformation tensor.[2]) When n = 1, we have the Newtonian case and K = η; when n < 1, the flow is "pseudo-plastic" (shear-thinning) and when n > 1, it is shear-thickening. Stability considerations forbid a negative value of n, so that n = 0 represents the extreme lower limit for n. In that case, we have slug (or plug) flow in the tube and the extrudate expansion is zero. Dimensional theory then shows us that

$$\chi = \chi(n) \tag{5}$$

in the creeping-flow limit. For $n = 0$ (plug flow) $\chi = 1$; for $n = 1.5$, $\chi \sim 1.18$.

Again, it is clear that no great change of expansion takes place due to change in viscosity with shear rate. In fact, the usual variation of viscosity with shear rate demands $1 > n$ (shear-thinning) and this yields somewhat less expansion than the Newtonian case.

Viscoelastic Effects

Introduction of memory effects may be accomplished in several ways. The Maxwell fluid has a relaxation time (λ), in addition to the viscosity and density, as a fluid parameter. Under creeping flow conditions, dimensional theory shows that

$$\chi = \chi(\lambda \bar{w}/R_o) \tag{6}$$

The group $\lambda \bar{w}/R_o$ is the Weissenberg number (Wi) (it is preferred to use this terminology and to reserve the Deborah number for Eulerian unsteady flows). In the limit of slow flow a viscoelastic fluid behaves as a second-order fluid, which has a connection between the stress tensor $\underline{T}$, the pressure p, and the first and second Rivlin-Ericksen tensors $\underline{A}$ and $\underline{B}$ in the form[3]

$$\underline{T} = -p\underline{I} + \eta\underline{A} + (\nu_1 + \nu_2) \underline{A} - \tfrac{1}{2}\nu_1\underline{B} \tag{7}$$

where $\underline{I}$ is the unit tensor and ν_1 and ν_2 are material parameters with dimensions of kg/m. Dimensional theory shows that in this case

$$\chi = \chi\left(\frac{\nu_2}{\nu_1}, \frac{\nu_1\bar{w}}{\eta R_o}\right) \tag{8}$$

In plane flows, we find ν_2 is irrelevant, and hence can be omitted from (8), so that χ is only a function of $\nu_1\bar{w}/\eta R_o$. The best available experimental evidence shows that ν_2/ν_1 is in any case roughly a constant (about -0.10) and is probably not a very important factor in die-swell. If we note that ν_1/η has the dimensions of time, then the equivalence of (6) and (8) is apparent when ν_2 is ignored. Also, if we note that the first normal stress difference N_1 is proportional to $\nu_1(\bar{w}/R_o)^2$, then the group $\nu_1\bar{w}/\eta R_o$ is proportional to (N_1/τ) evaluated at the tube wall; thus the

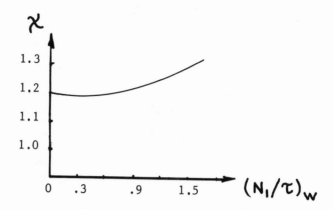

Fig. 2. χ as a function of $(N_1/\tau)_w$ for the plane case with second-
order fluid.[3]

equivalence of the present formulation to those based on "recoverable
shear" $(N_1/2\tau)_w$ is demonstrated. Results for a limited range of
$(N_1/\tau)_w$ are shown (for the plane case) in Fig. 2; these computations are based on the second-order fluid model and did not converge
for the axisymmetric case or for higher Wi values in the plane case.
Further calculations by the author and others for the convected
Maxwell liquid case yield results similar to those in Fig. 5;
there is a slight initial contraction, then an expansion.

There is evidence that the expansion is beginning to increase
rapidly for $(N_1/\tau)_w > 1$. Thus, as expected, we see viscoelasticity
as a cause of die-swell.

Thermal Effects

In the case where the fluid is not isothermal, several more
dimensionless groups appear. In order to have any significant
changes in swelling, we need to couple the thermal and mechanical
fields by permitting the viscosity η to vary with temperature.
Thus, we let

$$\eta = \bar{\eta}\exp\{-\alpha(T - T_w)\} \qquad\qquad (9)$$

where T is the temperature
      α is a positive constant
      T_w is the die-wall temperature
      $\bar{\eta}$ is the viscosity at T_w.

Note that this means the fluid is still Newtonian, since $\dot{\gamma}$ does not appear in (9), Phuoc and Tanner[4] have shown that large jet expansions (~ 70%) can occur due to thermal effects.

Wall Slip

This factor reduces swelling.

AN INELASTIC THEORY OF SWELLING

The above effects can be explained quite well by a new theory of swelling[5] based on the variation of fluid "viscosity" across the die. If we divide the fluid into two parts, one in tension (outer layer) and one in compression (inner layer), then it can be shown[5] that for the plane case the die swell above the Newtonian swelling is given by

$$\chi - \chi_{Newt} = [(\eta_o/\eta_i) - 1]t_o \qquad (11)$$

where η_o is the elongational viscosity for the outer layer and η_i is the same quantity for the inner layer; t_o is the fractional thickness of the outer layer at exit. From (11) when $\eta_o > \eta_i$, we get extra swelling. Similar results hold for the axisymmetric case.

We have presented a new mechanism for extrudate swelling, additional to (a) the Newtonian swelling apparently caused by residual stress or pressure distributions[2], and (b) the elastic recoil mechanism. Since the new mechanism can occur in inelastic fluids, it is proposed to call it the "inelastic swelling mechanism". What proportion each mechanism contributes to typical polymer swelling remains to be discovered.

ACKNOWLEDGMENTS

I thank the Australian Research Grants Committee for supporting the investigation, Mrs. S. Edwards for typing the paper and Mr. T. Shearing for making the drawings.

REFERENCES

1. K.R. Reddy and R.I. Tanner, Computers and Fluids, 6, 83 (1978).
2. R.I. Tanner, R.E. Nickell and R.W. Bilger, Comp. Methods in Appl. Mech. & Engng., 6, 155 (1975).
3. K.R. Reddy and R.I. Tanner, J. of Rheology, 22, 661 (1978).
4. H.B. Phuoc and R.I. Tanner, Thermally-Induced Extrudate Swell, J. Fluid Mech. in press (1980).
5. R.I. Tanner, J. Non-N. Fluid Mech., 6, 289 (1980).

TIME DEPENDENCY OF EXTRUDATE SWELL OF MOLTEN POLYETHYLENE

J.M. Dealy and A. Garcia-Rejon*

Department of Chemical Engineering
McGill University, 3480 rue Universite
Montreal, PO, Canada   H3A 2A7

INTRODUCTION

Extrudate swell is of interest to rheologists because it is
one of the very few readily measured properties that provide
information on the elastic response of a melt to shearing at high
strain rates. For a capillary, which is the flow normally used
for rheological characterization, there is only one swell ratio,
$B \equiv D/D_o$, and this is usually represented as a function of wall
shear rate or wall shear stress for steady flow. In fact, the
swell ratio is also time-dependent, so that the swelling of an
element of extrudate depends on the time which has elapsed since
that element left the die. This introduces an element of
uncertainty into the meaning of reported values of swell measured
in the laboratory, and White and Roman[1] have pointed out that
measured values depend strongly on the details of the technique
used.

Aside from its importance as a rheological phenomenon, the
time dependency of extrudate swell plays an important role in
industrial processes in which the extrudate flows freely from the
die for a short time before being subjected to a post extrusion
forming process. Henze and Wu[2] have discussed the importance of
time dependent swell in the blow molding process.

Except in the case of extrusion from a simple capillary die,
the swelling behavior cannot be described in terms of a single
ratio. For example, in the case of extrusion from an annular
gap, as occurs in the blow molding process, two ratios are

* Facultad de Quimica, Universidad Nacional Autonoma de Mexico

63

necessary. One possible choice, involving the use of a diameter swell, B_1, and a thickness swell, B_2, is illustrated in Figure 1. Alternative choices include the area swell and the weight swell.

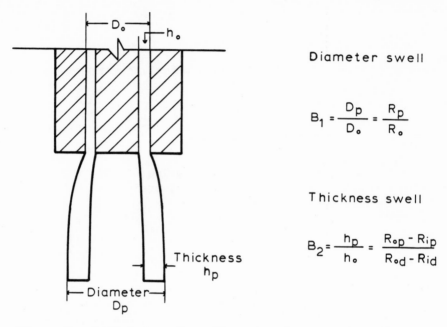

Diameter swell

$$B_1 = \frac{D_p}{D_o} = \frac{R_p}{R_o}$$

Thickness swell

$$B_2 = \frac{h_p}{h_o} = \frac{R_{op} - R_{ip}}{R_{od} - R_{id}}$$

Figure 1.   Definition of swell ratios for extrudate from an
            annular die.

EXPERIMENTAL TECHNIQUES

     The methods which have been commonly used to measure extrudate swell do not lend themselves to the study of time dependent swell because of the roles played by temperature variation and draw down due to gravity. For example, if the extrudate is cooled rapidly, the swell is frozen at its value at some ill-defined time in the cooling cycle. Annealing allows an estimate of the equilibrium swell but provides no basis for establishing the B(t) curve. The use of a laser beam, as in the Monsanto Automatic Die Swell Detector[3] is not well suited for long-term studies of isothermal swell where the extrudate is not suspended in air.

     The apparatus used in the present study was a modification of that described by Utracki et al.[4]. An Instron Capillary Rheometer was used for extrusion, with the extrudate flowing into a thermostatted glass test tube containing oil having the same density as the melt to prevent sagging under the influence of gravity. After extruding until the behavior becomes independent of die residence time, the extrudate is cut off and a test specimen

collected in a test tube and allowed to reach its equilibrium
dimensions. Photographs taken through a window in the thermo-
statting chamber allow the determination of the time dependency
of the swell.

The die used for the capillary studies has a diameter of
0.13 cm and an L/D of 40. For the annular flow studies, the die
had outer and inner diameters of 0.613 cm and 0.5 cm while the
length was 7.5 cm. The swells of a Newtonian fluid leaving a die
with approximately this value of D_o/D_i have been calculated by
Crochet and Keunings[5] to be as follows: B_1 = 1.055; B_2 = 1.17

For the annular die, the transparency of the melt, together
with a fortuitous matching of the refractive indexes of the oil
and melt, permitted the direct determination of the inside diameter
of the tubular extrudate. This allows a much more accurate determination of the thickness swell than is possible by use of the
pinch-off mold technique originally devised by Sheptak and Beyer[6],
the disadvantages of which have been discussed by Kalyon et al.[7].

Because of the design of the barrel and heating mantle of the
Instron capillary rheometer attachment, it was not possible to
observe the very rapid swell which occurs at the die exit. The
time dependency reported here is that which occurs over a time
scale much larger than that associated with this nearly
instantaneous swell.

Data for two high density polyethylene blow molding resins are
presented. Number 22 is SCLAIR 59C manufactured by DuPont of
Canada, while number 26 is grade 40054 manufactured by Dow Canada.
Complete rheological characterizations of these materials have been
reported by Garcia-Rejon et al.[8]. All studies were carried out at
170 ^{o}C.

DISCUSSION OF RESULTS

The swell ratio for capillary flow is shown in Figures 2 and 3
at several shear rates. The time is that which has elapsed since
the specimen was cut. Thus, it does not take into account the
variable history of various elements of the specimen. The
uncertainty introduced by this simplification is not significant,
however, as the time scale involved in sample collection was large
compared to the nearly instantaneous swell at the die exit but
small compared to the time scale over which the long term swell
occurs.

The involvement of two distinct time scales in the swelling
process is a universal feature of the results. The rapid swelling

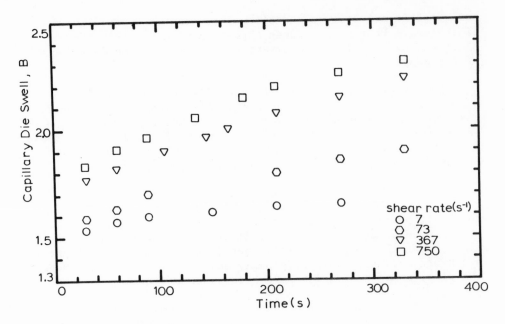

Figure 2.   Capillary extrudate swell as a function of time for
            resin no. 22 at 170 °C.

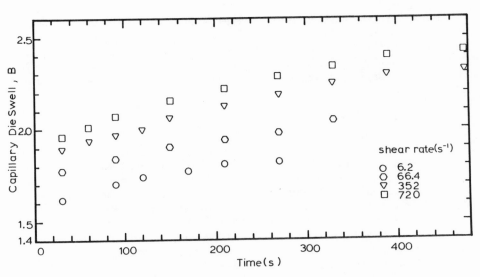

Figure 3.   Capillary extrudate swell as a function of time for
            resin no. 26 at 170 °C

at the die exit occurs over a period of less than a second, while
the long-term approach to equilibrium occurs over a period of 5 to
10 minutes. The rapid early swell accounts for roughly 70 to 80%
of the ultimate swell. The same general phenomena can be seen
in the case of the annular die data shown in Figures 4 and 5.

Henze and Wu[2], used the following equation for weight swell in
their model of parison behavior.

$$B(t) = B_\infty - (B_\infty - B_0)^{-t/\lambda}$$ (1)

where B_∞ and B_0 are the ultimate and instantaneous swell and λ is
a recovery time. This equation fitted our data fairly well, but
the values of λ so obtained varied randomly with shear rate over
a rather wide range. When we plotted ln B versus ln t we found
most of the data to be approximately linear with a slope which was
insensitive to shear rate. A good representation of these lines
was found to be given by equation (2).

$$B = A(\tau_w)^C t^M$$ (2)

Of course this empirical expression cannot accommodate the observed
limiting behavior at very small and very large times.

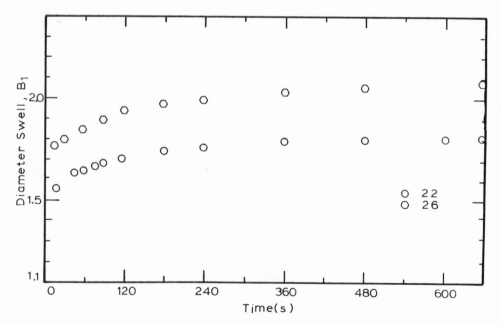

Figure 4.   Diameter swell for annular die extrudate as a function
            of time at 170 $^\circ$C for 2 resins.

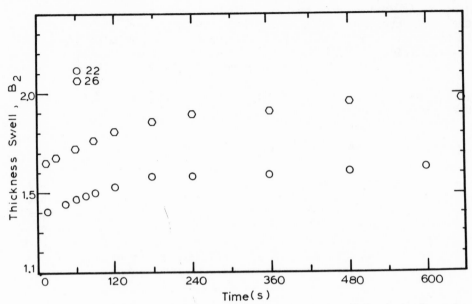

Figure 5.   Thickness swell for annular die extrudate as a function
of time at 170 °C for 2 resins.

REFERENCES

1. J.L. White and J.F. Roman, "Extrudate swell ...", J. Appl.
    Polym. Sci., 20:1005 (1976); 21, 869 (1977).
2. E.D. Henze and W.C.L. Wu, "Variables affecting parison
    diameter swell and their correlation with rheological
    properties", Polym. Eng. & Sci., 13:153 (1973).
3. Data sheet G-14, Monsanto Co., Akron, Ohio, December 1976.
4. L.A. Utracki, Z. Bakerdjian and M.R. Kamal, "A method for the
    measurement of the true die swell of polymer melts", J. Appl.
    Polym. Sci., 19:481 (1975).
5. M.J. Crochet and R. Keunings, "Die swell of a maxwell fluid-
    numerical prediction", J. Non-Newt. Fl. Mech., in press.
6. N. Sheptak and C. Beyer, SPE Journ., 21(2):190 (1965).
7. D. Kalyon, V. Tan and M.R. Kamal, "The dynamics of parison
    development in blow molding", SPE Tech. Papers, 25:991 (1979)
    (to appear in Polym. Eng. & Sci.)
8. A. Garcia-Rejon, J.M. Dealy and M.R. Kamal, "Rheological
    comparison of four blow molding resins", Can. J. Chem. Eng.,
    in press.

DESIGN OF POLYMER MELT EXTRUSION DIES TO AVOID

NON-UNIFORMITY OF FLOW

H.A.A.HELMY * AND R.A.WORTH **

* BONE CRAVENS DANIELS LTD. SHEFFIELD S9 4LS

**MANCHESTER POLYTECHNIC, MANCHESTER M1 5GD

INTRODUCTION

A well designed annular extrusion die used in the manufacture of tubular film, pipe or cable covering, should provide a tube of melt that is uniform thermally, geometrically and kinematically. The mandrel of such a die must be supported in some way, either by a spider in an 'in-line' die or by a flow deflector in a 'cross-head' die. There may be two problems associated with the subsequent disturbance to the flow; firstly weld lines are formed in the extrudate where the flow separates then rejoins around the spider legs or flow deflector (1) as shown in figures 1 and 2. The weld introduces an area of weakness, which may cause premature failure in pressurised pipe. By using a compression zone in the die exit region it is possible to minimize the effect, depending on the residence time of the melt in the compression region. Another method of reducing the weld line effect is to use mixing devices in the die

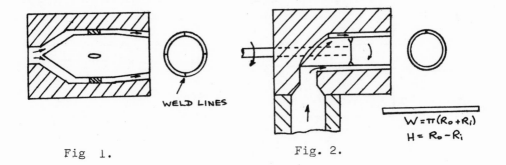

WELD LINES

$$W = \pi(R_o + R_i)$$
$$H = R_o - R_i$$

Fig 1.   Fig. 2.

such as breaker plates, spreader plates (2) or spiral spreaders.

The molecular diffusion mechanism across the weld is often inadequate, particularly in the case of ultra-high molecular weight polymers, and it seems that more positive mixing is needed to overcome the weld line problem.

The second problem is that non-uniform velocity and temperature profiles are set up within the die. In cross-head dies the variations in velocity, which result in uneven wall thickness in the extrudate may be due to either the non-uniform die resistance (flow lengths being greater on the outside of the 90° bend) or the elastic nature of the melt (residual stresses). In the latter case the effect may be reduced by introducing a relaxation zone in the die, so that the stresses developed in the cross-head can relax before the melt leaves the die. A suitable design procedure was described by Parnaby and Worth (3). The problem of non-uniform die resistance can be alleviated by having a choke ring in the die so that the unbalance in the cross-head region is compensated by that in the choke. Pearson (4) showed how cross-head die geometry can be designed to provide uniform die resistance.

The non-uniform velocity profile is often accompanied by temperature non-uniformities, which are equally undersirable, and which may be due to variations in shear history and residence time of the melt, or to asymetrical heat transfer. Localized areas of high temperature in blown film create low resistance to deformation, and hence bad film thickness distribution. This effect is illustrated in table 1 for 12 micron nominal thickness HDPE film where the variation in film thickness in die A can only be attributed to temperature non-uniformity. Other experiments have shown that a temperature variation of $\pm$ 3.5 $^{\circ}$C will cause a gauge variation of $\pm$ 12% in polypropylene film.

Table 1 Gauge Variation in Blown Film

DIE	% THICKNESS VARIATION PRIOR TO STRETCHING.	% GAUGE VARIATION AFTER STRETCHING.
A	$\pm$ 2.6%	$\pm$ 20%
B	$\pm$ 11%	$\pm$ 26%

A method of achieving uniform flow in annular dies and improving the weld line distribution is to use spiral mandrels. The latter are particularly suited to film blowing and large diameter pipe dies.

A design procedure for the spiral mandrel is described and a method
for assessing temperature non-uniformities due to differential shear
histories is proposed. The thickness uniformity of both film and
pipe produced by spiral mandrel dies has been measured on a commercial
film plant and pipe line and the results are presented.

SPIRAL MANDREL DIE

A typical spiral mandrel die is shown in figure 3a. Melt from a
common central entry point is fed to the start of a series of helical
channels of decreasing depth. These channels connect to an annular
channel of increasing clearance between the die body and the mandrel.
The combination of reduction in spiral depth with increase in annular
clearance allows an increasing proportion of melt to leave the spiral
and flow vertically over the land into the adjacent spiral above.
This results in layering of the flows from all the spiral grooves
which eliminates weld lines and improves flow uniformity. The key to
successful design is the correct balance between spiral and leakage
flows.

The first comprehensive analysis of flow in a spiral die was by
Procter (7) who derived a non-Newtonian model for flow in the die and
demonstrated the effect on flow uniformity of varying the design
parameters. The latter as used in the die flow calculations are
shown in figure 3b.

A computer programme has been developed to predict flow
distribution, shear rates and pressure distribution both in the
spiral channel and clearance passage. The programme functions by
dividing the die into a number of flow elements. One flow element

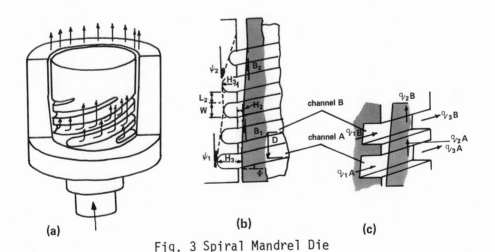

(a)   (b)   (c)

Fig. 3 Spiral Mandrel Die

is shown in figure 3c. The flow q_1 inlet to the spiral element is
divided into two streams, axial or leakage flow q_2 and spiral flow q_3.
Using a power law model $\dot{\gamma} = K \tau^n$ where τ = shear stress and $\dot{\gamma}$ = shear
rate, k and n are the melt rheological constants, then

$$q_2 (x) = K_2 G_2^n (x) \Delta p_2^n (x) \ldots\ldots\ldots\ldots\ldots (1) \quad \text{and}$$
$$q_3 (x) = K_3 G_3^{nl}(x) \Delta p_3^{nl}(x) \ldots\ldots\ldots\ldots\ldots (2)$$

where G_2 and G_3 are flow resistance constants and are functions of
the dimensions of each flow elements and also of the melt constant n
(for an element of circular cross section of radius R and length L
$G = (\dfrac{\pi}{3 + n})^{1/n} \ (\dfrac{R \ 3/n + 1}{2L}), \Delta p_2$ and Δp_3 are the pressure drops
along the clearance passage and the spiral passage respectively in
the flow element shown in figure 3c.

Equations(1) and (2) can be coupled through the relationship
between Δp_2 and Δp_3 which is obtained by assuming (as a first
approximation) a constant rate of pressure drop along the spiral
channel. This is regarded as a reasonable assumption due to the
slow rate of change of spiral channel area. A further condition for
calculation is that q_1 (x)= q_2(x) + q_3(x)$\ldots\ldots\ldots\ldots\ldots$(3).

The programme proceeds from one element to the next and
calculates values for q_2(x) and q_3(x) for each element. The leakage
flow from one spiral element entering the spiral element above can
be considered to be redistributed again as a ratio of the flow
resistances in the clearance and along the spiral.

Based on the set of calculated values of leakage flow q_2(x)
a new set of spiral flows q_3(x) is calculated and hence a new value
for spiral pressure drops Δp_3(x). Using the latter a more accurate
set of land flows q_2(x) are calculated and compared with the
original set calculated previously. The design method continues
until a small prescribed difference between the leakage flows from
two consecutive loops is achieved. Figures 4a, b and c show a
comparison between the flow uniformity as predicted by the design
method and that measured experimentally. The effect of the more
non-Newtonianity of the melt is reflected both on flow uniformity
of the extruded tube and even in a more drastic way on the film
thickness after blowing. The higher blow up ratio i.e. more
transverse stretching in the HDPE case coupled with the shear
sensitive nature of the high molecular weight HDPE used is
responsible for the bad gauge variations in the example in figure 4b.
Figure 5 shows the possible error encountered in using a constant
rate of pressure drop in the spiral. Although the flow from the die
showed reasonable uniformity after the first approximation it
certainly exhibited signs of deterioration as judged by the premature
emptying of the spiral channel.

Although flow uniformity is of prime importance in die design,
pressure drop across the die is also a very important factor which

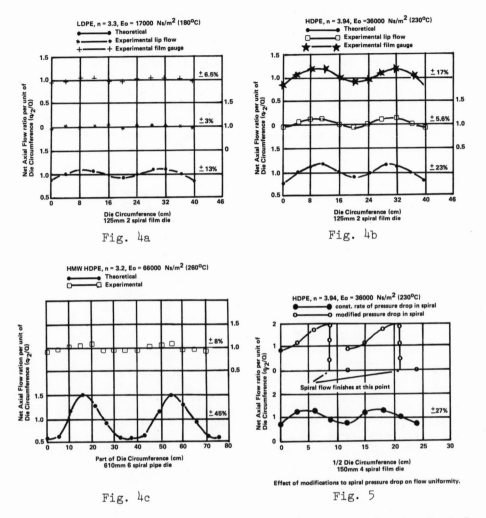

Fig. 4a

Fig. 4b

Fig. 4c

Fig. 5

must not be overlooked. A high pressue die means reduced output from
the extruder and also high melt temperature. Pressure drops in dies
should be typically in the range of 21-28 MN/m^2 for LDPE and 35-40
MN/m^2 for HDPE.

The design method described above proved to be a successful
tool for designing spiral dies. The use of a more accurate pressure
drop in the spiral is an improvement on previous published work (7).
A further improvement to the method is to incorporate a measure of
the amount of shear (shear rate x residence time) as experienced by
the different flow elements to ensure uniform shear history. In the
present method this is monitored by printing out the shear rate at
each element. The requirement for monitoring the shear experienced
by different flow elements is very important in the film blowing

application particularly when dealing with shear sensitive polymers
such as HMW HDPE and Polypropylene.

A second useful solution to the problem of extrudate non-
uniformity is the use of a rotating die core, which distributes the
flow around the circumference of the annulus. A degree of laminar
mixing is also provided to overcome the weakness of the weld line.
The paper describes a suitable rotating die design, and the effect
on weld line formation is investigated theoretically and
experimentally.

ROTATING DIE SYSTEM

Figure 2 illustrates schematically a rotating die assembly,
based on a cross-head die with a parallel die land region where the
core can be rotated by a drive shaft supported in bearings at the
rear of the die. A weld line is formed at the tip of the flow
deflector, which without die rotation would persist to the exit of
the die, producing weakness in the extruded pipe. To illustrate
weld line formation theoretically the isovels in the die have been
plotted assuming that the annulus can be approximated by a
rectangular channel, as shown in figure 2, then translating the
results to the cylindrical co-ordinates of the annular die.
The velocity distributions for a Newtonian fluid (5) have been used
to plot the isovels shown in figure 6a.

When the die core is rotated a tangential velocity field is
created, so that any element of fluid within the die will be rotated
through an angle depending on the tangential velocity of the element
and its residence time in the die (6).

Modified isovel distributions are shown in figure 7a for a die
speed of 5 r.p.m. The weld line now forms an oblique line across
the pipe wall and consequently when the die is rotated at high
speeds the weld line will form a spiral around the annulus, the
isovels tending towards concentric circles.

The velocity distributions within the die were investigated
experimentally by examinations of this section of carbon black
pigmented polyethylene pipe. Photographic prints, using sections
of 40 microns thickness as negatives, are reproduced in figures
6b and 7b, which suggests that the carbon black particles tend
to segregate along the isovels which exist in the cross-head
section of the die. This is thought to be due to the chain-like
structure of carbon black, where chains which cross the shear
stress gradient are destroyed, but chains lying along lines of
constant shear stress are maintained.

The experimental findings support the theoretical results,
and show that when the die rotation speed is increased the weld

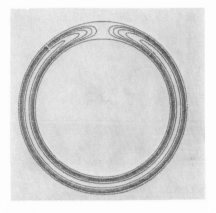   

a) THEORETICAL   b) EXPERIMENTAL

Fig.6   ISOVEL DISTRIBUTIONS FOR STATIONARY DIE.

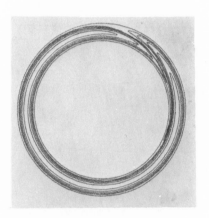   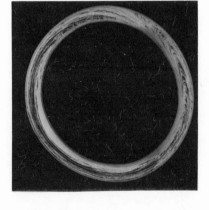

a) THEORETICAL   b) EXPERIMENTAL

Fig.7   ISOVEL DISTRIBUTIONS AT 5 r.p.m.

line begins to spiral around the pipe wall, thereby increasing the
area available for cohesion at the weld, and suggesting that the
strength of the pipe will be improved.

The laminar mixing provided by the rotating die core should
also encourage temperature and velocity uniformity at the die orifice.

The use of the rotating die is particularly suitable for the
production of glass fibre reinforced thermoplastic pipe where it is
unlikely to distribute the fibres across the weld line using spider
or cross-head type dies.

REFERENCES

1. U. Kleindienst, Kunststoffe, 63, 5, 423 (1973).
2. G.Prall, Modern Plastics, p.118 (Nov.1970).
3. J.Parnaby and R.A.Worth, Trans.I.Chem.E.
    52, 4, 368 (1974).
4. J.R.A.Pearson, Trans. J.Plastics I,
    31, 95, 125 (1963).
5. J.M.McKelvey, 'Polymer Processing', Wiley, (1962).
6. R.A.Worth, 'Modifications to Weld Lines in
    Extruded Pipe using a Rotating Die System',
    Polymer Eng. Sci, in press.
7. B.Procter, SPE Journal, 28, 2, 41 (1972).

BROKEN SECTION METHOD FOR ANALYZING MOLTEN FLOW IN EXTRUSION DIE OF PLASTIC NET

Katsuhiko Ito, Yasuhiko Kato*, Shuzo Kimura**

College of Engineering, Hosei University,
*Daicel Chemical Ind., Ltd., Central Research Laboratory,
**Dai-Nippon Plastics Co., Ltd., Matudo Factory,
Kajino-cho, Koganey, Tokyo 184, Japan
*Ohui-machi, Iruma-gun, Saitama 354, Japan
**Midoridai, Matudo-city, Chiba 271, Japan

INTRODUCTION

This paper describes the analytical results of the molten flow in the extrusion die of the plastic net by the broken section method.

The die has the inner and outer lips of which the series of narrow channels are located at both sides of sliding contact surfaces with equal intervals. The lips rotate in the opposite directions.

The separation and the join of the series of narrow channels correspond to the strand and the cross area of the net.

From the start of cross joining of the inner lip with the outer lip to the complete cross joining, nine profiles of channels are considered with the time dependent behavior.

In these profiles of channels the interesting isovels are given, and the large variation of volumetric flow rate is found.

BROKEN SECTION METHOD

Considering a channel, as shown in Figure 1, it is assmed that the polymer melt flow in the channel is isothermal, laminar and steady, and is free of slip along the walls. Further, the polymer melt is assumed to be incompressible.

If v_i is the flow velocity of polymer melt in the i-th section along its flow direction which is taken as the Z-axis, v_i is expressed as in eq. (1),

$$v_i = v_{i\ max}\left[1-\left|\frac{2}{Hi}\left\{Y_i-g(x_i)\right\}-1\right|^{n+1/n}\right] \qquad (1)$$

where $v_{i\,max}$ is the maximum velocity in the i-th section, H_i is the channel height, and n is the flow index.

$v_{i\,max}$ is obtained as follows,

$$v_{i\,max}=-\frac{H_i}{2}\left(\frac{n}{n+1}\right)\left[\frac{H_i}{2\,\eta^0|\dot\gamma^0|^{1-n}}\left(\frac{dp}{dz}\right)\right]^{1/n} \tag{2}$$

where $\dot\gamma^0$ is the shear rate in a standard state (e.g., $\dot\gamma^0=1\,sec^{-1}$), η^0 is the non-Newtonian viscosity at the sear rate $\dot\gamma^0$, dp/dz is the pressure gradient in the flow direction.

The value Y_i of the ordinate that gives a certain velocity V is obtained from eq. (1) as follows,

$$Y_i=\frac{H_i}{2}\left\{1\pm(1-\frac{V}{v_{i\,max}})^{n/n+1}\right\}$$
$$+\,g(x_i) \tag{3}$$

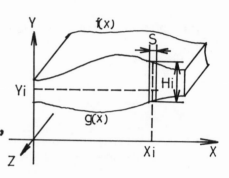

Fig. 1 Common Die

Since the maximum value of $v_{i\,max}$ in the channel, V_{max}, corresponds to the maximum height $H_{i\,max}$, eq. (2) is simply re-written as follows,

$$V_{max}=-\frac{H_{i\,max}}{2}\left(\frac{n}{n+1}\right)\left[\frac{H_{i\,max}}{2\,\eta^0|\dot\gamma^0|^{1-n}}\left(\frac{dp}{dz}\right)\right]^{1/n} \tag{4}$$

From eq. (2) and eq. (4) are derived eq. (5) and eq. (6).

$$v_{i\,max}/V_{max}=(H_i/H_{i\,max})^{n+1/n} \tag{5}$$

then,

$$v_{i\,max}=(H_i/H_{i\,max})^{n+1/n}\cdot V_{max} \tag{6}$$

If K is taken as the ratio of V to the maximum velocity V_{max}, K is given as in eq. (7).

$$K=V/V_{max} \tag{7}$$

From eq.(3), the value Y_i is written as function of only H_i and n.

$$Y_i=\frac{H_i}{2}\left[1\pm\left\{1-K(\frac{H_{max}}{H_i})^{n+1/n}\right\}^{n/n+1}\right]+g(x_i) \tag{8}$$

The volumetric flow rate Q_i at the i-th section is given by integrating eq. (1).

$$Q_i = \left(\frac{n+1}{2n+1}\right) \cdot S \cdot H_i \cdot v_{i\,max} \qquad\qquad (9)$$

where S is the width of each section.
The overall flow rate Q_o is a simple sum of the volumetric
flow rate Q_i as in eq. (10).

$$Q_o = \sum_i Q_i \qquad\qquad (10)$$

APPLICATION TO DIE LIP OF HALF CIRCULAR TYPE

The half circular lip, as shown in Figure 2, is chosen as the
typical die lip used for the plastic net. The isovels for nine
selected profiles are obtained by using eq. (8).
From the start of cross joining of the inner half circular lip
with the outer half circular lip to the complete cross joining,
nine profiles of the channels are studied.
For the calculation it is given
that the diameter of the half
circular lip is one and is divided
into 100 equal sections. As for K
are selected the eleven values
between 0 and 1 at equal intervals.
Then, for certain n the isovels
are obtained, as shown in Figure 3.
The volumetric flow rate Q_i
and the overall flow rate Q_o are
calculated by using eq. (9) and
(10). For the analysis of the
practical cases the values of the
variables are selected as follows,

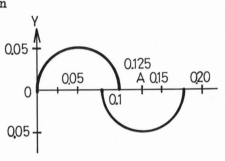

Fig. 2   Half Circular Type Die

diameter of the lip = 0.1 cm
n   = 0.4673
$\dot{\gamma}^o$ = 1 sec^{-1}
η^o = 6.8×10^4 poise
dp/dz   = 4.905×10^7 dyne/cm$^2 \cdot$ cm

The results are given in Figure 4 and 5.

RESULTS

Figure 6 shows the cross sections of the extrude products.
The deviation for the cross sections from the calculated results
is partly due to the stretch by the winder.

In the isovels, when the inner and outer half circles cross
join about a quater of diameter, it is shown that the position

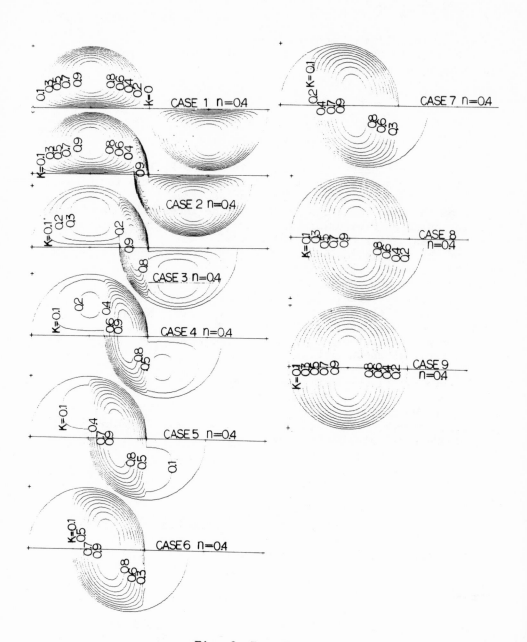

Fig. 3   Isovels

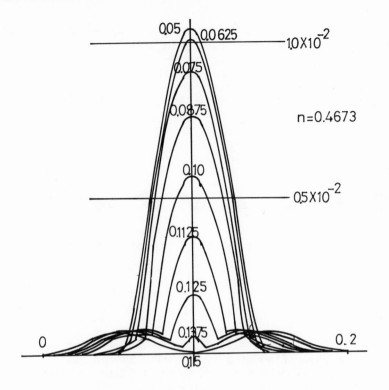

Fig. 4   Volumetric Flow Rate Q_i

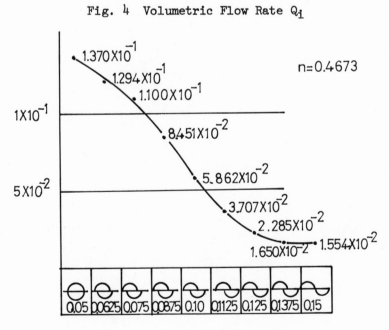

Fig. 5   Overall Flow Rate Q_o

having the maximum velocity moves from the center of the half
circle to the center of the cross join area.

The variation of the volumetric flow rate between the start
of cross joining and the complete cross joining is very large.
The flow rate of the latter is ten times as large as the flow rate
of the former.

Fig. 6 Cross Sections of Extrude Products

REFERENCE

K.Ito, M.Ishida, and N.Kikuchi, "Broken Section Method for
    Analyzing Flow of Polymer Melts in Die Having Cross
    Sections of Varying Heights", Applied Polymer Symposium,
    20: 99 (1973)

RHEOLOGICAL BEHAVIOUR, EXTRUSION CHARACTERISTICS AND VISCOUS DISSIPATION IN FIBER REINFORCED POLYMER MELTS

B. Anders Knutsson
James L. White
Polymer Engineering
University of Tennessee
Knoxville, TN   37916 USA

Kent B. Abbas

Telefon AB L M Ericsson
S-126 25 Stockholm, SWEDEN

INTRODUCTION

In this paper we shall consider the influence of glass fibers on the flow characteristics of polycarbonate. Viscous dissipation during the flow of fiber filled suspensions will also be discussed.

Earlier studies of glass fiber reinforced thermoplastics have been limited largely to polyolefin and polystyrene melts (1-5). Papers concerning rheological properties of polycarbonate have appeared (6-7), but emphasize generally shear viscosity for unreinforced grades. Degradation of polycarbonate and its influence of the rheological properties has been investigated by Abbås (8).

Fiber damage during flow has been investigated by Czarnecki and White (9) for polystyrene compounds and by O'Connor (10). Fiber motions and orientations in flow of thermoplastics are investigated by some authors (11-14).

MATERIALS

The polycarbonates and the glass fiber reinforced compounds are listed in Table 1. The grades investigated were all produced by Bayer AG, West Germany for commercial purposes. These grades are commonly used in electronic applications. The materials also contain thermal stabilizers, release agents and in two cases flame retardants.

TABLE 1

MATERIAL	SUPPLIER	GLASS FIBERS (weight %)	SYMBOL	$\overline{M}_w$	$\overline{M}_w/\overline{M}_n$	NOTE
Makrolon 2805	Bayer AG	0	■	29300	2,44	
Makrolon 6555	— „ —	0	▲	28900	2,43	Flame ret.
Makrolon 9410	— „ —	10	●	28600	2,33	— „ —
Makrolon 8324	— „ —	20	▼	30400	2,41	
Makrolon 8035	— „ —	30	◆	30700	2,34	
Makrolon 8344	— „ —	40	★	27900	2,51	

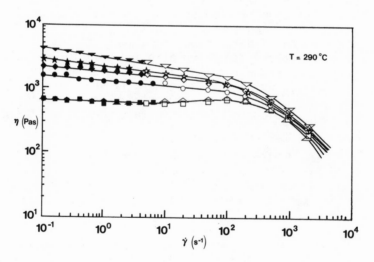

FIGURE 1.   Shear viscosity versus shear rate. (For symbols see
            Table.)

Molecular weight distributions of the polycarbonates were measured in a Water Associates Gel Permeation Chromatography (GPC) Model 200. Weight average molecular weights and the ratios of weight to number average molecular weight are also summarized in Table 1.

RHEOLOGICAL BEHAVIOUR

Capillary measurements were carried out in an Instron Capillary Rheometer using a series of dies with D = 1.47 mm and length/diameter ratios 5, 10, 20, 30 and 40. The shear stress was obtained from a Bagley plot (15) of total pressure vs. die L/D. The measurements were carried out at 290°C.

A Rheometrics Mechanical Spectrometer in the cone-plate mode was used to determine shear viscosities and principal normal stress differences at shear rates in the range $10^{-1} - 10^1$ s^{-1}. Shear measurements were done at 250, 275, 290 and 300 °C. Owing to very low normal forces it was only possible to determine the principal normal stress difference accurate at 250°C.

In Fig. 1 the shear viscosities are plotted vs. shear rate for all grades. The data for the cone-plate and the capillary instruments agree. Generally the fiber reinforced melts exhibit higher viscosities than the unfilled melts. There is however not a monotonic increase with loading level. This can be explained by different fiber length distributions and molecular weights.

The unreinforced polycarbonates behave quite Newtonian to shear rates of 150 s^{-1}. By adding fibers to the melt, the flow behaviour become more non-Newtonian. The viscosity data in Fig. 1 are not treated for pressure effects. The pressure was shown to have a large influence of viscosity by Baumann and Steinsiger (6). Due to the relative large capillary diameter in our experiments we did not find the influence of pressure on viscosity significant. We observed however that the Bagley plots turned to become non-linear at higher pressures (50 MPa). The pressure effect was less for the reinforced compounds than for the unfilled melts.

The temperature dependence of the viscosity is shown in Fig. 2, where we plot zero viscosity vs. reciprocal temperature. The activation energy will decrease by adding fibers, although the molecular weight seems to have a larger influence.

The principal normal stress difference (N_1) is plotted vs. shear stress (σ_{12}) in Fig. 3. We see that the values of N_1 at fixed σ_{12} are increased by the presence of fibers. This agrees

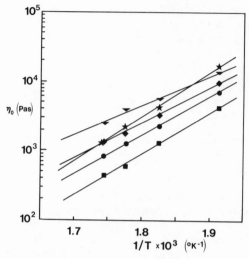

FIGURE 2.   Zero viscosity versus reciprocal temperature. (For symbols see Table 1.)

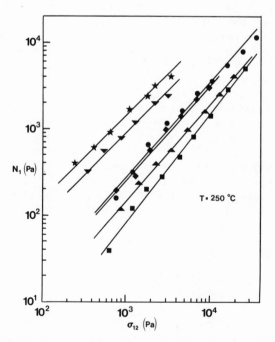

FIGURE 3.   Principal normal stress difference versus shear stress. (For symbols see Table 1.)

with the experiments of Czarnecki and White for polystyrene melts
(9). The magnitude and slope are however lower than the results
of Han (16) and Oda et al (17) concerning polystyrenes and polyolefins. In a viscoelastic fluid in the Newtonian region one
expects the principal normal stress difference to be proportional
to the square of the shear stress.

EXTRUSION CHARACTERISTICS

Extrudate swell was measured at 290°C using a die diameter
of 1.47 mm and L/D = 40 in an Instron Capillary Rheometer. The
extrudate swell vs. shear rate is shown in Fig. 4. The swell for
unreinforced polycarbonate is not large. It is however completely
surpressed by the addition of the glass fibers.

Surface morphology was examined in a scanning electrone
microscope. A series of photos taken, which showed the surface
to become more smooth when the shear rate was increased. The
glass fiber length seemed to exhibit a major influence on surface
smoothness.

Fiber orientation during flow was determined by monitoring
cross-sections from slit extrudates in a scanning electron
microscope. The ratio between the highly oriented zone and the
total extrudate thickness vs. shear rate is shown in Fig. 5.
We obtain completely oriented extrudates at moderate flow rates.

Fiber damage was investigated by using a microscope. The
polymer was solved away and the lengths of the fibers could be
measured. Fiber length measurements were carried out for virgin
material and for samples extruded at an apparent shear rate of
1922 s^{-1} through a die of diameter 1.47 mm and L/D = 20 at 290°C.

During shear flow the glass fibers will undergo a shear
stress induced buckling, which was first described by Forgacs
and Mason (18). From their theory we can predict a critical
fiber length/diameter in our compounds to be of order 10 at shear
rates of order 2000 s^{-1}. The experimental results agreed quite
well to the theory.

VISCOUS DISSIPATION

Due to the increasing viscosity caused by the glass fibers,
we believe the viscous dissipation to be increased by the
addition of fibers to a polymer melt. If the flow were assumed
to be adiabatic and incompressible, the bulk temperature rise is

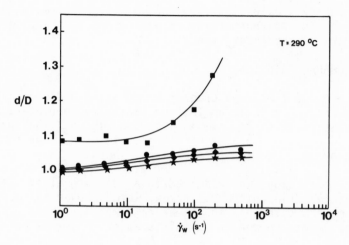

FIGURE 4.   Extrudate swell versus wall shear rate.   (For symbols
            see Table 1.)

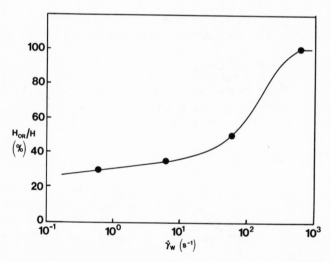

FIGURE 5.   Average orientation ratio versus wall shear rate.
            (For symbols see Table 1.)

given by

$$\Delta T = \frac{\Delta P}{\rho c} \tag{1}$$

The product of ρc will increase by adding fibers to the melt, but not that much as the pressure loss increases. Thus giving a higher bulk temperature increase, indicating viscous dissipation to be larger for compounds than for unreinforced melts.

Using the physical equations of continuity, momentum and energy and rheological relationships it is possible to derive an expression for the temperature field. For a temperature dependent viscosity, no analytical solution may be obtained. The solution of such problems has been considered by several persons (19-23).

By adding fibers, the melt is no longer isotropic but we have a compound highly dependent on the fiber orientation. The energy equation has the form:

$$\rho c \left(\frac{\partial T}{\partial t} + \underline{v} \cdot \nabla T\right) = \nabla \cdot \underline{\underline{k}}\nabla T + \underline{\underline{\tau}} : \nabla\underline{v} - T\left(\frac{\partial P}{\partial T}\right)_\rho (\nabla \cdot \underline{v}) \tag{2}$$

where we have considered the thermal conductivity tensor

$$\underline{\underline{k}} = \begin{vmatrix} k_1 & 0 & 0 \\ 0 & k_2 & 0 \\ 0 & 0 & k_2 \end{vmatrix} \tag{3}$$

in the coordinate system of an oriented element along its principal axis. It may be shown that

$$k_{zz} = k_2 + (k_1 - k_2) \overline{\cos^2\theta}_{1,z}$$

$$k_{yy} = k_1 - (k_1 - k_2) \overline{\cos^2\theta}_{1,z} \tag{4a,b}$$

where $\theta_{1,z}$ is the angle the fiber makes to the flow direction (z). Local convection effects due to fiber rotation during flow may also enhance the thermal conductivity.

Assuming a steady state the energy equation can be simplified (if metal fibers are not used) to the Graetz-Brinkman asymptote. The temperature distribution for long dies is given by

$$T(y) = \int \frac{1}{k_{yy}} \left\{ \int K \left(\frac{\partial v}{\partial y}\right)^{n+1} dy \right\} dy \qquad (5)$$

From this we can see that a higher viscosity will increase the temperature. The thermal conductivity of the glass fibers is of order five times larger than the polymer melt.

REFERENCES

1. D. P. Thomas and R. S. Hagman, Society of the Plastics Industry, Reinforced Plastics Division Preprints, 3-C (1966).
2. J. M. Charrier and J. M. Rieger, Fiber Sci. Tech., 7, 161 (1974).
3. Y. Oyanagi and Y. Yamaguchi, J. Soc. Rheol. Japan, 3, 69 (1975).
4. Y. Chan, J. L. White and Y. Oyanagi, J. Rheology, 22, 507 (1978).
5. Y. Chan, J. L. White and Y. Oyanagi, Polym. Eng. Sci., 18, 268 (1978).
6. G. F. Baumann and S. Steinsiger, J. Polym. Sci., Pt A-1, 3395 (1963).
7. M. Yamada and R. S. Porter, J. Applied Polym. Sci., 18, 1711, (1974).
8. K. B. Abbas, Proceeding 2nd Int Symposium Degradation and Stabilization of Polymers, Dubrovnik, Yug. (1978).
9. L. Czarnecki and J. L. White, J. Appl. Polym. Sci., (in press).
10. J. E. O'Connor, Rubber Chem. Technol., 50, 945 (1977).
11. M. W. Darlington and P. L. McGinley, J. Mat. Sci., 10, 910 (1975).
12. L. A. Goettler, R. I. Leib and A. J. Lambright, Rubber Chem. Technol., 52, 838 (1974).
13. K. Yoshida, G. Budiman, Y. Oyayama and T. Kitao, Seni-Gakkaishii, T-225 (1975).
14. S. Wu, Polym. Eng. Sci., 19, 638 (1979).
15. E. B. Bagley, J. Appl. Phys., 28, 624 (1957).
16. C. D. Han, "Rheology in Polymer Processing," Academic Press, NY (1976).
17. K. Oda, J. L. White and E. S. Clark, Polym. Eng. Sci., 18, 25 (1978).
18. H. L. Goldsmith and S. G. Mason in "Rheology: Theory and Applications," Vol. 4, p. 85 Academic Press, NY (1967).
19. J. E. Gerrard, F. E. Steidler and J. K. Appeldorn, Ind. Eng. Chem. Fundam., Vol. 4, No. 3 (1965).
20. R. A. Morette and C. G. Gogos, Polym. Eng. Sci., 8, (1968).
21. H. T. Kim and E. A. Collins, Polym. Eng. Sci., 11, (1971).
22. H. W. Cox and C. W. Macosko, Am. Inst. Chem. Eng., J-20, (1974).
23. R. Kohler, Kunststoffe, 68, 6, (1978).

TIME-DEPENDENT RHEOLOGICAL BEHAVIOR OF POLYMERIC SYSTEMS

A. I. Isayev, C. A. Hieber, R. K. Upadhyay and S. F. Shen

Sibley School of Mechanical and Aerospace Engineering
Cornell University
Ithaca, N.Y. 14853

INTRODUCTION

The capability to describe time-dependent rheological material behavior is a necessary element of any new proposed viscoelastic equation. Recently, Leonov[1,2] published a rheological equation, based upon irreversible thermodynamics, which describes the behavior of elastic polymeric fluids over a wide range of elastic strain. This model has subsequently been used for predicting flow orientation in injection molding[3], where it was shown that the time-dependent rheological behavior is important for predicting frozen-in birefringence in molded parts. In the present paper, by using the Leonov equation together with numerical methods, the following two time-dependent problems are considered: (1) transient simple shear flow with impulsive change from zero shear rate to non-zero constant shear rate with subsequent relaxation following cessation of flow; (2) oscillatory shear flow in the linear and non-linear regimes.

TRANSIENT SIMPLE SHEAR FLOW

For experimental determination of stresses built up during transient shear flow, rotational devices have usually been employed in which the stresses are measured by mechanical means, having limited rigidity. Recently, however, in papers by Gortemaker, et al.[4] and Osaki, et al.[5], optical methods have been used for measuring polymeric behavior in transient flows. It has been shown[4] that mechanical devices can give incorrect information about transient flow, particularly concerning the duration of the process. Accordingly, in checking out the predictions of the Leonov theory below, comparison will be made with the optical measurements[4,5] concerning birefringence and extinction angle during flow and relaxation.

We consider here the one-dimensional shear flow between two plates which results when one plate is impulsively given a velocity U at time $t = 0^+$ while the other plate is maintained fixed with a distance b between the two. The relevant set of equations according to the Leonov model[1,2] is then given as

$$\rho \frac{\partial u}{\partial t} = \sum_{k=1}^{N} \eta_k \, s \, \frac{\partial^2 u}{\partial y^2} \Big/ (1-s) + \sum_{k=1}^{N} \frac{\eta_k}{\theta_k} \frac{\partial C_{12,k}}{\partial y}$$

$$\partial C_{11,k}/\partial t - 2C_{12,k}(\partial u/\partial y) + \left(C_{11,k}^2 + C_{12,k}^2 - 1 \right)/(2\theta_k) = 0$$

$$\partial C_{12,k}/\partial t - C_{22,k}(\partial u/\partial y) + \left(C_{11,k} + C_{22,k} \right) C_{12,k}/(2\theta_k) = 0$$

$$C_{11,k} \, C_{22,k} - C_{12,k}^2 = 1 \tag{1}$$

with boundary and initial conditions:

$$u(y, t = 0) = 0 \;, \; u(0, t > 0) = 0 \;, \; u(b,t) = U$$

$$C_{11,k}(y,0) = 1 = C_{22,k}(y,0) \;, \; C_{12,k}(y,0) = 0 \tag{2}$$

The stress field is given by

$$\underset{\approx}{\sigma} + p\underset{\approx}{\delta} = \sum_{k=1}^{N} \frac{\eta_k s \dot\gamma}{1-s} \begin{pmatrix} 0 & 1 & 0 \\ 1 & 0 & 0 \\ 0 & 0 & 0 \end{pmatrix} + \sum_{k=1}^{N} \frac{\eta_k}{\theta_k} \begin{pmatrix} C_{11,k} & C_{12,k} & 0 \\ C_{12,k} & C_{22,k} & 0 \\ 0 & 0 & 1 \end{pmatrix} \tag{3}$$

which has been used in writing the first equation in (1), with p being the isotropic pressure, $C_{11,k}$, $C_{22,k}$ and $C_{12,k}$ being components of the elastic strain tensor, η_k and θ_k being the shear viscosity and relaxation time of the kth mode and s being a rheological parameter $(0 \leqslant s \leqslant 1)$.

In order to solve (1) subject to the conditions in (2), the model parameters, η_k, θ_k (k = 1 , ..., N) and s, must first be specified. These constants have been evaluated for an 8% solution of polystyrene in Aroclor 1248 on the basis of the steady shear viscosity data presented by Osaki, et al.[5] and for a polystyrene melt on the basis of the frequency dependence of the storage G' and loss G" moduli in the linear range, as presented by Gortemaker, et al.[4]. In both cases, a least-square procedure has been developed to fit the data with the following results: $\eta_k = 9.7 \times 10^4$, 1.57×10^4 poise, $\theta_k = 41.2$, 6.0 sec, s = .007 for polymer solution and $\eta_k = 2.28 \times 10^5$, 5.8×10^5, 2.24×10^5, 1.96×10^4 poise, $\theta_k = 171.0$, 9.7, 0.57, 0.024 sec, s = 0.0012 for polymer melt.

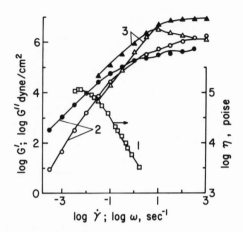

Fig. 1.   Apparent viscosity vs $\dot{\gamma}$($\square$); G'($\triangle$,$\bigcirc$) and G'' ($\blacktriangle$,$\bullet$) vs
ω. Theoretical fits: curves; data: 8% polystyrene in
Aroclor at 35°C[5] ($\square$); polystyrene at 170°C[4] ($\bullet$,$\bigcirc$); poly-
butadiene (M_w = 1.41 x 10^5; M_w/M_n = 1.1) at 25°C[12] ($\blacktriangle$,$\triangle$).

Resulting plots are shown in Fig. 1.

     An explicit finite-difference scheme has been used to solve
the full set of equations in (1) with both developing velocity and
elastic strains. For the polymer solution with U = .0428 cm/sec
and b = 0.2 cm, such that U/b = 0.214 sec^{-1}, it has been found that
a uniform shear flow is reached within about 40 microsec. During
this time, there is essentially no change in any of the $C_{11,k}$, $C_{22,k}$
or $C_{12,k}$. Accordingly, in all following calculations, the developing $C_{ij,k}$ components have been determined in the presence of a uniform shear flow, $\gamma = \partial u/\partial y$ = constant. In particular, a fourth-
order Runge-Kutta scheme has been used to integrate the second and
third equations in (1), in conjunction with the final algebraic
equation in (1). Similarly, upon impulsively stopping the flow,
the $C_{ij,k}$ have been solved with $\partial u/\partial y$ = 0 and the initial conditions corresponding to the steady-state values for the prescribed
γ. In order to relate the resulting predictions with the measurements[4,5] of birefringence, Δn, and extinction angle, χ, stress-
optical laws have been employed[4].

     Figure 2 shows a comparison between the theoretical and experi-
mental dependence of Δn and χ upon time at different shear rates
for the polymer solution during deformation and relaxation. Similar
comparisons are shown in Fig. 3 for the polymer melt during deformation. It is seen from the presented results that the Leonov theory
is in good agreement with the experiments over a large range of
deformation, $\gamma = \dot{\gamma}t$.

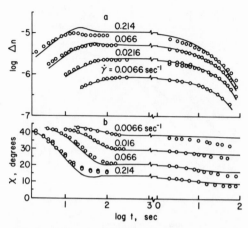

Fig. 2. Δn (a) and χ (b) vs time following impulsive shear flow
         (left curves) and during relaxation (right curves) at
         different γ̇. Theory: curves; O –data[5].

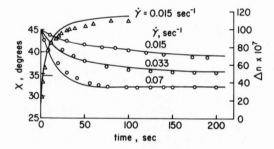

Fig. 3. Δn (a) and χ vs time following impulsive shear flow at
         different γ̇. Theory: curves; O , Δ –data[4].

OSCILLATORY SHEAR FLOW IN THE LINEAR AND NON-LINEAR REGIMES

Under oscillatory shear flow, with $\gamma(t) = \gamma_0 \sin \omega t$ and $\dot{\gamma}(t) = \dot{\gamma}_0 \cos \omega t$ where γ_0 is the deformation amplitude and $\dot{\gamma}_0 = \omega \gamma_0$, we introduce the following dimensionless quantities $\tau \equiv \omega t$, $\Omega_k \equiv \theta_k \omega$, $\Gamma_k \equiv \theta_k \dot{\gamma}_0 \cos \tau$, $\Gamma_{0,k} \equiv \theta_k \dot{\gamma}_0$. In the case of linear oscillatory shear flow $(\gamma_0 \ll 1)$, the set of equations for $C_{ij,k}$ in (1) gives the following stationary solutions for $C_{12,k}$ and $C_{11,k} - C_{22,k}$:

$$C_{12,k} = \gamma_0 (A_k \sin \tau + B_k \cos \tau) + 0 \, (\gamma_0^3)$$

$$C_{11,k} - C_{22,k} = \gamma_0^2 (A_k + C_k \sin 2\tau + D_k \cos 2\tau) + 0 \, (\gamma_0^4)$$

(4)

From the theory it follows that the shear stress and normal stresses include odd and even harmonics, respectively, with the steady component of the first normal-stress difference, $(\sigma_{11} - \sigma_{22})_o$, being equal to $\gamma_o^2 G''$. Further, it is noted that C_k and D_k are related to A_k and B_k as follows:

$$C_k(\Omega_k) = B_k(\Omega_k) - \frac{1}{2} B_k(2\Omega_k), \quad D_k(\Omega_k) = -A_k(\Omega_k) + \frac{1}{2} A_k(2\Omega_k) \quad (5)$$

These relations are also predicted by many known theories (see references in [6,7]) and have been corroborated by experiment[6-9].

In the case of nonlinear behavior (γ_o no longer small), the $C_{ij,k}$ equations in (1) have been normalized and solved numerically for given values of γ_o and Ω_k. It is noted that the problem is identical for each mode, and hence need be solved but once in the $\gamma_o - \Omega$ plane. Accordingly, the subscript $(\ )_k$ will be omitted in the following. It has been found useful to base the initial guesses for C_{11} and C_{12} at $\tau = 0$ upon the above small-amplitude results and to then integrate forward in τ to $\tau = 2\pi$ by means of a fourth-order Runge-Kutta scheme. By comparing the initial and final values for both C_{11} and C_{12} and by using a Newton-Raphson type procedure, the initial values can be iterated upon until the initial and final values for C_{11} and C_{12} agree to within the desired accuracy. Typically, this numerical scheme has converged in one to five iterations, depending upon the size of γ_o. Calculations have been carried out in the range $\log \gamma_o = -2.00$ to 1.25, in increments of .25, and $\log \Omega = -2.0$ to 3.0 in increments of .5. The corresponding Fourier components of the converged numerical solutions have then been obtained by quadrature. These results indicate, e.g., that the amplitude of the third harmonic of C_{12} at $\gamma_o = 10$ can be as large as 30% of the first, with this maximum percentage occurring at $\Omega \approx 1$. However, after combining the contributions from the various modes, it is found that the relative amplitude of the third harmonic of σ_{12} is considerably smaller. Indeed, this small effect of the higher harmonics in the nonlinear regime has also been observed experimentally. For example, Philippoff[10], with polymer solutions, and Vinogradov et al.[7], with elastomers, have noted that the contribution of higher harmonics on measured shear stress is less than 5% for values of γ_o as large as 7.2 and 5.0, respectively.

In the following, comparison will be made between the above theory and experimental results in the nonlinear γ_o range for shear stress and the steady component of the first normal-stress difference (no work seems to be available for the 2ω component of the latter stress). In particular, Philippoff[10] has measured shear stress under oscillatory flow with large γ_o for a 2% solution of polyisobutylene in Primol. Plots of the resulting data are shown

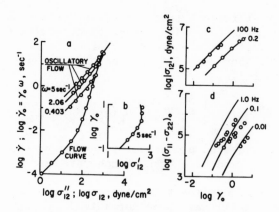

Fig. 4.   Flow curve and σ_{12}'' vs $\dot{\gamma}_o$ at different ω (a); σ_{12}' vs γ_o at
$\omega = 5$ sec^{-1} (b); $|\sigma_{12}|$ (c) and $(\sigma_{11} - \sigma_{22})$ (d) vs γ_o at
different $\omega/2\pi$. Theory: curves; O -data from [10,11] (a,b)
and [7,12] (c,d).

in Fig. 4a in terms of σ_{12}'', the amplitude of shear stress out of
phase with deformation, versus shear-rate amplitude, $\dot{\gamma}_o = \gamma_o\omega$, at
various frequencies. In contrast to the small-amplitude range, in
which each frequency corresponds to a distinct curve, in the large-
amplitude range the results seem to approach a common asymptote
independent of frequency. This asymptote was compared by Philippoff
with the steady-shear correlation between σ_{12} and $\dot{\gamma}$, whereby he concluded that, for polymer solutions, the steady-shear flow curve is
an upper bound for the σ_{12}'' versus $\dot{\gamma}_o$ behavior in the large γ_o region.
Also shown on Fig. 4a are corresponding predictions which are seen
to corroborate the above conclusion. In fact, the theoretical de-
pendence for σ_{12}'' ($\dot{\gamma}_o$) at different frequencies does merge with the
theoretical flow curve of the polymer solution at large $\dot{\gamma}$. The
model parameters used for this polymer solution have been obtained
by fitting the flow curve over a large shear rate range, as presented by Philippoff and Stratton[11]; resulting values are as follows:
η_k = 9.98 x 10^3, 1.39 x 10^3, 9.4 x 10^1 poise, θ_k = 171, 10.7, 0.35
sec, s = 0.0014. Further, Fig. 4b shows theoretical and experimental
results for σ_{12}', the shear-stress component in phase with deformation, versus γ_o at $\omega = 5$ sec^{-1}. Both the data and theory are seen
to indicate a much stronger non-linear behavior in σ_{12}' than in
σ_{12}''.

Vinogradov, et al.[7,12] have conducted measurements of the shear
stress amplitude $|\sigma_{12}|$ and steady component of the first normal-
stress difference, $(\sigma_{11} - \sigma_{22})_o$,in oscillatory flow at large deformation amplitudes for a series of polybutadienes and polyisoprenes
with narrow molecular weight distributions. For these polymers, the
non-linearity in the shear stress is small and, as a critical value

of γ_o is approached, there is observed a spurt (polymer detaches from measuring surface) with a corresponding sharp decrease in the shear stress. On the other hand, even before the critical value of γ_o is reached, these polymers exhibit a significant departure from a quadratic dependence of $(\sigma_{11} - \sigma_{22})_o$ upon γ_o. Evidently, the present theoretical results do not describe a spurt in the polymer but they do indicate the experimentally observed weak non-linear dependence of $|\sigma_{12}|$ upon γ_o (Fig. 4c) and a significant non-quadratic dependence of $(\sigma_{11} - \sigma_{22})_o$ versus γ_o (Fig. 4d). Model parameters for the polybutadiene used in this calculation were obtained from data[12], by fitting values of G' and G" versus ω in linear range (Fig. 1). Resulting parameters are: η_k = 2.69 x 10^5, 5.3 x 10^5, 7.84 x 10^5, 1.92 x 10^5 poise and θ_k = 16.2, 1.31, 0.13, 0.0072 sec, s = 0.000555. It is seen from Fig. 4d, however, that the measurements indicate a stronger deviation from a quadratic dependence than is predicted by the theory, with the discrepancy increasing with frequency. A possible cause of this discrepancy with the data is the use of a simple classic potential from network-theory elasticity in deriving the equations in (1) and (3). Further work should perhaps be continued with a potential of more realistic form.

This work is part of the Cornell Injection Molding Project, supported by NSF under Grant DAR78-18868.

REFERENCES

1. A. I. Leonov, Nonequilibrium thermodynamics and rheology of viscoelastic polymer media, Rheol. Acta, 15:85 (1976).
2. A. I. Leonov, E. H. Lipkina, E. D. Paskhin and A. N. Prokunin, Theoretical and experimental investigation of shearing in elastic polymer liquids, Rheol. Acta, 15:411 (1976).
3. A. I. Isayev and C. A. Hieber, Toward a viscoelastic modelling of the injection molding of polymers, to appear in Rheol. Acta.
4. F. N. Gortemaker, M. G. Hansen, B. de Cindio, H. M. Laun and H. Janeschitz-Kriegl, Flow birefringence of polymer melts: application to the investigation of time-dependent rheological properties, Rheol. Acta, 15:256 (1976).
5. K. Osaki, N. Bessho, T. Kojimoto and M. Kurata, Flow birefringence of polymer solutions in time-dependent field, J. Rheol., 23:457 (1979).
6. H. Kajuira, H. Endo and M. Nagasawa, Sinusoidal normal stress measurements with the Weissenberg rheogoniometer, J. Polym. Sci., Phys. Ed., 11:2371 (1973).
7. G. V. Vinogradov, A. I. Isayev, D. A. Mustafayev and Y. Y. Podolsky, Polarization-optical investigation of polymers in fluid and high-elastic states under oscillatory deformation, J. Appl. Polym. Sci., 22:665 (1978).

8. H. Endo and M. Nagasawa, Normal stress and shear stress in a viscoelastic liquid under oscillatory shear flow, J. Polym. Sci., A-2, 8:371 (1970).
9. L. A. Faitelson and A. I. Alekseenko, Normal stresses during periodic shear deformation, Polym. Mech., 9:275 (1973).
10. W. Philippoff, Vibrational measurements with large amplitudes, Trans. Soc. Rheol., 10:1:317 (1966).
11. W. Philippoff and R. A. Stratton, Correlation of the Weissenberg rheogoniometer with other methods, Trans. Soc. Rheol., 10:2:467 (1966).
12. G. V. Vinogradov, A. I. Isayev and E. V. Katsyutsevich, Critical regimes of oscillatory deformation of polymeric systems above glass transition and melting temperature, J. Appl. Polym. Sci., 22:727 (1978).

PHASE BEHAVIOR AND RHEOLOGICAL PROPERTIES OF AMORPHOUS

POLYMERS PLASTICIZED BY CRYSTALLINE SOLIDS

M. T. Shaw and J-H. Chen

Department of Chemical Engineering and
Institute of Materials Science
University of Connecticut
Storrs, Connecticut 06268

INTRODUCTION

Many high-Tg thermoplastics are difficult to process because
of the close proximity of the glass transition and degradation
temperatures. For these polymers only a few degrees of latitude
in processing temperature are available to the fabricator, causing
a high rate of rejection of the finished parts. Common problems
include short shots and weak welds because of low temperatures,
or splash and poor color because of excessively high temperatures.
To increase the processability of high performance polymers, processing aids can be added. These include plasticizers, which
lower melt viscosity but also adversely affect the high temperature performance of the resin. Because of the close relationship
between melt viscosity and glass transition -- as described, for
example, by the WLF relationship -- it seems evident that true
plasticizers are not desirable as processing aids unless they can
be removed from the polymer after fabrication. Removal of the
plasticizer might be, conceivably, by its polymerization, vapori-
zation, or solidification as a separate phase. The latter was
chosen as the subject of this work.

Fig. 1 demonstrates schematically the possibilities with the
proposed system consisting of amorphous polymer plus cyrstalline
solid. In this figure, the phase boundary (solid line) intersects
a curve representing the glass transition temperature of the
system if the plasticizer were a liquid. Because a system of
composition Φ_1 separates on cooling into a phase of pure crystalline plasticizer and a metastable glassy solution low in plasticizer (the abscissa of point B) the Tg is enhanced by an amount

99

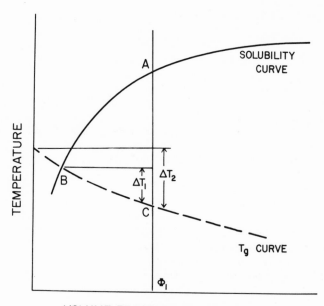

VOLUME FRACTION PLASTICIZER

Fig. 1. Schematic phase diagram for system consisting of amorphous polymer and a crystalline plasticizer. (See text
for explanation of labeled points.)

ΔT_1, as shown in the figure. The lower the solubility, the
greater ΔT_1 will be. The melt viscosity, on the other hand, will
be related to the Tg represented by point C, as long as the system
is in the single phase region about point A. Thus the mixture
could be processed at a temperature ΔT_2 below that required for
pure polymer.

     The simplistic picture painted here will be muddied in prac-
tice by the failure of the system to attain equilibrium. This
problem is perfectly illustrated by the early studies of Tobolsky
and Takahashi (1) who found that "elastic" sulfur -- polymeric
sulfur plasticized by monomeric S_8 sulfur -- was a metastable
solution, the S_8 gradually crystallizing out in the rubbery matrix.
Quenching below the glass transition effectively halts all crystallization; for this reason the point B in Fig. 1 is considered to
represent the final composition of the matrix. In recent studies,
Narkis et al. (2,3) noted the tendency of systems containing
crystalline flame-retarding additives to form metastable solutions.
The practical application of crystalline substances as fugitive
processing aids may thus require a annealing period to complete
the phase separation if it is particularly slow. Blyer (4), on
the other hand, reports a system (ABS/amide wax) that does not

appear to be troublesome in this respect.

Because of the vast number of possible systems containing an amorphous polymer and a crystalline plasticizer, this study was limited to a search for some extremes in behavior and properties. Parameters considered to be important were the difference between the Tg of the polymer and the melting point of the plasticizer, the interaction energy between the two components, the heat of fusion of the plasticiser, and the molar volume of the plasticizer. Phase behavior, viscoelastic properties, mechanical properties and the rheology of the mixtures were determined.

EXPERIMENTAL

Materials

Selected for this work were three amorphous polymers: polystyrene (PS), phenoxy (PHE), and polysulfone (PSF). The first two have Tg's near 100°C, while PSF has a Tg near 180°C. Polysulfone has the very polar sulfone group, while PHE features a pendant hydroxyl group capable of hydrogen bonding. PS is an essentially non-polar hydrocarbon. The solubility parameters, all calculated in a consistent fashion using the group contribution scheme of Fedor (5), are listed in Table 1. All resins were used as received except for the PSF, which was freed of monomer and oligomers.

Table 1. Polymer Components

Polymer	Tg, °C	Solubility Parameters, $(J/cm^3)^{1/2}$
PS	100	20.3
PHE	100	25.1
PSF	180	26.9

The plasticizers were selected on the basis of melting point, heat of fusion, molar volume, and solubility parameter. These properties are listed in Table 2.

From these polymers and plasticizers, the following systems were chosen: PS/WAX, PS/ANTH, PHE/WAX, PSF/DZPS, PSF/BPTA, PSF/DCAQ, and PSF/ANTH. Using a two-roll mill, mixtures from 0 to 40 volume %, in increments of 5 volume %, were prepared.

Table 2.　Crystalline Plasticizers

Name	Abbrevia- tion	ΔH_f, kJ/mole	T_m, °C	Molar Volume cm^3/g-mole	Solubility Parameter $(J/cm^3)^{1/2}$
1,8,9-anthra- cenetriol	ANTH	6.5	178 – 180	116	37.8
bisamide wax (Advawax 200)	WAX	56	165	623	19.7
2,5-bis(4- pyridyl)-1,3,4- thiadiazole	BPTA	30	240	121	29.7
1,5-dichloro- anthraquinone	DCAQ	26	245– 247	152	28.4
Dibenzothiophene sulfone	DZPS	22	231– 233	140	24.7

Phase Diagrams

Phase diagrams of the polymer/plasticizer systems were determined to guide subsequent experiments and to aid in understanding
the results. Potentially each phase diagram will consist of two
curves: the boundary between the single-phase and two-phase
liquid regions, and the boundary between the single-phase liquid
region and the two-phase region where solid plasticizer constitutes one of the phases (the solubility of the polymer in the
crystalline plasticizer is assumed to be negligible). Although
it is not an equilibrium feature of the system, the glass transition temperature vs. composition curve is also important to the
system's behavior, and is included on all the phase diagrams.

The phase boundaries were determined by Thermal Optical
Analysis (TOA) using both the depolarized and axial light intensity modes (6). By using a slow temperature scanning rate (6°C/
min), the difference between the heating and cooling curves was
minimized; however, the temperatures reported are those where the
last crystals disappeared on heating. All temperatures were
calibrated against ultra-pure standards. During all runs, the
specimen was observed, using one of the occulars of the binnocular
microscope, to ascertain the nature of the phase changes.

Glass transition temperatures were determined using a DuPont Dynamic Mechanical Analyzer (DMA), Model 981. The samples were heated at 5°C/min, about the same rate as the optical determination. The Tg was taken as the temperature at the maximum in the damping peak, while the dynamic modulus was calculated from the resonant frequency using a suggested relationship (7). The positions of the β-relaxation peaks in the PHE and PSF were also noted.

Mechanical Properties

The elongation at break, tensile strength, and the elastic modulus of all the samples were measured with an Instron tensile machine at 5.1 cm/min. Sample dimensions were 0.64cm x 0.64 mm with a 5.1 cm gauge length. Five strips from each material were tested.

Rheological Properties

Viscosities for selected compositions were determined with a Rheometrics Mechanical Spectrometer, using cone and plate fixtures in steady shear. The temperatures were selected based on the phase behavior and are designated by a cross on the phase diagrams. The shear-rate range commenced at 0.025 s^{-1} and continued to 25s^{-1}, if instabilities did not intervene.

RESULTS AND DISCUSSION

Phase Behavior

A typical phase diagram of the normal type is given in Fig. 2a. This type, qualitatively predicted by the basic Flory-Huggins lattice model, obtained for the systems PS/ANTH, PS/WAX, PHE/WAX, PSF/BPTA and PSF/DCAQ. The dotted lines on diagram are predicted phase boundaries using the equation

$$\ln \phi_1 + (1 - \phi_1) + \mu(1 - \phi_1)^2 = \frac{\Delta H_f}{R} \left(\frac{1}{T_m} - \frac{1}{T}\right) \tag{1}$$

where ϕ_1 is the volume fraction of plasticizer; ΔH_f and T_m are its heat of fusion and melting temperature, respectively; and μ, the interaction parameter, is given by

$$\mu = \mu_s + \frac{V_1(\delta_1 - \delta_2)^2}{RT} \tag{2}$$

In this equation V_1 is the molar volume of the plasticizer; δ_1 and δ_2 are the solubility parameters of plasticizer and polymer, respectively; and the entropic contribution, μ_s, was set equal to 0.4, a typical value for solvent/polymer systems. In cases where

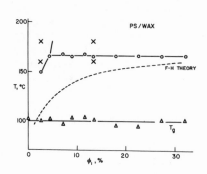 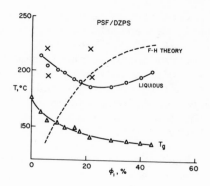

Fig. 2. Phase diagram and glass transition temperature for
system: (a) PS/WAX, and (b) PSF/DZPS.

two phases are predicted above T_m, the usual expressions for the
chemical potentials of the components were used (8). While much
is known (9) about the inaccuracies of the simplified expression
for the interaction parameter given by Eq. 2, it does permit one
to do calculations with very little experimental information about
the components.

Two of the systems studied -- PSF/DZPS and PSF/ANTH -- exhibited unusual phase behavior of the type shown in Fig. 2b. In
the case of the latter, the phase boundary was in the liquid
region, and could well be an example of the minimum solution
temperature type of phase behavior often found with polymer-
solvent systems (10). In the case of PSF/DZPS, the phase boundary
is below the melting point of the DZPS. This does not exclude a
liquid-liquid phase equilibria above this point, but the microscope observations did not reveal clearly two liquid phases. One
is tempted to cite the eutectics found, for example, by Koningsveld
(10) with two crystalline substances, one a polymer. As far as
we know, PSF does not crystallize, but the possibility that it
was not freed completely of crystallizable substances must always
be considered.

Relaxation Transitions

Typically, the observed glass transition of the polymer/
plasticizer systems decreased with plasticizer content except in
the cases of PS/WAX and PHE/WAX where the very low solubility of
the plasticizer resulted in little change in Tg. The decrease of
Tg beyond the intersection of the phase boundary and the Tg curve
(point B in Fig. 1) is good evidence for metastability. Presum-

ably, annealing would raise the Tg to the expected level, but this imposes an economic penalty.

Fig. 3 shows the variation of the prominent secondary transition in PHE and PSF with plasticizer content. The antiplasticization of this Tg by the crystalline plasticizers was often sufficient to raise the modulus of the mixture in spite of the decreases in Tg. This is illustrated in Fig. 4 for the system PSF/DZPS, where the initial rise in modulus might be interpreted as due to a combined filler and antiplasticization effect; the drop as due to the lowering of the glass transition; and the final, gradual rise as evidence for the accumulation of further plasticizer in a second, solid phase. The tensile strength, as might be expected, followed a similar pattern.

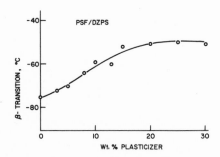

Fig. 3. Antiplasticization of the β-transition in PSF with DZPS.

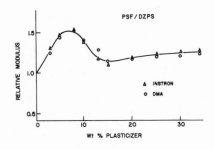

Fig. 4. Behavior of the modulus in the PSF/DZPS system.

Melt Viscosity

The melt viscosities were measured at pairs of temperatures and compositions which would bridge the liquidus line of the phase diagram, as illustrated by the crosses in Fig. 2. The viscosities at a shear rate of 0.025 s^{-1} are listed in Table 3 for the selected temperatures and compositions. Beside each viscosity are noted the phases that were presumed to be present in the sample at its viscosity was measured. The system PSF/ANTH could not be included because of the rapid sublimation of the plasticizer.

Table 3. Shear Viscosities of Plasticized Systems at 0.025 s^{-1}

| System | Temperature,°C | Viscosity,kPa·S,and Phases Present with ϕ_1 | |
		Low	High
PS/ANTH	145	1.4, M[a]	1.8, M+S
	171	0.15, M	0.37, M
PS/WAX	160	0.56, M	0.36, M+S
	180	0.20, M	0.14, M+L
PHE/WAX	137	8.4, M	2.0, M+S
	154	3.0, M	0.31, M(+L)
PSF/BPTA	194	24, M	3.5, M+S
	214	6.0, M	0.92, M
PSF/DCAQ	193	3.2, M	0.56, M+S
	215	4.4, M	0.72, M
PSF/DZPS	195	9.2	8.0
	215	14	37

[a] M = Melt, S = Crystalline plasticizer, L = Liquid Plasticizer.

PSF/ANTH had what might be termed the expected behavior at 145°C: the addition of plasticizer to form a second, solid phase, increased the viscosity. However, this increase carried over to the melt state at 171°C, which cannot be explained in terms of

distinctly lower Tg at higher plasticizer content. The expected
change with plasticizer content is that shown in Table 3 for
PSF/BPTA and PSF/DCAQ at 214 and 215°C, respectively. The wax-
containing systems both showed a decrease in viscosity with increasing solid plasticizer concentration. Since the phase assign-
ments are quite certain in these systems, we are forced to conclude that the wax can indeed act as a solid lubricant. The
behavior of PSF/BPTA and PSF/DCAQ at low temperatures can be
explained by the strong decrease of Tg with plasticizer content;
an effect that is absent in the wax-containing systems.

The most unusual observation of all is the rather large increase in viscosity with temperature for the systems PSF/DCAQ and
PSF/DZPS. The increase is particularly strong in the latter at
high plasticizer contnet (25%). The melts are also far more
pseudoplastic than at lower plasticizer levels.

To check on the possibility of instrument problems or crosslinking reactions, the samples were rerun in oscillating shear
using a DMA modified for this purpose. The changes in G" with
temperature are shown in Fig. 5 for both low and high plasticizer
concentrations (5 and 25wt%). G" goes through a maximum in both
cases, the peaks starting approximately at the phase boundary

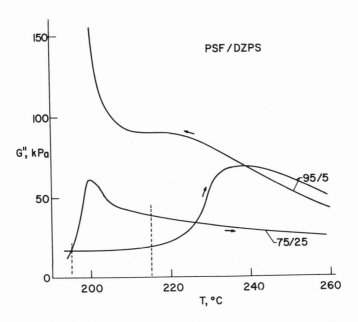

Fig. 5.   Change of loss modulus with temperature for PSF containing 5 and 25% DZPS.

shown in Fig. 2b. Qualitatively correct are the relationships of
G" and the viscosity measured in the cone and plate at 195 and
215°C (dotted lines on Fig. 5). Possibly, the systems decompose
in the neighborhood of 200°C into a melt phase of polymer with
little plasticizer (and of high viscosity) and solid plasticizer.
Another explanation is that the systems are metastable at low
temperatures and precipitate plasticizer at the "phase boundary".
The latter explanation is supported by the cooling curve (Fig. 5)
which shows the high viscosity developing near Tg in the usual
manner and only a slight tendency, evidence by the shoulder,
toward redissolution.

CONCLUSIONS

The behavior of the system amorphous polymer/crystalline
plasticizer can be exceedingly complex. Simplified lattice theory
of polymer solutions, well qualitatively correct in a number of
cases, does not provide a useful guideline for predicting phase
behavior. A notable failure of the group contribution method was
with the bisamide wax, which is far less soluble in polymers than
predicted. This wax appears to be capable of lubricating the
polymer melt, reducing the measured melt viscosity even as a solid.
Noted were some possible examples of lower critical solution
behavior, giving melts with viscosities which increased with
temperature.

ACKNOWLEDGEMENT

This work was supported by the University of Connecticut
Research Foundation.

REFERENCES

1.   A. V. Tobolsky and N. Takahashi, J. Polym. Sci. Part A,
     2, 1987 (1964).
2.   M. Narkis, A. Siegmann, A. Dagan, and A. T. DiBennedetto,
     J. Appl. Polym. Sci., 21, 989 (1977); M. Narkis, A. Siegmann,
     M. Puterman, and A. T. DiBennedetto, J. Polym. Sci., Polym.
     Phys. Ed., 17, 225 (1979).
3.   A. Siegmann, M. Narkis, and A. Dagan, Polymer, 15, 499 (1974).
4.   L. L. Blyer, Polym. Eng. Sci., 14, 806 (1974).
5.   D. W. Van Krevelen, "Properties of Polymers" 2nd ed., p. 143
     Elsevier, Amsterdam, 1976.
6.   J. A. Reffner, Ph.D. Thesis, University of Connecticut, 1975.
7.   DuPont Co., Wilmington, DE, "Instruction Manual, 980 Dynamic
     Mechanical Analyzer", 1977.
8.   R. L. Scott and M. Magat, J. Chem. Phys., 13, 172 (1945).
9.   H. Tompa, "Polymer Solutions", Butterworth, London, 1956.
10. D. D. Patterson, J. Paint, Technol. 45, 495 (1968).
11. R. Koningsveld, Br. Polym. J., 7, 435 (1975).

GELATION AND FUSION CHARACTERISTICS OF PVC RESINS IN PLASTISOL BY

DETERMINATION OF THEIR VISCOELASTICITY

N. Nakajima and D.W. Ward

B.F. Goodrich Chemical Division
Avon Lake Technical Center
Avon Lake, Ohio 44012

A method of characterizing gelation and fusion behavior by viscoelastic measurements has been developed previously[1]. The present work is an application of this method for finger-printing the gelation and fusion profiles of commercial dispersion resins. The Rheometrics Mechanical Spectrometer was used in the dynamic oscillatory mode at 1 Hz with a programmed increase of temperature. The elastic modulus, G', loss modulus, G'' and the complex viscosity $|\eta^*|$ were recorded as functions of temperature. Two temperature profiles were used: one was $20^\circ C/min$. to simulate a process, i.e., a heating rate in an oven, and the other was $5^\circ C/min$., a slower heating for a better characterization of gelation behavior.

The viscoelastic data with Resin A are shown in Fig. 1, where G', G'' and $|\eta^*|$ are recorded as functions of time. With the increase of temperature, the viscosity and moduli initially decrease, indicating that the system remains a suspension of PVC solids in DOP. At slightly above $50^\circ C$, the gelation process begins and the viscoelastic properties increase very rapidly with temperature. The maximum values are reached between 130 and $170^\circ C$, indicating the end of gelation. Above this temperature the viscoelasticity of the material decreases for two reasons; one is the normal temperature-dependence, primarily attributable to the thermal expansion, and the other is caused by the melting of micro-crystallites. Finally the data are seen to approach their equilibrium values at $195^\circ C$.

In Fig. 2 gelation and fusion curves are drawn separately and extended with the hypothetical curves; first, G' increases very rapidly during gelation and reaches the highest value at the completion of gelation, (real curve). Thereafter, G' would stay at the high level, if there were no melting of crystallites, (hypothetical curve). On the other hand G' of the gelled material would be high already, (hypothetical curve), and starts decreasing as the

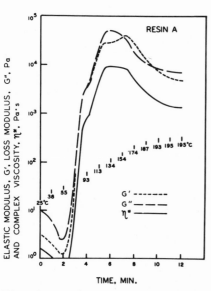

Fig. 1   Gelation and Fusion of
          Plastisol with Resin A.

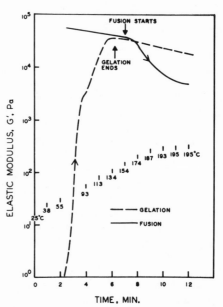

Fig. 2   Schematic Curves of
          Gelation and Fusion.

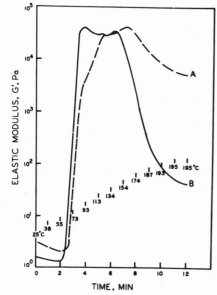

Fig. 3   Gelation and Fusion of
          Plastisols with Resins
          A and B,

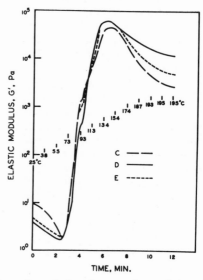

Fig. 4   Gelation and Fusion of
          Plastisols with Resins
          C, D and E.

crystallites start melting, (real curve). The experimental data of
gelation and fusion follow the arrows as indicated.

The gelation and fusion behavior of Resins A and B are shown in
Fig. 3 as G'. The only basic difference between these resins is the
molecular weight, Table I. The gelation of Resin B starts at a few
degrees lower than that of Resin A, and the maximum in G' is reached
at about 85-90°C for the former and at about 125-130°C for the latter.
This shows that with the lower molecular weight resin gelation not
only starts at a lower temperature, but it completes at much lower
temperatures.

The "gel-point" (SPI-VD-T5) of these resins were different only
by 3°C, being 77°C for Resin A and 74°C for Resin B. The gel-point
may be compared to some intermediate stage in the progress of gelation. If we take $G'=3X10^2$ Pa, Fig. 3, the corresponding temperature for two samples are 77°C for Resin A and 73°C for Resin B.
These temperatures are very similar to the gel point. Yet, the information, which is limited to a particular temperature only, is very
misleading. It becomes obvious, when both temperature and the corresponding modulus are shown. For example, at 73°C the plastisol
with Resin A has attained $G'=8X10$ Pa, whereas that with Resin B is
already at $G'=3X10^2$ Pa. At 77°C it is more dramatic, $G'=3X10^2$ Pa
with Resin A and $G'=5X10^3$ Pa with Resin B. This shows that at each
corresponding temperature the gelation has progressed much further
with Resin B than with Resin A. The difference is seen to become
more significant as the gelation progresses.

In spite of the clear dependence of gelation on the molecular
weight, the primary cause of the above difference is in morphology,
which is influenced by the polymerization condition designed to give
different molecular weights. When the fusion proceeds, the PVC crystallites begin to melt; this process is complete at 195°C and the
values of G' are those of the plasticized melt, which are proportional to the molecular weight of the polymer chain[3].

The data shown in Fig. 4 are G' of Resins C, D and E. These
resins are of a different type from Resins A and B. Among the three
resins, the only basic difference is again the molecular weight.
However, the gelation behavior is much less dependent on the molecular weight than what was observed with Resins A and B. Whether this
is a reflection of the difference in the resin type or the dependence
on the molecular weight diminishes at the higher molecular weight is
the subject of the future study. A significant dependence on molecular weight is observed at the fused state, as expected.

Fig. 5 is a comparison of Resins A and E with respect to G':
these resins have the same molecular weight, I.V.=1.2, but belong to
different types. The gelation of Resin E takes place at approximately
10°C higher than that of Resin A. The fusion curves of these resins
become identical at those temperatures higher than 175°C; this might
indicate that the crystallites have by-and-large melted and the viscoelasticity is a function of molecular weight only[3].

Although the difference in the gelation behavior were observed
in Figures 1-5, the increase in viscosity and moduli are too fast for

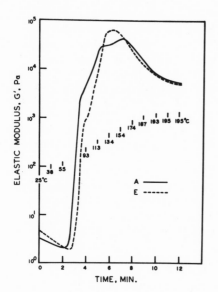

Fig. 5   Gelation and Fusion of
          Plastisols with Resins
          A and E.

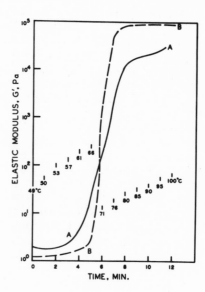

Fig. 6   Gelation Rate of
          Plastisols with Resins
          A and B.

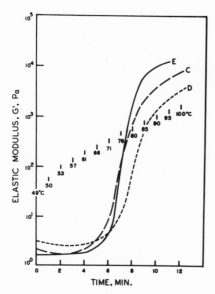

Fig. 7   Gelation Rate of
          Plastisols with Resins
          C, D and E.

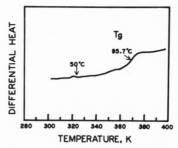

Fig. 8   DSC Thermogram of
          Resin D.

precise comparison. Therefore, a slower heating rate was applied as
described previously. Comparisons of the gelation behavior are shown
in Figs. 6 and 7, where the effect of molecular weight and that of
the resin type are shown as before. The real advantage of the slower
heating rate is that the differences in the gelation rate are clearly
seen as differences in the slope of the curves. This enables us to
identify the initiation of gelation and the gelation rate for each
resin. The former is expressed as the temperature at which G' begins
to rise, and the latter as the slope of G' curve taken G'=100 Pa.
Also the temperatures at which G'=5 Pa and 100 Pa are read off the
curves, because the former is the point at which the onset of gelation is near, and the latter is the value of G' when the slope of G'
vs. temperature is near its maximum. The results are summarized in
Table III. The temperature of the start of gelation and those at
fixed value of G' differentiate two types of Resins, whereas the
gelation rate of G'= 100 Pa is approximately proportional to the
reciprocal of molecular weight, with an exception of Resins E. However, the overall implication is that the differences in morphology
must be influencing the gelation behavior.

The start of gelation is when the plasticizer begins to penetrate into the solid PVC. This can happen only when there is a
significant increase in free volume and an increase in the segmental
motion of the polymer chain so that the plasticizer molecule can
diffuse into the polymer matrix. In short, the glass-rubber transition must commence at this temperature. Normally, the glass transition temperature, Tg, of PVC homopolymer is to be in the range of
80-100°C, which depends upon the molecular weight and polymerization
temperature[4]. However, the gelation started in these experiments at
50-60°C, as shown in Figs. 1-7 and Table III. The dynamic mechanical
properties reported in the literature[5] show that the glass transition
actually starts at about 40-60°C, the exact temperature depending
upon the sample history.

The differential scanning calorimetry, DSC, can be performed
with the plastisol samples as well as with the resin powders. However, no clear cut change in the slope of DSC trace was found in the
temperature range of 50-60°C, Fig. 8. A significant change of slope
took place only within plus minus several degrees of Tg. Evidently,
the mechanical measurements are more sensitive in detecting the start
of the transition from the glassy to the rubbery state than are the
thermal methods. Therefore, we propose that the viscoelastic measurements described in the present study be used for characterizing the
glass-transition behavior, when the sample is a powder and can not be
molded to a test specimen for the dynamic mechnical measurement. A
partial support of the above proposal may be found in Table III,
where the Tg of the present samples are listed along with the parameters describing the gelation behavior.

REFERENCES
1. N. Nakajima, D. W. Ward and E. A. Collins, Polym. Eng. Sci.,
       19 210 (1979).

2. C. A. Daniels and E. A. Collins, J. Macromol. Sci.-Phys.,
    B10(2) 287 (1974).
3. N. Nakajima, D. W. Ward and E. A. Collins, J. Appl. Polym. Sci.,
    20 1187, (1976).
4. E. A. Collins, C. A. Daniels and C. E. Wilkes, "Polymer Handbook"
    2nd ed., J. Brandrup and E. H. Immergut, eds., J. Wiley and
    Sons, New York, 1975, v-44.
5. G. Pezzin, G. Ajroldi and C. Carbuglio, J. Appl. Polym. Sci.,
    11, 2553 (1967).

TABLE I   PVC DISPERSION RESINS

Designation	Product Name	Molecular Weight Expressed as I.V.[a], dl/g
A	Geon [R] 121	1.20
B	Geon [R] 120 X 241	0.74
C	Geon [R] 128	1.10
D	Geon [R] 120 X 271	1.45
E	Geon [R] 120 X 279	1.20

[R] products of BFGoodrich Chemical Division
a I.V.: inherent viscosity in cyclohexanone at 30°C [2].

TABLE II   PLASTISOL FORMULATIONS

	Gelation and Fusion Study, Formulation I 20°C/min.	Gelation Study, Formulation II 5°C/min.
Heating Rate		
Formulation, parts by weight		
Resin	100	100
Plasticizer, DOP	60	60
Epoxydized Soybean Oil	5	--
Barium, Cadmium, Zinc Stabilizer	3	--

TABLE III   GELATION BEHAVIOR

Resin	IV	Temperature Start of Gelation	G'=5Pa	G'=100Pa	Gelation Rate d log G'/dt at G'=100Pa, min⁻¹	Tg
A	1.20	52°C	67°C	69.5°C	0.9	82.0,81.1,80.9
B	0.74	51	62	70	4.2	81.3
C	1.10	57	69	76.5	1.1	91.6,90.3
D	1.45	59	72	80.5	1.05	95.7
E	1.20	60	71.5	76	1.6	91.0

PROCESS RHEOLOGY IN THE POLYMER MANUFACTURING FOR

SYNTHETIC FIBRES

Dicoi Ovid

Zimmer AG
Borsigallee No. 1 6000
Frankfurt/Main, West Germany

The object of this paper is to develop a background for optimum design and operation of the polycondensation reactor based on real time measuring the flow behaviour of molten polymers.

The capillary rheometer illustrated in Fig. 1 consists of a metering pump (3) developing flow rates from 6 to 300 cm3/s and an interchangeable capillary block (4) with the capillary tube (5). The two blocks are heated at the temperature T4 that is close to the level of T2 and T3. The polymer melt from the main line at the temperature T1 and pressure P1 flows through the inelt valve (1) into the pump (3) and enters in the inlet reservoir where pressure P2 and temperature T2 are measured. After passing the capillary tube (5) the pressure P3 and temperature T3 are measured and the polymer melt is either returned to the main line (on-line operation) or removed from the system (off-line operation). The overall accuracy and repeatability of the measurements are 1 percent of full scale and the reponse time is less than 1 min. for the illustrated configuration.

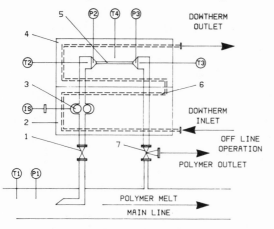

Figure 1. Process Capillary Rheometer

The principal factors that have to be taken into consideration by real time measuring of viscosity are: the capillary tube and reservoirs size and geometry, the metering pump speed and flow rate, the pressure measurement, the pressure in the main line and the isothermal flow through the capillary tube. The energy losses due to sudden contraction and enlargement by entering and leaving the capillary tube are neglected. The chosen capillary tubes have a length to radius ratio of at least 100 and the pressure sensors are installed very close to the capillary edges. The isothermal flow of polymer melt through the capillary tube depends on the actual viscosity of the polymer and the shear rate. In the case of a disk ring continuous polycondensation reactor (CPR) for manufacturing of polyethylene terephthalate (PET) the polymer viscosity is maintained constant so that the isothermal flow is attained at so called "isothermal shear rate" (see Fig. 2).

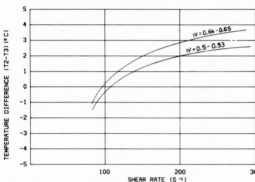

Figure 2. The Temper. Difference (T2-T3) Between Inlet and Outlet of Capillary Tube vs. Shear Rage

The melt viscosity is affected by the fluctuation of the polymer pressure in the process line. Fig. 3 illustrates the strong dependency of the apparent viscosity of PET melt from the pressure in the main line for off-lines and on-line operations of the capillary rheometer.

A few on-line capillary rheometers are provided with a second discharge metering pump assuring a linear proportionality between apparent viscosity and shear stress. The discharge pump keeps pressure P3 equal to the melt vapor pressure.

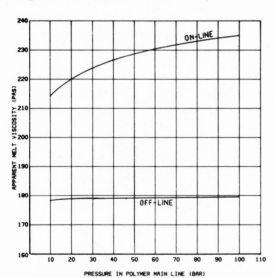

Figure 3. The Apparent Melt Viscosity vs. Pressure in Polyester Main Line

The optimum operating point of the capillary rheometer for a given polymer viscosity is achieved when the shear rate assured an isothermal flow at significant pressure drop and the speed of metering pump is low enough to keep at minimum the polymer leaks. The calculation of the apparent melt viscosity and the solution viscosity are carried out by means of a process computer. From the measured pressures P2 and P3 and pump speed is computed the actual apparent melt viscosity (Va). The solution viscosity (SV) and the reference apparent melt viscosity (Vr) are calculated by solving the following set of equations:

$$Vr = A*(Va)^B*Exp(C*P1)*Exp(D+E/T+F*Ln(SV))*f(r) \qquad (1)$$

$$SV = G*Exp(H*Vr) \qquad (2)$$

where A,B,C,D,E,F,G are constants.
   f(r) is a function of the shear rate

For a narrow range of the pressure in the main line (P1) and at constant shear rate the solution viscosity may be given by:

$$SV = Exp(A+B/T+C*Ln (Va)) \qquad (3)$$

The real time measuring of the flow behaviour of molten polymers supplies practical information about the rate of poly-condensation reaction as function of process variables (tempera-ture, residence time, pressure), the optimum conditions for carrying out a reaction in practice, and for reactor design.

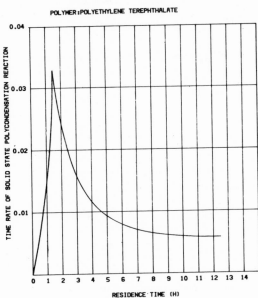

POLYMER:POLYETHYLENE TEREPHTHALATE

Figure 4. Time Rate of Solid State Polycondensation Reaction vs. Residence Time

The rate of polycon-densation reaction may be expressed as the time rate of change of the concentra-tion of the polymer end groups (EGC) and is written as follows:

$$r = \frac{d(EGC)}{dt} \qquad (4)$$

The end group concen-tration can be on-line com-puted from the molecular weight and the last is ob-tained from the solution viscosity according to the following equations:

$$(EGC) = 2*10^6/MW \qquad (5)$$

$$MW = A*(SV)^B \qquad (6)$$

As example is given in Fig. 4 the reaction rate of solid state polycondensation of the PET as function of the residence time.

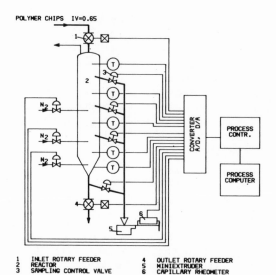

POLYMER CHIPS IV=0.65

```
1   INLET ROTARY FEEDER        4   OUTLET ROTARY FEEDER
2   REACTOR                    5   MINIEXTRUDER
3   SAMPLING CONTROL VALVE     6   CAPILLARY RHEOMETER
```

Figure 5. Continuous Solid State Polycondensation Reactor for Manufacturing of High Molecular Weight Pet

This function can be obtained by operating of a continuous solid state polycondensation reactor as illustrated in Fig. 5. The automatic sampling system supplies polymer to the chip viscosity measurement system. The progression of the solid state polycondensation is controlled by means of monitoring the increase in intrinsic viscosity. The chips are melted in the mini-extruder and the melt passes the capillary rheometer (off-line operation).

During the polycondensation reaction for manufacturing ofPET the molecular weight increases while the end group concentration decreases and at the same time results ethylene glycol (EG). The flow rate of split EG may be estimated from the rate of decreasing of end group concentration and polymer flow rate.

Then the reactor hold-up is computed by integrating of equation (7)

$$\frac{d\ (Hold\text{-}up)}{dt} = \begin{matrix} prepolymer \\ into\ reactor \end{matrix} - \begin{matrix} end\ polymer \\ out\ of\ reactor \end{matrix} - \begin{matrix} split\ EG \\ out\ of\ reactor \end{matrix} \quad (7)$$

The average residence time (RT) in the CPR is estimated as follows:

$$RT = \frac{Reactor\ Hold\text{-}up}{Polymer\ flow\ rate} \quad (8)$$

The partial pressure of EG in reactor has a direct influence on the rate of the polycondensation reactor and can be computed from the rate of decreasing end group concentration.

The on-line measurement of the flow behaviour of PET melt enables to improve the operation and design of the disk ring continuous polycondensation reactor. The main goal when manufacturing PET is to produce polymer of constant quality.

Small changes in the feeding conditions (e.g. flow rate, intrinsic viscosity, concentration in EG) of the prepolymer or in the flow rate of the end polymer may lead to fluctuation of the residence time and EG vapour pressure over the polymer melt with negative repercussion on the product quality.

In Fig. 6 is illustrated the CPR with the entire process control to minimise the difference between set polymer viscosity and actual measured viscosity by optimisation of operating conditions.

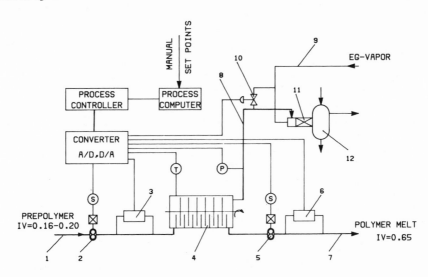

1- PEPOLYMER MAIN LINE.
2- PREPOLYMER METERING PUMP
3- PREPOLYMER ON-LINE CAPILLARY RHEOMETER
4- DISK RING POLYCONDENSATION REACTOR
5- POLYMER DISCHAGE PUMP
6- POLMER ON LINE CAPILLARY RHEOMETER
7- POLYMER MELT LINE
8- VAPOR MAIN LINE
9- EG-VAPOR
10- CONTROL VALVE
11- EJECTOR
12- CONTACT CONDENSER

S-SPEED
T-TEMPERATURE
P-PRESSURE

Figure 6. Continuous Polycondensation Reactor for Manufacturing of Polyethylene Terephthalate

Two on-line capillary rheometers are provided for measuring of the viscosity of prepolymer and the polymer, feed and discharge metering pumps with variable speed for measuring the flow rates, vacuum ejector with EG vapour as motive fluid, A/D and D/A, converters , process controller for digital control functions and a process computer for data logging and estimation of the new set points.

According to on-line measurements of the prepolymer and polymer flow rate and viscosity, and the temperature and pressure in the reactor are estimated the residence time, EG partial vapour pressure in the reactor and the deviation of actual viscosity from the set point.

The new vacuum set point is computed from an optimization algorithm that minimise the viscosity deviation. This control system is provided to improve the performance of the 16 disk ring CPR with a capacity of 4.2 t/h each where the absolute deviation of the intrinsic viscosity is less than 0.015 for a viscosity range from 0.6 to 0.65.

The design of the disk ring CPR may be improved by on-line measuring of the apparent melt viscosity as function of shear rate and time at constant temperature. It is well known however that the variation of apparent melt viscosity with shear rate and time delivers data about the history of the polymer. Keeping constant conditions we expect to obtain on-line information about the molecular weight distribution of the polymer melt in different places inside the reactor. This data enables to minimise the total residence time, the dead space in the reactor with a view to preventing partial degradation of polymer, to improve the plug flow of the polymer in reactor and to reduce the entrainement and sublimation of the oligomers and polymer.

However, the development of a capillary rheometer for on-line measuring of apparent melt viscosity as function of a variable shear rate (10 to 1500 s^{-1}) requires information about the relationship between isothermal flow, capillary length, shear rate, polymer degradation and time response.

MULTIPHASE FLOW IN POLYMER PROCESSING

C.D. Han

Department of Chemical Engineering
Polytechnic Institute of New York
Brooklyn, New York 11201

INTRODUCTION

Two types of multiphase flow may be distinguished on the basis of their degree of phase separation. One type is dispersed multiphase flow, in which one or more components exist as discrete phase, dispersed in another component forming the continuous phase. The other type is stratified multiphase flow, in which two or more components form continuous phases separated from each other by continuous boundaries.

Depending on the nature of the discrete phase, the bulk flow properties of a dispersed multiphase system may vary. That is, the bulk flow properties of a dispersion can be different, depending on whether rigid particles, deformable droplets, or gas bubbles are suspended in the continuous polymeric medium. All three kinds of dispersed system are widely used in the polymer processing industry. As summarized in Figure 1, the industry uses not only two-phase (liquid-liquid, liquid-solid, liquid-gas) systems, but also three-phase (liquid-liquid-solid, liquid-solid-gas) systems and four-phase (liquid-liquid-solid-gas) systems.

Depending on the geometry of the die through which a system of fluids may be forced to flow, stratification may occur either in a rectangular duct, or in a circular tube, or in an annular space. Figure 2 summarizes various applications of the stratified multiphase flow, often referred to as 'Coextrusion' in the polymer processing industry. The number of layers, the layer thickness, and the manner in which the individual components are arranged, control the bulk flow properties and the ultimate physical/mechanical properties of a stratified system. There are

121

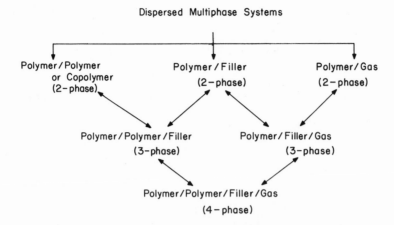

Fig. 1.   Schematic showing various ways of producing dispersed
          multiphase systems encountered in the polymer processing
          industry.

situations where one may have a combination of dispersed and
stratified flow, a practice frequently encountered in the polymer
processing industry. For instance, a polymer stream containing
gas bubbles or additives may be coextruded with another polymer.
There are numerous other examples of industrial importance.

DISPERSED MULTIPHASE FLOW IN POLYMER PROCESSING

A rapidly growing interest in engineering thermoplastics has
stimulated the polymer industry to develop dispersed multiphase
polymeric systems. They comprise many polymeric materials being
used in industry, including reinforced plastics, mechanically
blended polymers and copolymers, and foamed thermoplastics. From

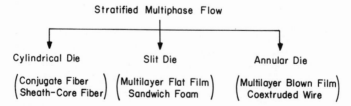

Fig. 2.   Schematic showing various applications of stratified
          multiphase flow encountered in the polymer processing
          industry.

the point of view of the availability of base materials, namely
polymer, solid reinforcing agent, foaming agent, an enormously
large number of combinations is possible for making dispersed
polymeric materials. In all such cases, one hopes to improve the
mechanical/physical properties of the finished product, utilizing
some additional component as a modifying agent.

Applications of dispersed multiphase systems to polymer processing operations include: (1) reinforced thermoplastics or
thermosets; (2) rubber modified plastics (e.g., high-impact polystyrene (HIPS) and acrylonitrile-butadiene-styrene (ABS) resins),
and mechanically blended thermoplastics and elastomers (e.g.,
blends of PVC with impact modifier, blends of styrene-butadiene
rubber (SBR) with natural rubber); and (3) foamed plastics or
rubbers as well as reinforced structural foams.

When a reinforcing component (e.g., an inorganic filler) is
used, improved mechanical/physical properties of the finished
product can only be obtained by achieving good adhesion between the
reinforcing component and the suspending polymeric medium. In
order to obtain good adhesion, multifunctional silanes or titanates
are often used as coupling agents.

In the processing of, for instance, heterogeneous polymer
blends and copolymers, both phases are usually non-Newtonian and
viscoelastic. The polymer processing industry makes use of blends
of two or more polymers of the same molecular structure, but of
different molecular weight distributions or of different molecular
structures. It is usually the case that when two polymers of
different structure are melt-blended, one polymer forms the discrete phase dispersed in the other, which is the continuous phase,
thus giving rise to a two-phase polymer system[1].

In dealing with either particle-filled polymers (composites)
or gas-charged polymers (foamed plastics), processing-morphology-
property relationships may be as complex as those for polymer
blends and copolymers of heterogeneous nature. For instance, in
processing highly filled thermoplastics or thermosetting resins,
the choice of particle size and its distribution, the shape of
particles, and wettability of the particles to the base polymer
matrix are important to control the bulk rheological properties
of the mixture and to obtain the finished product with consistent
quality in terms of mechanical/physical properties.

Figure 3 shows the effect of filler particles ($CaCO_3$) on the
shear viscosity of a non-Newtonian viscoelastic molten polypropylene. It is seen that the solid particles increase the melt
viscosity very rapidly as the shear rate is decreased, giving
rise to a yield stress as the shear rate approaches zero. However,

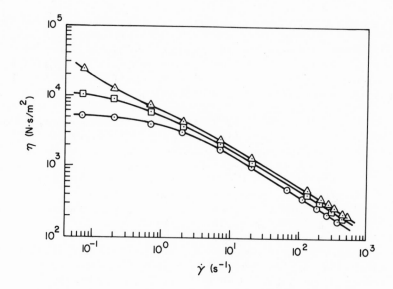

Fig. 3.   Viscosity versus shear rate of polypropylene (PP) (T=200°C)
filled with calcium carbonate (CaCO$_3$), with and without a
silane coupling agent: (⊙) pure PP; (Δ) PP/CaCO$_3$ = 50/50
by wt.) without coupling agent; (□) PP/CaCO$_3$ = 50/50 (by
wt.) with coupling agent (1 wt. %).

the addition of an additive (coupling agent) reduces the melt viscosity of the CaCO$_3$-filled polypropylene.

In a foam processing operation, gas bubbles resulting either
from the decomposition of a chemical blowing agent or from the injection of a gas remain in the molten polymer while it is being
processed. Foam products are obtained by means of either extrusion
or injection molding. Examples are polyolefin foam, PVC foam,
polycarbonate structural foam, etc. Foam formation is a very complicated process, in which one has to consider not only the formation of gas bubbles, but also the growth of gas bubbles inside the
processing equipment. In a foamed plastic, the shape and size of
the cells determine the physical properties to a large degree.
It may be surmised that the cell size and its distribution would
depend on the choice of melt temperature, injection pressure, injection speed, the rate of cooling of the mold, and the amount of
the blowing agent used. Consequently, all these processing
variables are expected to affect the foam density as well as the
mechanical properties of the finished foam product.

Difficulties arise in analyzing the rheological data of

multiphase systems, in general. This is because their rheological properties are influenced by many factors, such as the particle size, its shape, and the concentration of the dispersed phase. In other words, the state of dispersion influences the rheological behavior of multiphase systems. The rheological properties that are important to the design of equipment for polymer processing operations are <u>viscosity</u> and <u>elasticity</u>. Information concerning these properties is essential, for instance, for determining pressure drops (and thus production rates), viscous heating, stress relaxation, etc.

In dispersed multiphase polymeric systems, there are many variables interrelated to each other, affecting the ultimate mechanical/physical properties of the finished product. Such relationships are pointed out schematically in Figure 4. For instance, the method of preparation (e.g., the method of mixing polymer blends, the intensity of mixing during polymerization of copolymers) controls the morphology (e.g., the state of dispersion, particle size and its distribution) of the mixture, which in turn controls the rheological properties of the mixture. On the other hand, the rheological properties very much dictate the choice of processing conditions (e.g., temperature, shear stress), which in turn strongly influence the morphology and therefore the ultimate mechanical/physical properties of the finished product.

STRATIFIED MULTIPHASE FLOW IN POLYMER PROCESSING

Recent developments in polymer processing technology have stimulated researchers to obtain a better understanding of the various problems involved in stratified multiphase flow, often referred to as 'Coextrusion.' Representative commercial products available in the market, produced by coextrusion, are conjugate (bicomponent) fiber, multilayer flat film, multilayer blown film, coextruded cables and wires, sandwiched foam composites, etc. Figure 5 gives schematics of the cross sections of some representative coextruded products.

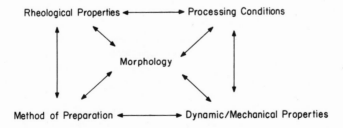

Fig. 4.   Schematic pointing out processing-morphology-property relationships in dispersed multiphase polymeric systems.

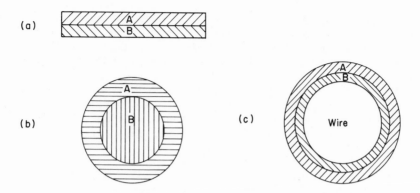

Fig. 5.   Schematic describing various shapes of the cross section
of the dies being used in coextrusion processes:
(a) rectangular die; (b) circular die; (c) annular die.

In order to have successful operation of coextrusion processes,
it is important to investigate the processability of given combinations of polymeric materials over a wide range of processing variables, and to evaluate the physical/mechanical properties (i.e.,
adhesion, tensile properties, permeability) of coextruded products
as affected by processing conditions. It is also important to
develop mathematical models simulating the various coextrusion
processes (e.g., flat- and blown-film coextrusion, wire coating
coextrusion), in terms of the rheological properties of the
individual polymers being coextruded, the variables involved with
die design, and the processing variables (i.e., melt extrusion
temperature, extrusion rate).

One of the most fundamental and important problems in coextrusion is to obtain an interface of the desired shape between the
individual components in the final product. It is reported[1] that,
under certain situations, the position of the interface migrates
as the fluids flow through an extrusion die, and that the direction of interface migration depends on the rheological properties
of the individual polymers involved.

It is also reported in the literature[2] that, under certain
flow conditions or with certain combinations of polymers, the
interface between the phases becomes irregular, resulting in
interfacial instability that must be avoided in obtaining products
of acceptable quality. Figure 6 shows a schematic displaying
irregular interfaces that may be observed in various coextrusion
processes.

It is desirable to correlate the processing conditions,

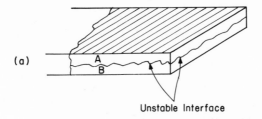

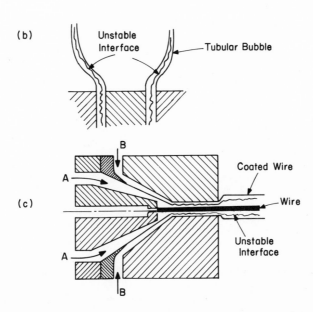

Fig. 6. Schematic describing irregular interfaces that may occur
         in coextrusion processes: (a) flat film coextrusion;
         (b) blown film coextrusion; (c) wire coating coextrusion.

at which interfacial instability begins, to the rheological
properties of the individual polymers being coextruded, for each
coextrusion process (e.g., flat- and blown-film coextrusion, wire
coating coextrusion). A better understanding of the phenomenon of
interfacial instability occurring in each coextrusion process will
help the industry concerned produce coextruded composites having
consistent quality.

     A combination of dispersed and stratified multiphase flow
has also been used to produce commercial products. For instance,
in producing a sandwiched foam product, the gas charged core component is either coextruded with the skin component through a
sheet-forming die or is coinjection-molded in a mold cavity,

giving rise to the product, as schematically shown in Figure 7. Instead of using a foaming agent, one may use a reinforcing agent (i.e., an inorganic filler) in the core component.

In this review lecture, we shall first discuss various rheological problems of multiphase flow as applied to polymer processing and, then, point out the need for further research on multiphase flow of polymeric liquids[3].

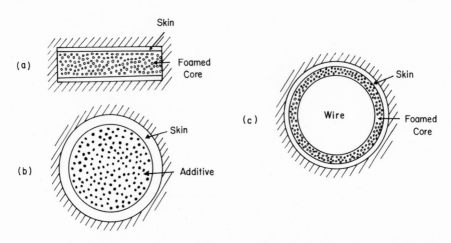

Fig. 7.   Schematic illustrating the cross sections of the products obtained by a combination of dispersed and stratified flows: (a) sandwiched foam sheet; (b) sheath-core rod with an additive in the core; (c) coextruded wire with a foamed core.

REFERENCES

1. C.D. Han, "Rheology in Polymer Processing," Academic Press, New York (1976).
2. C.D. Han and R. Shetty, Studies on Multilayer Film Coextrusion. II. Interfacial Instability in Flat Film Coextrusion, Polym. Eng. Sci., 18:180 (1978).
3. C.D. Han, "Multiphase Flow in Polymer Processing," Academic Press, New York (in press).

INSTABILITIES AND DISTURBANCES ON INDUSTRIAL MELT SPINNING.

G.Colombo, G.Manfrè[*], S.Stellino

SNIA Viscosa S.p.A.
Centro Sperimentale "F.Marinotti"
20031 Cesano Maderno (MI), Italy

INTRODUCTION

To rationalize the melt spinning process of synthetic fibres, many approaches have been devoted in the recent years[1,2,3]. Our recent investigation[4] has included also the inertia forces as our model will be applied even for high spinning speeds. The results of this model are reliable only in the very upper part of the spinning path, as the applied constitutive equation neglects the elasticity.

Generally speaking, the results of the model has visualized that the real instabilities in this part of the spinning path can be attributed to melt-fracture, oscillations of die-swelling and draw-resonance; the last it has been described and analysed in details in the literature[5,6,7,8,9]. But usually this kind of instabilities, predicted by the mentioned authors, do not occur in the real conditions of industrial processes simply for high productivity and good quality reasons.

There exist other kinds of instabilities occurring in the real industrial processes, which we had to face in order to rationalize deeply and more completely what happens in the first part of the drawing zone, where the temperature gradient along the fibre is normally small and on other hand a very high part of total fibre attenuation takes place.

---

[*] present address: Centro Ricerche FIAT
                      10043 Orbassano (TO), Italy

INSTABILITIES AND DISTURBANCES

To be precise it is necessary to distinguish two different kinds of spinning instabilities:

a) <u>process instabilities</u> due to critical conditions of steady-state spinning, nearly intrinsic to a process. Among the other the main are: draw-resonance, melt-fracture, breakage due to capillary phenomena. Usually this critical conditions are avoided or at least the amplitude of the instability is confined to quite low value. In fact our experience leads to say that it is not possible to avoid completely the process instabilities and so it is our opinion that the art of spinning is to damp them properly as smallest as possible.

b) <u>sudden instabilities</u>, which lead to unsteady spinning due to sudden variations either of process variables or of molten polymer parameters. The intrinsic variation of molten polymer can be caused by: clusters, gels, degraded polymer, heterogeneous inclusions, and so on. These variations basically can be approached as a sudden change of viscosity of the molten polymer flowing from the spinneret.

The main aim of the present work is to rationalize this last kind of instabilities, called more properly <u>disturbances.</u>

GOVERNING EQUATIONS

The governing equations of the melt spinning process are:

continuity

$$\frac{\partial R}{\partial t} + V\frac{\partial R}{\partial x} = -\frac{R}{2}\frac{\partial V}{\partial x} \quad ,$$

momentum

$$\rho\left(\frac{\partial V}{\partial t} + V\frac{\partial V}{\partial x}\right) = \frac{2\,\chi}{R}\frac{\partial R}{\partial x}\frac{\partial V}{\partial x} + \chi\frac{\partial^2 V}{\partial x^2} + \frac{\partial \chi}{\partial x}\frac{\partial V}{\partial x} \quad ,$$

energy

$$\frac{\partial T}{\partial t} + V\frac{\partial T}{\partial x} = -2h\frac{T-T_a}{\rho C_p\,R} \quad .$$

The equations are those used in our steady-state spinning model with the further terms relative to the derivatives of variables R, V and T in function of time t. The heat transfer coefficient applied[2] is:

$$h = 4.73 \ 10^{-5}\left(\frac{V}{\pi\,R^2}\right)^{1/3}\left\{1 + \left(\frac{8V_a}{V}\right)^2\right\}^{1/6}$$

and the applied constitutive equation relative to elongational viscosity is:

$$\chi = 3\eta_o = A\,\exp\left(\frac{B}{T}\right)$$

which takes into account only its variation with temperature.

Table 1 Shows the values of A and B coefficients in addition to the values of C_p and ρ, regarded as constants, for nylon 6 (N6) and polyethilenetherephtalate (PET), which are the commercial polymers for melt spinning treated in the present paper.

Table 1.- Constants of N6 and PET polymers.

polymer	A (poise)	B ($^{o}K^{-1}$)	ρ (g/cm^3)	C_p (cal/g ^{o}K)
N6	$5.92 \cdot 10^{-2}$	$5.38 \cdot 10^3$	1.05	0.45
PET	$1.13 \cdot 10^{-6}$	$1.17 \cdot 10^4$	1.33	0.30

SOLUTIONS

The mathematical approach used to compute the governing
equations is the numerical method at finite differences without
linearization. The steady-state equations have been solved assuming
that the solutions are asymptotical limits of a transient process.
The starting point is to impose to the model a rough solution at
t=0, that can be obtained assuming $T_a=T_0$, that is to say:

$$V = V_0 \, \exp\left[\frac{x \, \ln(V_L/V_0)}{L}\right] \; .$$

After this we perturb the system taking the real T_a of the process,
with the following boundary conditions:

$$\begin{array}{lll} \text{at} \quad x=0 & T=T_0 , \quad V=V_0 , \quad R=R_0 , \\ \text{at} \quad x=L & V=V_L . \end{array}$$

The method of computation has been prepared in such a way that
the computer works up to find the steady-state solution of the whole
spinning path with the real process conditions. If the process
conditions are that the steady-state solution is not possible, the
computer stops after a certain amount of time.

The same procedure has been carried out to find the solution
for the case of perturbed spinning conditions. Of course the best
starting point is to begin at t=0 with the steady-state solution,
obtained earlier, modifying the parameter which we have decided to
perturb for a certain amount of time.

Because of the chosen parameter perturbed, the occurring
phenomena can be visualized in two ways: the first is to see what
happens at a given point of the spinning path during the time, the
second is to give an instantaneous picture of the whole spinning
path, either during or after the perturbation time. The computer
used for calculations is an IBM Series/1.

RESULTS

Among the several spinning approached cases in our investigation we present here only two cases which we think are representative.
The boundary conditions of these two cases are shown in table 2.

Table 2.- Boundary conditions in the melt
spinning of N6 and PET.

polymer	T_0 (°K)	V_0 (cm/s)	V_L (cm/s)	R_0 (cm)
N6	533	11.2	1166	0.0195
PET	558	11.2	1166	0.0195

These are the main conditions in addition to the constants shown
in table 1.

The first result worth to mention is the steady-state solutions
relative to N6 and PET. Figure 1 shows the radius R versus the
spinning way x and figure 2 the elongational velocity gradient
versus x.

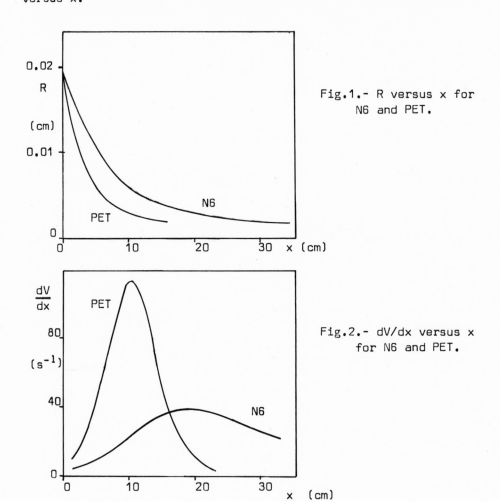

Fig.1.- R versus x for
N6 and PET.

Fig.2.- dV/dx versus x
for N6 and PET.

It is evident that the differences in the drawing zone are mainly related to the specific heat C_p and the constant B,the last related to the activation energy.

The perturbed cases show what happens when there exists in the polymer a sudden change in viscosity due mainly to inhomogeneities of the molten state flowing from the spinneret. We have decided to present the results only of N6, but of course the method can be applied to any melt spinning. The results of PET show the same behavior of N6.

We think that the sudden increase of viscosity can be related with a well known phenomenon in the industrial spinning process, called "nubs". The perturbed conditions assumed here are relative to a (5 ms of time) sudden viscosity increase 1.5 times bigger. The results are shown in figure 3, which pictures the final radius R_f in function of time t; the radius was taken at a distance from the spinneret,where attenuation of fibre radius is completed.

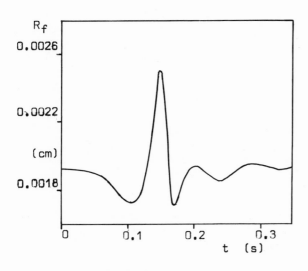

Fig.3.- Disturbance of R_f versus t at a chosen point of the spinning way. Viscosity increased 1.5 times for 5 ms.

It can be seen that the fibre final radius has a slight decrease before the nub occurs. After occurring of the maximum disturbance of the fibre radius, corresponding to the passage of the nub, the radius oscillations damp towards the steady-state spinning like that before the perturbation.

The second perturbation presented here is relative to a sudden decrease of viscosity, which is another common phenomenon verifying in the industrial process, called "thin filaments", usually leading to more probable breakage of the fibre. The results of our compu-tation are shown in figure 4 and are relative to a sudden decrease of viscosity for a time of 5 ms and a value 0.7 times lower.

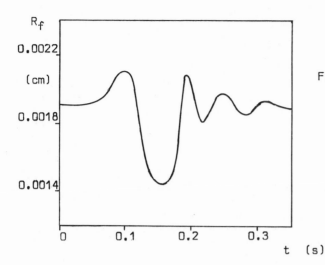

Fig.4.- Disturbance of R_f versus t at a chosen point of the spinning way. Viscosity decreased 0.7 times for 5 ms.

Even the figure 4 pictures the final radius R_f in function of time t, taken at a distance from the spinneret where the fibre presents no attenuation any more. The sudden decrease of the final radius is preceded by a slight increase and it is followed by the damped oscillations.

CONCLUSIONS

It has been presented a model to study both the steady-state and the perturbed solutions of equations governing the melt spinning process, applied particularly to N6 and PET.

The reason to approach the perturbed state in addition to our preceding model of steady-state has become a necessity. In fact we have remarked that the process instabilities, usually mentioned in the literature, are normally avoided, although not completely, in the actual industrial spinning process. Usually the industrial spinning conditions are already optimized in such a way that quality and productivity are as highest as possible. Therefore we start to compute the disturbances produced by a sudden perturbation of process parameters, which in our opinion the main is the polymer viscosity change.

At first glance the results show the remarkable variation of final fibre diameter in function of a sudden change of viscosity, which can be seen in terms of bad quality and also low productivity (breakage).

But on the other hand we have to point out that so far we have not measured neither the real viscosity change nor its time, so any attempt in the future will be to compute the disturbances with the real values of this kind of sudden change in order to improve the production.

Acknowledgment: The authors thanks SNIA Viscosa to have let the delivery of this paper.

SYMBOLS

h	heat transfer coefficient $(cal/cm^2 \ s \ ^oK)$
t	time (s)
x	axial coordinate parallel to the spinning direction (cm)
C_p	specific heat $(cal/g \ ^oK)$
L	lenght of the whole spinning path (cm)
R	fibre radius (cm)
R_f	final fibre radius at completed attenuation (cm)
R_o	initial fibre radius (cm)
T	fibre temperature (^oK)
T_a	room temperature (^oK)
T_o	initial fibre temperature at spinneret exit (^oK)
V	fibre speed (cm/s)
V_a	transverse speed of the cooling air (cm/s)
V_o	initial fibre speed (cm/s)
V_L	final fibre speed (cm/s)
dV/dx	elongational velocity gradient (s^{-1})
η_o	newtonian shear rate viscosity $(poise)$
ρ	polymer density (g/cm^3)
χ	elongational viscosity $(poise)$

REFERENCES

1. S.Kase, T.Matsuo - J.of Polymer Science, part A, 3, 2541 (1965)
2. S.Kase, T.Matsuo - J.of Applied Polymer Science, 11, 251 (1967)
3. A.Ziabicki - Applied Polymer Symposium, 6, 1 (1967)
4. G.Manfrè, S.Stellino - Paper delivered at Convegno-Scuola "Cristallizzazione dei Polimeri", Gargnano (BS) Italy, May 14.17 1979, Proceedings p.245
5. Y.T.Shah, J.R.A.Pearson - Industrial Engineering and Chemistry, Fundamentals, 11, 150 (1972)
6. S.Kase - J.of Applied Polymer Science, 18, 3279 (1974)
7. J.L.White, Y.Ide - Applied Polymer Symposium, 27, 61 (1975)
8. R.J.Fisher, M.M.Denn - A.I.Ch.E. Journal, 22, 236 (1976)
9. J.C.Hyun - A.I.Ch.E. Journal, 24, 418 (1978)

ANALYSIS ON THE STRATIFIED MULTI-PHASE FLOW OF

POLYMER MELTS IN WIRE COATING

Osamu Akita and Katsuhiko Ito

College of Engineering, Hosei University
3-chome, Kajino-cho, Koganei, Tokyo, JAPAN

INTRODUCTION

Theoretical studies on the wire coating process have been made by investigators until now.[1,2] These studies are almost related to the single phase flow between a coating die wall and a travelling wire. Then, the authors expand these studies to the multi-phase flow of non-Newtonian fluid in a wire coating die regarding polymer melts as power law fluids.

THEORETICAL

In order to analyse the concentrically stratified multiphase flow in wire coating process, several fundamental assumptions are postulated as follows.

1. Polymer melts are incompressible non-Newtonian fluids which obey power law.
2. The flow is so viscous that the inertial and gravitational effects are negligible.
3. The flow is steady, and the pressure gradient and the travelling velocity of the wire are invariable.
4. The flow field formed in wire coating die is axially symmetric around the wire and the flow is one dimensional in the direction along the wire axis.

For the convenience of the mathematical treatment, a cylindrical coordinate system in which z coordinate axis is taken in the wire direction is introduced as illustrated in the figure. The momentum balance in z direction is described as

137

$$- \frac{dP}{dz} + \frac{1}{r} \frac{d}{dr} (r\tau) = 0$$

(1)

where $\frac{dP}{dz}$ is the pressure gradient and τ is the shearing stress in z direction.   Integrating Eq.(1) with respect to r and taking into account the boundary condition $\tau = \tau_w$ at $r = r_w$ where τ_w is the shearing stress acting on the wire surface and r_w is the wire radius as shown in the figure, the shearing stress acting on the fluid element is solved as follows.

$$\tau = \frac{r}{2} \frac{dP}{dz} + \frac{1}{r} \left(\tau_w r_w - \frac{r_w^2}{2} \frac{dP}{dz} \right)$$

(2)

If there exists a cylindrical surface $r = \overset{*}{r}$ in the flow field between $r = R$ and $r = r_w$ where the velocity gradient vanishes, the velocity gradient changes its sign across the surface.   In this case, the constitutive equation for power law fluid should be evaluated for each case as

$$\tau = K_i \left(\frac{dv}{dr} \right)^{n_i} \qquad \text{for} \qquad \frac{dv}{dr} \geq 0 \qquad (3)$$

$$\tau = -K_i \left(-\frac{dv}{dr} \right)^{n_i} \qquad \text{for} \qquad \frac{dv}{dr} < 0 \qquad (4)$$

where, K_i and n_i are material constants for i th phase.
Substituting Eq.(2) into Eq.(3) or (4) to give

$$\frac{dv}{dr} = \phi_i \left(\frac{\overset{*}{\chi}^2}{\chi} - \chi \right)^{\frac{1}{n_i}} \qquad \text{for} \left(-\frac{dP}{dz} \right) \geq 0 \text{ and } \chi \leq \overset{*}{\chi} \qquad (5)$$

$$\frac{dv}{dr} = -\phi_i \left(\chi - \frac{\overset{*}{\chi}^2}{\chi} \right)^{\frac{1}{n_i}} \qquad \text{for} \left(-\frac{dP}{dz} \right) \geq 0 \text{ and } \chi > \overset{*}{\chi} \qquad (6)$$

$$\frac{dv}{dr} = -\overset{\backprime}{\phi}_i \left(\frac{\overset{*}{\chi}^2}{\chi} - \chi \right)^{\frac{1}{n_i}} \qquad \text{for} \left(-\frac{dP}{dz} \right) < 0 \text{ and } \chi \leq \overset{*}{\chi} \qquad (7)$$

$$\frac{dv}{dr} = \overset{\backprime}{\phi}_i \left(\chi - \frac{\overset{*}{\chi}^2}{\chi} \right)^{\frac{1}{n_i}} \qquad \text{for} \left(-\frac{dP}{dz} \right) < 0 \text{ and } \chi > \overset{*}{\chi} \qquad (8)$$

where, $\chi = r/R$, $\overset{*}{\chi} = \overset{*}{r}/R = \sqrt{\chi_w^2 + \frac{2\tau_w}{R} \left(-\frac{dP}{dz} \right)^{-1} \chi_w}$ and $\chi_w = r_w/R$

$$\phi_i = \left\{ \frac{R}{2K_i} \left(-\frac{dP}{dz} \right) \right\}^{\frac{1}{n_i}} \qquad , \qquad \overset{\backprime}{\phi}_i = \left\{ \frac{R}{2K_i} \left(\frac{dP}{dz} \right) \right\}^{\frac{1}{n_i}}$$

It is further defined that

$$\Lambda_\chi^i = \int_{\chi_i}^\chi \left(\frac{\overset{*}{\chi}^2}{\chi} - \chi \right)^{\frac{1}{n_i}} d\chi$$

(9)

$$\overset{\sim}{\Lambda}{}^{i}_{\chi} = \int_{\chi_i}^{\chi} (\chi - \frac{\overset{*}{\chi}{}^2}{\chi})^{\frac{1}{m_i}} d\chi$$

(10)

where χ_i denotes the interface between $(i - 1)$ th phase and i th phase, and for $i = 1$ $\chi_1 \equiv \chi_w$. For the case that $- dp/dz \geq 0$, the wire travel velocity V is determined from the boundary condition that $v = 0$ at the die wall $r = R$ $(\chi = 1)$

$$V = - R(\phi_1 \Lambda'_{\chi_2} + \phi_2 \Lambda^2_{\chi_3} + \cdots + \phi_k \Lambda^k_{\chi} - \phi_k \int_{\overset{*}{\chi}}^{\chi_{k+1}} (\chi - \frac{\overset{*}{\chi}{}^2}{\chi})^{\frac{1}{m_k}} d\chi -$$
$$\phi_{k+1} \overset{\sim}{\Lambda}{}^{k+1}_{\chi_{k+2}} - \cdots - \phi_m \overset{\sim}{\Lambda}{}^m_1)$$

(11)

Hence, the fluid velocity v, at χ in i th phase is obtained as follows.

$$v = V + R(\phi_1 \Lambda'_{\chi_2} + \cdots + \phi_k \Lambda^k_{\overset{*}{\chi}} - \phi_k \int_{\overset{*}{\chi}}^{\chi_{k+1}} (\chi - \frac{\overset{*}{\chi}{}^2}{\chi})^{\frac{1}{m_k}} d\chi -$$
$$\phi_{k+1} \overset{\sim}{\Lambda}{}^{k+1}_{\chi_{k+2}} - \cdots - \phi_i \overset{\sim}{\Lambda}{}^i_\chi)$$

(12)

This situation is illustrated in the figure. For the case that $- dp/dz < 0$, the wire travel velocity $\tilde{V}$ is determined by

$$\tilde{V} = R(\overset{\sim}{\phi}_1 \Lambda'_{\chi_2} + \overset{\sim}{\phi}_2 \Lambda^2_{\chi_3} + \cdots + \overset{\sim}{\phi}_k \Lambda^k_{\overset{*}{\chi}} - \overset{\sim}{\phi}_k \int_{\overset{*}{\chi}}^{\chi_{k+1}} (\chi - \frac{\overset{*}{\chi}{}^2}{\chi})^{\frac{1}{m_k}} d\chi -$$
$$\overset{\sim}{\phi}_{k+1} \overset{\sim}{\Lambda}{}^{k+1}_{\chi_{k+2}} - \cdots - \overset{\sim}{\phi}_m \overset{\sim}{\Lambda}{}^m_1)$$

(13)

then, the velocity v at χ in i th phase is

$$v = \tilde{V} + R(-\overset{\sim}{\phi}_1 \Lambda'_{\chi_2} - \overset{\sim}{\phi}_2 \Lambda^2_{\chi_3} - \cdots - \overset{\sim}{\phi}_k \Lambda^k_{\overset{*}{\chi}} + \overset{\sim}{\phi}_k \int_{\overset{*}{\chi}}^{\chi_{k+1}} (\chi - \frac{\overset{*}{\chi}{}^2}{\chi})^{\frac{1}{m_k}} d\chi +$$
$$\overset{\sim}{\phi}_{k+1} \overset{\sim}{\Lambda}{}^{k+1}_{\chi_{k+2}} + \cdots + \overset{\sim}{\phi}_i \overset{\sim}{\Lambda}{}^i_\chi)$$

(14)

Eq.(12) or (14) represents the case that the surface $\chi = \overset{*}{\chi}$ $(r = \overset{*}{r})$ lies in k th phase and χ lies in i th phase such that $i > k$, and at χ in any other phase, v can be obtained in a similar manner. If there does not exist $\overset{*}{\chi}$ between $\chi = \chi_w$ and $\chi = 1$, then the velocity distribution is represented by continuously decreasing or increasing function of χ . In this case, the substitution of Eq.(3) or (4) into Eq.(2) gives

$$\frac{dv}{dr} = \phi_i \left[\left\{ \chi_w^2 + \frac{2\tau_w}{R} (-\frac{dP}{dz})^{-1} \chi_w \right\} \frac{1}{\chi} - \chi \right]^{\frac{1}{m_i}}$$

for $\quad (-\frac{dP}{dz}) \geq 0, \tau_w \geq 0.$

(15)

$$\frac{dv}{dr} = \overset{\vee}{\phi}_i \left[\chi - \left\{ \chi_w^2 + \frac{2T_w}{R} \left(-\frac{dP}{dz} \right)^{-1} \chi_w \right\} \frac{1}{\chi} \right]^{\frac{1}{n_i}}$$

for

$$\left(-\frac{dP}{dz} \right) < 0 \,, \; T_w \geq 0 \,.$$
(16)

$$\frac{dv}{dr} = -\phi_i \left[\chi - \left\{ \chi_w^2 + \frac{2T_w}{R} \left(-\frac{dP}{dz} \right)^{-1} \chi_w \right\} \frac{1}{\chi} \right]^{\frac{1}{n_i}}$$

for

$$\left(-\frac{dP}{dz} \right) \geq 0 \,, \; T_w < 0 \,.$$
(17)

$$\frac{dv}{dr} = -\overset{\vee}{\phi}_i \left[\left\{ \chi_w^2 + \frac{2T_w}{R} \left(-\frac{dP}{dz} \right)^{-1} \chi_w \right\} \frac{1}{\chi} - \chi \right]^{\frac{1}{n_i}}$$

for

$$\left(-\frac{dP}{dz} \right) < 0 \,, \; T_w < 0 \,.$$
(18)

The following integrals are defined similarly as Eqs.(9) and (10).

$$\Pi_\chi^i = \int_{\chi_i}^{\chi} \left[\left\{ \chi_w^2 + \frac{2T_w}{R} \left(-\frac{dP}{dz} \right)^{-1} \chi_w \right\} \frac{1}{\chi} - \chi \right]^{\frac{1}{n_i}} d\chi$$
(19)

$$\overset{\approx}{\Pi}_\chi^i = \int_{\chi_i}^{\chi} \left[\chi - \left\{ \chi_w^2 + \frac{2T_w}{R} \left(-\frac{dP}{dz} \right)^{-1} \chi_w \right\} \frac{1}{\chi} \right]^{\frac{1}{n_i}} d\chi$$
(20)

Thus, the wire travel velocities V_a, V_b, V_c, V_d are calculated for respective conditions as follows.

$$V_a = -R \sum_{k=1}^{m} \phi_k \Pi_{\chi_{k+1}}^k \qquad \text{for } \left(-\frac{dP}{dz} \right) \geq 0 \,, \; T_w \geq 0 \,. \quad (21)$$

$$V_b = R \sum_{k=1}^{m} \phi_k \overset{\vee}{\Pi}_{\chi_{k+1}}^k \qquad \text{for } \left(-\frac{dP}{dz} \right) \geq 0 \,, \; T_w < 0 \,. \quad (22)$$

$$V_c = -R \sum_{k=1}^{m} \overset{\vee}{\phi}_k \overset{\approx}{\Pi}_{\chi_{k+1}}^k \qquad \text{for } \left(-\frac{dP}{dz} \right) < 0 \,, \; T_w \geq 0 \,. \quad (23)$$

$$V_d = R \sum_{k=1}^{m} \overset{\vee}{\phi}_k \Pi_{\chi_{k+1}}^k \qquad \text{for } \left(-\frac{dP}{dz} \right) < 0 \,, \; T_w < 0 \,. \quad (24)$$

Where $\chi_{m+1} \equiv 1$ corresponds to the die wall $r = R$, and the velocity distribution at χ in i th phase is calculated for each condition:

$$v = V_a + \left\{ \sum_{k=1}^{i-1} \phi_k \Pi_{x_i}^k + \phi_i \Pi_x^i \right\} R$$

$$\text{for } \left(-\frac{dP}{dz}\right) \geq 0, \tau_w \geq 0. \quad (25)$$

$$v = V_b - \left\{ \sum_{k=1}^{i-1} \phi_k \hat{\Pi}_{x_i}^k + \phi_i \hat{\Pi}_x^i \right\} R$$

$$\text{for } \left(-\frac{dP}{dz}\right) \geq 0, \tau_w < 0. \quad (26)$$

$$v = V_c + \left\{ \sum_{k=1}^{i-1} \overset{*}{\phi}_k \hat{\Pi}_{x_i}^k + \overset{*}{\phi}_i \hat{\Pi}_x^i \right\} R$$

$$\text{for } \left(-\frac{dP}{dz}\right) < 0, \tau_w \geq 0. \quad (27)$$

$$v = V_d - \left\{ \sum_{k=1}^{i-1} \overset{\vee}{\phi}_k \Pi_{x_i}^k + \overset{\vee}{\phi}_i \Pi_x^i \right\} R$$

$$\text{for } \left(-\frac{dP}{dz}\right) < 0, \tau_w < 0. \quad (28)$$

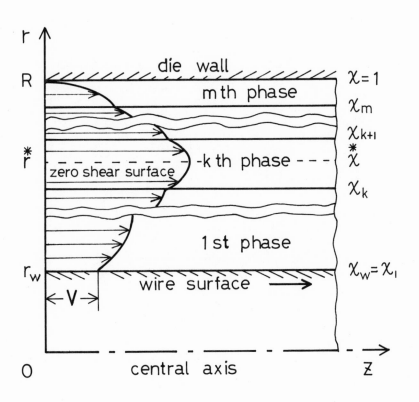

CONCLUDING REMARKS

The velocity distribution is solved in this paper providing the pressure gradient and the wire surface stress. The zero-shear surface $r = \overset{*}{r}$ can be determined from Eq.(2) without any other information since the shearing stress τ vanishes where so that τ_w is preferably employed as a boundary condition rather than the wire travelling velocity. The shearing stress τ_w acting on the wire can be calculated from the wire radius, the drawing force, and the contacting wire length with polymer melts in the coating die. The volumetric flow rate can be easily obtained by integrating the velocity field in a usual way.

REFERENCES

1. C. D. Han and D. Rao, "Study on Wire Coating Extrusion" Polym. Eng. Sci. Vol. 18, No. 13, 1019 (1978).

2. M. Kasajima and K. Ito, "Posttreatment of Polymer Extrudate in Wire Coating" Appl. Polym. Symp., No. 20, 221 (1972).

MICROSTRUCTURAL ORIENTATION DISTRIBUTIONS

INJECTION MOLDED POLYETHYLENE ARTICLES

Musa R. Kamal* and Francis Moy

Chemical Engineering Department
McGill University
3480 University Street
Montreal, Quebec, Canada, H3A 2A7

INTRODUCTION

A number of theories and related techniques have been developed in recent years for investigating the microstructure of partially crystalline polymers. Many of these techniques have been employed to study the anisotropy of molded articles (1-6). In the present work, a multiplicity of techniques is employed to give a detailed characterization of the microstructure (crystallinity and orientation distributions) of injection molded polyethylene parts. Furthermore, these results are employed in conjunction with an existing mechanical model to calculate the distribution of tensile moduli in the moldings. Comparison between experimental and calculated results shows good agreement.

EXPERIMENTAL

Two injection molding polyethylene resins (Ex. 1 and Ex. 2) were employed. Injection temperatures of 176°C and 205.9°C were used in conjunction with a cooling temperature of 35°C. Moldings (12.7cm x 6.35cm x 3.18mm) were obtained with a reciprocating screw 2 1/3 oz Metalmec machine. Thin sections of 40 microns thickness were obtained using a large Reichert microtome.

Sonic velocity was measured with a Pulse Propagation Meter (PPM-SR) supplied by H.M. Morgan Co. Inc., Norwood, Mass. A Reichert Zetopan-Pol polarizing microscope was used in conjunction with a Berek-type compensator (Carl Zeiss Company) to measure birefringence. In wide angle X-ray measurements, samples were irradiated with Ni-filtered CuK_2 radiation in a plane normal to the

* To whom all correspondence should be addressed

X-ray plane of the specimen.   The azimuthal intensity distributions
were measured using Joyce-Loebl Automatic Recording Microdensito-
meter Model MkIII C.S.   The Instron Mechanical Tester was employed
to determine tensile properties.

RESULTS AND DISCUSSION

Figures 1-3 show that the degree of crystallinity (density)
increases with distance from the surface.   This effect is due to
the lower cooling rates near the center.   The degree of crystallinity tends to be unaffected or marginally increased as the melt
temperature is increased.

For an unoriented sample, the measured sonic modulus is expressed as

$$\frac{3}{2E_u} = \frac{x_c}{E_{t,c}^o} + \frac{(1 - x_c)}{E_{t,am}^o} \tag{1}$$

where E_u is the measured sonic modulus of the unoriented sample in
Pascals (Pa), x_c is the crystalline volume fraction, $E_{t,c}^o$ and $E_{t,am}^o$
are the intrinsic lateral moduli in Pascals (Pa) of the crystalline
and of the amorphous phases of the polymer, respectively.

By the use of Equation (1) in conjunction with compression
molded Ex. 2 samples, the values of $E_{t,am}^o$ were determined to be
0.570×10^{10} Pa and 0.43×10^{10} Pa, respectively.   Sakurada et al.
(7) and Horio (8), each using the same stress - X-ray diffraction
technique, obtained values of $E_{t,c}^o$ for polyethylene ranging from
0.22×10^{10} to 0.42×10^{10} Pa.

For an oriented sample, the sonic modulus is expressed as:

$$\frac{3}{2} \left(\frac{1}{E_u} - \frac{1}{E_{OR}}\right) = \frac{x_c f_{cR}}{E_{t,c}^o} + \frac{(1 - x_c)f_{am}}{E_{t,am}^o} \tag{2}$$

where E_{OR} is the sonic modulus of the oriented sample measured in
Pascals; f_{cR} and f_{am} are defined orientation functions for the
crystalline and amorphous phases, respectively.

Increasing the melt temperature tends to increase the sonic
modulus, although the change shown by resin Ex. 2 is marginal.
Resin Ex. 1 exhibits minimum orientation near· the surface, whereas
resin Ex. 2 exhibits maximum orientation near the surface.   Also,
except near the surface, resin Ex. 1 tends to exhibit higher sonic
modulus (total orientation) throughout the molding.

The patterns of birefringence variation are similar to those
observed with sonic modulus.   The measured birefringence can be

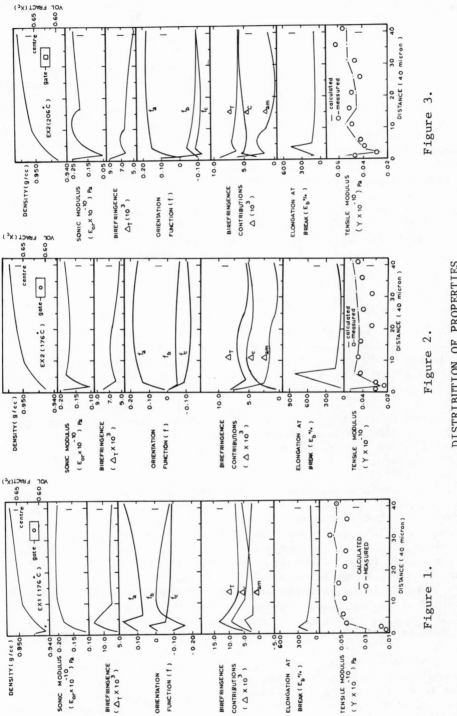

Figure 1.

Figure 2.

Figure 3.

DISTRIBUTION OF PROPERTIES

expressed (9) as

$$\Delta_T = x_c f_{cR} \Delta_c^o + (1 - x_c) f_{am} \Delta_{am}^o + \Delta_f \qquad (3)$$

where Δ_T is the total birefringence of the sample, x_c is the crystalline volume fraction, f_{cR} and f_{am} are the orientation functions of the crystalline and amorphous phases, respectively; Δ_c^o and Δ_{am}^o are the intrinsic birefringence of a perfectly oriented crystalline and amorphous phase, respectively; Δ_f is the form birefringence and is assumed to be negligible (10,11). By coupling birefringence measurements with X-ray measurements, it was found that $\Delta_c^o = 0.045$ and $\Delta_{am}^o = 0.064$. These values are in reasonable agreement with results reported by other workers (10, 12).

The crystalline contribution to the total orientation, as measured by birefringence, may be completely described in terms of the orientation functions defined as (13, 14)

$$f_a = (3 \overline{\cos^2 \alpha} - 1)/2 \qquad (4a)$$

$$f_b = (3 \overline{\cos^2 \beta} - 1)/2 \qquad (4b)$$

where α and β are the angles between the x-axis (the flow direction) and the crystallographic a and b axes respectively. The c-axis orientation function, f_c, is obtained from the orthogonality relationship that exists among the three mutually perpendicular directions for an orthorhombic system.

Generally, the crystallographic a-axis tends to be oriented in the flow direction. The degree of orientation tends to increase from a minimum near the surface to a maximum towards the core region, regardless of resin or melt temperature.

By combining the results of density (crystallinity), sonic modulus, and x-ray diffraction, it is possible to obtain information regarding the orientation of the amorphous phase by the use of equation (2). Generally, the degree of amorphous orientation is a maximum at or near the surface, which is consistent with the measurement of total orientation measured by both the sonic modulus and birefringence techniques discussed earlier. Furthermore, resin Ex. 1 tends to exhibit higher amorphous orientation than resin Ex.2 molded at the melt temperature of 176^oC. Increasing the injection melt temperature appears to have marginal effects on the degree of amorphous orientation of resin Ex. 2.

Figures 1 to 3 show that, near the surface, increasing the melt temperature tends to decrease the percent elongation at break. Furthermore, resin Ex. 1 exhibits much lower ε_b values near the surface than those observed for resin Ex. 2. Also, it is observed that the percent elongation at break towards the core region of the molding, appears to be independent of differences between resins and of melt temperature.

Finally, the experimental tensile modulus distributions for the three cases are compared to those estimated theoretically by employing a mechanical model proposed recently by Seferis and Samuels (15):

$$J = 1/E_{oR} = x_c(1-f_{cR})A_{cR}+(1-x_c)(1-f_{am})A_{am} \tag{5}$$

$$A_{cR} = \frac{2}{3E_c^o} \quad \text{and} \quad A_{am} = \frac{2}{3E_{am}^o} \tag{6}$$

where J and E_{oR} are the tensile compliance and modulus respectively. The values of x_c, f_{cR}, f_{am}, E_c^o and E_{am}^o are obtained from the results of density, birefringence and X-ray measurements. The agreement between the experimental and calculated distributions is surprisingly good, as shown in Figures 1-3. Furthermore, the theoretical model predicts accurately the qualitative differences between the two resins and the influence of temperature, especially in the critical area near the mold wall. Thus it is felt that techniques to predict and control crystallinity and orientation distributions in injection molded parts would contribute substantially to the prediction and control of the ultimate properties of these articles.

REFERENCES
1. G. Menges and G. Wubken, SPE ANTEC, 19, 519 (1973)
2. W. Woebcken and B. Heise, Kunststoffe, 68, 99 (1978)
3. Z. Bakerdjian and M.R. Kamal, Polym.Eng. and Sci., 17, 96 (1977)
4. V. Tan and M.R. Kamal, SPE ANTEC, 21, 339 (1976)
5. V. Tan and M.R. Kamal, J.Appl.Polym.Sci., 22, 2341 (1978)
6. F.H. Moy and M.R. Kamal, SPE ANTEC, 25, 108 (1979)
7. I. Sakurada, T. Ito and K. Nakamae, J.Jap.Soc. Test. Mat., 11, 683 (1962)
8. M. Horio, Symp. at Polytechnic Institute of Brooklyn, Sept. 7, 1963
9. R.S. Stein, J.Polym.Sci., C34, 709 (1959)
10. W.T. Mead, C.R. Desper and R.S. Porter, J.Polym.Sci., Polym. Phys., 17, 859 (1979)
11. Y. Fukui, T. Asada and S. Onogi, Polym. J., 3(1), 100 (1972)
12. K. Nakayama and H. Kanetsuna, J.Mater.Sci., 10, 1105 (1975)
13. R.S. Stein, J.Polym.Sci., C34, 709 (1959)
14. J.J. Hermans, et al., Rec.Trav.Chim., 65, 427 (1946)
15. J.C. Seferis and R.J. Samuels, Polym.Eng.Sci., 19, 975 (1979)

MELT DEFORMATION DURING PARISON FORMATION

AND INFLATION IN EXTRUSION BLOW MOLDING

Musa R. Kamal*, Dilhan Kalyon and Victor Tan

Chemical Engineering Department
McGill University
3480 University Street
Montreal, Quebec, Canada, H3A 2A7

INTRODUCTION

A significant part of the work associated with the analysis of the extrusion blow molding process has concentrated on the development of techniques to characterize parison swell and draw-down behaviour (4-5) and on establishing empirical correlations between operating variables and parison behaviour (6-8). Other studies have attempted to relate parison behaviour to material properties (9-16). On the other hand, a limited amount of work has been reported on parison inflation, involving mainly simplified methods to predict the distribution of the bottle thickness in simple mold geometry (17-19).

Recently, the authors (20) reported experimental data obtained with high speed cinematography to study parison formation and development during extrusion blow molding. The present work extends the application of high speed cinematography to the study of the parison inflation stage. Moreover, a first order mathematical treatment is proposed to permit the calculation of the relevant dimensions during the parison formation, development, and inflation stages of the extrusion blow molding process. This treatment may also be extended to allow the estimation of distribution of bottle thickness.

EXPERIMENTAL

Two commercial blow molding grade high density polyethylene resins, designated as Resins D and E, were employed in the study.

* to whom all correspondence should be addressed

These are the same two resins employed in the parison study reported earlier (20).   An Impco, model A13-R12, reciprocating screw
extrusion blow molding machine was employed in conjunction with a
specially constructed transparent mold.   The cylindrical mold was
22 cm long and 10.16 cm in diameter.   Two 16 mm Bolex (H16) movie
cameras were employed simultaneously for the continuous filming of
the parison, with one facing the front view of the transparent mold
and the other monitoring the bottom view of the mold via an in-
clined mirror.

RESULTS AND DISCUSSION

A.   Parison Extrusion
      For extrusion of the parison through an annular die, it is
necessary to define two important quantities associated with the
transient swelling characteristics of the polymer.   The transient
diameter swell, known as $B1(t)$, is defined as $B1(t)=D_o/D_{o,d}$, where
D_o and $D_{o,d}$ are the outside diameters of the parison and die
respectively.   The transient thickness swell, $B2(t)$, is defined
as $B2(t)=H/G$, where H and G represent thicknesses of the parison
and the die gap respectively.

      In the absence of sag, due to gravity, the area of a differen-
tial ith element of the parison is

$$A_{no\ sag}^{i}(z,t) = \pi \left| D_o^2(z,t) - D_i^2(z,t) \right| /4$$

where D_i is the inside diameter of the parison.   Moreover,

$$D_o(z,t) = B1(t)\ x \cdot D_{o,d} \qquad \text{and}$$

$$D_i(z,t) = B1(t)\ x\ D_{o,d} - B2(t)\ x\ 2G$$

      It can be shown that, in the absence of sag, the area of a
differential ith element of the parison is

$$A_{no\ sag}^{i}(z,t) = \pi \left| B1(t)\ x\ D_{o,d} - B2(t)\ x\ G \right| \left| B2(t)\ x\ G \right|$$

      Ajroldi (10) has shown that the tensile compliance, $J(t)$,
could be used for computing the effect of sag, as follows:

$$\ln W^{i}(t) = \ln W_r^{i}(t) - g\ell\rho J(t) \{\Sigma W^{i}(t)\}/W^{i}(t)$$

where $W^{i}(t)$ is the weight of the ith pillow after it has recovered
for time t, $W_r^{i}(t)$ is the weight of the ith pillow at time t in the
absence of sag, ρ is the density, and ℓ is the length of the
pillow, and

$$\Sigma W^{i}(t) = W^{i}(t)/2 + \sum_{k=1}^{k=i-1} W^{k}(t)$$

Here, we denote the leading end as i = 1.

It is proposed here that, in the place of $W^i(t)$, one employs the cross-sectional area of the parison A^i. Thus

$$\ln A^i_{sag}(t) = \ln A^i_{no\ sag}(t) - g\ell\rho J(t)\{\Sigma W^i(t)\}/W^i(t)$$

Therefore, the length of the ith element, $\Delta L^i(t)$, at time t, is given by

$$\Delta L^i_{sag}(t) = \dot{W}(\Delta t)/\rho\ A^i_{sag}(t)$$

where W is the extrusion rate, and Δt represents the time period required to extrude the ith element.

Hence, the total length of the parison $L^T_{sag}(t)$ is given by

$$L^T_{sag}(t) = \sum_{i=1}^{i=N} \Delta L^i_{sag}(t)$$

where N is the total number of increments extruded in time t.

The diameter profiles and the weight distributions obtained in Reference (20) are employed to estimate the transient swelling functions B1(t) and B2(t). In this work, the values of creep compliance J(t) reported by Plazek and co-workers, Figure 1 of Reference (21), were employed.

Typical results are shown in Figure 1 for the diameter profiles and the thickness distribution for the parison based on Resin D at 200°C. The excellent agreement reflects the accuracy of the two transient swelling functions. The deviation at the leading end is due to the actual experimental conditions which, due to intermittent operation, allow the material at the leading end to relax substantially at the tip of the die before extrusion. Thus this portion of the parison exhibits low swell. Fortunately, this portion of slightly sheared material becomes the tail flash of the bottle.

B.   The Inflation Stage
In this section, a first order mathematical procedure is proposed for the analysis of the inflation stage to estimate the variability of bottle thickness in both the axial and the circumferential directions. Typical cinematography results showing the effects of extrusion temperature and air pressure on the effective inflation rate of resin E are shown in Figure 2. They indicate that the inflation pressure has a marked influence on the rate of inflation.
    (1)   Parison Shape During Inflation
The shape of the parison at the end of extrusion tends to be circular. During the clamping stage, the bottom portion of the

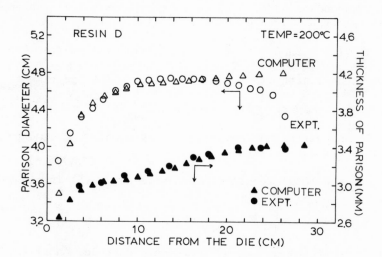

Figure 1.   Comparison of Calculated and Experimental Parison
            Diameter and Thickness Distributions

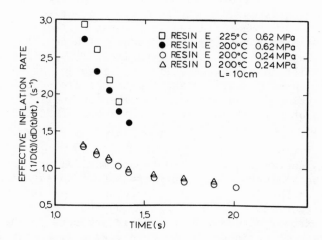

Figure 2.   Compariosn of the Effective Inflation Rates of
            the Two Resins

parison becomes flat, while the bulk of the parison resembles a
toothpaste tube.   High speed cinematography shows that, upon the
introduction of air, the middle portion of the parison returns to
the original circular shape, while at other axial positions it
takes on an elliptical shape.   Therefore, it is assumed that, at
an axial distance, z, from the die, the ellipse major and minor
axes are a(z,t) and b(z,t) respectively, at time t.

The profile made by the outer surface of the parison may be
estimated employing the shape characteristics proposed by Fukase
and co-workers (18).   In this work a correction was applied to
allow for the redistribution of mass near the pinch-point as a
result of clamping.

Figure 3 shows experimental data relating to the progression
of the inflation stage for both resins D and E, as obtained by
cinematography.   The diameters for resin E are greater than for
resin D at all times.   The Figure also compares calculated and
experimental inflation profiles for resin D.   Good agreement is
noted between calculated and experimental results.

(2)   Estimation of Extension Ratios and Bottle Thickness

The extension ratio in the radial direction, $\lambda_{rj}^i(t)$, which is
related to the change in parison thickness, may be estimated from
the angular deformation ratio, $\lambda_{\theta j}^i(t)$, and the axial deformation
ratio, $\lambda_{zj}^i(t)$.   A mass balance yields:

$$\lambda_{rj}^i(t) = 1/\lambda_{zj}^i(t) \ \lambda_{\theta j}^i(t)$$

The axial deformation ratio, $\lambda_{zj}^i(t)$, is obtained from the
ratio of the total length of the vertical arc, as estimated by the
Fukase model, at times $t + \Delta t$ and t.

In order to estimate $\lambda_{\theta j}^i(t)$, the standard hoop stress for a
circular thin shell of radius R and thickness H is modified with a
shape factor, $F_j^i(t)$, in order to estimate the hoop stress, σ_j^i,
along a point (i,j) on the elliptical perimeter of the parison.
Thus,

$$\sigma_j^i(t) = P(t) \ R(z,t) F_j^i(t)/2H(z,t)$$

The details of the mathematical procedure for the estimation of
$F_j^i(t)$ will be described elsewhere, due to space limitations.

Figure 4 compares calculated and experimental results for the
circumferential distribution of the bottle thickness at two locations.   The agreement is reasonable, showing a maximum deviation
of about 8%.

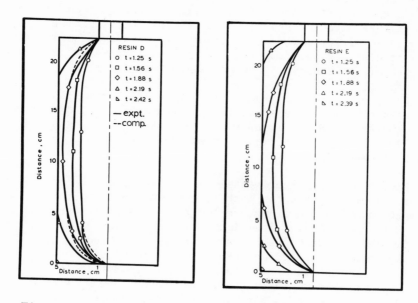

Figure 3.   Inflation Characteristics of Resins D and E

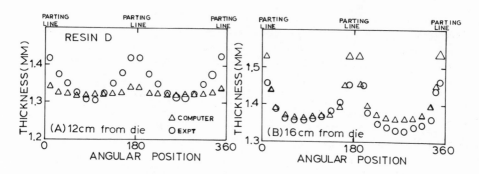

Figure 4.   Calculated and Experimental Circumferential Bottle
             Thickness Distribution

REFERENCES

1. R. Bird, C. Armstrong and O. Hassager, "Dynamics of Polymeric Liquids", John Wiley, New York (1977)

2. R. Mendelson, F. Finger and E. Bagley, J. Polym. Sci., 35, 178 (1971)

3. M. Harig, S. Lidarikis and J. Vlachopoulos, Trans. Soc. Rheology, 16, 670 (1972)

4. N. Sheptak and C. Beyer, SPE Journal, 190 (1965)

5. K. Chao and W. Wu, SPE Journal, 190 (1965)

6. N. Wilson, M. Bentley and B. Morgan, SPE Journal, 26, 34 (1970)

7. K. Blower and N. Standish, Polym. Eng. Sci., 13, 3, 222 (1973)

8. E. Henze and W. Wu, Polym. Eng. Sci., 13, 2, 153 (1973)

9. F. Cogswell, Plastics and Polymers, 391 (1970)

10. G. Ajroldi, Polym. Eng. Sci., 18, 10, 742 (1978)

11. T. Clifford, SPE Journal, 25, 32 (1969)

12. J. Miller, Trans. Soc. Rheology, 19, 3, 341 (1975)

13. J. Schaul, M. Hannon and K.F. Wissburn, Trans. Soc. Rheology, 19, 3, 351 (1975)

14. F.N. Cogswell, P. Webb, J.C. Weeks, S. Maskell and P. Rice, Plastics and Polymers, 340 (1971)

15. R.A. Worth and J. Parnaby, Trans Instn Chem. Engrs, 52, 368 (1974)

16. R. Pritchatt, J. Parnaby and R. Worth, Plastics and Polymers, 55 (1975)

17. A. Dutta, SPE Technical Papers, Vol. 25, 913 (1979)

18. H. Fukase, A. Iwaaki and T. Kunio, SPE Technical Papers, Vol. 24, 650 (1978)

19. C.J.S. Petrie, "Polymer Rheology and Plastics Processing", Proceedings of Conference, edited by P.L. Clegg et al., Plastics and Rubber Inst., London, pp. 307-318, 1975

20. D. Kalyon, V. Tan and M.R. Kamal, SPE Technical Papers, Vol. 25, 991 (1979)

21. D.J. Plazek, N. Raghupathi, R.F. Kratz and W.R. Miller, Jr., J. Appl. Polym. Sci., Vol. 24, 1309 (1979)

HIGH MODULUS POLYETHYLENE OBTAINED WITH INJECTION MOULDING

J. Kubat and J.A. Manson

Chalmers University of Technology
Gothenburg, Sweden

(Abstract)

There have been substantial efforts to produce high modulus
materials based on polyethylene using techniques such as
compression moulding, solid state extrusion, various drawing
procedures, flow controlled crystallization etc. In our laboratory
a commercial processing technique, i.e. injection moulding, has
been employed to produce specimens of linear high molecular weight
polyethylene containing, even though to a lesser degree, high
modulus structure of a similar kind as obtained with the above
methods. The injection moulding machine is designed to operate
with injection pressures up to 500 MPa, to be compared with the
normally used range of 150-200 MPa.

A polyethylene specimen produced by high pressure injection
moulding has the character of a composite, consisting of a core
and a surrounding outer layer. This layer contains spherulites of
the normal type, while the core consists to a large degree of
high-strength structures, i.e. probably a combination of extruded
chain and shish kebab structures, resulting in improved stiffness
and strength. Their formation is probably due to the combined
action of the elevated pressure and the shear deformations occuring
during the packing of the mould. The stiffness and strength of high
pressure injection moulded specimen are improved by ca.200-300 per
cent compared with the corresponding values obtained at normal
injection pressures. This is still far below what is obtained with

the techniques mentioned above. However, by varying the processing condition the amount of high strength structures can be increased which further improves the mechanical properties. It is shown that a lowering in melt temperature from 250°C to 170°C increases the tensile modulus from ca. 4 GPa to 7.5 GPa (injection pressure 500 MPa). The tensile strength varies in a similar way.

The significance of the shear deformation of the melt is substantiated by the observation that increasing the thickness of the specimens results in lower stiffness and strength values. The results obtained at lower melt temperatures show that the high strength structures obtained are similar to those reported for solid state extrusion.

The high strength structures obtained with high pressure injection moulding are further characterized using differential thermal analysis and wide angle X-ray diffraction studies.

FLOW OF MOLTEN POLYMERS USED IN

THE SYNTHETIC FIBRE INDUSTRY THROUGH GRANULAR BEDS

Z. Kembłowski, M. Michniewicz, J. Torzecki

Institute of Chemical Engineering
Łódź Technical University
Łódź, Poland

INTRODUCTION

The paper is concerned with some fundamental aspects of filtration of molten polymers, used in the synthetic fibre industry. The process of filtration of spinnable polymers is of great importance to the quality of the fibres produced. Because of the leading position of polyester and polyamide fibres in the world production of man-made fibres, we investigated the rheological properties and flow through granular beds of molten poly (ethylene terephthalate) - PET, and polycaproamide - PA6.

In our previous papers[1-4] the results on flow of various rheologically complex fluids through granular beds were presented. Among the investigated media there were also molten polymers, but their rheological properties were independent of the residence time in molten state. In order to investigate the flow of molten PET and PA6 through granular beds we had to take into account their rheological instability in molten state, caused by the thermal, hydrolytic and oxygen degradation.

DETERMINATION OF RHEOLOGICAL PROPERTIES

The characteristics of the experimental media is given in Table 1. The rheological properties were investigated at different temperatures using two capillary rheometers: an Instron 3211 plug rheometer and a constant-pressure rheometer RK-3 of own construction[5] (the rheometer RK-3 enabled the measurements in the region of low

shear rates to be made). The polymer samples were carefully dried and kept in the atmosphere of nitrogen in
order to avoid (as far as possible) the hydrolytic and
oxygen degradation.

Table 1. Experimental media, their inherent
         viscosity $[\eta]$ and weight-average
         molecular weight M_W

Medium Symbol	$[\eta]$	M_W
PET-1	0.672	43900
PET-2	0.639	40800
PET-3	0.611	38200
PET-4	0.545	32300
PA6	1.259	33700

Results Obtained for Molten PET

        In the first part of the investigations the depen-
dence of zero-shear-rate viscosity η_o of molten PET on
residence time in molten state was determined. An example
of such results for PET-1 is presented in Fig. 1 for the
temperature range 270-300°C. One can see that in the
temperature range 275-295°C there is a region in which
the value of η_o remains constant at such time intervals
which enable the processing of the melt without noticeable thermal degradation. These time intervals vary from
ca 120 minutes at 275°C to ca 25 minutes at 295°C.

        In the second part of the investigations the depen-
dence of shear-dependent viscosity η of molten PET on
the nominal shear rate $\dot{\sigma}_n$ in the region of thermal stability of the melt was determined. An example of such
results for PET-1 is given in Fig. 2. A range of Newtonian
viscosity at lower shear rates as well as a range of
shear-dependent viscosity at higher shear rates is clearly visible on this graph. Similar results were obtained
for PET-2, PET-3 and PET-4.

        Using the Mendelson superposition procedure[6,7] a
master curve was obtained in coordination system

$$\frac{\eta}{a_T b_M} \quad \text{vs.} \quad a_T b_M \dot{\sigma}_n$$

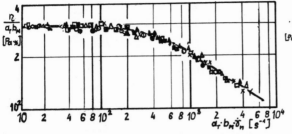

Fig.3 Master flow curve for
molten PET

	275 °C	280 °C	285 °C	290 °C	295 °C
PET-1	x	Y	✝	✱	⅄
PET-2	△	▲	◮	◮	◮
PET-3	□	■	◪	◪	◫
PET-4	○	●	◐	◑	◫

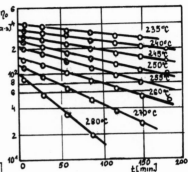

Fig.4 Zero-shear rate
viscosity of
molten PA6 vs. resi-
dence time in molten
state

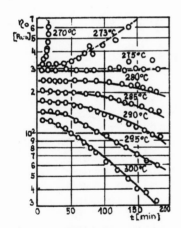

Fig.1 Zero-shear-rate
viscosity of
molten PET-1 vs. resi-
dence time in molten
state.
--poor reproducibility

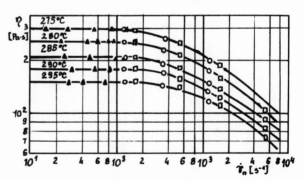

Fig.2 Shear-dependent viscosity of
molten PET-1 vs. shear rate
at different temperatures
▲ RK-3   L=76.426mm, D=0.770mm
O Instron L=51.036mm, D=1.252mm
◻ Instron L=25.542mm, D=0.765mm

where a_T - temperature shift factor
      b_M - molecular weight shift factor.
The master curve for all experimental data concerning
molten PET is presented in Fig. 3. The curve was described by the Carreau model[8], which for molten polymers
may be presented in the following form

$$\eta = \eta_0 \left[1 + (\Lambda \dot{\sigma}_n)^2\right]^{-N} \tag{1}$$

where η_0, Λ and N are the rheological parameters of the
model. The rheological parameters of the curve presented
in Fig. 3 are described by the following dependences

$$\eta_0 = 294 \, a_T b_M \quad [Pa \cdot s] \tag{2}$$

$$\Lambda = 0.00295 \, a_T b_M \quad [s] \tag{3}$$

$$N = 0.167 \tag{4}$$

where

$$a_T = exp(-19.386 + 10625/T) \tag{5}$$

and

$$b_M = 7.836 \, [\eta]^{5.16} = 5.090 \; 10^{-17} M_W^{3.51} \tag{6}$$

with T being the absolute temperature of the medium.

Results Obtained for Molten PA6

     In Fig. 4 the dependence of the zero-shear-rate
viscosity η_0 of molten PA6 on residence time in molten
state is presented for the temperature range 235-280°C.
One can see that for molten PA6 there is no region in
which the value of η_0 remains constant. In order to correlate the experimental data, in a similar way as in the
case of molten PET, one had to take into account the dependence of the rheological parameters of the Carreau
model (1) on the residence time in molten state t. Theoretical considerations which are outside the scope of this
paper resulted in the following formulae

$$\eta_0 = A \, exp\left(\frac{B}{RT} - ct \, exp \, \frac{\beta}{RT}\right) \tag{7}$$

$$\Lambda = \frac{\alpha}{\varrho T} \, exp\left(\frac{B'}{RT} - c't \, exp \, \frac{\beta'}{RT}\right) \tag{8}$$

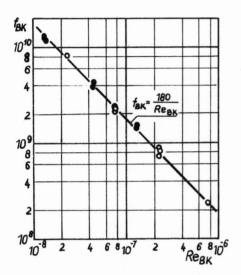

Fig.5 Friction factor vs.
      Reynolds number for
molten PET at different
temperatures (bed height
10-42mm)
   O - d_p = 0.175mm
   ● - d_p = 0.102mm

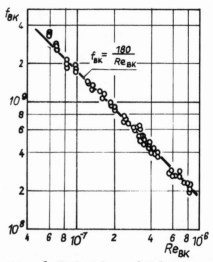

Fig.6 Friction factor vs.
      Reynolds number for
molten PA6 at different
residence times and dif-
ferent temperatures (bed
height 10-35mm, d_p=0.210mm)

$$N = \frac{1}{2} \left(1 - \frac{1}{n_o + mT} \right) \tag{9}$$

The numerical values of the constants occurring in eqs.
(7), (8) and (9) were determined experimentally for the
investigated molten PA6 and are presented in Table 2.

Table 2. Numerical values of constants occurring in eqs.
(7), (8) and (9)

A [Pa·s]	B [J/mol]	c [s⁻¹]	β [J/mol]	α [kg·s·K/m³]
$6.456 \cdot 10^{-7}$	$8.631 \cdot 10^4$	$2.37 \cdot 10^7$	$-1.146 \cdot 10^5$	832

B' [J/mol]	c' [s⁻¹]	β' [J/mol]	n_o [−]	m [K⁻¹]
0	$1.97 \cdot 10^9$	$-1.351 \cdot 10^5$	18.64	−0.0317

FLOW THROUGH GRANULAR BEDS

The above presented dependences enable the prediction
of rheological behaviour of the investigated molten PET
and PA6 during the flow through granular beds. Because
of the low shear rates which occurred in the experimental
beds, the media behaved as Newtonian fluids. Therefore
the pressure drop data were correlated, according to the
classical approach of Blake-Kozeny-Carman[4], in the coordinate system

$$f_{BK} \quad vs. \quad Re_{BK}$$

The friction factor f_{BK} and Reynolds number Re_{BK} are
defined as follows

$$f_{BK} = \frac{\Delta p}{\rho v_o^2} \frac{d_p}{l} \frac{\varepsilon^3}{1-\varepsilon} \tag{10}$$

$$Re_{BK} = \frac{v_o d_p \rho}{\eta_o (1-\varepsilon)} \tag{11}$$

where Δp - pressure drop due to friction, ρ -fluid density, v_o - superficial velocity, d_p - effective

particle diameter, l - bed height, ε - bed porosity. The value η_0 was determined for every investigated case according to dependence (2) or (7).

The experimental data concerning pressure drop during the flow through granular beds are presented for molten PET and PA6 in Fig. 5 and Fig. 6 respectively, and are well described by the Blake-Kozeny-Carman equation

$$f_{BK} = \frac{180}{Re_{BK}} \tag{12}$$

This confirms the applicability of the adopted approach to the correlation of data, concerning polymers which are unstable in molten state because of degradation.

REFERENCES

1. Z. Kembłowski, J. Mertl, Chem.Eng.Sci. 29:213 (1974)
2. Z. Kembłowski, J. Mertl, M. Dziubiński, Chem.Eng.Sci. 29:1343 (1974)
3. Z. Kembłowski, M. Dziubiński, Rheol.Acta 17:176 (1978)
4. Z. Kembłowski, M. Michniewicz, Rheol.Acta 18:730 (1979)
5. Z. Kembłowski, M. Dziubiński, H. Fidos, Włókna Chemiczne 4:138 (1978)
6. R.A. Mendelson, Trans.Soc.Rheol. 9:53 (1965)
7. R.A. Mendelson, Polym.Lett. 5:295 (1967)
8. P.J. Carreau, Trans.Soc.Rheol. 16:99 (1972)

VISCOELASTIC PROPERTIES OF PHENOL-FORMALDEHYDE OLIGOMERS AND POLYMERS IN THE PROCESS OF CROSS-LINKING

Yu.G.Yanovsky, V.I.Brizitsky, G.V.Vinogradov

Institute of Petrochemical Synthesis of the USSR
Academy of Sciences, Moscow, USSR

(Abstract)

A specific aspect of processing composites based on reactive oligomers is that products have to be structurized during their fabrication. It is therefore necessary to know the mechanism and kinetic aspects of the oligomer-polymer transition to optimize curing and obtain products exhibiting excellent strength characteristics.

This work involves studies on the viscoelastic properties of reactive oligomers (PFO) in the process of structurization, as well as examination of the effect of alloying agents on this process. The method of low-amplitude forced harmonic motion in the range of straining frequencies of 10^{-3} to 10 sec^{-1} and temperatures of 120 to 180°C was used. This method has enabled us to measure continuously the viscoelastic characteristics of the system in the process of its structurization, when the values of G' and G" varied in the range of 10^2 to 10^{11} dyn/cm^2, that is from the fluid to cured state. Also measured were the induction period, the onset and completion of the structurization process, and the duration of the process as a whole. The temperature dependence of cured phenol-formaldehyde polymers (PFP) is given.

It has been established that the cure rate of alloyed PFP exceeds that of the starting polymers. The addition of alloying agents yields materials with greater values of G' and lesser loss values. Comparison of the G' values and strength of PFP reveals a correlation between the values of G' and strength characteristics. Corresponding to greater values of G', which characterize a greater degree of cure, are greater values of the compression and bending strengths.

RUBBER

DYNAMIC MECHANICAL PROPERTIES OF THERMOPLASTIC URETHANE

ELASTOMERS BY THERMALLY STIMULATED CREEP

Talal El Sayed, Daniel Chatain and Colette Lacabanne

Laboratoire de Physique des Solides
Université Paul Sabatier
118, Route de Narbonne
31077 Toulouse Cédex (France)

In the chemistry of urethanes, major emphasis in recent years, has been placed upon efforts to develop aliphatic isocyanates to impart light stability, and improved stability toward hydrolysis (1). The first commercial aliphatic diisocyanate used was 1,6 hexamethylene diisocyanate (HDI). A series of elastomers was prepared in the melt from polyester diol (ES) $-M_\omega$ = 2,000-, from HDI and from 1,4 butane diol (BDO) for mole ratio of ES/HDI/BDO : 1/1/0, 1/2/1, 1/3/0, 1/4/3 and 1/5/4 (2). The dynamic mechanical properties of these thermoplastic urethane elastomers were investigated over the temperature range −200°C to 100°C using the Thermally Stimulated Creep (TSCr)technique (3)-. It is the purpose of this paper to investigate the molecular relaxation mechanisms of thermoplastic urethane elastomers by studying the effect of microphase segregation and intermolecular bonding on their dynamic mechanical properties.

EXPERIMENTAL

In TSCr experiments, the sample was subjected to a shear stress; then the temperature was lowered in order to freeze molecular motions and the stress was removed. Next, the sample was heated at a linear rate so that the mobile units can reorientate. The time derivative of the strain $\dot{\gamma}$ is recorded versus the temperature of the sample. Fig.1 shows the overall TSCr spectrum of thermoplastic urethane elastomer 1/5/4 as example.

SECONDARY RELAXATIONS

For all of the polymers, a double relaxation peak is observed at −150°C (γ) and −110°C (β).

ɤ-Relaxation

The magnitude of this TSCr peak is a decreasing function of
the organization of hard segments : Annealed samples have a relatively higher peak. From data reported in the literature (4-6), this
relaxation can be attributed to local motions of methylene sequences.

β-Relaxation

Increasing hard segment content produces increased magnitude
of this βTSCr peak. Annealing at high temperature, results in the
disappearance of this peak which is restored by annealing for 12
hours at room temperature. Several explanations have been proposed
for analogous relaxations in polyurethanes (4,7). Here, it is
assigned to the motion of CO and/or NH groups to which water molecules are associated by hydrogen bonding.

GLASS RELAXATIONS

Thermoplastic polyurethane elastomers possess two glass
temperatures : a lower one $Tg(\ell)$ which correspond to the α (ℓ) loss
peak and an upper value $Tg(u)$ which is responsible of the $\alpha(u)$
shoulder on the high temperature side of the $\alpha(\ell)$ peak.

$\alpha(\ell)$ Lower Glass Relaxation

The $\alpha(\ell)$ TSCr peak is observed in all of the urethanes at
-38°C. Increasing hard segment content produces increased breadth
over which the TSCr peak occurs while reducing its magnitude. In
addition, for a given sample, a quench results in an increasing
of the TSCr peak while an annealing has an opposite effect.
This peak has been resolved into elementary processes whose retardation times, τ_i , obey the "compensation" equation :

$$\tau_i = \tau_c \exp \{E/k \ (1/T - 1/T_c) \}$$

where E is the activation energy,
and   k is the Boltzmann constant.
At the "compensation" temperature T_c, all the processes would have
the same retardation time τ_c : in other words, the activation
entropy is a linear function of the activation energy.

Fig. 2 shows as example, the Arrhenius plot of the glass
relaxation for the thermoplastic urethane elastomer 1/5/4 for the
four processes constituting the $\alpha(\ell)$ component : τ_c= 30 secondes
and T_c = -37°C. By changing the hard segment concentration, a
constancy of the lower glass transition has been observed.

This relaxation is often designated as the glass transition
of soft segments (4-13) and it is assigned to microbrownian

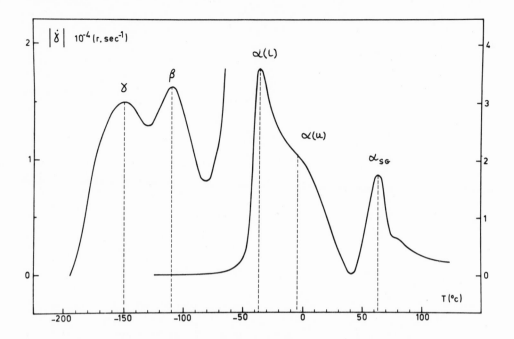

Figure 1.   Overall TSCr spectrum of thermoplastic urethane elastomer
            1/5/4 between liquid nitrogen temperature and 100°C.
            The applied stress was 1007 N cm^{-2} for the left scale,
            and   197 N cm^{-2} for the right hand scale.

motions of soft segments in the amorphous regions.

α(u) Upper Glass Relaxation

     The α(u) TSCr peak appears as a shoulder on the high tempera-
ture side of the preceding peak, in the temperature range -25 to
-7°C (cf Fig.1). This peak becomes prominent as the concentration
of hard segments is increased and its magnitude increases by quenching. A fine analysis of this component shows that it is consti-
tuted by a discrete distribution of retardation times following
Arrhenius equations. Fig. 2 shows the four retardation times
isolated in the elastomer 1/5/4. Several interpretations have been
proposed for this mode (4, 14). From the above data, it is assigned
to microbrownian motions of soft segments submitted to constraints
from aggregates.

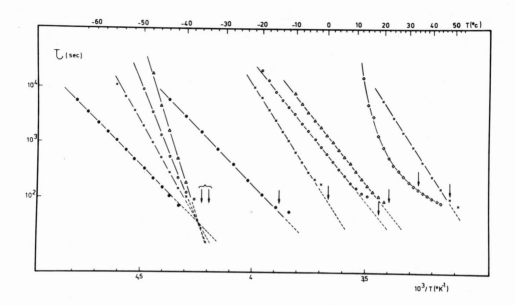

Figure 2.   Arrhenius plot of the elementary TSCr peaks isolated
             in the thermoplastic urethane elastomer 1/5/4.
             The arrows indicate the temperature of the TSCr maximum.

SUB-GLASS RELAXATION

     A TSCr peak α_{sg} is observed at $\sim 50°C$ in thermoplastic urethane
elastomers (cf. Fig.1). It is characterized by a retardation time
following a Vogel equation (cf. Fig. 2), so a discontinuity in the
free volume expansion might be expected. This α_{sg} peak is only
observed in samples annealed for 12 hours at room temperature. In
other urethanes, an analogous relaxation has been previously
reported in the same temperature range (4, 5, 11, 13, 15, 16).
From TSCr findings, this relaxation is interpreted as due to the
hydrogen bonding dissociation in amorphous regions. Then, the
micro-brownian motions result in a fluidification of amorphous
regions.

REFERENCES

1. K.C. Frisch and S.L. Reegen, "Advances in Urethane Science
    and Technology", Vol 2, Technomic Publishing, West port (1973)
2. T. El Sayed, Thesis, Toulouse (1979)
3. J.C. Monpagens, D.G. Chatain, C. Lacabanne and P. Gautier,
    J. Polym. Sci. Phys. Ed. 15, 7.67 (1977)
    J.C. Monpagens, Thesis, Toulouse (1977)
4. D.S. Huh and S.L. Cooper, Polym. Eng. Sci. 11(5), 369 (1971)
5. H.N. Hg, A.E. Allegrezza, R.W. Seymour and S.L. Cooper,
    Polymer 14, 255 (1973)
6. P. Hedvig, "Dielectric Spectroscopy of Polymers", Adam Hilger
    Bristol, p. 243 (1977)
7. C.G. Seefried Jr., J.V. Koleske and F.E. Critchfield, J. Appl.
    Polym. Sci. 19, 2493 (1975)
8. N.S. Schneider, C.S. Kaik Sung, R.W. Matton and J.L. Illinger,
    Macromolecules 8(1), 62 (1975)
9. Yu.S. Lipatov, "Advances in Urethane Sciences and Technology",
    Vol. 4, K.C. Frisch and S.L. Reegen Ed., Technomic Publishing,
    Westport, p.1 (1976)
10. J.L. Illinger, "Polymer Alloys", D. Klempner and K.C. Frisch
    Ed., Plenum Press, New York, p. 313 (1977)
11. J. Ferguson and N. Ahmad, European Polymer J. 33, 859 (1977)
12. R.R. Aitken and G.M.F. Jeffs, Polymer 18, 197 (1977)
13. G.A. Senich and W.J. Macknight, Polymer Prepr. 19(1), 11 (1978)
14. S. L. Cooper, J.C. West and R.W. Seymour, "Encyclopedia of
    Polymer Science and Technology", Suppl. n°1, John Wiley and
    Sons, p. 521 (1976)
15. R.W. Seymour and S.L. Cooper, Macromolecules, 6, 48 (1973)
16. C.H.M. Jacques, "Polymer Alloys", D. Klempner and K.C. Frisch
    Ed. Plenum Press, New York, p. 1 (1977)

RUBBER NETWORKS CONTAINING UNATTACHED POLYMER MOLECULES AND CARBON BLACK

Chen Chih-Chuan[*], Ole Kramer   John D. Ferry

Department of Chemistry   Department of Chemistry
University of Copenhagen   University of Wisconsin
Universitetsparken 5   Madison, Wis. 53706
DK-2100 Copenhagen Ø, Denmark   U.S.A.

INTRODUCTION

Rubber networks containing unattached polymer molecules were studied previously[1-5] in order to test the reptation model of de Gennes[6]. For a linear unattached polymer, de Gennes predicted that the time to completely renew its conformations would depend on molar mass to the third power. Experimentally the relaxation time was found to depend on molar mass to a power of 3.0 - 3.5.[1,2]

The presence of unattached polymer molecules in a rubber network is responsible for some interesting properties of such networks. The relaxation of the unattached polymer molecules gives rise to the appearance of an additional loss peak which may be tuned to any desired frequency by choosing the proper molar mass of the unattached molecules.

Although embedded in a stable three-dimensional network, the unattached polymer molecules can move freely inside the network and across the boundaries of the material. It means that these networks exhibit adhesion and self-healing properties similar to those of un-crosslinked elastomeric materials.

The present work was undertaken to investigate whether the presence of carbon black in such networks would change the relaxation

---

[*]Present address: Department of Organic Chemistry
                   South China Industrial Institute
                   Canton, China

rate of the unattached polymer molecules. It was also hoped that this work would give new information on the possibility of physical adsorption of rubber molecules onto the surface of the carbon black particles. Physical adsorption has been mentioned as a possible reinforcement mechanism[7] in carbon black filled rubbers.

Table 1.   Polymer Characterization

Sample No.	$\overline{M}_n$-1 kgmol	$\overline{M}_v$-1 kgmol	Ethylene Wt %	Diene wt %
V-2504 (EPDM)	43	137	52.0	4.6
2303 B (EPM)	100	217	51.2	0

Table 2.   Composition of Sample Mixtures*

Sample code	Polymer-Composition		Vulcanization parts per 100 EPDM			Carbon black parts per 100 polymer
	V-2504 (EPDM)	2303 B (EPM)	TMTD	MBTS	S	
D_1	100	0	1.0	0.5	1.5	0
D_2	50	50	1.0	0.5	1.5	0
D_3	50	50	1.0	0.5	1.5	50
D_4	50	50	1.0	0.5	1.5	50**
D_8	100	0	1.0	0.5	1.5	50

*All mixtures contained in addition 3.0 parts zinc oxide and 1.0 part stearic acid per 100 parts of polymer.

**Graphitized carbon black

EXPERIMENTAL

The ethylene-propylene terpolymer (EPDM, V-2504) and ethylene-propylene copolymer (EPM, 2303 B) were kindly supplied by Dr. G.J. Ziarnik of the Exxon Chemical Company, New Jersey. Polymer characterization data are given in Table 1.

The EPDM and EPM polymers were blended by dissolving the polymers in cyclohexane (about 5% by weight). The solvent was removed by evaporation in large trays.

The compounds were mixed on a mill. They were cured the following day for 35 minutes at 433 K as sheets about 1 mm thick. The composition of the sample mixtures is shown in Table 2. The carbon black was Regal 600 (N 219) which together with graphitized Regal 600 was kindly supplied by Dr. B.B. Boonstra of the Cabot Corporation, Billerica, Massachusetts.

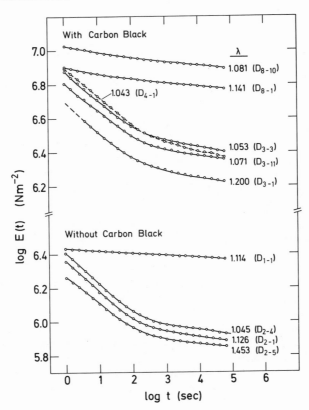

Figure 1. Double logarithmic plot of apparent Young's stress-relaxation modulus at 323 K versus time. Sample codes are identified in Table 2. Samples D_2, D_3, and D_4 contain unattached polymer molecules.

Strips (80 mm by 4 mm) were cut from the cured sheets. Before testing, the strips were conditioned for 24 hours at 323 K in an attempt to remove any effects which could have been introduced by handling of the samples.The cross-sectional area of each sample was calculated from microscope measurements of width and thickness.

Stress-relaxation at times ranging from 1 s to about 10^5 s was measured at 323 K in a new apparatus which is going to be described elsewhere. Extension ratios were calculated from measurements of the distance between two fiducial marks with a travelling microscope before and after stretching of the sample. Dynamic measurements at 323 K and frequencies of 3.5, 11, 35, and 110 Hz were made on a Rheovibron model DDV-II.

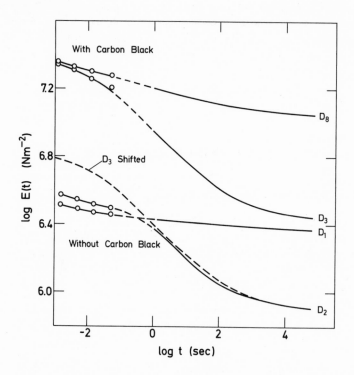

Figure 2. Double logarithmic plot of Young's stress-relaxation modulus E(t) at 323 K extrapolated to zero strain versus time. Sample codes are identified in Table 2. Samples D_2 and D_3 contain unattached polymer molecules. For comparison, D_3 is also shown after vertical shifting.

RESULTS

The apparent Young's stress-relaxation modulus, E(t), was calculated from the neo-Hookean formula

$$E(t) = 3f(t)/A_o(\lambda-\lambda^{-2}) \tag{1}$$

where $f(t)$ is the time dependent force, A_o is the unstrained cross-sectional area, and λ is the stretch ratio in simple extension. Representative results are shown in Figure 1.

It may be seen from Figure 1 that the apparent stress-relaxation modulus, $E(t)$ as calculated from eq. 1, depends strongly on stretch ratio, λ. This is especially true of carbon black filled compounds. The relaxation modulus at infinitely small strains is obtained by linear extrapolation to zero strain. The extrapolated stress-relaxation results are shown in Figure 2 together with the Rheovibron results. Sample D_4 with graphitized carbon black is omitted in Figure 2. It follows D_3 in the steep portion of the curve and falls slightly below at the short and long time ends, both.

DISCUSSION

For samples without carbon black (D_1 and D_2 in Figure 2), the behavior is the same as found in the previous study[2]. By comparison to pure EPDM (D_1), the presence of 50% EPM (D_2) has very little effect at short times while the modulus at long times is decreased by 65-75%. As pointed out in the previous study[2], this result indicates that the equilibrium modulus of crosslinked EPDM primarily is due to chain entangling. Pearson and Graessley[8] has shown this more directly in a recent study, using the Langley method.

For samples with carbon black (D_8 and D_3 in Figure 2), the moduli are of course found to be much higher than for samples without carbon black. Vertical shifting of the curve for sample D_3 in Figure 2 shows that the presence of carbon black does not change the relaxation behavior for log t > 0 which is the region in which the unattached polymer molecules relax. Thus, no new relaxation mechanism has appeared, and therefore practically no bonds between carbon black and the polymer molecules could have been broken in this time region.

Unless physical bonds between EPDM and the carbon black surface are so strong that they may be considered permanent, the results presented here seem to exclude physical bonds as a possible reinforcement mechanism, at least at longer times.

Diene rubbers, containing unattached polymer molecules and carbon black, should give interesting additional information. However, such systems cannot be made by sulfur crosslinking, and they are also difficult to make by endlinking.[5]

ACKNOWLEDGEMENTS

The authors wish to thank Dr. L.L. Chapoy at the Technical University of Denmark for permission to use the rubber compounding equipment, and Dr. A. Holm at Risø National Laboratory for loan of the Rheovibron. The work was supported, in part, by Nato Research Grant No. 935.

REFERENCES

1. Kramer, O.; Greco, R.; Neira, R.A.; Ferry, J.D. J.Polym.Sci., Polymer Phys. Ed. 1974, 12, 2361-74
2. Kramer, O.; Greco, R.; Ferry, J.D.; McDonel, E.T. J.Polym.Sci., Polymer Phys. Ed. 1975, 13, 1675-85
3. Greco, R.; Taylor, C.R.; Kramer, O.; Ferry, J.D. J.Polym.Sci., Polymer Phys. Ed. 1975, 13, 1687-94
4. Kramer, O.; Ferry, J.D. In "Chemistry and Properties of Cross-linked Polymers"; Labana, S.S., Ed.; Academic Press: New York, 1977; p. 411
5. Nelb, G.W.; Pedersen, S.; Taylor, C.R.; Ferry, J.D., submitted for publication in J. Polym. Sci.
6. de Gennes, P.G. J.Chem.Phys. 1971, 55, 572-
7. Kraus, G. In "Science and Technology of Rubber"; Eirich, F.R., Ed.; Academic Press: New York, 1978; p. 357
8. Pearson, D.S.; Graessley, W.W. Macromolecules, in press.

ON DETERMINATION OF CONSTITUTIVE EQUATIONS IN FINITE ELASTICITY

H.C. Strifors

The Royal Institute of Technology, Stockolm, Sweden

(Abstract)

In this contribution results are reviewed from experimental determination of elastic constitutive equation for a rubber material, being assumed isotropic and incompressible. With these results as a starting point, general problems concerning determination of response functions in finite elasticity are discussed.

The experimental technique is based on a controllable inhomo geneous deformation, namely, extension and torsion of a solid circular, cylindrical bar. The longitudinal stretch and twist of specimen are measured together with the resultant normal force and torque. By a numerical method, employing least square approximations, the parameters of a given mathematical form of the constitutive equation are computed with aid of experimental data from a large number of deformations. Admitting a considerable variety of combinations of the invariants of the deformation tensor as the basis for determination of response functions, this experimental technique may be considered attractive.

Particular forms of the response functions, containing different number of parameters, have been compared with each other in regard of the predictions furnished for the resultant force and torque on the test specimens in the test situation. These comparison suggest that a more complex form of the constitutive equation does not necessarily improve predictions of the behavior of a body on the whole given by, e.g., the Mooney-Rivlin type of equation.

DERIVATION OF THE PLAZEK TIME-CHAIN CONCENTRATION SHIFT FACTOR

FOR ELASTOMERS

R.F.Landel, T.J.Peng

(Abstract)

In 1966 Plazek discovered that the creep compliance for an elastomer crosslinked to various extents, after normalization to the equilibrium compliance, could be superposed by a shift factor a_x . The latter was found to vary with chain concentration ν as $a_x = (\nu_o / \nu)^{13.4}$, where ν_o is a reference state. Using the concept of a stress-moderated potential barrier to diffusional steps, it will be shown that the a_x factor arises naturally but that its proper form is exponential, i.e., $a_x = \exp(k\nu)$. This result has been experimentally confirmed.

ON THE TRANSITION FROM LINEAR TO NON-LINEAR VISCOELASTIC BEHAVIOUR OF NATURAL RUBBERS

B. Stenberg and E. Östman

Department of Polymer Technology
The Royal Institute of Technology
Stockholm, Sweden

INTRODUCTION

The theoretical basis for the analysis of the linear visco-elastic behaviour of polymers has been discussed excellently by Yannas[1-4]. Linear viscoelastic behaviour of polymers can be determined from isochronous stress-strain diagrams.

Yannas[3] has shown that the limiting strain increases very sharply when T_g is exceeded. Strain limits observed below T_g are usually reported in terms of the strain function $\ln\lambda$ where λ is the ratio of the extended to the original specimen length, while observations several degrees above T_g are reported in the form $1/3(\lambda^2-\lambda^{-1})$ which is familiar from the equation of state for ideal rubbers. The latter strain function is negligibly different from $\ln\lambda$ at small strains. In Yannas' reviews[3,4], the reported strain limits are shown as $1/3(\lambda^2-\lambda^{-1})$; below T_g the strain limits of linearity fall in the range 0.1-10%, whereas the limits in the rubbery state fall in the range 10-100%.

The linear viscoelastic behaviour of rubbers has earlier been discussed by Yannas[3,4], Tobolsky[5] and Bartenev and Lyalina[6,7].

EXPERIMENTAL

The recipes of the vulcanisates studied are shown in table 1. SAF is a super abrasion furnace black which is a reinforcing carbon black which has a higher specific area than the non reinforcing Thermax MT black. The cross-link density was found by swelling measurements to be approximately identical for materials A-C. The relaxation behaviour was studied in air in a thermostated relaxometer at 298K. Strips 10 cm long, 3 mm wide and 1 mm thick

Table 1.   Recipes of materials studied.

Material	A	B	C
Natural rubber	100	100	100
N 110 (SAF)		50	
Thermax MT			50
Stearic acid	1	1	1
ZnO	3	3	3
TMTD	8	8	8

Curing temperature: 430K
Curing time: 16 min.

were used. These were made by Skega AB, Ersmark, Sweden. Since the humidity can influence the relaxation rate, as reported by Derham[8], this was checked with a hygrometer. During the measurements, the humidity varied within fairly narrow limits. In accordance with Yannas[3], the strain measure $1/3(\lambda^2-\lambda^{-1})$ was used, where λ is the extension. The force was calculated on the actual cross-sectional area.

The measurements for the determination of ε_L were made on specimens which had not been previously strained. Stress and strain values obtained at 100s for materials A-C were registered and plotted as shown in figure 1.

In the second part of the investigation the influence of prestrain was studied. The specimen was stretched to the desired prestrain for 60s and then unloaded for 15 min before the isochrone at 100s was registered. This process was repeated in steps of 5% until a prestrain of 25% was reached.

RESULTS

Isochronous diagrams were drawn from stress relaxation measurements at 100s as illustrated in figure 1 for material A, for which the limiting strain (ε_L) is 14%. In figure 2, ε_L for the filled materials B and C was found to be 7 and 3.5% respectively. The results from the prestrain experiments are shown in figure 3. ε_L was found to increase with increasing prestrain for all materials.

DISCUSSION

From figures 1 and 2 it is obvious that ε_L is lower for the filled vulcanisates (B and C) than for the unfilled material A. This difference can probably be understood from the concept of strain amplification[9]. The filler will be essentially undeformed and the strain must be borne by the elastomer phase alone. The overall strain (macroscopic strain) is less than the strain in the rubber phase. ε_L is lowest for the vulcanisate with the low reinforcing filler Thermax MT.

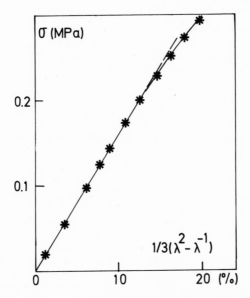

Figure 1.   Isochrone at 100s for material A.

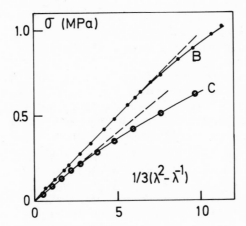

Figure 2.   Isochrones at 100s for materials B and C.

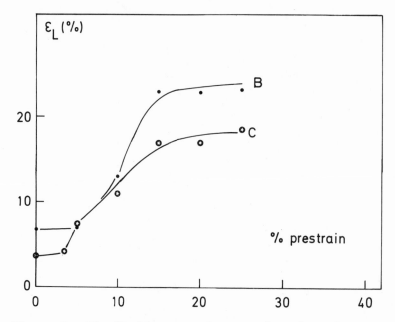

Figure 3.   The limiting strain as a function of prestrain.

The deviation from linearity is towards the strain axis, and this means that at the transition from linear to non-linear visco-elastic behaviour the vulcanisates became softer.   When a simple network is strained without the occurrence of scissions, however, the network will become stiffer as the strain increases.   If the results are discussed in terms of a distribution of ordered supra-structures, the deviation from linearity towards the strain axis can be explained as resulting from the successive breakdown of such stiffer suprastructures, which has been discussed by, for instance, Blockland and Prins[10] and Bartenev and Lyalina[6,7,11]. When ε_L is exceeded, more and more of the superstructures are de-composed.   The material then becomes softer.

The influence of prestraining on ε_L can also be interpreted in terms of suprastructures.   Prestrains lower than the limiting strain ε_L for the unstrained material will not change this value. When, however, the prestrain is higher than the initial ε_L, a higher ε_L willbe obtained.   This is seen in figure 3 for materials B and C.   The same tendency holds also for material A, when the prestrains exceed 14%.   During the prestraining a certain amount of the suprastructures is decomposed.   In materials B and C this decomposition takes place in the filler as a breakdown of filler aggregates and as a rupture from filler particles of rubber macro-molecules.

A modulus-prestrain relationship constructed from isochrones at different prestrains shows that the modulus for material C is independent of prestrain. This must mean that Thermax MT acts mainly hydrodynamically. For material B, the modulus decreases linearly with increasing prestrain. The decrease is probably a result of breakdown of suprastructures in the filler and of ruptures from filler particles of rubber macromolecules.

REFERENCES
1. I.V. Yannas and A.C. Lunn, J.Macromol.Sci.-Phys., B4(3):603 (1970)
2. I.V. Yannas, N-H. Sung and A.C. Lunn, J.Macromol.Sci.-Phys., B5(3):487 (1971)
3. I.V. Yannas, J.Macromol.Sci.-Phys., B6(1):91 (1972)
4. I.V. Yannas, J.Polymer Sci., Macromolecular Reviews, 9:163 (1974)
5. A.V. Tobolsky, "Properties and Structure of Polymers", Wiley, New York (1960)
6. G.M. Bartenev and N.M. Lyalina, Plaste u. Kautschuk, 16:901, (1969)
7. G.M. Bartenev and N.M. Lyalina, Polymer Sci. U.S.S.R., 12:420 (1970)
8. C.J. Derham, J.Polymer Sci.Phys., 13:1855 (1975)
9. M. Shen, in "Science and Technology of Rubbers", F.R. Eirich ed., Chap. 4, Academic Press, New York (1978)
10. R. Blockland and W. Prins, J.Polymer Sci., 7:1595 (1969)
11. G.M. Bartenev and N.M. Lyalina, Plaste u. Kautschuk, 24:741 (1977)

EFFECTS OF CURATIVES AND ANTIDEGRADANTS ON FLOW PROPERTIES OF UNCURED RUBBER COMPOUNDS

J.L. Leblanc

Monsanto Europe S.A., Louvain-la-Neuve, Belgium

(Abstract)

Whilst it is well known that changes in polymers plasticizers and fillers can produce large effects in flow properties of uncured rubber compounds, the influence of other compounding ingredients such as curatives and antidegradants is less well defined.

Recently developed capillary Rheometers with extrudate swell detector allow rheological properties of rubber to be analyzed in a more precise manner and the effects of chemicals can be observed. Using new processability testing techniques, it has been possible to determine the effects of antidegradant or cure system changes on the flow properties of rubber compounds. This paper discusses recent results and their practical significance.

ANALYSIS OF EXTRUDATE SWELL BEHAVIOUR OF RUBBER COMPOUNDS USING

LASER SCAN DETECTOR

J.L.Leblanc

Monsanto Europe S.A., Louvain-la-Neuve, Belgium

(Abstract)

Controlling the elastic response of filled rubbers, as it manifests itself in extrudate swell, is a major problem in rubber processing.

Classical extrudate swell measurements such as the gravimetric or the shrinkage techniques, are difficult, complex and time consuming methods. Using laser scan detector, new instruments have been recently developed which allow the extrudate swell behaviour of rubber compounds to be analyzed in a quick and reliable manner.

This paper discusses the requirements for suitable analysis of the extrudate swell of rubber compounds. Due to the specific elastic response of rubbers, these requirements differ from those for thermoplastics. Recent results obtained using typical industrial formulations are presented and their practical significance is discussed.

SOLIDS

VISCOELASTIC RESPONSE OF A SOLID POLYMER AFTER YIELDING

Giovanni Rizzo , Giuseppe Titomanlio

Istituto di Ingegneria Chimica

Viale delle Scienze , 90128 Palermo (Italy)

EXPERIMENTAL RESULTS

The viscoelastic behavior of solid polymers after yielding has been recently the subject of some experimental studies. In particular both stress relaxation and constant force creep tests were performed in tension and in compression on Mylar[1] (a biaxially oriented polyethylene terephthalate manufactured by du Pont) and polymethylmetacrylate[2] (Plexiglass) respectively; furthermore both tensile and compressive tests were carried out on polycarbonate Lexan[3,4].

All data indicate that creep and stress relaxation depend both on the sample loading rate α in the last strain interval just prior to the test and on the strain $e = 1/l_0$ (where l_0 is the sample length or heigth in the unoriented configuration and 1 is its value at the beginning of the test). Furthermore the dependence is similar for all the materials investigated. In particular larger deformation rates during the sample loading give rise to faster creep or stress relaxation during the subsequent test; two creep or stress relaxation curves, relative to different values of α , but to the same strain e with respect to the non-oriented configuration, can be superimposed by means of a time shift factor proportional to α . In other words, after yielding, the dominant relaxation time λ is proportional to the inverse of α and to a function of the strain :

$$\lambda \propto f(e)/\alpha \qquad (1)$$

As for the dependence of the dominant relaxation time λ on the strain e , the data collected in tension on Mylar indicate that for e larger than one λ does not depend significantly on

the strain e ; the data collected on Plexiglass in compression
show that,for 0.3<e<0.75 , λ , and thus f(e) , is proportional
to the inverse of e . Lexan was studied both in tension and in
compression covering for e the range 0.3-1.7 and for α the
range 0.4-110 hr^{-1} ; for tensile tests strip samples were cut from
sheets prepared by compression molding and subsequent rolling, the
latter being performed in order to avoid necking during the sample
loading. Furthermore compressive tests were performed also at e>1
on samples which had been previously elongated up to necking (which
occurred with a strain e=1.7). This whole body of data confirms
the indications obtained from Mylar and Plexiglass : f(e) is a
slightly increasing function of e for e>1 and coincides with
1/e for small values of e .

    In order to show the degree of superposition which may be ob-
tained by shifting the creep and stress relaxation curves by a factor proportional to λ one should consider the dimensionless time :

$$\theta = t\alpha \,/f(e)\propto\ t/\lambda \tag{2}$$

    Some of the data taken on Lexan are reported in Figs 1 and 2
versus θ . In evaluating θ , f(e) has been substituted by an
invariant strain measure which satisfies the above mentioned requirements; this measure is the magnitude of $\underline{C_u^{-1}}$, the inverse of the
Cauchy strain tensor[5] of the present material configuration with
respect to the unoriented one.

    The creep data are considered in Fig.1,where the ratio between
the creep lenght (or height) variation Δl and the sample lenght
(or height) l_{ot} when the test load is reached, is plotted versus
θ . At small times all curves are very close. At larger times,
the divergence of the tensile curves from the compressive ones is
due to the fact that, the tests being performed under constant
load, the former are stress increasing and the latter stress decreasing , as creep deformation proceeds.

    As for stress relaxation, in analogy to what suggested by
Kubàt et al.[6,7], the ratio $\bar{\sigma}$ between the relaxed stress $\sigma_o-\sigma$
and the relaxable one[3,4] $\sigma_o-\sigma_\infty$ is plotted in Fig. 2 versus the
same dimensionless time θ . Also in this case, as for Fig. 1,
the superposition may be considered satisfactory, in view of the
amplitude of the operating conditions (e,α , e_i , tension, compression).

SOME SUGGESTIONS FOR A RHEOLOGICAL MODEL

    A quantitative rheological description of the viscoelastic
behavior observed after yielding is not available in the literature.
A generalized Maxwell model is here considered as a basis for such

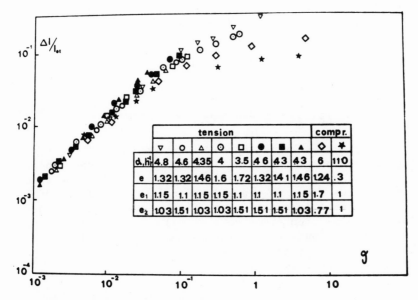

	tension								compr.	
	▽	○	△	⊙	□	●	■	▲	◇	✻
d,hr	4.8	4.6	4.35	4	3.5	4.6	4.3	4.3	6	110
e	1.32	1.32	1.46	1.6	1.72	1.32	1.41	1.46	1.24	.3
e_1	1.15	1.1	1.15	1.15	1.1	1.1	1.1	1.15	1.7	1
e_2	1.03	1.51	1.03	1.03	1.51	1.51	1.51	1.03	.77	1

Fig.1 Creep curves. e_1 is the initial (after rolling or necking)
strain along the loading direction and e_2 along a direction
perpendicular to loading. The original data are reported
in ref.2.

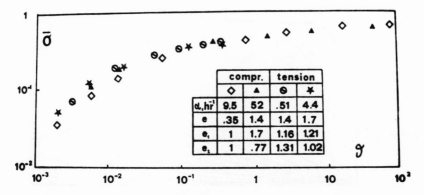

	compr.		tension	
	◇	▲	❂	✻
d,hr	9.5	52	.51	4.4
e	.35	1.4	1.4	1.7
e_1	1	1.7	1.16	1.21
e_2	1	.77	1.31	1.02

Fig.2 Master stress relaxation curve. e_1 is the initial (after
rolling or necking) strain along the loading direction and
e_2 along a direction perpendicular to loading. The original data are reported in ref.4.

a description: schematic assumptions on the type of non-linearity involved allow an expression to be obtained for the dominant relaxation time λ as a function of α and e .

Let us consider the deformation of a sample by elongational kinematics. As the deformation proceeds, the stress of the ith element changes according to the equations :

$$\tau_i^{11} + \lambda_i \; (\; \frac{\delta \tau_i^{11}}{\delta t} - 2\alpha \; \tau_i^{11} \;) = 2\alpha \eta_i \tag{3}$$

$$\tau_i^{22} + \lambda_i \; (\; \frac{\delta \tau_i^{22}}{\delta t} - \alpha \tau_i^{22}) = -\alpha \eta_i \tag{4}$$

Starting from a stress-free situation, stress changes elastically with deformation at small strains whereas non-linear phenomena take place at larger strains. In particular the smallest relaxation times are soon reduced substantially because of the effect of stress. At material yielding, a dramatic decrease affects most of the relaxation times down to values which, as shown by the data, are proportional to the inverse of the deformation rate α ; as the deformation proceeds, new elements (with larger relaxation times) undergo their own yielding.

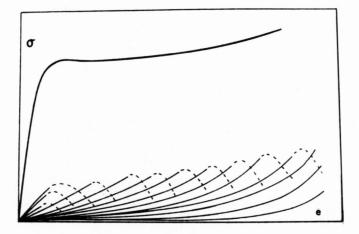

Fig.3   Schematic representation of the model. ── total stress; ── elastic stress contribution of ith element (before its yielding); --- relaxable stress contribution of ith element.

In the case when, after the yielding of each element, the relaxation time goes on decreasing because of further deformation
each stress-strain curve would have a maximum as indicated in Fig. 3.
Schematically, at a given strain, the total stress would be made
of contributions due to some elements which are elastically deformed,
to others whose stress is rapidly decreasing and to those which are
at maximum stress.

Let us assume that the last elements give rise to the greatest
contribution to the relaxable part of stress and thus correspond
to the dominant relaxation time. Subtracting Eq. 4 from Eq. 3 and
imposing the condition of maximum contribution of the ith element
to the total stress, i.e. $d(\tau_i^{11}-\tau_i^{22})/dt= 0$, give :

$$\frac{\tau_i^{11} - \tau_i^{22}}{G_i} - \lambda_i \alpha \frac{2\tau_i^{11} + \tau_i^{22}}{G_i} = 3\alpha\lambda_i \tag{5}$$

where $G_i = \eta_i/\lambda_i$.

As long as the yielding of each element may be considered a
process rapid enough to neglect the strain interval during which
it takes place with respect to the actual strain, one may write :

$$\frac{\tau_i^{11}}{G_i} = e^2 - 1 \qquad\qquad \frac{\tau_i^{22}}{G_i} = \frac{1}{e} - 1 \tag{6}$$

at yielding of the ith element.

Substitution of Eqs 6 into Eq. 5 gives for the relaxation
time :

$$\lambda = \frac{e^2 - 1/e}{2e^2 + 1/e} \cdot \frac{1}{\alpha} \tag{7}$$

Eq. 7 gives the right dependence of the relaxation time upon
the deformation rate. As far as the effect of strain e is concerned, Eq. 7 reproduces the behavior observed both for small values
of e and for e sufficiently larger than one. For e close to
one, Eq. 7 cannot be taken into account because the assumption that
there is a single dominant relaxation time cannot be accepted near
the total yielding, when numerous elements yield at deformations
very close to each other; furthermore Eqs 6 cannot be considered
valid for values of e close to unity, when the strain interval,
during which yielding takes place, becomes comparable to the actual
strain.

The qualitative agreement of Eq. 7 with experimental observations indicates that a generalized Maxwell model can describe the viscoelastic behavior observed after yielding of solid polymers in the case when non-linearities are introduced, which reproduce for each element the behavior hypothesized above, i.e. a rapid decrease of the stress after yielding.

A possibility is that the relaxation times decrease according to a differential equation; some forms for it involving the free volume concept have already been proposed. The driving force should be a stress measure. In such a case a decrease of the relaxation time for each element after yielding could be due to the fact that λ_i does not follow the stress τ_i instantaneously. If the driving force should take into account, not only τ_i , but also the total stress, an increase in this would cause a further decrease in λ_i .

REFERENCES

1. G. Titomanlio and G. Rizzo, Master curves of viscoelastic behavior in the plastic region of a solid polymer, J. Appl. Polym. Sci., 21:2933 (1977)
2. G. Titomanlio, Compressive viscoelastic behavior of polymethyl-metacrylate and effect of temperature rise due to plastic deformation, Acta Polimerica, in press
3. G. Titomanlio and G. Rizzo, Compressive large deformation visco-elastic behavior of polycarbonate, Polymer, in press
4. G. Titomanlio and G. Rizzo, Experimental study of stress relax-ation and creep behavior of polycarbonate after yielding, in: "Proceedings 10th EPS Conference on Macromolecular Physics, Noordwijkerhout, 21-25 april 1980"
5. G. Astarita and G. Marrucci, Principles of non-newtonian fluid mechanics, Mc Graw-Hill, London (1974)
6. J. Kubàt, J. Petermann and M. Rigdahl, Internal stresses in polyethylene as related to its structure, Mat. Sci. Eng., 19:185 (1975)
7. J. Kubàt, J. Petermann and M. Rigdahl, Internal stresses in cold-drawn and irradiated polyethylene, J. Mater. Sci., 10:2071 (1975)

DYNAMIC MECHANICAL ANALYSIS OF PC-SAN BLENDS

Giancarlo Locati and Giampaolo Giuliani

Assoreni - POLI/TEPO

20097 S. Donato Milanese - Italy

INTRODUCTION

Dynamic Mechanical Spectroscopy is a powerful tool for characterizing polymeric materials that, when extended over a broad temperature and frequency range, affords valuable structural and morphological information. When applied to polymer blends Dynamic Mechanical Spectroscopy gives further insight to the knowledge of the materials revealing structural and compositional effects. A wide temperature and frequency scanning, hovever, cannot be dove by a single instrument or by a single geometry because of the large variations of viscoelastic properties experienced by polymers as a consequence of temperature and frequency variations. In this paper the results of an investigation on the influence of the two phase structure of polycarbonate SAN blends on dynamic mechanical spectra, performed with three different instruments, are presented.

EXPERIMENTAL

Two commercial thermoplastic polymers were used; a polycarbonate (PC) an injection moulding grade; styrene-acrylonitrile (SAN) copolymer containing about 30% weight of acrylonitrile. Both the polymers were in the form of chopped extrudates. Blends of different composition were prepared by mixing 45 g of the polymers in a Brabender mixer at 200°C for 5 minutes. No additives were added. The samples for the dynamic measurements were compression molded at 210°C. The dynamic instruments utilized through this work were: a Rheovibron model

DD V II, a Du Pont Dynamic Mechanical Analyzed (DMA)
and a Rheometrics.
Films O.8 mm thick were used for Rheovibron measurements;
sheets 1.1 and 3.0 mm thick, were used respectively for
DMA and Rheometrics experiments. The scanning rate was
5°C/min in DMA experiments. The geometry adopted for
Rheometrics was plate-plate.

RESULTS

        Fig. 1 shows a plot of tan δ vs. T at 3.5 Hz for
some selected PC/SAN blends, as obtained by Rheovibron.
In the low temperature region a peak is evident, the intensity of which is nearly proportional to the amount
of PC present in the blend. Around 100°C blends containing less than 50% PC present an upturn whose position,
too, depends on the amount of PC. In the intermediate
region of temperature tan δ vs. T follows a flat course
with the absolute values following a somewhat disordered
pattern. The overall picture does not differ substantially at 110 Hz except for the fact that 75 and 85% PC blends
do not show an upturn following the first peak located
at about 100°C.
Attempts to scan the temperature region below the glass
transition of SAN by DMA give erratic results which are
to be attributed to the clamping system. Where DMA shows
its interest is in the region where the glass transitions
of the components occur, that is the temperature region
between 80 and 170°C. In this region we were able to record two complete peaks also in the cases where only a
hint of the second one was suggested by the Rheovibron
spectrum, that is for blends containing more than 75%
PC (fig. 2). The possibility of observing two peaks even
when they are only about 20°C apart was explored
with blends PVC/SAN. It turned out that the DMA technique resolves two peaks also in this case, fig. 3.
As for as the intensity of the peaks is regarded the
technique is not quantitative because the second peak
is influenced by the variations of mechanical properties
brought about by the first transition. The situation is
somewhat similar to that found in DSC or DTA measurements.
The rubbery and flow regions of blends were explored by
Rheometrics, fig. 4 and 5. An interesting feature emerges from log G' and log G" vs. log ω plots. At low frequencies a plateau begins to appear for blends containing 25% PC. As a consequence the viscoelastic values
G' and G" of blends are higher than corresponding values
of PC and SAN over an extended range of frequency.

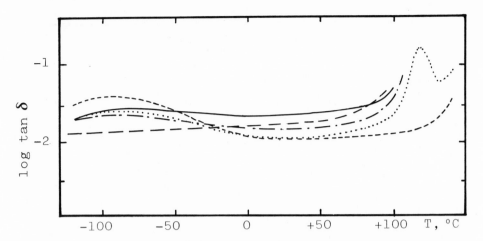

Fig. 1: Typical tan δ curves for PC/SAN blends.
Rheovibron data at 3.5 Hz.

— — SAN; —— 25/75 PC/SAN; —·—50/50 PC/SAN;

...... 75/25 PC/SAN; - - - PC

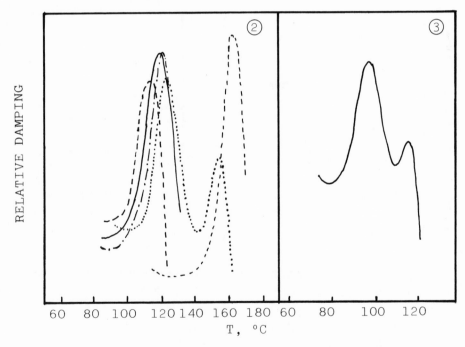

Fig. 2: Relative damping for PC/SAN blends. DMA.

Fig. 3: Relative damping for a 50/50 blend PVC/SAN.

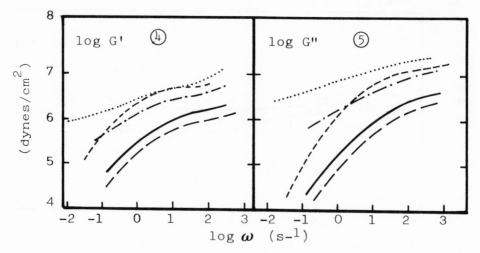

Fig. 4 and 5: log G' and log G" vs. log ω for PC/SAN
             blends. Data obtained by Rheometrics.
             T = 190°C

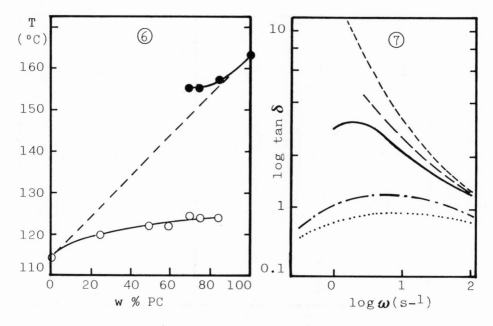

Fig. 6: Tg of PC/SAN blends as a function of composi-
        tion. Data obtained by DMA.

Fig. 7: log tan δ vs. log ω for PC/SAN blends at 230°C

DISCUSSION

The overall picture emerging from different techniques confirms the well known incompatibility of the two materials over the entire composition range[1]. Each technique, however, reveals particular aspects of the biphasic character of the blends. Perhaps, the most clear indication of biphasicity comes out from DMA spectra. This technique allows a surprisingly good resolution of the mechanical transitions into two peaks, also when the pure component peaks are as close as $20^{o}C$. Incompatibility is not complete, however. Within each phase a partial mixing of the two components occur, as can be inferred from the relative shifts of the two main peaks as a function of composition, fig. 6. This is confirmed also by Rheovibron measurements which show a progressive shift of the SAN glass transition with increasing PC content, fig. 1. The Rheometrics data, fig. 4 and 5, show that blends containing more than 25% PC have a viscoelastic behaviour which cannot be simply traced to that of the parent polymers by viscoelastic models or theorics. The tendency of G' and G" curves to approach a plateau at low frequencies is reminescent of what found in other composite systems such as ABS, acrylonitrile-butadiene copolymers or filled polymers[2]. Another way to look at the same fact is to plot $\tan \delta$ as a function of frequency (fig. 7).Instead of the monotonic decrease of $\tan \delta$, expected and found for simple polymers, a maximum is observed in the case of blends, the origin of which is clearly related to the tendency of the functions G' and G" to reach a plateau both at low and high frequencies.

REFERENCES

1. R.J. Peterson, R.D. Corneliussen, and L.T. Rozelle, Polym. Prepr. Amer. Chem. Soc., Div. Polym. Chem. 10, 385 (1969).
2. H. Münstedt, Proc. of the VIIth Int. Congress on Rheology, Gothenburg Aug 23-27, 1976, pag. 496.

DYNAMIC MECHANICAL PROPERTIES OF AROMATIC

POLYAMIDEIMIDE DEGRADED IN NO_2 ATMOSPHERE

Hirotaro Kambe and Rikio Yokota

Institute of Space & Aeronautical
Science, University of Tokyo
Komaba, Meguro-ku, Tokyo 153

INTRODUCTION

The degradation of aliphatic polyamide as nylon 66 in an NO_2 atmosphere has been investigated by Jellinek,[1] based on the molecular weight changes determined by the change of solution viscosity. The aromatic polyamideimides are also accessible to NO_2, but are slightly soluble in polar solvents as N-methyl pyrrolidone(NMP) and the solutions obtained are not readily stable to estimate the degree of degradation from the usual viscosity measurements. Therefore, we investigated the degradation of aromatic polyamideimide by following the change of dynamic mechanical properties in an NO_2 atmosphere.

EXPERIMENTAL

Materials

We used an aromatic polyamideimide(PAI) with the following structure as the sample:

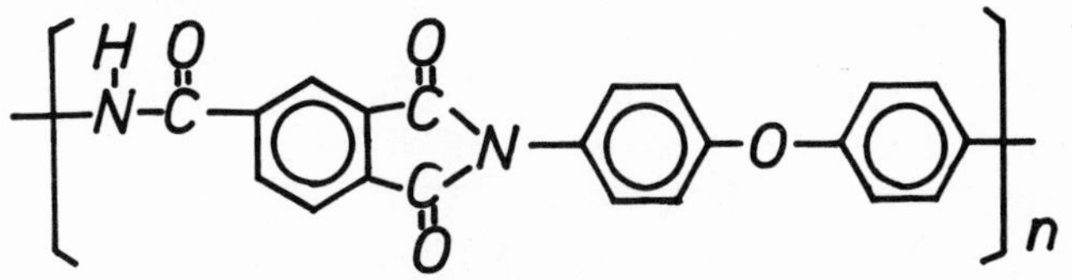

A commercial film of the sample polymer was supplied by Toray Co. The thickness of the sample film was 50μm. By a previous experiment on superstructure of PAI, as revealed by small angle X-ray scattering,[2] we treated Toray film in vacuum at 290°C for 2h to

211

gave a stable molecular aggregation to the sample film. These films were used for tensile dynamic mechanical measurements as a sample.

The degradation was also monitored by chemical changes appeared in IR spectra. For this purpose, we prepared a thin film of 5-7µm by casting from a solution. IR monitor films were also thermally treated with the same conditions as the Toray film.

Sample films of mechanical measurements were degraded in an atmosphere of 100mmHg NO_2 at 50°C for several days, together with IR monitor films. Details of the experimental procedure have been reported in a previous paper.[3]

A glass-reinforced composite of PAI for DMA measurements was prepared by immersing a glass cloth in 10% solution of PAI in NMP, dried in a hot air at 60°C, and thermally treated at the same conditions. The composite involved ca. 35% fiber component. The composite was degraded in 100mmHg NO_2 at 60°C for several days.

Mechanical properties of the degraded IR monitor films were measured by preparing a sandwich of degraded film between steel plates as a sample for DMA.

Dynamic Mechanical Measurements

The dynamic mechanical measurements were conducted with tensile and flexural modes of oscillation. Tensile oscillations were measured by a Toyo-Baldwin Rheovibron DDC IIC and flexural oscillations by a du Pont 980 Dynamic Mechanical Analyzer(DMA). Tensile oscillations were measured at 3.5Hz in a flowing nitrogen at a uniform heating rate of 3°C/min. From these measurements we could obtain temperature dependences of tensile storage modulus E' and loss modulus E". Flexural oscillations of glass-reinforced composites and also of sandwiched samples were measured by DMA. The results were expressed as the temperature dependences of resonant frequency and damping capacity, which correspond to E' and E", respectively.

RESULTS AND DISCUSSION

In Fig. 1, the changes of temperature dependences of E' and E" are shown after various periods of degradation. The E' curve for the unreacted PAI film showed glass transition at 290°C and a corresponding loss peak(α) was found in E" curve. After 3 days degradation, we found a new loss peak(α') at a little lower temperature than α peak. The α' peak gradually grew up and shifted for the lower temperature in the course of degradation.

By the thermal treatments at 230, 260, and 305°C, α' peak shifted again for the high temperature, as shown in Fig. 2. It was confirmed[3] by IR spectroscopy and mechanical scraping and solvent

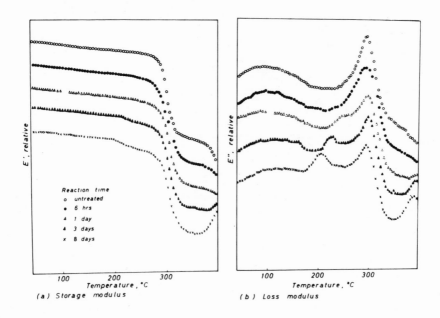

Fig. 1. Tensile dynamic mechanical properties of PAI films
degraded in 100mmHg NO₂ at 50°C for various periods.

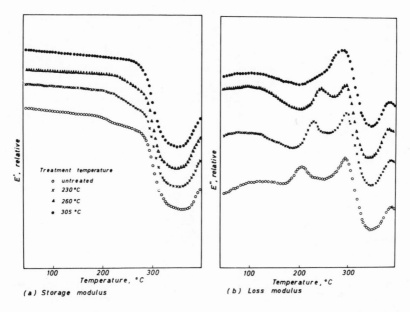

Fig. 2. Effect of thermal treatments on tensile dynamic mechanical
properties of PAI film degraded for 8 days in 100mmHg NO₂
at 50°C.

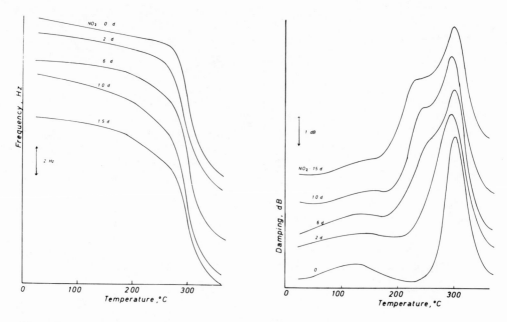

Fig. 3. DMA curves of glass-reinforced PAI degraded in 100mmHg NO₂ at 60°C for various periods.

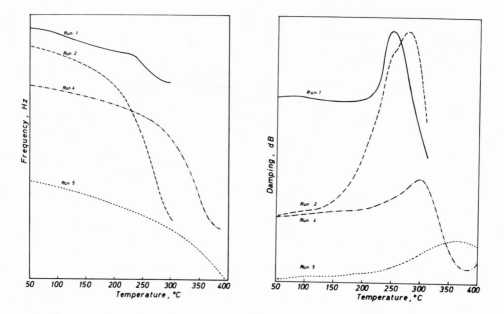

Fig. 4. Effect of thermal treatments on DMA curves of degraded PAI IR monitor film sandwiched between steel plates.

extraction of the degraded films that the appearance of α' peak by NO$_2$ degradation of PAI was owing to the formation of degraded surface layers on the sample film. From the changes of IR spectra during degradation, it was found that imide ring structures are intact and amide linkages are attacked by NO$_2$. Degradation of IR monitor films was so complete that the films became so brittle as to measure tensile properties by Rheovibron.

By using DMA, we can measure flexural oscillations of somewhat more rigid specimen. Then, we prepared a glass-cloth reinforced PAI for this technique. DMA curves obtained for the glass-reinforced PAI degraded in 100mmHg NO$_2$ at 60°C for various periods are shown in Fig. 3. Changes of temperature dependences of frequency and damping factors were found in coincidence with those observed for film samples, as in Fig. 1. The degradation of glass-reinforced PAI in NO$_2$ was not so complete as we observed for IR monitor films.

IR monitor films were not measurable with Rheovibron after degradation. However, if we put the degraded IR monitor film sandwiched between two steel plates, we can measure flexural vibrations of completely degraded PAI, as shown in Fig. 4. By the thermal treatment of degraded films, we observed similar effects as those in Fig. 2.

The α' peak observed for PAI films degraded in NO$_2$ may be originated from the glass transition of degraded layers on the surface of films. The softening of degraded films was observed by thermomechanical analysis(TMA) at the corresponding temperature range. By the thermal treatments of degraded films, the degraded layers recover some of its original structure and α' peak shifts for higher temperature side and finally merges into the α peak.

REFERENCES

1. H. H. G. Jellinek, Reaction of Polymers with Pollutant Gases, in "Aspects of Degradation and Stabilization of Polymers," H. H. G. Jellinek, ed., Elsevior, Amsterdam (1978).
2. S. Isoda, M. Kochi, R. Yokota, and H. Kambe, Rept. Progr. Polym. Phys. Japan, 20:219 (1977).
3. H. Kambe and R. Yokota, To be published in Polym. Eng. Sci.

THE APPEARANCE OF NONLINEAR VISCOELASTICITY IN GLASSY POLYMERS

Jan-Fredrik Jansson

Dept. of Polymer Technology
The Royal Institute of Technology
S-100 44 Stockholm   Sweden

INTRODUCTION

Glassy polymers show linear or approximately linear visco-elastic behaviour only for small uniaxial tensile stress-strains as can be observed by the deviation from linearity in the iso-chronous stress/strain creep diagram.

The transition to marked non linear viscoelasticity appears at about the same strain level (< 1%) for most glassy polymers.

Above this level the creep rate and the stress dilatation in-crease. Ferry and Stratton[1] proposed, based on this observation, that the origin of the non linearity emanates from stress activated molecular mechanisms, at least as far as no irreversible deforma-tion is concerned, resulting in a shift of the relaxation time spectrum towards shorter times. The shift was interpreted in terms of free volume and thus associated to the stress dilatation of the material, which has been observed to accelerate above the limit of linear viscoelasticity.

Due to the fact that the superposition principle has no vali-dity for non linear viscoelastic polymers, it is not possible, however, to analyse the noticed changes in the relaxation time spectrum from a molecular point of view.

More recently the effect of dilatation on the viscoelastic relaxation times has been examined by for instance Sternstein[2] and Matsuoka[3]. According to Matsuoka the dilatational effects are pro-posed to be the prime reason for the stress dependent, non linear viscoelastic behaviour in polymeric solids.

The appearance of non linear viscoelasticity in glassy polymers has also been related to fracture initiation, based on the
observation that the strain for total fracture, the strain at
first "visible" craze and the strain limit of linear viscoelasticity
trends to proceed towards approximately the same value for long
creep time[4].

Based on results reported in the literature and on a series of
studies carried out during the last years we now propose a molecular model for the appearance of marked non linear viscoelasticity
in glassy polymers and for the connection between non linear viscoelastic behaviour and fracture initiation.

DEFINITION OF THE STRESS/STRAIN LIMIT OF LINEAR VISCOELASTICITY

Although the transition to marked non linear viscoelasticity
is reasonably distinct in glassy polymers the change in behaviour
is gradual. Therefore the stress/strain limit has to be related
to an arbitrarily chosen deviation from the linear behaviour, which
in turn has to be related to the accuracy of the measurements.

When comparing results by different authors it seems obvious
that the more accurately the measurements have been done, the
lower stress/strain limit has been detected.

In the creep measurements reported in the following the accuracy is determined predominantly by the uncertainty of the strain
measurements and is better than ± 0.05% strain in the tensile direction.

Therefore in the following discussion the limit of linear
viscoelasticity in uniaxial creep is defined as the point where
the deviation of the isochronous curve from the prolonged straight
line for small stress/strains exceeds the accuracy of the measurements.

INFLUENCE OF RELAXATION MECHANISMS IN THE GLASSY REGION

The transition to marked non linear viscoelastic behaviour
has been studied for PC, PVC, PMMA and PEMA[5,6], all of which have
similar types of secondary relaxation processes in their glassy
regions. For PC and PVC the β-relaxation is supposed to be due
to local main chain motions, the detailed nature of which, however, is not fully understood at present. The β-mechanism of PMMA
and PEMA which appears around room temperature is suggested to be
caused by coupled motions in the main chains and the voluminous
side groups. Necessary conditions for the rotation of the side
groups are either special conformations in the main chains or a
simultaneous rotation around the covalent bonds in the main chains.

Fig. 1 shows the temperature dependence of the stress limits
of linear viscoelasticity for the four materials. According to
Yannas[7] the stress limit will decrease at temperatures above the

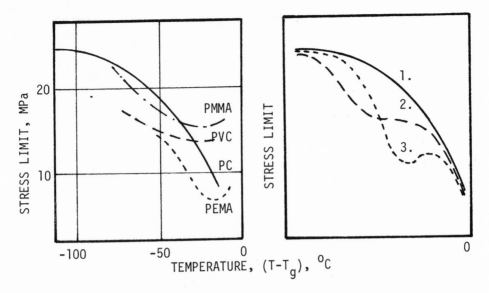

Fig.1 The influence of secondary transitions on the stress limit
       of linear viscoelasticity in glassy polymers
       1. absent or very distant β-transition
       2. wide range between the α- and β-transitions
       3. β-transition very close to the α-transition

glass temperature, T_g. The stress limit of linear viscoelastic behaviour, therefore, goes through a local maximum for temperatures
close to the α-transition and through a minimum at temperatures
between the α- and β-transitions. In addition to this, Jansson[8]
has reported that the β-relaxation of PEMA is more easily activated
by increasing stress than the glass transition mechanism (i.e. more
rapidly shifted towards shorter times). Based on this, Bertilsson
and Jansson[6] have proposed the influence, shown in Fig. 1, from the
actual types of secondary transitions on the stress limit of linear
viscoelasticity.

INFLUENCE OF ANTIPLASTIZATION

     By adding small amounts of compatible substances to for in-
stance PVC, a pseudo-crosslinking effect is obtained, which causes
severe hindrances to the local main chain motions associated with
the β-relaxation, although a shift of the α-relaxation towards
lower temperatures, shorter times, is obtained simultaneously.

     The β-transition is suppressed and disappears at a certain
amount of added substance. The phenomenon is called antiplastization and is also accompanied by increase in density.

     The effect of antiplastization on the creep compliance and
stress limit of linear viscoelasticity is shown in Fig. 2[9,10,11].

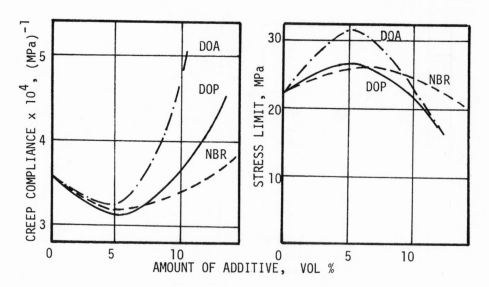

Fig.2   The influence of antiplasization on the creep compliance
        and stress limit of linear viscoelasticity at 1000 s, 23°C,
        for PVC.
        DOA: di-octyl-adipate, DOP: di-octyl-phtalate,
        NBR: acrylonitrile-butadiene-rubber

Especially for the plasticizer DOA (di-octyl-adipate) a remarkable
increase in the stress limit is obtained without any corresponding
decrease in creep compliance. Similar effects have also been observed from adding small amounts of compatible polymers, e.g. NBR
and PCL[10,11] to PVC.

INFRARED DICHROISM OF GLASSY POLYCARBONATE AT SMALL STRAINS

The changes of the average main chain conformations and degree
of orientation of polymers can be detected by infrared dichroism.
The method can be made very sensitive and has shown for PC that no
conformational changes or only very small ones ($\Delta R < 2 \cdot 10^{-3}$) appear
for short times (10^4-10^5 s.) in stress relaxation below a strain
level which corresponds to the strain limit of linear viscoelasticity[12]. Above this level the dichroic ratio increases very rapidly to a value at 1.5% strain, of more than ten times the value at
0.7% strain.

Based on these data it seems reasonable to suggest that the
transition from linear to non linear viscoelastic behaviour in
glassy polycarbonate is marked by the onset of significant rotation
around backbone bonds.

INFLUENCE OF COOLING RATE AND VOLUME RECOVERY

By cooling a polymer rapidly through its glass transition a state of unequilibrium is introduced. The resulting excess of volume slowly recovers with time at a rate which is dependent on the temperature compared to the temperatures of the glass and main secondary transitions. The excess of volume results in an increasing creep rate and creep compliance.

Recent studies[13] on PMMA have shown, however, contrarily to what can be expected from the excess of volume, that both the stress and the strain limits of linear viscoelasticity increase with cooling rate (increasing excess of volume), although the creep compliance increases. This indicates that the cooling rate does not only influence the over all average density but also the distribution of molecules in the material. Thus, if the rapid cooling gives a more even distribution of molecules in the sample, the load will be distributed over a large number of molecules. The stress activated deformation processes causing the non linear behaviour therefore appear at larger macroscopic stress/strains.

CONCLUSIONS

It seems obvious that the transition from linear or approximately linear to marked non linear viscoelastic behaviour in glassy polymers is caused by stress/strain activated rotation around backbone bonds and changes in conformations of the main chains. It is not possible, however, to decide at the moment whether these motions originate from
- changes in the nature of the mechanisms acting in the linear region
- an acceleration of the mechanisms acting in the linear region or
- the introduction of new mechanisms that will not appear even for very long times at low stresses.
It is of course very difficult to differentiate between these types of processes and it seems most likely that the phenomena causing the transition to non linear behaviour include all three mechanisms.

The non linear behaviour starts in the region of the glassy structure which contains the smallest number of molecules (the low density regions). The distribution of molecules is determined by for instance annealing and the cooling procedure through the glass transition of the polymer, giving more evenly distributed molecules at high cooling rates.

The onset of orientation of the molecules in these weak regions will cause an "opening" of the material and the formation of crazes. Thus the fracture initiation in glassy polymers proceeds through a number of stages, starting with the stress activation of the non linear viscoelastic deformation mechanisms in the low density regions.

The reported studies are parts of a research program on Mechanical Long Term Properties of Polymers, sponsored by the Swedish Board for Technical Development (STU). The author wishes to thank Prof. B. Rånby, the head of the Department of Polymer Technology, who has made these studies possible.

REFERENCES

1. J.D. Ferry and R.A. Stratton, Kolloid-Z., 171, 107 (1960)
2. S.S. Sternstein and T.C. Ho, J.Appl.Phys., 43, 4370 (1972)
3. S. Matsuoka, H.E. Bair and C.J. Aloisio, J.Polym.Sci. Symp., 46, 115 (1974)
4. G. Menges and H. Schmidt, Plastics and Polym., Febr., 13 (1970)
5. H. Bertilsson and J-F. Jansson, J.Appl.Pol.Sci., 19, 1971 (1975)
6. H. Bertilsson and J-F. Jansson in H. Bertilsson, Thesis: On the Transition to Marked Nonlinear Viscoelasticity in Solid Polymers, The Royal Institute of Technology, Stockholm, Sweden (1977)
7. I.V. Yannas, J.Polym.Sci.Macromol. Rev., 9, 163 (1974)
8. J-F. Jansson, Angew.Makromol.Chem., 37, 27 (1974)
9. H. Bertilsson and J-F. Jansson, J.Macromol.Sci.Phys., B14(2), 251 (1977).
10. G. Bergman, H. Bertilsson and Y.J. Shur, J.Appl.Pol.Sci. 21, 2953 (1977)
11. N. Sundgren, G. Bergman and Y.J. Shur, J.Appl.Polym.Sci., 22, 1255 (1978)
12. J-F. Jansson and I.V. Yannas, J.Pol.Sci.Phys., 15, 2103 (1977)
13. M. Robertsson and J-F. Jansson, VIIIth Int. Congr. Rheology, Naples, Italy, 1980.

THE INFLUENCE OF COOLING RATE ON THE TRANSITION TO MARKED

NONLINEAR VISCOELASTICITY IN POLY(METHYLMETHACRYLATE)

Mats E. Robertsson and Jan-Fredrik Jansson

Dept. of Polymer Technology
The Royal Institute of Technology
S-100 44 Stockholm   Sweden

INTRODUCTION

For amorphous glassy polymers a transition exists from approximately linear to marked nonlinear behaviour. The transition
has been suggested to be due to stress activated changes in the
deformation mechanisms[1,2,3]. In poly(methylmethacrylate),(PMMA),
as well as in other polymers both the α- and β-mechanisms have
been shown to be essential for the appearance of the nonlinearity[1,2]. The transition is accompanied by an accelerated stress dilatation and is supposed to be influenced by the available free
volume in the material[4].

It is known that amorphous materials at temperatures substantially below their glasstransition deviate from thermodynamic e-
quilibrium due to the kinetic nature of the glasstransition[5].
Their volume, enthalpy and entropy etc are larger than in the
state of equilibrium. However owing to the limited main chain
mobility that still exists in the glassy region the materials
slowly approach their equilibrium[6]. The slow and gradual approach
to equilibrium is often called physical aging[6].

In this paper we report initial studies of uniaxial creep
measurements at room temperature on PMMA with different thermal
histories. Our intention is to obtain a deeper insight into the
phenomena connected with the transition from approximately linear
to marked nonlinear viscoelasticity and into the phenomena connected with the concept of non equilibrium in glassy amorphous polymers. The conclusion[7] that the linear viscoelastic limit is the
limit below which crazing cannot occur in creep experiments, im-
plicates that the phenomena mentioned above affect the long term
strength.

EXPERIMENTAL

The material used was a commercially available PMMA supplied in the form of 2 mm thick sheets. Its glasstransition temperature was 113.5°C, determined with a differential scanning calorimeter (Perkin Elmer DSC-2), as that temperature at which the change in heat capacity is one half of its maximum value. The measurement was performed with a heating rate of 10°C/min directly after heating to above T_g and cooling to below T_g with the same rate. From the sheet ordinary dumbbell-shaped specimens were machined out (approximately according to ASTM D 638 Type II). The specimens were then subjected to different annealing procedures and cooling programs (see results) in a special oven provided with a very accurate temperature regulation (Heto 02 PG 000). The changes in weight during the annealings were negligible. Density measurements were performed in a density column (according to ASTM D 1505-68) at 1 hour and 24 hours after the annealing procedure had been completed to 23°C. Uniaxial creep measurements were performed in a creep equipment with two specimen sites. The temperature was thermostatically adjusted to 23.0°C±0.1°C, and controlled with a calibrated platinum resistance thermometer mounted close to the specimens. A polycarbonate sample tested in the apparatus between crossed polarizers confirmed that the uniaxial stress was uniformly distributed in the narrow measurement section of the specimen. The longitudinal and transversal strains were detected simultaneously by extensometers (Instron G-51-11M and Instron 2640-005). They were very carefully calibrated (Instron 2602-004) and mounted so that their weights could not interfere with the measurements. The extensometers were directly connected to a carrier amplifier (Peekel 581 DNH) the output of which was connected to a 2-channel recorder. Two specimens of each material were tested following the loading programme illustrated in figure 1. The strains at 300 s were determined as ln (L/L_0). The

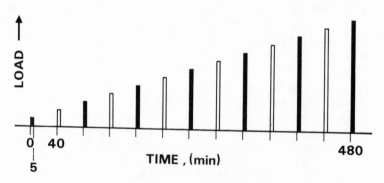

Fig. 1   Principal loading programme
         Specimen 1 = filled load periods
         Specimen 2 = unfilled load periods

stresses were calculated from the loads and the original cross
sectional areas of the specimens. For stress/strains below the
limit of linear viscoelasticity the accumulated remaining strain
at the end of the last recovery period, measured directly on the
amplifier by the zero-balance method, was less than ± 0.003%.

RESULTS AND DISCUSSION

    Figure 2 shows the isochronous stress vs strain plots obtai-
ned from uniaxial tensile creep experiments on PMMA at 23°C and
300s. The samples were cooled with different cooling rates, -0.6,
-6.0 and -350°C/h from 120°C. Each curve is based on 12-13 data
points from two different specimens (see also fig. 1). Straight
lines corresponding to the linear part of the curves have been
determined with linear regression analysis using at least 5 points.
The correlation coefficient of the determination was better than
0.9999 indicating an almost perfect linear fit. The linear limit
is arbitrarily chosen as the value of the prolonged regression
line where the experimental strain is 0.01% higher at constant
stress. Obviously, it is impossible to obtain evidence for the
existence of true linear viscoelastic behaviour by means of mechanical experiments only, since they always have a limited degree of resolution. As the linearity can only be determined with-

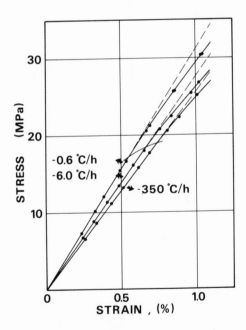

Fig. 2   Isochronous stress strain diagrams from uniaxial tensile
         creep at 23°C and 300s for PMMA cooled with different
         cooling rates (see figure) from 120°C.

in this resolution, we are discussing the transition from approximately linear to marked nonlinear behaviour. Moreover, the stress applied on a specimen is not entirely homogeneously distributed over the polymer chains. A small fraction of the bonds will be highly stressed even at low macroscopic stress levels, which has been shown for different solid polymers by IR-spectroscopy[8]. If the appearance of nonlinear viscoelasticity is supposed to be due to stress activated deformation mechanisms[1-3], it is thus clear that a small fraction of the nonlinear mechanisms will be activated already at low macroscopic stress levels.

In figure 3 the tensile creep compliance in the linear viscoelastic range at 23°C and 300s is plotted against log (cooling rate). The creep compliance decreases with decreasing cooling rate. This is in concordance with the behaviour observed[6] for glassy amorphous materials which have been quenched from temperatures above T_g. Thus for glassy amorphous polymers between T_g and T_β there is at least a qualitative similarity between decreasing cooling rate when passing the glasstransition and increasing aging time after a quenching through the glasstransition. When altering the cooling rate from -350°C/h to -0.6°C/h the compliance decreases approximately 20%. The connected density increase caused by rearrangements on the molecular level is about 0.2% when measured with a density column.

Both the stress and strain limits of the linear viscoelasticity decrease when lowering the cooling rate as can be seen in figure 4. Even the dilatation, defined by $\Delta V/V_0 = 1-\exp(\varepsilon_1 + \varepsilon_2)$ for an isotropic material where ΔV=volume change, V_0=original unstrained volume, ε_1=longitudinal strain and ε_2=transversal strain,

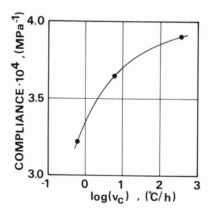

Fig. 3   Tensile creep compliance at 23°C and 300s in the linear viscoelastic range for PMMA vs cooling rate, (V_c) starting at 120°C

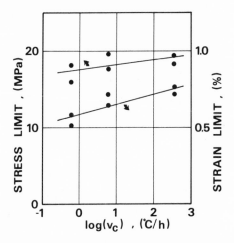

Fig. 4 Tensile creep stress and strain limits of approximately
        linear viscoelasticity at 23°C and 300s for PMMA <u>vs</u>
        cooling rate, (V_c) starting at 120°C.

at the limit of linear viscoelasticity decreases from about 0.2%
to about 0.1% when changing the cooling rate from -350°C/h to
-0.6°C/h. These results are at first sight rather unexpected,
since we have an excess of free volume in fast cooled amorphous
polymers below T_g and since free volume is supposed to facilitate[4]
the nonlinear deformation mechanisms. The results therefore indicate that it is not the absolute magnitude of the free volume
that determines the onset of nonlinear deformation mechanisms, but
rather the distribution of free volume in the material. This hypothetical difference in density distribution on the molecular
level between slowly and fast cooled specimens might also affect
the local stress distribution, when a macroscopic stress is imposed.

    For all specimens the creep periods are 300s followed by re-
covery periods of 15 x 300s (see also fig 1). The accumulated
longitudinal strain at the end of each recovery period did not
exceed ±0.003% up to stress and strain values of 1.25 times the
linearity limits for the slowly cooled materials. The fast cooled
material (-350°C/h), however, showed a marked decrease in accumulated strains when the linearity limit was exceeded, indicating a
small volume contraction, e.g. a mechanically enhanced aging. At
25% above the linearity limit the accumulated deformation was
-0.012 ±0.003%. These results show that the large dilatation at
the linear limit of the rapidly cooled material is accompanied by
irreversible structural changes in the marked nonlinear region.

REFERENCES

1. H. Bertilsson and J-F. Jansson, J.Appl.Polym.Sci., 19, 1971 (1975)
2. H. Bertilsson, Thesis, Royal Inst. of Technol., Stockholm, (1977)
3. J-F. Jansson and I.V. Yannas, J.Polym.Sci., Polym.Phys. Ed., 15, 2103 (1977)
4. S. Matsuoka, H.E. Bair and C.J. Aloisio, J.Polym.Sci. Symp., 46, 115 (1974)
5. A.J. Kovacs, Fortschr.Hochpolym.Forsch., 3, 394 (1964)
6. L.C.E. Struik, Polym.Eng.Sci., 17(3), 165 (1977)
7. G. Menges, Kunststoffe, 63, 95, 173 (1973)
8. S.N. Zhurkov, V.E. Korsukov, J.Polym.Sci., Polym.Phys.Ed., 12, 385 (1974)

VISCOELASTIC MODELS IN THE RHEOLOGY OF HYBRID POLYMERIC COMPOSITES OF (PHASE-IN-PHASE)-IN-PHASE TYPE

Horia Paven and Viorica Dobrescu

Central Institute of Chemistry
202 Splaiul Independentei
Bucharest, Romania

INTRODUCTION

The intrinsic academic interest as well as the increasing application of polymeric composites has largely motivated the significant development of both theoretical methods and intensive experimental studies aiming at describing their mechanical behaviour in terms of the properties of individual components, volume fraction, morphology and some characteristic parameters[1-3].

The results of numerous attempts to express quantitatively the elastic and viscoelastic properties of polymeric composites provide a lot of equations for the elastic and complex moduli, respectively. The priority utilization of "elastic-viscoelastic" correspondence principle rises in this case some special problems[4-9].

Although the phase-in-phase type composites have been extensively dealt with in literature, the hybrid (phase-in-phase)-in-phase composites focus ever more attention, given their higher qualities and the possibility of a better control of their properties. Evidently, the difficulties encountered are more important than in the case of ordinary composites, so that the utilization of mechanical models method and the development of the pertinent rheological equations of state (RES) seem to be necessary step in the better understanding of mechanical behaviour of (phase-in-phase)-in-phase composites, too.

This contribution presents an extension of author's method of directly deducing the composite RES on the basis of viscoelastic operators method rather than the composite moduli, taking into account the actual viscoelastic-type behaviour of polymeric materials.

THEORY

The mechanical behaviour of polymeric materials can be described in the framework of the linear viscoelasticity by the RES of the form:

$$P\sigma = Q\varepsilon \ , \qquad P = \sum_{r=0}^{m} p_r D^r \ , \qquad Q = \sum_{r=0}^{n} q_r D^r \qquad (1)$$

where P and Q are the characteristic viscoelastic operators, p_r and q_r represent the rheological parameters, $D = d^r/dt^r$ is the r-th order time derivative, and σ and ε are the stress and strain, considered, for the sake of simplicity, for the uni-dimensional case.

In the case of two phase, phase-in-phase, composites the following RES were obtained[10] according to Reuss approximation (uniform stress) and Voigt approximation (uniform strain), respectively:

$$[(1-v)P_a Q_b + vP_b Q_a]\sigma = [Q_a Q_b]\varepsilon \qquad (2)$$

and

$$[P_a P_b]\sigma = [(1-v)P_b Q_a + vP_a Q_b]\varepsilon \qquad (3)$$

where P_a, Q_a and P_b, Q_b stand for the characteristic viscoelastic operators for "a" and "b" components, respectively and v for the volume fraction of "b" in "a". These RES correspond to some limiting morphologies and evidently point, in the viscoelastic case, to the possibility of a dissimilarity, quantitative to the least, between the mechanical behaviour of components, on one side, and that of composite, on the other side and a difference between the behaviours of the two types of morphologies[11].

The two RES (2) and (3) yield directly the well known laws of mixting for the viscoelastic modulus (M_R^*, M_V^* for Reuss and Voigt approximations, respectively; M_a^* and M_b^* are the same quantities for components):

$$M_R^* = \frac{M_a^* M_b^*}{(1-v)M_b^* + vM_a^*} \qquad (4)$$

and

$$M_V^* = (1-v)\ M_a^* + v\ M_b^* \qquad (5)$$

As it is known these equations are usually obtained rather on the basis of correspondence principle than as a natural consequence of the RES.

In order to obtain the composite RES for a (phase-in-phase)-in-phase model i.e. in case in the phase "a" there is a phase-in-phase composite ("b"/"c"), the results above are applied first to ("b"/"c") phase-in-phase composite and then to (phase-in-phase)-in-phase resulting composite:"a"//"b"/"c". A set of four composite RES is thus obtained:

$$[(1-V)P_a Q_b Q_c + (1-v)VP_b Q_a Q_c + vVP_c Q_a Q_b]\sigma = [Q_a Q_b Q_c]\varepsilon \qquad (6)$$

$$[(1-v)(1-V)P_a P_c Q_b + v(1-V)P_a P_b Q_c + VP_b P_c Q_a]\sigma =$$
$$= [(1-v)P_c Q_a Q_b + vP_b Q_a Q_c]\varepsilon \quad , \qquad (7)$$

$$[(1-v)P_a P_b Q_c + vP_a P_c Q_b]\sigma = [(1-v)(1-V)P_b Q_a Q_c +$$
$$+ v(1-V)P_c Q_a Q_b + VP_a Q_b Q_c]\varepsilon \quad , \qquad (8)$$

$$[P_a P_b P_c]\sigma = [(1-V)P_b P_c Q_a + (1-v)VP_a P_c Q_b + vVP_a P_b Q_c]\varepsilon \qquad (9)$$

The "a"//"b"/"c" composite is described in terms of viscoelastic operators of components P_a, Q_a, P_b, Q_b, P_c, Q_c, of volume fraction v of component "c" in "b" and volume fraction V of composite ("b"/"c") in component "a". Eqs. (6)-(9) show that the mechanical behaviours of various morphologies may generally differ, like in the case of phase-in-phase composite, quantitatively as well as qualitatively. This set of RES allows immediate deduction of expresions giving the viscoelastic (or complex moduli). Thus, the following rules of mixing are obtained:

$$M_1^* = \frac{M_a^* M_b^* M_c^*}{(1-V)M_b^* M_c^* + (1-v)M_a^* M_c^* + vVM_a^* M_b^*} \qquad (10)$$

$$M_2^* = \frac{(1-v)M_a^* M_b^* + vM_a^* M_c^*}{(1-v)(1-V)M_b^* + v(1-V)M_c^* + VM_a^*} \qquad (11)$$

$$M_3^* = \frac{(1-v)(1-V)M_a^* M_c^* + v(1-V)M_a^* M_b^* + VM_b^* M_c^*}{(1-v)M_c^* + vM_b^*} \qquad (12)$$

$$M_4 = (1-V)M_a^* + (1-v)VM_b^* + vVM_c^* \qquad (13)$$

wich, in case of elastic components give the values of elastic moduli. It is worth noting that in case of hybrid composite under consideration there are two concentrations which determine transformation of "curves of mixing" into "surfaces of mixing".

The numerical study of the above relationships supplies interesting data regarding the dependence of composite modulus on corresponding component moduli and on v and V parameters.

To exemplify a particular case is presented next: a (phase-in-phase)-in-phase hybrid composite, selected from among the almost two hundred analyzed by the authors on the basis of seven types of linear viscoelastic behaviours. The composite is of H//KV/KV type i.e. it is characterized by Hooke (H) and Kelvin-Voigt (KV) RES:

$$\sigma = q_{H,0}\varepsilon$$

$$\sigma = (q_{KV,0}^{(1)} + q_{KV,1}^{(1)}D)\varepsilon$$
$$\sigma = (q_{KV,0}^{(2)} + q_{KV,1}^{(2)}D)\varepsilon \tag{14}$$

Utilization of Eqs. (6)-(9) yields the following RES:

$$\left\{ \left[(1-v)q_{KV,0}^{(1)}q_{KV,0}^{(2)} + (1-v)Vq_{H,0}^{(2)}q_{KV,0}^{(1)} + vVq_{H,0}^{(1)}q_{KV,0}^{(1)} \right] + \right.$$
$$+ \left[(1-V)(q_{KV,0}^{(1)}q_{KV,1}^{(2)} + q_{KV,0}^{(2)}q_{KV,1}^{(1)}) + (1-v)Vq_{H,0}^{(2)}q_{KV,1}^{(1)} \right] +$$
$$+ vVq_{H,0}^{(1)}q_{KV,1}^{(1)} \right] \; D + \left[(1-V)q_{KV,1}^{(1)}q_{KV,1}^{(2)} \right] D^2 \right\} \sigma =$$
$$= \left[q_{H,0}^{(1)}q_{KV,0}^{(1)}q_{KV,0}^{(2)} + q_{H,0}^{(1)}(q_{KV,0}^{(1)}q_{KV,1}^{(2)} + q_{KV,0}^{(1)}q_{KV,1}^{(2)})D + \right.$$
$$+ q_{H,0}^{(1)}q_{KV,1}^{(1)} q_{KV,1}^{(2)} D^2 \right] \varepsilon \quad , \tag{15}$$

$$\left\{ \left[(1-v)(1-V)q_{KV,0}^{(1)} + v(1-V)q_{KV,0}^{(2)} + Vq_{H,0} \right] + \left[(1-v)(1-V)q_{KV,1}^{(1)} + \right. \right.$$
$$+ v(1-V)q_{KV,1}^{(2)} \right] \; D \right\} \sigma = \left\{ \left[(1-v)q_{H,0}q_{KV,0}^{(1)} + vq_{H,0}q_{KV,0}^{(2)} \right] + \right.$$
$$+ \left[(1-v)q_{H,0}q_{KV,1}^{(1)} + vq_{H,0}q_{KV,1}^{(2)} \right] \; D \right\} \varepsilon , \tag{16}$$

$$\left\{ \left[(1-v)q_{KV,0}^{(2)} + vq_{KV,0}^{(1)} \right] + \left[(1-v)q_{KV,1}^{(2)} + vq_{KV,1}^{(1)} \right] \; D \right\} \sigma =$$
$$= \left\{ \left[(1-v)(1-V)q_{H,0}q_{KV,0}^{(2)} + v(1-V)q_{H,0}q_{KV,0}^{(1)} + Vq_{KV,0}^{(1)}q_{KV,0}^{(2)} \right] + \right.$$
$$+ \left[(1-v)(1-V)q_{H,0}q_{KV,1}^{(2)} + v(1-V)q_{H,0}q_{KV,1}^{(1)} + V(q_{KV,0}^{(1)}q_{KV,1}^{(2)} + \right.$$
$$+ q_{KV,0}^{(2)}q_{KV,1}^{(1)}) \right] D \; + \left[Vq_{KV,1}^{(1)}q_{KV,1}^{(2)} \right] \; D^2 \right\} \varepsilon, \tag{17}$$

$$\sigma = \left\{ \left[(1-V)q_{H,0} + (1-v)Vq_{KV,0}^{(1)} + vVq_{KV,0}^{(2)} \right] + \right.$$
$$+ \left[(1-v)Vq_{KV,1}^{(1)} + vVq_{KV,1}^{(2)} \right] \; D \right\} \varepsilon. \tag{18}$$

Eqs. (15)-(18) give explicit dependence of rheological parameters of composite on those of components and on concentrations (volume fractions). They also reveal significant differences in mechanical behaviour of composites of various morphologies. It is interesting to notice that utilization of components with elastic (Hooke) and viscoelastic (Kelvin-Voigt) behaviours yields for the hybrid composite a viscoelastic behaviour, which is either of Kelvin--Voigt – eq. (18) or Zener-Poynting-Thomson – eq. (16) or more complicated type – eqs. (15) and (17). Thus it turns out that the form of morphologies lead in this case to qualitatively rheological behaviours.

The concentration dependence of actual composite morphology and its importand effect upon the mechanical response of composite draw

the attention and also accounts for the interesting properties of
hybrid composites of (phase-in-phase)-in-phase type.

CONCLUSIONS

The method of viscoelastic operators may be succesfully extended
to hybrid polymeric composites of (phase-in-phase)-in-phase type.
The RES thus obtained allow the deduction of relations for viscoelastic moduli which gives the laws of mixing.
The strong morphology dependence of hybrid composite mechanical
properties causes both quantitative and qualitative differences in
the case of polymers whose behaviour is viscoelastic.
A highly accurate evaluation of mechanical behaviour of hybrid
polymeric composites requires improvement of present methods of investigation and utilization of mechanical models which, from this
point of view justify the interest shown to them.

1. L.E. Nielsen, "Mechanical Properties of Polymers and
      Composites", Marcel Dekker, New York (1974).
2. J.A. Manson and L.H. Sperling, "Polymer Blends and
      Composites", Plenum, New York (1976).
3. Yu.A. Lipatov, 1976, Relaxation and Viscoelastic
      Properties of Heterogeneous Polymeric Composites,
      Adv. Pol. Sci., 22;1.
4. H.H. Kausch, 1977, Mikromechanik mehrphasiger
      Polymersysteme, Angew. Makromol. Chem. 60/61: 139.
5. M. Takayanagi, H. Harima, Y. Iwata, 1963, Viscoelastic
      Behaviour of Polymer Blends and Its Comparison with
      Model Experiments, Mem. Fac. Eng., Kyushu Univ., 23:1.
6. Z. Hashin, 1965, Viscoelastic Behaviour of Heterogeneous
      Media, J. Appl. Mech., 32 E: 630.
7. R.A. Schapery, 1967, Stress Analysis of Viscoelastic
      Composite Materials, J. Composite Materials, 1:228.
8. R.M. Christensen, 1969, Viscoelastic Properties of
      Heterogeneous Media, J. Mech. Phys. Solids, 17:23.
9. R.A. Dickie, 1973, Heterogeneous Polymer-Polymer
      Composites. I. Theory of Viscoelastic Properties and
      Equivalent Mechanical Models, J. Appl. Pol. Sci., 17:45
10. H. Paven, 1978, Method for Determining the Linear
      Viscoelastic Behaviour Laws, Materiale Plastice, 15:163.
11. H. Paven and V. Dobrescu, 1980, Model Rheological
      Equations of State in the Linear Viscoelasticity of
      Polymeric Composites, to be published.

THERMAL PROPERTIES OF COMPOSITES

J.C.Seferis

Department of Chemical Engineering
University of Washington, Seattle, WA 98195, U.S.A.

(Abstract)

Mechanical properties of composite materials have been the object of numerous papers in the last years while little attention has been devoted to the thermal properties.

In this paper the composite mechanics analysis is utilized to predict thermal properties of composite systems and semicrystalline polymers.

STRESS RELAXATION OF GLASS-BEAD FILLED GLASSY AMORPHOUS POLYSTYRENE

F.H.J. Maurer

DSM, Central Laboratories
Geleen, Netherlands

INTRODUCTION

Recently much progress has been made in describing the creep and stress relaxation behaviour of unfilled amorphous polymers in the glassy region with respect to the thermal and mechanical history.[1-5] Struik[3] demonstrated convincingly that the momentary creep curve exhibits a universal shape over a rather large temperature range below T_g and above T_β . In accordance with these results, Booij and Palmen[5] proposed a formula for the momentary shear stress relaxation curve. At small strains the curve locations on the time scale are governed only by the thermal history, but at larger stresses and strains also by the mechanical history.

The influence of a high modulus filler on the relaxation curve at small strains may be accounted for by means of the quasi-elastic method. Neglecting the influence of the time dependence of the filler modulus-matrix modulus ratio on the composite modulus, the momentary shear stress relaxation modulus of a filled polymer can be presented by the equation:

$$G(t,t_a,q,T,\varphi,G_f/G_{om},\nu_m) = G_o(T,\varphi,G_f/G_{om},\nu_m)(t/t_r)^b \exp.-(t/t_r)^{0.4} \quad (1)$$

Thus the relaxation behaviour at small strains is characterized by two parameters, viz. elastic constant G_o, which depends on temperature T, the volumefraction of filler φ, the ratio between filler modulus and matrix elastic constant G_f/G_{om}, and the Poisson ratio ν_m, and the characteristic mechanical relaxation time t_r, which depends on annealing time t_a, cooling rate q, and temperature T. The value b represents a small number (~ 0.01), which is related to the mechanical

damping at relatively low temperatures or short times and is not
expected to be strongly influenced by filling.

The purpose of this work is to show that small strain-stress
relaxation data of filled and unfilled amorphous polystyrene can be
presented by means of Eq. 1, and that at larger strains, when filled
samples exhibit strong non-linearity, the same equation can still be
applied.

MATERIALS AND EXPERIMENTAL

Composites of polystyrene (Hostyren N4000) and untreated glass
beads (Sovitec, $\emptyset < 35\,\mu$ (95 %)) were blended on a two-roll mill at
190 $^{\circ}$C for 7 minutes. The resulting mill sheets were pressed to
sheets of 0.16 cm and 0.30 cm thickness under standard conditions at
190 $^{\circ}$C. Before blending the iron-particles present were removed and
the glass beads were dried for 48 hours at 120 $^{\circ}$C. Volumefractions φ
were determined to 9.5 % and 19.1 % $\pm$ 0.2 % by burning off and calculation. In a torsiondamping experiment at 0.2 cps no influence of
the glass beads on the loss maximum G" was observed.

Stress relaxation experiments in torsion were performed at 30, 50,
70, and 80 $^{\circ}$C with the aid of an apparatus developed at DSM's Central
Laboratory[5]. The dimensions of the specimen for stress relaxation in
torsion were 4.1x0.9x0.16 cm^3. In tension we used a 0.30 cm thick
Iso R 527 type 1 specimen. Before testing, each sample was heated for
half an hour at 115 $^{\circ}$C, about 20 $^{\circ}$C above T_g, to obtain the equilibrium volume, and quenched to the measuring temperature. After seve-
ral time lapses t_a (approximately $\frac{1}{2}$, 1, 2, 4, 6 and 23 hrs) a
maximum shear strain of 2.2x10^{-3} was applied. No small strain relaxation experiment lasted longer than 0.1 times the annealing time.
Stress relaxation and stress-strain curves in tension were measured
by means of a Universal Zwick 1474 tensile tester with closed loop
control. Measurements were made in the strain control mode with the
aid of a weight-compensated Zwick extensometer at 23 $^{\circ}$C. To reach
the desired strain level without any overshoot we used a strain
rate of 0.003 s^{-1}. Stress relaxation measurements in uniaxial tension
at a strain of 0.001 were made after annealing times of $\frac{1}{2}$, 1, 2
and 4 hrs at 23 $^{\circ}$C. Higher strain levels (0.005 and 0.007) were
applied after an annealing time of $\frac{1}{2}$ hr. During the resulting large
strain-stress relaxation experiment, we intermittently applied a
small additional strain of 0.0005 in order to determine the small
strain stress relaxation properties. Specimens were stored above
P_2O_5 to prevent moisture take-up.

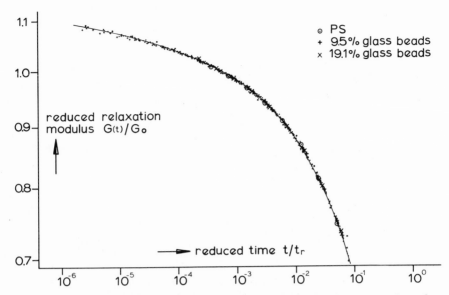

Fig. 1. Reduced relaxation modulus in torsion vs reduced time for
filled and unfilled polystyrene.

EXPERIMENTAL RESULTS

Stress relaxation in torsion

     All relaxation curves for filled and unfilled polystyrene, at
different temperatures T and different aging times t_a, were shifted
horizontally and vertically on a log G vs log t plot to the drawn
curve of Fig. 1. which satisfies Eq. 1., with b = -0.007. The accuracy of the shifting procedure supports the suitability of Eq. 1. In
Fig. 2., the characteristic relaxation time t_r is presented for
various temperatures and is found to be independent of the filler
volumefraction. The elastic constant G_o is slightly temperature-
dependent, independent of t_a and enhanced by filling as is shown in
Fig. 3. Predictions of the corrected van der Poel theory[6,7] are shown
by drawn lines (G_f/G_{om} = 25, ν_f = 0.25, ν_m = 0.37) and are in accordance with the experimental results.

Stress relaxation in tension

     In contrast to the nearly linear stress-strain curve in uniaxial
tension of unfilled polystyrene at 23 °C, the filled samples show below
a strain of 0.005 a typical knee, previously observed for several
glass-bead filled glassy amorphous polymers.[8] Below this knee, small
strain relaxation measurements were performed, and above it these
measurements were superimposed on large strain-stress relaxation.

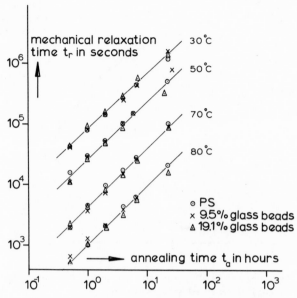

Fig. 2. Characteristic mechanical relaxation time t_r in torsion vs annealing time t_a at various temperatures T.

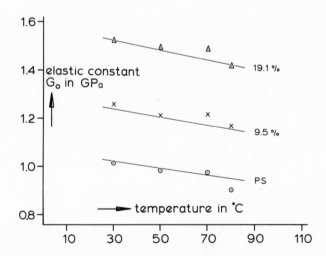

Fig. 3. Elastic constant G_o at various temperatures T for polystyrene and polystyrene filled with 9.5 % and 19.1 % glass beads by volume.

Table 1. Elastic Constants E_o at various Strain Levels

Strain after $t_a = \frac{1}{2}$ hr	0	0.0050	0.0070
Additional strain (intermittently)	0.001	0.0005	0.0005
E_o in GPa			
Polystyrene	2.830 ± 1 %	2.803 ± 2 %	2.780 ± 2.5 %
9.5 % glass beads	3.459 "	3.003 "	2.751 "
19.1 % glass beads	3.956 "	3.049 "	2.631 "

In the same manner as the shear relaxation results, the tensile
relaxation data were successfully analysed by means of Eq. 1. with a
corresponding tensile modulus E and with b = -0.008. The values determined for the elastic constants E_o, which were found to be unaffected
by annealing time t_a, are presented in Table 1. For the filled
samples E_o was strongly influenced by the strain level. The relaxation times are shorter at larger strains, both for unfilled and filled
polystyrene (Fig. 4.). The influence of the glass beads on relaxation
time seems to be a strain magnification effect. Shortly after application of the large strain a coupling between large strain and the
aging process becomes clear.

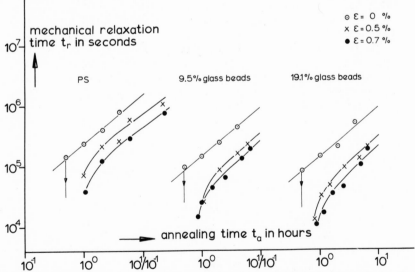

Fig. 4. Characteristic mechanical relaxation time in tension of stress
relaxation at small strain additional to different high
strain levels.

CONCLUSIONS

It has been shown that momentary small strain-stress relaxation data in shear can be described by Eq. 1. Both for unfilled polystyrene and polystyrene filled with untreated glass beads up to about 20 % by volume. The characteristic relaxation time and the aging rate $\partial \log t_r / \partial \log t_a$ are unaffected by filling. The elastic constant can be described by the van der Poel's theory. In spite of the strong non-linearity of the filled samples at larger strains in tension Eq. 1 can still be applied. Both relaxation times and elastic constants are then affected by filling and strain level.

ACKNOWLEDGEMENTS

The author is much indebted to Mr. J.W.A. Sleijpen and Mr. B.H.J. Henket for performing the shear stress relaxation measurements and to Mr. J. Drost for performing the tensile relaxation measurements and for drawing the figures. The stimulating discussions he had with Mr. H.C. Booij were greatly appreciated.

REFERENCES

1. R.A. Schapery, Polym. Eng. Sci. 9:295 (1969).
2. F.A. Myers, F.C. Cama and S.S. Sternstein, Ann. N. Y. Acad. Sci. 279:94 (1976).
3. L.C.E. Struik, "Physical aging in amorphous polymers and other materials", Elsevier, Amsterdam (1978).
4. S. Matsuoka, H.E. Bair, S.S. Bearder, H.E. Kern and J.T. Ryan, Polym. Eng. Sci. 18:1073 (1978).
5. H.C. Booij and J.H.M. Palmen, Polym. Eng. Sci. 18:781 (1978).
6. C. van der Poel, Rheol. Acta 1:198 (1958).
7. J.C. Smith. J. Res. Nat. Bur. of Standards, 78A:355 (1974).
8. R.E. Lavengood, L. Nicolais and M. Narkis, J. Appl. Polym. Sci. 17: 1173 (1973).

FAILURE PROPERTIES OF FILLED ELASTOMERS

AS DETERMINED BY STRAIN ENDURANCE TESTS

Y. Diamant, Z. Laufer and D. Katz

Department of Materials Engineering
Technion-Israel Institute of Technology
Haifa

INTRODUCTION

Determination of the resistance of a material to failure when held at constant strain for an extended period of time is of great practical importance. For elastomers the failure properties are usually described by the well known Smith's "failure envelope". This failure envelope is accepted to be independent of the kind of experiment used for its determination and is applicable in all cases in which the material remains basically (chemically and physically) unchanged during the period of investigation. In reality almost all elastomers fail as a result of their stress or strain history. During this history the load changes and the material has to withstand different stresses for long periods of time. Simultaneously the material ages and internal structural changes may occur. These aging effects are probably accelerated due to the fact that the material is kept deformed and its internal energy is higher than in the case of an unstrained system. Thus, the real failure properties of a material which is being used can not be probably represented by a failure envelope determined by testing of unaged materials only.

In this study efforts were made to determine the failure properties of two composite polymeric systems by investigation of their strain endurance (S.E.) and by stepwise increase of strain on some of the S.E. experiments.

EXPERIMENTAL

A simple device in which dumbbell shaped specimens were subjected to a constant strain for a specified period of time or until

243

failure, was used for S.E. tests. The strain levels were about 70
to 100 percent of the ultimate strain which was determined by constant strain rate measurement conducted on the same type of specimens in an Instron testing machine. Usually the specimens were
kept strained for a period of 30 days in the S.E. experiment. The
testing apparatus was sealed and desicated to avoid humidity effects
and tension was applied by screw operated jaws while turning the
screws by hand at a uniform and rather slow straining rate. The
specimens which did not fail in the S.E. tests, were removed and
strained to failure in an Instron testing machine at a rate of
strain of 0.7 min^{-1} for evaluating the effect of the prestraining
on the mechanical properties of the investigated materials. The
S.E. tests and the straining of the specimens removed from the S.E.
tests were performed at ambient temperature. The data presented in
the figures and the tables are an average of results obtained by
testing 2-3 specimens.

The effect of stepwise increase in strain magnitude in the S.E.
tests, was also studied. In these experiments the specimens were
initially strained to 0.7 of their ultimate strain and then the
strain was increased in steps of 0.05, 0.10 or 0.15 of the ultimate
strain until failure occurred. In each strain level the specimens
were held for about 100 hours.

Two different compositions of particulate filled elastomeric
polymers,   designated material A and material B were investigated
in this study.

Experimental Results

Uniaxial tension tests were performed at various temperatures
(in the range of -50°C to +100°C) and rates of strain for determination of the failure envelope (see fig.1). A difference in the
failure properties of the two elastomers can be observed in the
part of the curve considered in the clockwise direction from the
point of maximum strain (data obtained at low strain rates and high
temperatures). While material A exhibits the usual shape of a
failure envelope material B showed in the above mentioned range of
the envelope, a constant strain almost equal to the maximum strain.
This significant difference, as will be shown later explains other
differences in the failure behaviour of the two materials.

In the S.E. tests a similar behaviour was observed in both
materials: specimens which did not fail during the first hour
withstood the imposed strain for the complete straining period.

Material A, with the ordinary failure envelope, showed usual
strain endurance capabilities and the specimens failed at about 80%
of the ultimate strain. On the other hand, specimens of material B
did not fail unless they were stretched to their ultimate strain.

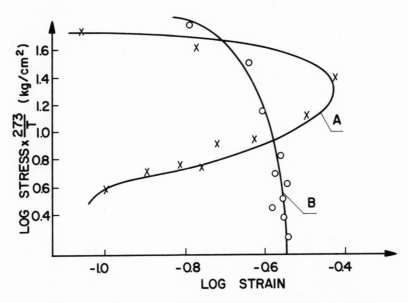

Figure 1:   Failure Envelopes of materials A and B

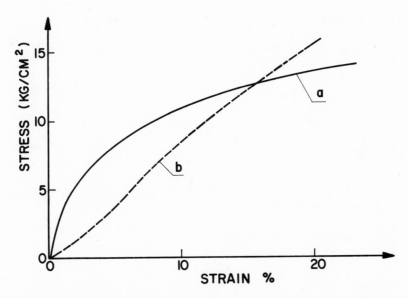

Figure 2:   Typical stress-strain curves of (a) A reference
            specimen, (b) A specimen which was prestrained in
            S.E. Test.

A typical stress-strain curve of a specimen which did not fail in a S.E. test is compared in figure 2 with that of a specimen which was not prestrained. Due to the softening of the material on straining, a decrease in its initial modulus is observed, while an increase in the maximum stress, accompanied in most of the cases by lowering of the ultimate strain, occurs.

Since the above results indicate that some internal changes in the structure of the material occur during the S.E. test, the effect of strain duration and the time after load release on the ultimate properties of previously strained specimens was investigated (table I and II).

The results indicate that the internal structure changes occur in the material during the first 24 hours and these changes are not reversible after the load release.

The stepwise S.E. experiments are described schematically in figure 3. The specimens were kept at each strain level for a certain time before more strain was imposed on them. Material A was almost not effected by this mode of strain application and failed at about 0.85-0.9 of the ultimate strain value. On the other hand material B exhibited an unexpected high S.E. capability and its failure occurred outside the boundaries determined by the failure

Table I: The effect of S.E. duration on the failure properties of Material B specimens (ambient temp., strain rate 0.07 min^{-1}).

S.E. duration days	Initial modulus kg/cm^2	Maximum stress kg/cm^2	Maximum stress %
1	51	19	27
3	38	17	20
10	42	19	23
30	46	17	17
Reference, no prestrain	82	15	29

Table II: The effect of time after load release on the failure properties of specimens strained previously for 14 days.

Rest time hours	Initial modulus kg/cm^2	Maximum stress kg/cm^2	Maximum stress %
Material A			
2	46	9.1	19
30	51	8.8	17
Material B			
2	43	18.6	23.1
30	47	18.7	22.5

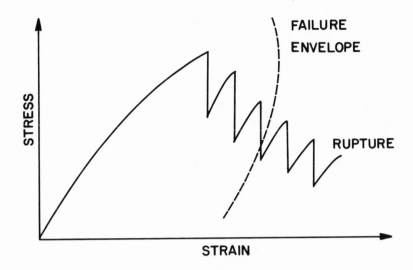

Figure 3: Schematic description of a stepwise S.E. experiment

envelope, namely at about 1.40 of the ultimate strain. For both
materials the S.E. capability was not influenced by the magnitude
of the strain increments.

DISCUSSION

     The two investigated materials exhibited a different strain
capability. Material A which had the usual shape of the failure
envelope showed no different behaviour even when the strain was
increased stepwise. Material B, characterised by a failure envelope
with a part of it showing a constant strain, could withstand higher
strains when stretched stepwise and left for several hours at each
strain level, reaching even beyond the limits of the failure envelope.
In the filled and highly strained elastomeric materials tested in
this study, internal structural changes apparently occurred during
the experiment due to application of high strains for relatively long
periods. This seems to be the reason, why the stress strain curves
of the strained material in the early stage of the experiment are
different from the curves after additional stretching and the ultimate behaviour of material B was not predictable by use of the failure envelope.

     Data in the two tables show, that if some alternation in the
internal structure of the systems occurred, it happened at the
beginning of the experiment and these changes were not reversible.
The idea that changes occur in the initial stages of the experiment
is supported by the fact that in cases in which rupture of the speci-
ment occurred, it happened shortly after strain application.

One of the possible explanations of strengthening of the material could be based on assumption, that orientation occurred during the stretching period, but no evidence for it was found in observations made by use of a scanning electron microscope.

In order to see if any changes occurred in the network structure of the polymer binder, swelling experiments were performed and no significant changes were observed in the swelling ratio and amount of solubles.

Further studies are taking place in order to investigate the possible changes occurring in the filler-matrix interphase of these composite materials as a result of wetting-dewetting processes.

For practical uses of filled elastomers, their failure properties will probably be more realistically represented, in most of the cases, by use of S.E. tests. Further studies of S.E. tests at various temperatures are in progress and the results of these investigations will be used for construction of a more general failure envelope.

AGING OF A STRUCTURAL ADHESIVE

D. Katz, A. Buchman and S. Gonen

Department of Materials Engineering
Technion-Israel Institute of Technology
Haifa, Israel

INTRODUCTION

The use of structural adhesives, in the form of a one component
mixture containing a thermosetting prepolymer and all the necessary
additives with an adequate amount of a suitable curing agent, often
spread on a reinforcing fabric is becoming more common in aircraft
and related industries. Most of the research and development work
deals with the nature and strength of the adhesive after bonding and
its behaviour and aging in different environmental conditions. To
the best of our knowledge, not much attention was paid to changes
occurring in the prepolymerized adhesive before its use, changes
which doubtlessly affect the performance of the final product.

EXPERIMENTAL

Materials

A commercial epoxy based adhesive reinforced by a nylon fabric,
"FM-73", made by "American Cyanamid" and kept in cold storage, was
used for this study. The adherents were Aluminium 20-24 plates of
1.7 mm thickness. Samples of the adhesive were aged for various
periods of time at room temperature before polymerization. The
properties of the unaged and aged adhesive were investigated
before polymerization and after polymerization in a teflon mold.
The properties of the bond created between aluminium plates by use
of the unaged and aged adhesive were also studied. Aging periods
were 3,8 and 16 days.

The Prepolymerized Adhesive

Differential thermal analysis (DTA) thermograms of samples

unaged and aged till 30 days were obtained by heating of 30 mg of
the prepolymer up to 120°C at a rate of 2°C/min, and keeping of the
samples at 120°C for 1 hour. The thermogram areas related to the
exothermic section occurring during polymerization, were measured
and compared.

Thermo-gravimetric analysis (TGA) of aged and unaged samples
were recorded while heating about 250 mg of prepolymer at a rate
of 1°C/min up to 200°C. The weight loss percentage was calculated.

The Polymerized Adhesive

Tension test: One layer of the aged and unaged adhesive film
of thickness 0.15 mm was molded between teflon plates. The samples
were polymerized by heating under load at a rate of about 2°C/min
up to 120°C, and then for one hour at 120°C (1). They were tested
at room temperature in an Instron machine, according to ASTM (2);
σ_B and ε_B were calculated.

Swelling and I.R. Spectroscopy: Specimens of the polymerized
unaged and aged adhesive were swollen till equilibrium in an acetone/
water solution (1:1 by volume). Weight percentages of the swollen
material and of extracts were calculated. The solubles were dissolved in absolute acetone and analysed by use of an I.R. spectro-
photometer.

Torsion Modulus vs. Temperature: 2 mm thick test samples prepared by polymerization of 6 layers of the adhesive film place one
upon the other were tested by use of a Gehman apparatus (3), in a
temperature interval of +20 - +160°C. Changes in the torsion modulus vs. temperature curve can be related to changes occurring in
the prepolymer during its controlled aging process.

Adhesive-Adherent System

Shear experiments: Bonding of aluminium plates was performed
by polymerizing one layer of adhesive film between two plates according to the procedure described previously and specimens with a
single lap joint configuration   were prepared. The dimensions
of the aluminium plates with edges trimmed at an angle of 45° were
10x2.5x0.17 cm; the thickness of the polymerized adhesive was 0.1 mm,
and the joint surface was 2.54x1.27 cm. Five specimens were prepared
from each aged adhesive, while 15 joints were made by use of the
unaged adhesive. The shear strength of the joints at room temperature was measured by means of an Instron machine at a strain rate
of 2 mm/min. Values of τ_B and fracture energy (E) were calculated.

Microscopic observations: The fracture surface of the samples
broken in shear experiments was examined by means of an optical and
a Scanning Electron Microscope (SEM).

Experimental Results and Discussion

The prepolymerized adhesive

Data from DTA thermograms show that the energy released per gram of prepolymer decreased with the increase in time of aging (figure 1). This decrease in curing energy can be attributed either to partial curing of the resin during aging or to side reactions in which the curing agent could be involved during the aging time and which prevented it from further participation in building of the polymer network. The TGA thermograms show no weight loss of the system during heating up to 200°C.

The Polymerized Adhesive

Tension Test: According to the results tabulated in table 1, aged samples seem to be more brittle (higher breaking stress lower breaking strain), than the unaged ones.

Swelling and I.R. spectroscopy: A decrease in the percentage of equilibrium swelling and in the amount of solubles in samples aged for long periods before polymerization (table 1) seems to indicate that higher degrees of crosslinking exist in the network of samples prepared from the adhesive aged before polymerization than in those formed from the unaged adhesive.

The I.R. spectra of the solubles extracted in different stages of aging of the adhesive samples, are shown in figure 2. The absorption bands at 3200,3300 cm^{-1}, typical for -NH groups, are smaller for aged samples. The absorption bands of the epoxy groups (1040-1050, 990, 870 cm^{-1}) are also decreasing in samples aged before polymerization.

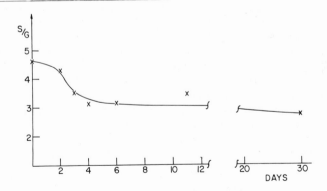

Fig. 1. Relative curing energies for unaged and aged adhesives (from DTA).

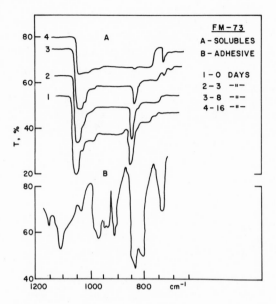

Fig. 2. I.R. spectra of solubles from aged and unaged adhesive
compared to I.R. spectra of the prepolymer.

Torsion modulus temperature relationship: The various transition temperatures characteristic for the unaged and aged adhesive
samples after polymerization, as derived from measurements of the
torsion modulus, G, vs. temperature, are shown in table   1. Unaged
samples, and those aged for a short period (3 days), had two transition temperatures $T_{(t)}1$, $T_{(t)}2$, while adhesives aged for longer
periods had only one transition. An increase in the aging period
of the prepolymer seems to lead to a rise in the transition temperatures of the resulting polymers, and to an augmentation in the torsion modulus values in their rubbery region $(G_{(10, R)})$. These facts
indicate that aging of the prepolymer results either in loss of a
volatile agent with plasticizing properties or in formation, during
polymerization, of different network structures with higher degrees
of crosslinking, or in both of those processes.

Adhesive-Adherent System

Shear: The values of ultimate shear stress and fracture energy
for the specimens are summarized in table no.1. The results show
a slight rise in τ_B of the joint in which the aged adhesive was used,
as compared with the joint made by use of an unaged adhesive, while
the joint fracture energy decreases with the increase of the aging
period of the adhesive before its use.

Table 1. Properties of the unaged and aged polymerized adhesive and of joints prepared from these adhesives.

Material	Properties	Aging Time (days)	0	3	8	16
Polymerized Adhesive	Swelling	Swelling (% wt)	41.0	35.2	23.8	24.7
		Solubles (% wt)	2.72	2.73	1.78	1.41
	Torsion	$T_{(t)}1$ (°C)	49	63	–	–
		$T_{(t)}2$ (°C)	89	90	99	100
		$G_{(10,R)} \times 10^{-7}$ (dyne/cm^2)	3.1	3.1	4.5	5.4
	Tension	σ_B (kg/cm^2)	178.4	275.0	289.0	316.0
		ε_B (%)	3.10	2.45	2.40	2.30
Joint	Shear	τ_B (kg/cm^2)	256	264	262	240
		E (kg/cm^2)	275	246	233	228

<u>Microscopic observations</u>: Investigation of the fracture surface of the joint showed that aging of the prepolymer caused a reduction in the concentration and the size of pores in the cured polymer; after 16 days of aging of the prepolymer no pores were detected. The fracture occurred mainly in the adhesive, although near the edges of the aluminium plates trimmed to 45°, the cohesive type fracture could be detected in the aluminium.

Investigation of the fracture surface by means of SEM shows that the mode of fracture of the adhesive layer of the unaged adhesive before polymerization was ductile and long fibres of the adhesive were formed during failure, although the experiment was carried out at room temperature. (Fig.no.3a). The pores on the fractured surface are round and crater shaped. The adhesive aged before use showed a brittle type of fracture with many sharp edges and cracks; no fiberlike structures formed during failure are found. The pores formed during polymerization are much smaller in the adhesive aged before polymerization than in the unaged one, and they

 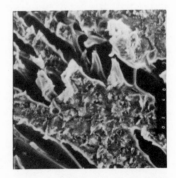

SEM photomicrograph of fracture surfaces of specimens prepared from an adhesive which was before polymerization.
3a) unaged, x 100   3b) aged for 7 days, x 150
(Reduced 10% for reproduction)

are elongated (Fig.3b). The wetting of the nylon fibres by the resin which is good in unaged adhesives seems to be insufficient in specimens prepared from the aged adhesive.

CONCLUSIONS

The results of the tests conducted on the prepolymer, polymer and the adhesive-adherent system indicate clearly that changes occur in the structure and properties of the adhesive as a result of aging of the prepolymer at room temperature before its use.

The adhesive stored at low temperature shows a certain amount of ductility after polymerization and formation of adhesive joints. This can be seen by observation of the fracture surface of the polymer and from data on its mechanical properties (high ε_B, low σ_B). Aging of the prepolymerized adhesive at room temperature, results in formation of a more brittle polymer, as shown by microscopy of the fractured surface of the polymer and by data related to its mechanical properties (lower ε_B, higher σ_B). Aging of the adhesive before polymerization leads to a decrease in the number and size of pores in the polymer. These phenomena become more pronounced with the increase in aging time of the prepolymerized adhesive. The above described observations can be explained by assuming that crosslinking of the prepolymer occurs during aging and this leads to a higher degree of crosslinking in the final cured polymer. It happens probably because of a slow increase in the viscosity of the reactive system and a more efficient utilization of functional groups (a higher crosslinking efficiency) which results in elimination of the plasticizing effect of longer network chains or/and branches with functional groups trapped during the short polymerization process of the unaged material due to the very fast increase in the viscosity of the system. Lower concentration of epoxy groups, as shown by I.R. spectra lower swelling percentages and lower polymerization energy evident from DTA thermograms of polymers obtained from aged prepolymerized adhesives, support the above assumption. The results obtained in TGA experiments show that no volatiles are lost during aging of the prepolymerized adhesive and the possibility of an eventual plasticizing effect of such materials on the final properties of the polymer obtained from an unaged adhesive has to be excluded. The occurrence of two transitions in the Gehman curve in the polymer made from an unaged adhesive, can be perhaps attributed to the existence of branches (with unreacted functional groups) attached to the basic network.

REFERENCES

1.   American Cyanamid Co. "FM-73 Brochure", London (1974).
2.   ASTM Standards, 26, D2141 (1972).
3.   ASTM Standards, 27, D1043 (1973).

THE STRENGTH OF ORIENTED SHORT FIBER REINFORCED PLASTICS

J. L. Kardos, J. C. Halpin, and S. L. Chang

Materials Research Laboratory and Department of
Chemical Engineering
Washington University
St. Louis, MO 63130

INTRODUCTION

Prediction of strength for short fiber reinforced plastic
systems is a complex but industrially crucial problem. Even in
the case of unidirectionally aligned fibers with tensile stress
applied in the fiber direction, failure may occur in the fibers,
in the matrix phase, or at the interface. Furthermore, failure
may take place in a tensile or shear mode and may be brittle or
ductile in nature.

Historically, the strength of materials approach to composite
strength prediction began with the work of Cox (1) in 1952, which,
although not exact, served as the basis for later developments of
what is now called the "shear lag" analysis. Later detailed
analyses by Dow (2) and Rosen (3) produced the same basic result.
In 1964 Cottrell (4) formulated the basic shear lag analysis based
on a single fiber and introduced the concept of a critical fiber
length, ℓ_c, above which the fiber's ultimate strength will be fully
utilized. One year later Kelly (5-7) extended this principle to
describe the behavior of short fiber reinforced metals and predicted
that up to 95% of continuous fiber composite properties should be
attainable with short fiber systems.

While there were other approaches (8) and occasional attempts
to modify the shear lag approach (9), it was not until the pioneer-
ing work of Chen in 1971 (10) that the interaction between neighbor-
ing short fibers was fully taken into account. Chen utilized a
finite element approach, which included a distortional energy
criterion, to calculate the strength of several uniaxially aligned

255

short fiber systems. He found that the composite strength reached
a plateau as the fiber aspect ratio was increased at constant
volume loading. Furthermore, this plateau occurred at only 80% of
the continuous fiber value for tungsten-copper composites and at an
incredibly low 55% and 60% of continuous values for boron-epoxy
and glass-epoxy respectively. Barker and MacLaughlin (11) and
Riley (12) also concluded that interacting fiber stress concentrations should reach a plateau at large aspect ratios.

In this paper we present the details of a new approach which
accounts for the large stress concentration penalties in a perfectly
aligned short fiber composite. The rudiments of this approach have
been outlined in an earlier publication based on a limited data
base (13) and we report a more thorough treatment here. Although
empirical, the method permits calculation of a strength reduction
factor which can then be utilized with an appropriate failure
criterion to calculate the strength of a wide range of short fiber
composite systems.

RESULTS

In examining Chen's work and our own laboratory results on
uniaxially aligned short fiber systems, two important characteristics are clear. One is that a plateau in strength is reached at
large fiber aspect ratios when the fiber volume fraction is held
constant. The aspect ratio, at which the knee in the curve appears,
depends on the type of fiber and matrix and is much larger than that
predicted by the shear lag analysis. Secondly, if one plots
strength versus aspect ratio at constant volume fraction of fibers
for different systems, a family of sigmoidal curves results, whose
plateau values for strength are much lower than those predicted by
shear lag analysis and whose values also depend on the particular
fiber-matrix combination used. If one could collapse all these
curves on a single master curve utilizing generalized normalization
parameters, then the stress concentration penalty for interacting
fiber ends could be accurately determined for any short fiber system.

The problem may be approached by considering how the curves
may be shifted both horizontally and vertically to produce a normalized master curve. In effect, use may be made of an aspect of
dimensional analysis in which division of one dimensionless group
by another can collapse the data if the groups are properly chosen.

To accomplish the vertical shift, a strength reduction factor
[SRF] is defined as the uniaxially aligned short fiber system
strength divided by the strength of an aligned continuous fiber
system of the same volume fraction fibers. Thus,

$$[SRF] = \bar{\sigma}_c / \sigma_R V_R \qquad\qquad (1)$$

wherein the matrix contribution to the rule-of-mixtures continuous fiber strength is neglected. σ_R is the fiber strength and V_R the volume fraction reinforcement. As the aspect ratio approaches unity, the [SRF] approaches that for a sphere-filled system, namely $[SRF]_o$. This lower bound can be evaluated by utilizing the results of Narkis et al. (14-16) for glass bead-filled thermoplastics and thermosets,

$$[SRF]_o = \frac{\sigma_m E_c (1-V_R^{1/3})}{\sigma_R V_R E_m} , \qquad (2)$$

where σ_m is the matrix strength.

The value for the [SRF] at large fiber aspect ratios, $[SRF]_\infty$, may be shifted vertically by utilizing the fiber-to-matrix stiffness ratio, E_R/E_m, which in fact controls the magnitude of the stress concentrations at the fiber ends (10-12). The best fit for this shift yields

$$[SRF]_\infty = 0.5 + (E_R/E_m)^{-0.87} \text{ for } E_R/E_m > 5 \qquad (3)$$

The horizontal shift parameter, β, can be developed by noting the parameters of importance in the shear lag analysis. The critical aspect ratio and the ratio of fiber strength to matrix shear strength are related by a constant. Thus the horizontal shift parameter is

$$\beta = (\ell/d)/(\sigma_R/\tau_m) \qquad (4)$$

where τ_m is the shear strength of the interface or matrix, whichever is lower.

The final equation for the master curve may be normalized and expressed as follows:

$$G = \frac{[SRF] - [SRF]_o}{[SRF]_\infty - [SRF]_o} = 1 - 0.97 \exp [-0.42\beta] \qquad (5)$$

The above equations, collectively denoted is the Halpin-Kardos Equations, predict the lowering of continuous fiber strength due to the discontinuous nature of the reinforcement. Figure 1 presents a non-linear regression fit of finite element calculations using Chen's approach (10). For each experimental data set, a finite element curve was calculated and expressed in terms of Halpin-Kardos equation parameters. Points from these individual curves are

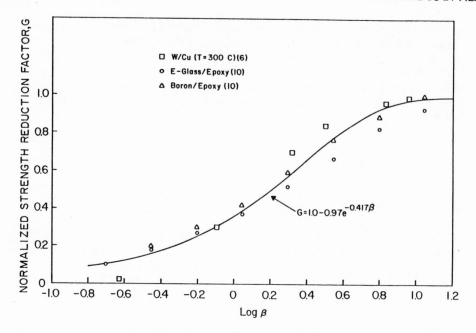

Figure 1.   Best fit of Halpin-Kardos equations to experiment-based
            finite element results.

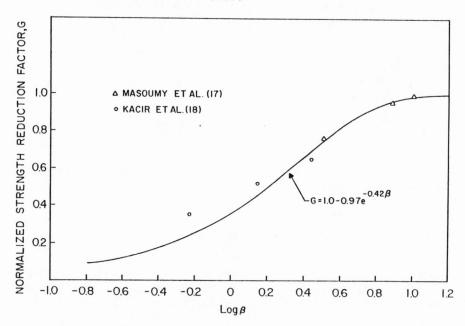

Figure 2.   Comparison of Halpin-Kardos equations with experimental
            data for E-glass /epoxy systems.

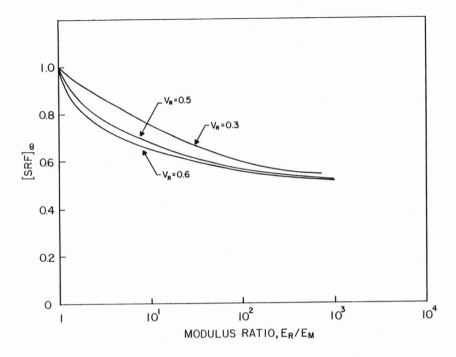

Figure 3.   Effect of fiber volume fraction on the modulus ratio -
             $[SRF]_\infty$ relation.

plotted in Figure 1 and the solid line represents a second non-
linear regression fit of the points to determine the best values
for the constants of the Halpin-Kardos equation. Figure 2 compares
the resulting Halpin-Kardos prediction (solid line) with experimental data for both ductile (17) and brittle (18) epoxy matrices
reinforced with E-glass fibers.

    The fiber volume fraction appears explicitly only in the
expression for $[SRF]_o$. Since all of the fitting was done using a
fiber volume fraction of 0.5, it is important to examine the effect
of fiber volume fraction on $[SRF]_\infty$. Since there was not enough well
characterized experimental data available at other volume fractions,
the finite element approach utilized by Chen (10) was used to calculate the effect of fiber loading on $[SRF]_\infty$. Figure 3 depicts the
results. It is clear that at ratios of E_f/E_m larger than 100, the
effect is quite small and may be neglected for practical design
considerations. Non-linear regression analysis yields the following
relation for $[SRF]_\infty$ at V_R = 0.5, in good agreement with the experimental fit (Eqn. 3).

$$[SRF]_\infty = 0.5 + 0.688 \ (E_R/E_m)^{-0.5} - 0.413 \ (E_R/E_m)^{-1}$$

$$+ \ 0.226 \ (E_R/E_m)^{-2} \tag{6}$$

The above equations may be collectively utilized to estimate allowable failure strains for aligned short fiber plies. In conjunction with laminated plate theory and a maximum strain failure criterion, these equations have led to the successful prediction of strength for systems whose in-plane fiber orientation distribution is known (17).

ACKNOWLEDGMENT

This work was supported by the Polymers Program of the National Science Foundation under Grant No. DMR78-12806. We thank Dr. Paul Chen for his very helpful discussions on the finite element calculations.

REFERENCES

1. H. L. Cox, Br. J. Appl. Phys., 3, 72 (1952).
2. N. F. Dow, General Electric Report R635D61 (1963).
3. B. W. Rosen, Mechanics of composite strengthening, in Fiber Composite Materials, American Soc. Metals, Metals Park, Ohio, 1965, Ch. 3.
4. A. H. Cottrell, Proc. Roy. Soc., Series A, 282A, 2 (1964).
5. A. Kelly, Fibre Reinforcement, in Strong Solids, Clarendon Press, Oxford, 1969, Ch. 5.
6. A. Kelly and W. R. Tyson, J. Mech. Phys. Solids, 13, 329 (1965).
7. A. Kelly and G. J. Davies, Metallurgical Rev., 10, 1 (1965).
8. J. O. Outwater, Jr., Mod. Plast., 33, 156 (1956).
9. J. K. Lees, Polym. Eng. Sci., 8, 195 (1968).
10. P. E. Chen, Polym. Eng. Sci., 11, 51 (1971).
11. R. M. Barker and T. F. MacLaughlin, J. Comp. Mater., 5, 492 (1971).
12. V. R. Riley, J. Comp. Mater., 2, 436 (1968).
13. J. C. Halpin and J. L. Kardos, Polym. Eng. Sci., 18, 496 (1978).
14. R. E. Lavengood, L. Nicolais, and M. Narkis, J. Appl. Polym. Sci., 17, 1173 (1973).
15. M. Narkis, Polym. Eng. Sci., 15, 316 (1975).
16. M. Narkis, J. Appl. Polym. Sci., 20, 1597 (1976).
17. E. Masoumy, L. Kacir and J. L. Kardos, submitted for publication.
18. L. Kacir, M. Narkis, and O. Ishai, Polym. Eng. Sci., 17, 234 (1977).

DEFORMATIONAL BEHAVIOR OF COMPOSITE PARTICLES

S.K. Ahuja

Xerox Corporation
Joseph C. Wilson Center of Technology
Rochester, New York 14644

INTRODUCTION

In recent years, a number of studies have been conducted on the fracture and deformation of single particles.[1,2,3,4] A previous study dealt with deformation of single, homogeneous, brittle, and ductile polymer spheres and related their contact stresses to the bulk stresses of polymers, polystyrenes, and polycarbonates.[4] In the present work, we have extended deformation of single particles to encapsulated, heterogeneous systems.

PREPARATION AND PROPERTIES OF THE MATERIALS

The encapsulated particles were obtained through spray drying.[5] Of the three types of encapsulated particles, the first two types had brittle, polystyrene shells and polyester cores. In the third type, the ductile polyamide shell was surrounded by ductile polyester. The polyester material, Dow Adipate or Dow Sebacate, was obtained as a reaction product of isopropylidene diphenoxy propanol and adipic acid or sebacic acid. The polyamide, Emerez 1540, was obtained as a reaction product of a dimeric acid with a linear diamine. The four encapsulated toners with their compositions are given in Table 1. The molecular weights, tensile moduli (E), shear moduli (G), compressive moduli (K), and Poisson ratios are given in Table 2.

The molecular weights were determined by gel permeation chromatography. The zero shear rate viscosities were measured with a Shirley-Ferranti viscometer. Tensile moduli and compressive moduli were determined by using an Instron and by shear moduli through the use of a Rheovibron.

261

Table 1. Material's Composition (by Weight)

Material	Composition by Weight
A	PS-2/Dow Adipate, PP578/carbon black 47.5/47.5/5.0
B	PS-2/Dow Adipate, PP582/carbon black 47.5/47.5/5.0
C	Emerez 1540/Dow Adipate PP583/carbon black 48/48/4.0
D	Styron 678/Dow Sebacate PP610/carbon black 47.5/47.5/5.0

Table 2.   Properties of Materials

Polymer	Temperature	M_w 10^{-3}	M_n 10^{-3}	ν	η $\times 10^{-7}$ poise	E $\times 10^{-8}$ dynes/cm^2	G 10^8 dynes/cm^2	K 10^{-8} dynes/cm^2
PP578 Dow Adipate	22°C	8.62	2.20	0.45	7.0	8.62	2.97	28.7
PP583 Dow Adipate	22°C	4.40	2.10	0.45	0.48	4.03	1.39	13.42
PP582 Dow Adipate	22°C	3.80	1.55	0.45	0.40	3.83	1.32	12.76
PP610 Dow Sebacate	22°C	5.68	2.38	0.45	0.02	1.5	0.51	5.0
Emerz 1540, polyamide	22°C	20.80	6.70	0.35	0.2	35.73	13.23	39.70
PS-2, polystyrene	22°C	22.0	6.5	0.33		187.6	56.51	184.0
Styron 678, polystyrene	22°C	250.0		0.33		201.6	60.72	197.6

COMPOSITE MODULI OF POLYMERS

There have been several models put forth to relate composite moduli to the volume concentrations of the components and the moduli of the individual components. Two recent reviews have dealt exhaustively with this subject.[7,8] The three different types of approximate formulas in use are (1) the Hashin and Shtrikman bounds,[9] (2) the model of Budiansky,[10] and (3) the law of mixtures.

$$G = G_m + \cfrac{\varphi_f}{\cfrac{1}{G_f - G_m} + \cfrac{6(K_m + 2G_m)\varphi_m}{5G_m(3K_m + 4G_m)}}$$

$$K = K_m + \cfrac{\varphi_f}{\cfrac{1}{K_f - K_m} + \cfrac{3\varphi_m}{3K_m + 4G_m}}$$

Highest Lower Bound, ------- Hashin Shtrikman

$$G = G_f + \cfrac{\varphi_m}{\cfrac{1}{G_m - G_f} + \cfrac{6(K_f + 2G_f)\varphi_f}{5G_f(3K_f + 4G_f)}}$$

$$K = K_f + \cfrac{\varphi_m}{\cfrac{1}{K_m - K_f} + \cfrac{3\varphi_f}{3K_f + 4G_f}} \quad \text{------- } \begin{array}{l} \text{Lowest Upper Bound,} \\ \text{Hashin Shtrikman} \end{array}$$

$$\frac{1}{G} = \frac{1}{G_m} + \left(1 - \frac{G_f}{G_m}\right)\frac{\varphi_f}{G + \beta(G_f - G)}$$

$$\frac{1}{K} = \frac{1}{K_m} + \left(1 - \frac{K_f}{K_m}\right)\frac{\varphi_f}{K + \alpha(K_f - K)}$$

$$\alpha = \frac{3K}{3K + 4G} \; ; \; \begin{array}{l} G, \, K, \, \varphi \text{ are shear modulus, bulk modulus and} \\ \text{volume fraction for matrix (m) and filler (f)} \end{array}$$

$$\beta = \frac{6(K + 2G)}{5(3K + 4G)} \quad \text{------------------------- Budiansky}$$

$$G = G_m + (G_f - G_m)\varphi_f$$

$$= (1 - \varphi_f)G_m + \varphi_f G_f \quad \text{--------------- Law of Mixtures}$$

On comparing composite compressive moduli with those obtained using different models (Table 3), it is clear about the applicability of the different models. The law of mixtures is applicable only where the ratio of the two shear moduli is very small, i.e., $G_f/G_m \simeq 3$. The Hashin-Shtrikman lowest upper bound agrees with the experimental results up to $G_f/G_m < 6.4$, but it fails when G_f/G_m is 14.4 or 39.5. The model of Budiansky appears to be applicable under the different ratios of G_f/G_m experimentally obtained in this study. This work thus shows that the model of Budiansky appears to be useful for a large class of composite materials.

PARTICLE DEFORMATION, CARBON BLACK, AND HEAT TREATMENT

A spherical particle supported on a flat, rigid substrate (glass) flattens when deformed by a flat, rigid probe (glass). The magnitude of force beyond which the particle flattens without significant increase in the applied force is termed the yielding force.

Table 3. Composite Moduli

Material	G_f/G_m	Temperature	V_1/V_2	K_H	K_M	K_B	K_E
	Shell/Core				10^{-8} dynes/cm^2		
PS-2/PP578	6.4	22°C	1/1	52.5	106.3	57.4	50.4
PS-2/PP582	14.4	22°C	1/1	25.2	98.3	61.2	61.9
Sty 678/PP610	39.5	22°C	1/1	10.4	103.3	51.0	51.6
Emerez 1540/ PP583	3.0	22°C	1/1	20.5	26.6	18.8	19.7

K_H = bulk modulus – Hashin and Shtrikman (1963)

K_M = bulk modulus – Law of Mixtures

K_B = bulk modulus – Budiansky (1965)

K_E = bulk modulus – Experimental

Figure 1 shows yielding force with particle size for particles with and without carbon content and particles subjected to heat treatment. The heat treatment consisted of subjecting spherical particles to 85°C (Tg + 20°C) in water and Triton X-100, non-ionic surfactant. The suspension was quenched with ice water. The particles are recovered by centrifugation, washed to remove sur- factant, and dried. The heat treatment reduced the number of cores and decreased the concentration of microvoids. The material A is an encapsulated, composite material with polystyrene, PS-2, as shell and polyester, Dow Adipate, as the core. The material A_c has the same polymer composition as A, but it contains 5% by weight of carbon black. The difference between curves A and A_c is in the same neighborhood as the difference between A_c and A_{cr}, a replicate of A. However, there is a significant difference between A and A_{ch}, a deformation curve of heat-treated particles. The decrease in the concentration of microvoids in the core would toughen the particles and increase the yielding force for various size parti- cles.

PARTICLE DEFORMATION, MOLECULAR WEIGHT, AND CHEMICAL STRUCTURE

Figure 2 shows the effect of increasing yielding force on in- creasing particle sizes with shell and core polymers differing in molecular weight and chemical structure. The material A consists of the polystyrene, PS-2, as shell and the polyester, Dow Adipate PP578, as core, The material B has the same molecular weight polystyrene as the shell with the lower molecular weight PP582 as

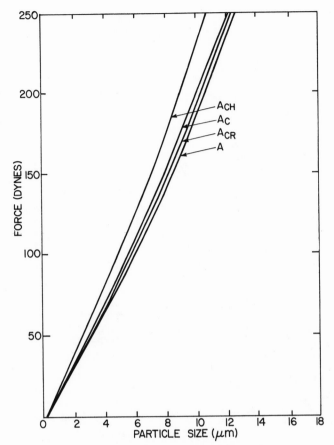

Fig. 1.   Composite particle deformation versus carbon black content
          or heat treatment of particles.

the core.   The material C has the polyamide, Emerez 1540, as the
shell and Dow Adipate PP583 with a molecular weight close to PP582
as the core.   The material D has the polystyrene, Styron 678, as
the shell and the polyester, Dow Sebacate PP610, as the core.   The
Styron 678 has about a decade higher molecular weight than PS-2,
whereas Dow Sebacate PP610 has a lower molecular weight and a softer
chemical structure (Tg) than Dow Adipate PP578.

     The curve A is less concave than curve B because of the higher
molecular weight of the core material, the higher modulus of the
core, and a stronger encapsulated particle.   The curve B is less
concave than curve C because the polystyrene shell in B has a higher
modulus than the polyamide shell in C.   The curve D is only slightly
more concave than curve A because the increase in the modulus of the
shell with Styron 678 versus PS-2 is compensated for by a decrease
in modulus with polyester Dow Sebacate PP610 versus polyester Dow
Adipate, PP578.

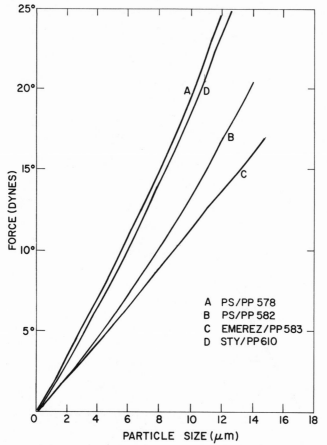

Fig. 2.   Composite particle deformation versus shell and core
           structure.

YIELD STRESSES OF ENCAPSULATED, COMPOSITE PARTICLES

An encapsulated, composite particle would yield when shear
stress becomes maximum. Thomas and Hoersch[6] have analyzed contact
stresses when two spheres come in contact under an application of
a normal force. The tensile stresses, σ_x, σ_y, and σ_z are maximum
at the surface, whereas, τ_{xz} shear stresses are maximum at a depth
of about half of the contact length a.

At z = 0.47a, τ_{xz} is maximum and is given by

$$\tau_{xz} = \frac{0.908}{\pi} \frac{\left[(1+\nu^*)(0.47\cot^{-1} 0.47-1) + 1.22\right]}{\left[\left(\frac{1-\nu^{*2}}{E^*} + \frac{1-\nu^2}{E}\right)\right]^{2/3}} \cdot \left(\frac{P}{R^2}\right)^{1/3},$$

$1/R = 1/R_1 + 1/R_2$, where, R_1 and R_2 are the radii of the two spheres, ν and ν^* are the Poisson ratios, E^* and E are the Young's moduli, and P is the yielding force. For flat glass on a sphere, $\nu = 0.26$ and $E = 1.8 \times 10^{10}$ dynes/cm^2.

The compressive yield stresses of single composite particles are compared with bulk compressive yield stresses, and these are shown in Table 4. It is clear that for the composite materials A, B, and C, the bulk yield stresses are not significantly dependent upon crosshead speed. In case of material D, where both the components are ductile, the yield stress is quite significantly dependent upon crosshead speed.

The single particle yield stresses show reasonably close-to-the-bulk yield stresses when compared for the same crosshead speed. The lowering of the average strength in single particles can be considered to be due to the presence of microvoids. As we have seen earlier, the heat-treated particles increased in strength because of the decrease in microvoid content.

Table 4. Composite Single Particle vs Bulk Yield Stresses

Material	Temperature	Single Particle Yield Stress (0.5 in./min.) 10^{-8} dynes/cm^2	Bulk Yield Stress 0.05 in./min. 10^{-8} dynes/cm^2	0.5 in./min.
PS-2/PP578	22°	1.4	1.3	1.6
PS-2/PP582	22°C	1.1	1.3	1.7
Sty 678/PP610	22°C	1.2	1.4	1.6
Emerez 1540/PP583	22°C	0.63	0.34	0.55

REFERENCES

1. R. H. Snow, Powder Technol., 5: 351-364 (1972).
2. H. Rumpf, F. Faulhaber, K. Schönert, and H. Umhauer, Paper A3, Second European Symposium on Size Reduction, Dechema Monograph, 57 (1967).
3. K. Steier and K. Schönert, Third European Symposium on Size Reduction, Dechema Monograph, 69 (1972).
4. S. K. Ahuja, Powder Technol., 1 (1977).
5. R. E. Wellman and R. G. Crystal, U.S. Patent No. 378899Y (1974).
6. M. R. Thomas and V. A. Hoerch, Bull. 46, Vol. XXXII, University of Illinois (1960).
7. W. E. A. Davies, J. Phys. D: Appl. Phys., 4: 318-328 (1971).
8. D. K. Hale, J. Mat. Sci., 11: 2105-2141 (1976).
9. Z. Hashin, J. Appl. Mech., 143 (1962).
10. B. J. Budiansky, J. Mech. Phys. Solids, 13: 223 (1965).

ACOUSTICAL RESPONSE OF PARTICULATE-LOADED VISCOELASTIC COMPOSITES

E. K. Walsh, J. W. Nunziato, and S. L. Passman

Univ. of Florida (EKW), and Sandia Laboratories

Gainesville, Florida and Albuquerque, N. M., U.S.A.

INTRODUCTION

In a recent study we proposed a general thermomechanical theory describing the behavior of multiphase mixtures which includes, for example, the response of a viscoelastic matrix with particulate inclusions.[1] Although the theory is a continuum theory, it considers the discrete nature of the phases of the mixture by associating with each material point a more complex structure to represent the interactions between phases. In particular the volume fraction of each phase is considered as an independent kinematic variable. This leads to additional force balance equations which relate to the interfacial force system and serve to describe the microstructural behavior of the mixture.

Here we consider a linear approximation of the theory and apply it to the propagation of small amplitude acoustic waves in a two-phase composite mixture of elastic particles in a viscoelastic matrix. We derive the dispersion relations for the wave speed and attenuation as functions of frequency for both longitudinal and shear waves. The results show that shear waves propagate without affecting the volume fractions or the temperatures of the phases and obey the propagation conditions for acoustic waves in a linear viscoelastic material with the relaxation behavior given by the shear relaxation modulus. Meanwhile the propagation condition for the longitudinal waves is influenced by the shear, longitudinal, and bulk relaxation behavior of the matrix material and includes a dissipative mechanism due to energy transfer associated with intergranular friction. It is interesting that the bulk relaxation function for the viscoelastic matrix also affects changes in the volume fractions.

269

In a later application of these results we will consider the effects of temperature on acoustic dispersion in two-phase composite materials by using time-temperature superposition for the viscoelastic phase. It is clear from the general results that the shear waves in the mixture would time-temperature superpose in the same manner as the viscoelastic matrix. However, the additional influences in the longitudinal waves as a result of multiple phases would affect the shifting of the dispersion relations. In particular the relaxation behavior associated with changes in the volume fraction may explain the differences in the activation energies used to determine the shift factor for shear and longitudinal waves in a study of a two-phase particulate-loaded viscoelastic composite.[2]

FIELD EQUATIONS AND CONSTITUTIVE EQUATIONS

We consider a two-phase mixture consisting of elastic particles in a linear viscoelastic matrix without voids. Diffusion between the constituents is neglected but account is taken of interaction between the phases through interfacial forces and energy exchange.

Neglecting diffusion, we describe the motion of the body by the function $x = \hat{x}(X,t)$, which gives the position x of the particle which occupied the position X in the reference configuration. Then $v = \dot{x} = \partial_t \hat{x}(X,t)$ is the velocity and $\dot{v}$ the acceleration of the particle. Furthermore, $F = \nabla \hat{x}(X,t)$ represents the deformation gradient and $L = \dot{F} F^{-1}$ the velocity gradient. The balance of mass for the ath constituent, $a = 1, 2$, is

$$\frac{\rho^{\circ}}{\rho} = \frac{\rho^{\circ}_a}{\rho_a} = \left| \det F \right| \tag{1}$$

where $\rho_a = \gamma_a \phi_a$ is the partial density of constituent a with γ_a the actual material density and ϕ_a the volume fraction of constituent a. In (1) ρ° represents the reference value of the density. The quantity ρ gives the mixture density and is related to the partial densities by, $\rho = \sum \rho_a$.* Also, we define a mixture volume fraction, $\phi = \sum \phi_a$. By assuming there are no voids in the mixture it is said to be saturated and $\sum \phi_a = 1$.

The balance of linear momentum for the mixture is given by

$$\rho \dot{v} = \text{div } T, \qquad T = T_1 + T_2 \tag{2}$$

where we have neglected external body forces. The volume fraction

*Here $\sum = \sum_{a=1}^{2}$

ϕ is an independent kinematic variable in the theory which motivates a balance law to describe the dynamical effects associated with changes in the volume fractions.[1] This equation, for the ath constituent, takes the form

$$\rho_a(k_a\dot{\phi}_a)^{\cdot} = \mathrm{div}\, \underset{\sim}{h}_a + v_a^+. \tag{3}$$

Here k_a is an equilibrated inertia, $\underset{\sim}{h}_a$, an equilibrated stress, and v_a^+, an equilibrated force interaction.

The balance of energy e_a for constituent a is given by

$$\rho_a\dot{e}_a = \mathrm{tr}(\underset{\sim}{T}_a^t\underset{\sim}{L}) + \underset{\sim}{h}_a\cdot\mathrm{grad}\dot{\phi}_a + \tfrac{1}{2}\rho_a k_a\dot{\phi}_a^2 - v_a^+\dot{\phi}_a + \rho_a r_a + e_a^+, \tag{4}$$

where r_a is the external heat supply and e_a^+ the energy interaction, both for constituent a.

The formulation of the constitutive assumption for the mixture is as follows. We assume that the free energy of the elastic particles ψ_1, taken as constituent 1, is determined by the volume fraction ϕ_1, the temperature θ_1, and the material density γ_1, all of constituent 1; i.e.,

$$\psi_1 = \hat{\psi}_1(\phi_1, \theta_1, \gamma_1). \tag{5}$$

The free energy of phase 2, the viscoelastic matrix, is given by

$$\psi_2 = \hat{\psi}_2(\phi_2, \theta_2, \gamma_2, (\gamma_2^t)_r, \underset{\sim}{H}, \underset{\sim}{H}_r^t) \tag{6}$$

where $\underset{\sim}{H} = (\mathrm{det}\, \underset{\sim}{B})^{1/3}\underset{\sim}{B}$ and $\underset{\sim}{B} = \underset{\sim}{F}\underset{\sim}{F}^t$. The notation $\underset{\sim}{H}_r^t$ refers to the restricted history of $\underset{\sim}{H}(t-s)$ and $\underset{\sim}{H}$ then represents the present value (similarly for γ_2).

Certain restrictions on these constitutive assumptions are obtained by requiring that the balance equations and the second law of thermodynamics be satisfied. These restrictions can be obtained from the more general results given previously.[1] Of particular significance to the present study is that the stress in the particles is hydrostatic and is given in terms of a pressure p_1, the equilibrated stresses $\underset{\sim}{h}_a$ are zero, and the constituent entropies are determined by the free energies.

LINEAR THEORY

In this section we linearize the above equations for the subsequent analysis of small amplitude acoustic waves. Toward this end we consider motions for which

$$|\underset{\sim}{u}| \; , \; |\text{grad } \underset{\sim}{u}| \; , \; |\dot{\underset{\sim}{u}}| \; , \; |\theta_a - \theta^\circ| \; , \; |\phi_a - \phi_a^\circ| \; , \; |\dot{\theta}_a| \; , \; |\dot{\phi}_a| << \delta,$$

where $\underset{\sim}{u}$ is the displacement vector and $\delta << 1$. For brevity, the explicit forms of the field equations will not be written out. However, the general form of the momentum equation is given as

$$\rho^\circ \ddot{\underset{\sim}{u}} = F_1(\phi_a^\circ, \; \underset{\sim}{g}_a, \; B_2(s), \; \lambda_2(s), \; \mu_2(s)). \tag{7}$$

In this equation $g_a = \text{grad } \theta_a$ are the constituent temperature gradients and $B_2(s)$, $\lambda_2(s)$, and $\mu_2(s)$ are the bulk, longitudinal, and shear relaxation functions for the viscoelastic phase.

Similarly, the interfacial force equation, written in terms of ξ_a, where ξ_a is related to ϕ_a by, $\phi_a^\circ = \phi_a(1 + \xi_a)$, can be written as

$$\phi_1^\circ \ddot{\xi}_1 = F_2(\rho_a^\circ, \; \overset{\circ}{\phi}_a, \; \overset{\circ}{k}_a, \; \underset{\sim}{u}, \; \theta_a, \; B_2(s)). \tag{8}$$

Note that, due to the constraint of saturation, only one equilibrated force equation is necessary. Further, it is of interest to note that the volume fraction behavior depends on the bulk stress relaxation function.

Finally, the heat equations become

$$\dot{\theta}_a = G(\rho_a, \; \gamma_a, \; \dot{\underset{\sim}{u}}, \; \theta_a, \; \dot{\xi}_1) \tag{9}$$

resulting in four equations in $\underset{\sim}{u}$, ξ_1, θ_1, and θ_2.

ACOUSTIC WAVES

To investigate the propagation of small-amplitude acoustic waves, we look for solutions of (7)-(9) of the form

$$\underset{\sim}{u} = \underset{\sim}{a}\bar{u}W, \; \xi_1 = \bar{\xi}_1 W, \; \theta_a = \theta_a^\circ + \bar{\theta}_a W, \tag{10}$$

where

$$W = \text{Re}\left\{ \exp\left[-\left(\alpha + \frac{i\omega}{U}\right) \underset{\sim}{m} \cdot \underset{\sim}{x} \right] \exp(i\omega t) \right\}. \tag{11}$$

The vector $\underset{\sim}{a}$ is the direction of displacement with $\underset{\sim}{m}$ the direction of propagation. Thus, for transverse waves $\underset{\sim}{a} \cdot \underset{\sim}{m} = 0$, for longitudinal waves $\underset{\sim}{a} \cdot \underset{\sim}{m} = 1$. The quantities $\bar{u}$, $\bar{\xi}_1, \bar{\theta}_a$ are the amplitudes. Calculating the necessary time and space derivatives of

(10) and substituting in (7)-(9) leads to four equations in the four amplitudes.

Transverse Waves

Substituting the condition $\underset{\sim}{a} \cdot \underset{\sim}{m} = 0$ in this latter system of equations leads to a reduced set of equations for the amplitudes $\bar{u}$, $\bar{\xi}_1$, $\bar{\theta}_1$, $\bar{\theta}_2$ for transverse waves. It is particularly noteworthy that only the equation for $\bar{u}$, which takes the form

$$\left\{ \rho^\circ \omega^2 + \phi_2^\circ \left[\mu_2(0) + \bar{\mu}_2'(\omega) \right] \left(\alpha + \frac{i\omega}{U} \right)^2 \right\} \bar{u} = 0 \qquad (12)$$

with $\bar{\mu}_2'(\omega)$ the Fourier transform of $\mu_2'(s)$, involves the quantities α and $\bar{U}$. With $\bar{u} \neq 0$ this is the propagation equation for small-amplitude progressive waves in a homogeneous linear visco-elastic material for which the dispersion relations for $U^2(\omega)$ and $\alpha(\omega)$ are known[3]. Remaining are three equations in the unknown amplitudes $\bar{\xi}_1$, $\bar{\theta}_1$, $\bar{\theta}_2$. For a non-trivial solution the determinant of the coefficients must vanish. However, with the coefficients independent of α and U it is always possible to find values of ω and material coefficients for which the determinant does not vanish, implying that the amplitudes are zero. That is, small-amplitude transverse progressive waves propagate without affecting the volume fractions or the temperatures of either phase.

Longitudinal Waves

Using the condition $\underset{\sim}{a} \cdot \underset{\sim}{m} = 1$ in the four equations for the unknown amplitudes yields a complete system of coupled equations. For a non-trivial solution to this system the determinant of the matrix of the coefficients must vanish. This leads to a determinantal equation which can be put in the form

$$-\rho^\circ \omega^2 \Gamma(\omega) = \left(\alpha + \frac{i\omega}{U} \right)^2 \Delta(\omega)$$

where Γ and Δ are complex functions of ω and involve the material coefficients and the equilibrium values of the constituent temperatures and densities. Expanding this equation and separating into real and imaginary parts leads to two equations which determine the dispersion relations for the longitudinal wave speeds and the attenuation in terms of the real and imaginary parts of $\Gamma(\omega)$ and $\Delta(\omega)$.

As expected, these results are similar in several aspects to those obtained earlier for acoustic waves in granular solids.[4] In particular, in the low frequency limit the wave speed is influenced by the multiphase nature of the mixture while the high frequency wave speed corresponds to the speed of acceleration waves in the same material. The results also show that there exists a resonant frequency based on the characteristic length of the mixture which serves to delineate two distinct regions of wave speed behavior. The attenuation is zero at the low frequency limit and increases with frequency to an upper limit associated with the energy transfer between the phases. It is important to note that the additional relaxation behavior associated with these results affects the manner in which time-temperature superposition would be applied to the mixture as compared to the matrix material alone.

ACKNOWLEDGEMENT

This study was supported in part by the U. S. Department of Energy and in part by the U. S. National Science Foundation. Sandia Laboratories is a facility of the U. S. Department of Energy.

REFERENCES

1. J. W. Nunziato and E. K. Walsh, On Ideal Multiphase Mixtures with Chemical Reactions and Diffusion, Arch. Rational Mech. Anal. (forthcoming)
2. H. J. Sutherland, Dispersion of Acoustic Waves by an Alumina-Epoxy Mixture, J. Comp. Mat. 13:35 (1979).
3. B. D. Coleman and M. E. Gurtin, On the Growth and Decay of One-Dimensional Acceleration Waves, Arch. Rational Mech. Anal. 19:239 (1965).
4. J. W. Nunziato and E. K. Walsh, Small-Amplitude Wave Behavior in One-Dimensional Granular Solids, J. Appl. Mech. 44:559 (1977).

THE ROLE OF CHAIN ENTANGLEMENTS AND CRYSTALS IN THE ORIENTATION PROCESS OF POLYMERS

Hans Gerhard Zachmann, Gerd Elsner, and
Harald Jürgen Biangardi

Abteilung Angewandte Chemie, Universität Hamburg,
Martin-Luther-King-Platz 6, D 2 ooo Hamburg 13,
and Institut für Nichtmetallische Werkstoffe,
Technische Universität Berlin

INTRODUCTION

If an amorphous polymer is oriented by drawing, some sort
of netpoints must be present in order to prevent the
motion of complete chains in different directions. Usually
chain entanglements act as such net points. With some polymers, also small crystals formed during drawing may be of
influence. In the following we report on investigations
performed on polyethylene-terephthalate in order to determine the influence of chain entanglements and small
crystals on the orientation. Polyethylene terephthalate
has a glass transition temperature of about 70° C. It can
be obtained in the amorphous state if quenched from the
melt, and it may be oriented without necking by drawing[1]
in the amorphous state at temperatures between 8o and
110° C.

UNIAXIAL DEFORMATION

Amorphous films of polyethylene-terephthalate obtained
from the company Kalle AG., Wiesbaden, with a molecular
weight M_w = 28.ooo and 3o nm thick were drawn at different
temperatures T_v with different drawing rates w.
The orientation obtained was characterized by the birefringence Δn_o of the samples. Fig.1 shows the birefringence
as a function of the drawing ratio for different drawing
rates and drawing temperatures. One can see that the
orientation is not only determined by the drawing ratio λ

but that it is influenced also markedly by the other drawing conditions.

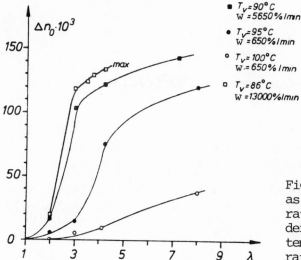

Fig.1: Birefringence Δn_o as a function of drawing ratio λ after uniaxial deformation. T_v = drawing temperature, w = drawing rate.

These results can be explained by the fact that, at the
beginning of the drawing, no stable netpoints are present.
Orientation occurs only because chain entanglements act
as temporary netpoints, which, due to segmental motion,
have only a comparatively small average life time; they
are formed, they disappear, they are formed again, and so
on. Therefore relaxation can take place during drawing.
With increasing drawing rate and decreasing drawing temperature, the relaxation effects become smaller, the number
of chain entanglements formed during drawing increases,
and the orientation obtained becomes better. The lowest
temperature and largest rate at which the material can
be drawn without obtaining necking ly at about T_v = 86°C
and w = 13.000 %/min.

During drawing also some crystallization takes place.
Fig.2 shows the degree of crystallization $x_c^{)}$ obtained
from density measurements assuming that the density of
the crystals is given by the values of Huisman and Heu-
vel[2] and that of the noncrystalline regions by 1.335 g/cm^3 .

The crystals formed act as permanent netpoints. Therefore,
at drawing ratios larger than 2, the orientation process
is influenced not only by the chain entanglements. More
information on the orientation mechanism can be obtained
if one determines the Herrmann orientation function of the
crystals, f_c, by x-ray wide angle scattering, and, sepa-

rately, the orientation function f_a of the noncrystalline regions. f_a is given by the difference of total orientation

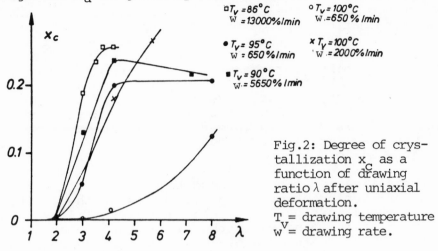

Fig.2: Degree of crystallization x_c as a function of drawing ratio λ after uniaxial deformation.
T_v = drawing temperature
w = drawing rate.

obtained by birefringence, and f_c. One sees that f_c is almost equal to 1, independent on total rientation, while f_a increases gradually with increasing birefringence. From this it follows that for $\lambda > 2$, the increase in orientation with increasing drawing ratio has two reasons:
(1) the increase of the orientation of the noncrystalline regions, (2) the increase of amount of crystals with almost perfectly oriented chains.

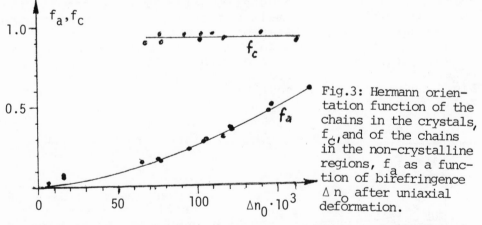

Fig.3: Hermann orientation function of the chains in the crystals, f_c, and of the chains in the non-crystalline regions, f_a as a function of birefringence Δn_0 after uniaxial deformation.

The existence of small crystals manifests itself also in the investigation of the relaxation after orientation. If no crystals are present the birefringence decreases upon annealing. In the presence of crystals no decrease takes place.

CRYSTALLIZATION DURING AND AFTER UNIAXIAL DEFORMATION

According to fig.2. the degree of crystallinity obtained
during deformation depends in a complicated manner on the
temperature, rate and ratio of drawing. However, if one
compares the results with those in fig.1, one sees that it
is mainly the birefringence which determines the degree of
crystallinity. The temperaure and rate of drawing are only
of minor influence;at the largest temperature and the lowest rate (T_V = 1oo$^{\circ}$C and w_V = 65o %/min.) the degree of
crystallinity is a little larger than at the same birefringence obtained with larger rates and lower temperatures.
Up to Δn_o= 2o.1o^{-3} almost no crystallization is obtained.

The crystallization process seems to proceed in the following manner: during orientation some parts of the chains
are oriented almost perfectly and form crystal-like bundles.
The number of the bundles is the larger the higher the
orientation is. These bundles act as crystal nuclei; therefore with increasing time, an additional crystallization
takes place during which other amorphous chains are attach-
ted to the bundles and, at the same time, oriented almost
perfectly. According to earlier investigations[4] the rate
of this crystallization increases markedly with temperature
and orientation. Due to this crystallization the degree
of crystallinity at the same birefringence is the larger,
the higher the temperature and the lower the rate of drawing.

Further details in the crystallization process can be obtained if one studies the small angle x-ray scattering during
an isothermal crystallization process following the drawing.
We were able to measure this scattering every ten seconds
by using the synchrotron radiation at DESY in Hamburg as a
powerful source and a vidicon system as a receiver
Fig.4 shows the azimuthual half width of the small angle[5]
reflection.

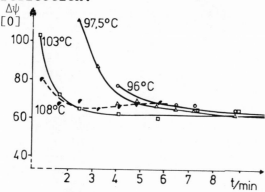

Fig.4: Azimuthual half
width $\Delta\psi$ of small angle
x-ray reflection as a
function of crystalli-
zation time t for un-
iaxially deformed mat-
erial with initial bire-
fringence Δn_o= 2o.1o^{-3}.
The parameter is the
temperature of crystal-
lization.

One can deduce from these results and from simultaneous
measurements of the long period that, with increasing crystallization time, the thickness of the crystals and non-
crystalline regions become smaller and the orientation of
the lamellae becomes better. At the same time, from wide
angle scattering, we deduce that the orientation of the
chains is almost perfect from the beginning. A model which
can explain these results is shown in fig.5. It is assumed
that, at the beginning of the crystallization process, the
lamellae are bended. With increasing time they become flat

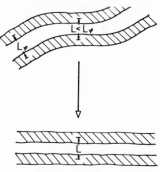

Fig.5: Form of crystal lamellae
at the beginning of crystallization (above) and at the end
of crystallization (below).

while the orientation of the chains is not changed.

BIAXIAL DEFORMATION

Also the dependence of biaxial orientation on the drawing
conditions has been investigated. The amorphous films of
polyethylene terephthalate described in the previous section were oriented first in z-direction at the temperature
T_{vz} = 92°C with a rate w_z= 4oo %/min. and a draw ration of
λ_z^{vz}= 4.4 and λ_z = 2.8 giving samples with a birefringence
of 95.1o^{-3} and 32.1o^{-3} respectively. Afterwards a second

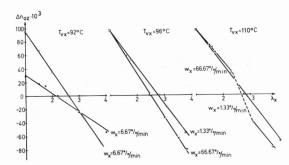

Fig.6: Birefringence
Δn_{oz} of biaxially deformed material as a
function of the draw
ratio λ_x at the second
deformation. T_{vx} = temperature and w_x^{vx}= rate
of drawing at the second
deformation.

orientation in x - directions, perpedicular to the
z-direction, was performed, with drawing temperature T_{vx},
drawing rate w_x and drawing ration λ_x. Fig.6 shows the

obtained birefringence measured in z-direction as a function of the drawing ratio λ_x with the other drawing conditions as a parameter. One sees that, with increasing
drawing ratio the birefringence becomes smaller, reaches
the value zero and becomes negative. Negative values
mean that the chains are oriented in x-direction.

Of special interest is the result that the birefringence
obtained is almost completely determined by the draw-
ratio λ_x and is almost not affected by the temperature
and rate of drawing. The situation here is different to
the situation of the first drawing (fig.1). The difference
is due to the fact that, during first drawing crystals
have been formed which now act as permanent netpoint.
Chain entanglements are only of minor importance for the
second drawing.

CONCLUSIONS

In order to derive quantitative relations between the
drawing conditions and the birefringence obtained in
amorphous polymers, one has to know the number and the
lifetime of chain entanglements. Such information can be
obtained from dynamic mechanical measurements as well as
from magic angle spinning NMR investigations[6].If crystal
nuclei are formed during drawing, one has also to know
the amount of such crystals. The orientation mechanism
in amorphous polymers is completely different from that
in crystalline polymers where, during drawing, crystal
blocks are separated and rearranged again. Therefore,
in crystalline polymers, the orientation obtained is
determined mainly by the drawing ratio while in amorphous
polymers also the drawing rate and the drawing termpera-
ture strongly influence the results.

1. Biangardi H.J., H.G.Zachmann,Progr.Coll.Polym.Sci.62,
   71 (1977)
2. 2. Huisman R., H.M.Heuvel, J.Appl.Polym.Sci.22,943(1978)
3. Stein R.S., F.H.Norris, J.Polym.Sci. 21,381 (1956)
4. G.Althen and H.G.Zachmann, Makromol.Chemie 180,2723(1979)
5. G.Elsner and H.G.Zachmann, IUPAC Symposium, Mainz 1979
6. R.Müller and H.G.Zachmann, Colloid and Polymer Sci.,
   in press

DEFORMATION INDUCED VOLUME RELAXATION IN A STYRENE-BUTADIENE

COPOLYMER

H.G. Merriman and J.M. Caruthers

School of Chemical Engineering
Purdue University
West Lafayette, Indiana 47907

INTRODUCTION

The viscoelastic behavior of a solid polymer in the vicinity of the glass transition temperature T_g depends upon the temperature, the level of deformation, and the thermal history (1-4). It has been suggested that all of these effects can be described by the free volume theory (5). The effect of temperature on the relaxation process is given by

$$\log \tau/\tau_0 = \log a_T = \frac{B}{2.303} \left(\frac{1}{f} - \frac{1}{f_o}\right) \tag{1}$$

where τ_0 is a relaxation time at a reference temperature T_0, τ is the same relaxation time at an arbitrary temperature T, a_T is the time-temperature shift factor, B is a constant usually equal to 1, f is the fractional free volume, and f_0 is the fractional free volume at the reference temperature. The fractional free volume is usually defined as

$$f = v_f/v_o = f_o + \Delta v/v_o \tag{2}$$

where v_f is the free volume, v_o is the total volume at T_0, $\Delta v = v - v_o$, and v is the volume at T. If the fractional free volume is a linear function of temperature, Equation 1 reduces to the WLF equation (1). Alternative forms of the free volume equation have been reviewed by Struik (2) and Kovacs, et al (6,7).

The effect of nonisothermal conditions on the linear viscoelastic behavior of solid polymers is described by an effective material time t* (8,9):

$$t^* = \int_0^t \frac{d\xi}{a_T[T(\xi)]} \tag{3}$$

where a_T is the time-temperature shift factor and $T(t)$ is the thermal history. The effective material time is the time it would take at the reference temperature to achieve the same deformation as is observed in time t in the nonisothermal experiment.

A solid polymer that is deformed near T_g exhibits nonlinear viscoelastic behavior for relatively small deformations (2-4). Moreover, a volume increase is observed, when a solid polymer is deformed at or below T_g. If the observed dilation results in an increase of the free volume, an effective material time for nonlinear relaxation processes can be computed by combining Equation 1 through Equation 3. It has been assumed the deformation induced dilation is equivalent to thermal effects. Thus, the nonlinear viscoelastic properties can be related to the linear viscoelastic behavior via the free volume theory (2-4).

EXPERIMENTAL

A styrene-butadiene random copolymer (85% styrene by weight) was dry blended with 0.1 weight % of benzoyl peroxide crosslinking agent. The mixture was compression molded and cured for 2 hours at 150°C. The sol fraction was less than 10% by weight. The glass transition temperature was determined with a dialatometer to be 43.3°C. All the tensile specimens used in this study were machined from a single compression molded sheet.

The stress relaxation response of the styrene-butadiene copolymer was measured in an Instron tensile tester. The specimen temperature was controlled within ± 0.1°C from 20°C to 90°C. The specimen was annealed for at least 2 hours at 60°C, almost 20°C above T_g, and then the temperature was lowered to the test temperature. The volume change upon deformation was determined by measuring both the axial and lateral strain. The axial strain was monitored with an LVDT. The time-dependent lateral strain was determined with a Hall effect proximity detector (10). Briefly, the Hall device is a semiconductor with an output voltage proportional to the intensity of a magnetic field. A Hall device was placed on one side of the specimen and a very small magnet was attached to the other; thus, the lateral dimension and thereby the volume was accurately measured as a function of time on a strip chart recorder.

RESULTS

The stress relaxation response at 50.1°C is illustrated in

Figure 1 for the styrene-butadiene random copolymer. The time-
dependent volume relaxation was determined with the Hall effect
device. Volume relaxation data at 50.1°C and 41.3°C are shown in
Figures 2 and 3 respectively. Linear and nonlinear stress relaxation data and volume relaxation data at other temperatures will be
presented elsewhere (11).

DISCUSSION

   The stress relaxation response illustrated in Figure 1 is
highly nonlinear even at relatively small strains. The modulus
decreases as the applied strain increases; however, at long times
the modulus at large strains begins to approach the linear viscoelastic response. The stress relaxation response is accelerated
by large strains; for example, at 50.1°C the modulus at 24 seconds is 10^7 Pa for an applied strain of 0.14, while at a strain of
0.017 the same modulus is reached at 220 seconds – almost ten
times longer. The acceleration of the viscoelastic processes with
large strains is consistent with the free volume theory.

   The deformation induces an increase in the specific volume
of the polymer   The volume increase is a maximum immediately
after the deformation and subsequently decreases with time. The
dilation increases with the level of applied strain. The volume
increase at short times depends upon the applied strain, but it is
relatively insensitive to the temperature. However, the volume

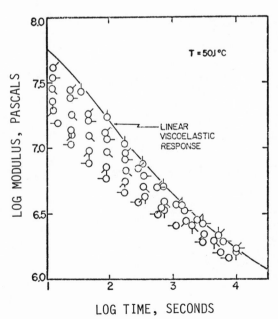

Figure 1. Stress relaxa-
tion modulus at 50.1°C
for various levels of
applied strain. Pips,
beginning with pip upward
and proceeding clockwise
at 45° intervals, indicate
the following strains:
0.0170, 0.0260, 0.0475,
0.0645, 0.105, 0.117,
and 0.140.

relaxation is more pronounced at lower temperatures.

Assuming that all the free volume created during deformation
is effective at short times, the fractional free volume in the
reference state can be determined by combining Equations 1 and 2.
The shift factor $\log a_\epsilon$ is analogous to the time-temperature shift
factor $\log a_T$. At short times $\log a_\epsilon$ is the horizontal shift
required to superpose nonlinear modulus data at 12 seconds with
the linear viscoelastic modulus. The $\log a_\epsilon$ shift factor is
shown in Figure 4 as a function of the volume increase at short
times. The shift factors predicted by the free volume theory
are computed from Equations 1 and 3 for B = 1 and various values
of f_o. As seen in Figure 4, the fractional free volume in the
undeformed state at 50.1° C is 0.20. The fractional free volume
at 41.3°C was determined to be 0.12 by a similar procedure. The
fractional free volume at T_g was predicted to be 0.113 by Simha
and Boyer (12).

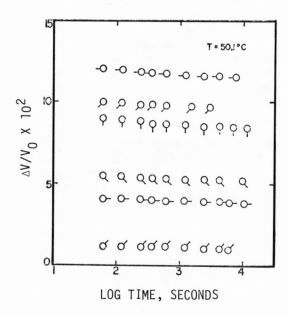

LOG TIME, SECONDS

Figure 2.   Volume relaxation at 50.1°C for various levels of
applied strain. Pips refers to the same strains as in
Figure 1.

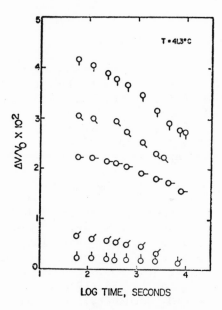

Figure 3.   Volume relaxation at 41.3°C for various levels of
            applied strain. Pips, beginning with pip upward and
            proceeding clockwise at 45° intervals, indicate the
            following strains: 0.0023, 0.0083, 0.0290, 0.0352
            and 0.0480.

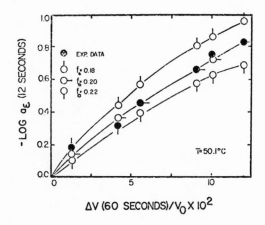

Figure 4.   Log a_ε as a function of the fractional free volume.

REFERENCES

1. J.D. Ferry, "Viscoelastic Properties of Polymers", John Wiley,
    NY, (1970).
2. L.C.E. Struik, "Physical Aging in Amphorous Polymers and
    Other Materials", Elsevier, Amsterdam, (1978).
3. S. Matsuoka, H.E. Bair, and C.J. Aloisio, J. Poly. Sci.: Poly.
    Symp., 46, 115 (1974).
4. S. Matsuoka, H.E. Bair, S.S. Bearder, H.E. Kern, and J.T. Ryan,
    Poly.Eng. and Sci., 18, 1073 (1978).
5. A.K. Doolittle, J. Appl. Phys., 22, 1471 (1951).
6. J.M. Hutchinson and A.J. Kovacs, "Symp. Structure of Non-
    Crystalline Materials", P.H. Gaskell, ed., Taylor and Francis
    Ltd, London (1977).
7. A.J. Kovacs, J.M. Hutchinsen and J.J. Aklonis, "Symp.
    Structure of Non-Crystalline Materials", ibid. (1976).
8. I.M. Hopkins, J. Poly. Sci., 28, 631 (1958).
9. L.W. Moreland and E.H. Lee, Trans. Soc. Rheol., 4, 233 (1960).
10. M. Okuyama, K. Yagil, S.C. Sharda, and N.W. Tschogel,
    Poly. Eng. and Science, 14, 38 (1974).
11. H.G. Merriman and J.M. Caruthers, to be published.
12. R. Simha and R.F. Boyer, J. Chem. Phys., 37, 1003 (1962).

A DEFORMATION ANALYSIS OF A POLYETHYLENE CRYSTAL SUBJECTED TO END

FORCES OF STRETCHING AND LATTICE EXPANSION

Jeffrey T. Fong

U.S. National Bureau of Standards
Center for Applied Mathematics
Washington, DC   20234   U.S.A.

INTRODUCTION

In a recent study by Hoffman[1,2] on a theory of flow-induced fibril formation in polymer solutions, it was proposed that each crystallite could be idealized as a rectangular parallelepiped of length ℓ and a square cross-section of side a, and that the four lateral surfaces could each assume the usual lateral surface free-energy density σ ($\sim$ 14 ergs/cm^2 for polyethylene). The end surface free-energy density σ_e, however, was assigned a much larger value, because, according to Hoffman[1], each end of the crystallite was subjected to two unique types of microstructure-induced forces of unknown magnitudes, namely, a longitudinal stretching due to the handling of the crystallite, and an equally biaxial expansion due to the crowding of the cilia at the crystal ends.

In a companion paper[3], I proposed a continuum mechanical model relating the crowding-induced free-energy at the end surfaces to the dilatation-induced strain energy in the bulk crystal. Several assumptions were made in deriving a suitable form of the quantity σ_e, and one of them was an assertion that the deformation within the crystal due to the stretching and biaxial expansion at its ends is approximately uniform. The purpose of this note is to show that, by an order of magnitude estimate based on a classical result in elasticity , the uniform deformation assumption for a polyethylene crystal as formed in a flow-induced undercooled solution environment, is reasonably accurate.

CASE I   CRYSTAL OF UNIT WIDTH FOR END LATTICE EXPANSION ONLY

Consider a rectangular parallelepiped of unit width, length ℓ,

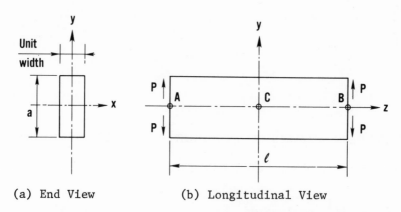

(a) End View   (b) Longitudinal View

Fig. 1.   Idealized Dimensions of a Polyethylene Crystal Subjected
           to Equal and Opposite Forces P at both end surfaces.

and height a. A system of Cartesian coordinates with its origin C
coinciding with the center of the parallelepiped is imposed as shown
in Fig. 1. Assuming the crystal is transversely isotropic with its
Young's moduli E_x, E_y, E_z estimated such that $E_z >> E_y = E_x$, it is required to estimate the magnitude and the direction of the deformation of an infinitesimal cube at the center C relative to the same
at the two ends A and B due to the application of the two equal and
opposite forces P at both ends of the crystal.

ANALYSIS BY SYMMETRY AND A BEAM-ON-ELASTIC-FOUNDATION MODEL

By symmetry, the deformation of a crystal of the dimensions
given in Fig. 1 , when subjected to end forces P of the lattice
expansion type, may be analyzed for one-half of the crystal (height
= a/2) such that the stresses along the x-z plane through the points
A,C, and B are, as a first approximation, uniaxial along the y-direction. This observation yields a beam-on-elastic-foundation model
for which the governing equation of the deflection y(z) of a line
infinitesimally above the symmetry line A-C-B assumes the form:

$$E_z \, I_z \, d^4y/dz^4 = - k \, y, \tag{1}$$

where I_z denotes the moment of inertia of the half-crystal beam of
depth a/2 and unit width, and k, the reaction of the fictitious
elastic foundation (provided by the other half-crystal) per unit
length along the z-direction when the displacement y equals unity.

To express k in terms of E_y and a, we observe that ky, by
definition, equals the vertical stress (reaction per unit area)
and 2y/a equals the vertical strain. Hence $E_y=(ky)/(2y/a)= ka/2$,
or alternatively, $k = 2 \, E_y/a$.

A CLASSICAL SOLUTION DUE TO HETENYI[4,5] AND TIMOSHENKO[6]

With the appropriate boundary conditions, eq. (1) admits a closed form solution for an infinitely long beam for arbitrary loadings. For beams of finite lengths, Hetenyi[4,5] applied the method of superposition and obtained closed form solutions for the displacements of an infinitesimal line above the line A-C-B as follows:

$$y_A = \frac{2 P \beta}{k} \frac{\cosh \beta\ell + \cos \beta\ell}{\sinh \beta\ell + \sin \beta\ell}, \qquad (2)$$

$$y_C = \frac{4 P \beta}{k} \frac{\cosh(\beta\ell/2) \cos(\beta\ell/2)}{\sinh \beta\ell + \sin \beta\ell}, \qquad (3)$$

where β is a parameter defined by

$$\beta = \sqrt[4]{k/(4 E_z I_z)}. \qquad (4)$$

Equations (2) through (4) appear in a standard engineering text written by Timoshenko[6], pp. 15-17. Substituting I_z of the half-crystal $(= a^3/96)$, and k in terms of E_y and a in eq. (4), the quantity $\beta\ell$ can be expressed in terms of the aspect ratio ℓ/a and the degree of anisotropy E_y/E_z as follows:

$$\beta\ell = 2.63 \ (\ell/a) \ \sqrt[4]{E_y/E_z}. \qquad (5)$$

By combining eqs. (2) and (3), we obtain a ratio of the two displacements y_C and y_A as a measure of the uniformity of the deformation throughout the crystal:

$$y_C/y_A = \frac{2 \cosh(\beta\ell/2) \cos(\beta\ell/2)}{\cosh \beta\ell + \cos \beta\ell}. \qquad (6)$$

POLYETHYLENE CRYSTAL WITH END LATTICE EXPANSION ONLY

Eq. (6) allows us to estimate the plausibility of the uniform deformation assumption for a specific crystal such as polyethylene. As shown by Hoffman[1], the aspect ratio ℓ/a for a polyethylene core fibril as observed in laboratory experiments is about 6. The degree of anisotropy, E_y/E_z, for a polyethylene crystal, may be estimated from a vibrational frequency spectrum model due to Broadhurst and Mopsik[7,8]. For our purposes here, we shall use the value of 3×10^{-3} as the quantity E_y/E_z. By eq. (5), we estimate $\beta\ell$ to be about 3.7. Substituting this value into eq. (6), we obtain the deformation ratio y_C/y_A to be equal to - 0.052.

The above estimate of the ratio y_C/y_A indicates that for a crystal subjected to lattice expansion forces at both ends alone, the central deformation is about 5% of the deformation at point A of the end of the crystal, but is opposite in sign. In other words, while the infinitesimal cube near A is in uniaxial tension due to the lattice expansion type of forces at the end surfaces, the point C at the center of the crystal is in uniaxial compression. For a polyethylene crystal of aspect ratio equal to 6 when subjected to end lattice expansion only, the uniform deformation assumption cannot be a viable assertion.

CASE II   COMBINED STRETCHING AND LATTICE EXPANSION AT CRYSTAL ENDS

It turned out[1],[2] that as a polymer core fibril was being formed, a certain amount of stretching at the end surfaces of a typical crystallite was unavoidable as part of the crystallization process. This introduced an extra set of forces at the crystal ends, namely, a longitudinal stretching force which could conceivably alter the estimate of the ratio y_C/y_A for zero stretching.

To determine this effect, we apply a generic result due to Timoshenko[6] (pp. 50–53) where the maximum deflection δ of a simply supported beam under the combined action of a tensile force S and a lateral load P', as shown in Fig. 2, is related to the maximum deflection δ_0 due to the lateral load P' alone as follows:

$$\delta \approx \delta_0/(1 + \alpha), \tag{7}$$

where α is a parameter defined by

$$\alpha = \frac{S\,\ell^2}{E_z\,I_z\pi^2}. \tag{8}$$

Eq. (7) was derived by Timoshenko[6] who retained the first term of a series expansion of the shape of a deflection curve of a beam. It was, nevertheless, a very good approximation, since Timoshenko went on to note that for a uniformly distributed lateral load, the error in eq. (7) at $\alpha = 1$ was about 0.3%. At $\alpha = 2$, he calculated the error to be 0.7%, and at $\alpha = 10$, it was 1.7%.

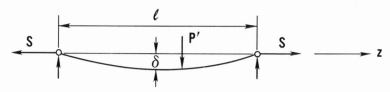

Fig 2.   Deflection Curve of a Simply Supported Beam of Length ℓ

Since the longitudinal stretching force, S , is relatively difficult to estimate or measure in the laboratory, it is necessary to convert eq. (8) into a form containing parameters which can at least be estimated. For example, we may introduce the parameter ε_z, the longitudinal strain term, to replace S, by assuming a simple linear relationship between the two quantities, i.e., $\varepsilon_z = 2S/(aE_z)$. Substituting $I_z = a^3/96$, and $E_z = 2S/(a\varepsilon_z)$ into eq. (8), we have

$$\alpha \simeq 4.8 \ \varepsilon_z \ (\ell/a)^2. \tag{9}$$

POLYETHYLENE CRYSTAL WITH STRETCHING AND LATTICE EXPANSION

Eqs. (5), (6), (7), and (9) can now be used to assess numerically the validity of the uniform deformation assumption for a specific polymer crystal, e.g., a polyethylene core fibril with an aspect ratio ℓ/a equal to 6 and an anisotropy ratio E_y/E_z equal to 3×10^{-3}. As an upper bound for the longitudinal strain ε_z, Zwijnenburg[9] measured the elongation strain to break for polyethylene to be about 3.6%. Consequently, by eq. (9), the parameter α has an upper bound of 6.2.

By using eq. (7) and the numerical estimate of the deflection of an infinitesimal line adjacent to A-C-B for the half-crystal subjected to lattice expansion only, we arrive at an estimate of the two extremes of the shape of the deflection curve of the half-crystal beam as shown in Fig. 3. As shown in previous sections, the central deflection for Case I loading (zero stretching) is 0.05 times the end deflection but is also opposite in sign. With maximum stretching to break and assuming linearity in the stress-strain relationship, the central deflection for Case II loading (maximum stretching) is 0.85 times the end deflection and is of the same sign. The shapes of the deflection curves for the two cases (Cases I & II) are given in Fig. 3 as the dotted and the solid curves, respectively.

It is interesting to note that this result, which is based on a continuum simple beam model, agrees with a qualitative argument

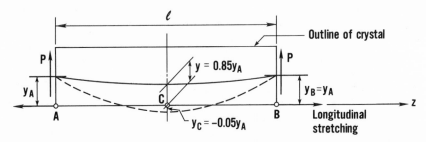

Fig. 3.   Deflection Curves of Crystal Centerline With or Without Longitudinal Stretching at Ends of the Crystal.

advanced by Hoffman[2] who postulated a molecular mechanism to explain the effect of the longitudinal stretching force. It is not difficult to estimate the upper bound of the magnitude of this longitudinal force for a polyethylene core fibril of realistic dimensions, and to show that the continuing existence of this force through some frictional or anchoring mechanism is not an inconceivable possibility. Without further experimental information of this specific conjecture, the result of this note indicates that the uniform deformation assumption used by Hoffman[1,2] and this author[3] is reasonable as a first approximation.

GENERALIZATION OF RESULTS FOR EQUI-BIAXIAL EXPANSION & STRETCHING

The above model for a crystal beam of unit width can be generalized not only to a square cross-section of width a and height a, but also to the general loading case of an equi-biaxial expansion at both ends of the crystal, if the assumption of a transversely isotropic elastic material holds. In other words, the qualitative result given in Fig. 3 applies equally well when we replace y's with the x's as far as the displacements are concerned. A more elaborate discussion of this result appears in Ref. 3.

REFERENCES

1. J. D. Hoffman, On the formation of polymer fibrils by flow-induced crystallization, Polymer, 20: 1071 (1979).
2. J. D. Hoffman, Theory of Flow-Induced Fibril Formation in Polymer Solutions, J. Res. Nat. Bur. Std. (U.S.), 84: 359 (1979).
3. J. T. Fong, On a Suitable Form of the Bundle-End Surface Free Energy of a Flow-Induced Polymer Core-Fibril, to appear in J. Res. Nat. Bur. Std. (U.S.).
4. M. Hetenyi, Analysis of Bars on Elastic Foundation, in:"Final Report of the Second International Congress on Bridge and Struct. Engrg., Berlin-Munich," (1936).
5. M. Hetenyi, "Beams on Elastic Foundation," The University of Michigan Press, Ann Arbor (1946).
6. S. Timoshenko, "Strength of Materials, Part II (Advanced Theory and Problems)," 3rd ed., Van Nostrand, Princeton (1956).
7. M. G. Broadhurst, and F. I. Mopsik, Normal Mode Calculation of Grueisen Thermal Expansion in n-Alkanes, J. Chem. Phys., 54: 4239 (1971).
8. M. G. Broadhurst, and F. I. Mopsik, Vibrational Frequency Spectrum for Polymers, J. Chem. Phys., 55: 3708 (1971).
9. A. Zwijnenburg, "Longitudinal Growth, Morphology, and Physical Properties of Fibrillar Polyethylene Crystals," Doctoral Thesis, University of Groningen (1978).

INDUCED ANISOTROPY OF THERMAL EXPANSIVITY

UNDER LARGE DEFORMATIONS*

Steven T. J. Peng

Jet Propulsion Laboratory
California Institute of Technology
Pasadena, CA,   91003, U.S.A.

INTRODUCTION

The induced anisotropic behavior of highly deformed polymer solids, which were initially isotropic, can result from orienting the links of the polymer chain in the direction of deformation. The degree of anisotropy may depend on the monomer structure, but it depends mostly on the degree of deformation and thus, orientation. In the case of thermal expansion it may be considered that the thermal expansion of polymer solid is due to the increased separation of neighborhood polymer chains. In the undeformed state, the large molecule is randomly coiled and orientation of the segment is averaged out, hence the thermal expansivity will be isotropic. In the deformed state, however, this will no longer be true, but will depend on the average deformation, i.e., the orientation along a given direction.

In what follows, we will show that starting with the following eq. (1) and (3), and using the assumption of the simplified network theory of Gaussian and non-Gaussian Chain, we obtain a simplified equation which describes the induced anisotropic behavior of thermal expansivity in terms of stretch ratios.

Induced Anisotropy of Thermal Expansivity

We consider that an isotropic, homogeneous, slightly compressible polymer solid before deformation will obey the isotropic behavior of thermal expansion. When the solid is deformed, the thermal expansivity is no longer isotropic and becomes anisotropic. Degree of anisotropy depends on the amount of deformation. Thus, it is

*This paper represents one phase of research carried out by the Jet Propulsion Laboratory, California Institute of Technology, Pasadena, CA, sponsored by NASA under Contract No. NAS7-100.

reasonable to assume that the induced anisotropic thermal expansivity can be expressed in the tensorial form

$$\beta^{ij} = \beta^{ij}(\gamma_{k\ell}, T) \tag{1}$$

Where T is the absolute temperature, β^{ij} is the contravariant component of the tensorial induced anisotropic thermal expansivity with respect to the imbedded curvilinear coordinate θ^i in the deformed state, and the strain tensor γ_{ij} is defined by

$$\gamma_{ij} = \tfrac{1}{2}(G_{ij} - g_{ij}) \tag{2}$$

where the tensor g_{ij} is a covariant metric of the imbedded curvilinear system θ^i of the undeformed body defined with respect to the material Cartesian system X_i. Similarly, G_{ij} is the corresponding covariant metric of the imbedded curvilinear system θ^i of the deformed body defined with respect to the spatial Cartesian system γ^i.

Noting that β^{ij} is a tensorial function of the tensor variable γ_{ij}, one may write that

$$\beta^{ij} = \frac{\partial G(\gamma_{k\ell}, T)}{\partial \gamma_{ij}} \tag{3}$$

where $G(\gamma_{k\ell}, T)$ is a scalar function of the tensor variable $\gamma_{k\ell}$. Since the thermal expansion depends on the distribution of chain orientation, the scalar function G is, in the essence, similar to the distribution function of the chain orientation which depends on the deformation gradient. The above statement is quite general. Since the polymer solid is initially isotropic and almost incompressible, the function $G(\gamma_{ij}, T)$ can be expressed by two principal invariants of γ_{ij}, i.e., I_1 and I_2, or by three principal values, γ_1, γ_2 and γ_3, i.e., $G = G(\gamma_1, \gamma_2, \gamma_3, T)$.

As stated in the introduction, we assume that the thermal expansion is due to the separation between neighborhood molecular segments caused by increased thermal excitation from rising temperature. Since in the undeformed state, the large molecular is randomly coiled and the orientation of the segments is averaged out, the thermal expansion will be isotropic. But, in the deformed state, this is no longer true, and the orientation of the segments will depend on the deformation, i.e., orientation along a given direction. The general calculation of the orientation is not necessary, however, for Treloar[1] has shown that the deformation of chain in a polymer network of Gaussian and non-Gaussian chains, can be resolved into the deformations of three equivalent chains, one lying along each principal coordinate direction. This implies that the contribution to the thermal expansivity can similarly be resolved into components along the principal coordinate directions, and so we therefore assume the scalar function $G(\gamma_1, \gamma_2, \gamma_3, T)$ can be reduced to three independent sets, i.e.,

$$G(\gamma_1, \gamma_2, \gamma_3) = g(\gamma_1, T) + g(\gamma_2, T) + g(\gamma_3, T) \tag{4}$$

Now what we need is the three principal values γ_α and three sets of principal directions $\vec{N}_\alpha (\alpha = 1,2,3)$ of the strain tensor γ_{ij}. It can be obtained from the usual characteristic equation in a material coordinate system. Note that we have

$$\gamma_\alpha = \tfrac{1}{2}(\lambda_\alpha^2-1) \text{ and } \gamma_\alpha = \gamma_{ij}N_\alpha^i N_\alpha^j \qquad (5)$$

Where λ_α are the principal stretch ratio.

With eqs. (4) and (5), eq. (3) becomes:

$$\beta^{ij} = \sum_{\alpha=1}^{3} \lambda_\alpha^{-1} g'(\lambda_\alpha) N_\alpha^i N_\alpha^j \qquad (6)$$

In the following, we will determine the function $g'(\lambda_\alpha)$ from the experimental data.

Experimental Determination of the Function $g(\lambda)$

For uniaxial deformation we identify the coordinate $\theta^i = X_i$, thus the strain tensor γ_{ij} becomes

$$\gamma_{ij} = \tfrac{1}{2}(\lambda_{(i)}^2 -1) \delta_{ij}^{(i)} \qquad (7)$$

The principal direction $\vec{N}^\alpha$ of simple extension is then given by

$$\vec{N}_i^1 = (1,0,0), \quad \vec{N}_i^2 = (0,1,0), \quad \vec{N}_i^3 = (0,0,1) \qquad (8)$$

Substituting eq. (8) into (6), we obtain

$$\beta_{11} = \lambda_1 g'^{-1}(\lambda_1) \equiv \beta_{||}, \text{ and} \qquad (9)$$

$$\beta_{22} = \beta_{33} = \lambda_2 g'^{-1}(\lambda_2) \equiv \beta_\perp \qquad (10)$$

Where λ_1 and $\vec{N}^1$ are parallel to the direction of stretch, λ_2 and $\vec{N}^2$ are perpendicular to the stretch. Since β_{11} is the heat conductivity parallel to the stretching direction, we denote $\beta_{||} = \beta_{11}$. Similarly, we denote the perpendicular to the direction of stretch by $\beta_{22} = \beta_\perp$

It was shown from the experimental data by Tjader and Protzman[2] that, to an excellent first approximation, the following relation is held

$$3\beta_0 = \beta_{||} + 2\beta_\perp \qquad (11)$$

Hence, from eqs. (9) and (10), we have

$$3 = \frac{g'(\lambda_1)}{\beta_0\lambda_1} + \frac{g'(\lambda_2)}{\beta_0\lambda_2} \qquad (12)$$

then we have eq. (12)

$$f(\lambda_1) + 2f(\lambda_2) = 3 \qquad (13)$$

where

$$f(\lambda) \equiv g'(\lambda)/\beta_0\lambda$$

Since the material is almost incompressible, i.e., $\lambda_1\lambda_2^2 \approx 1$ for simple extension, we have $\lambda_2 \approx \lambda_1^{-\frac{1}{2}}$. Hence, eq. (13) becomes

$$f(\lambda_1) + 2f(\lambda_1^{-\frac{1}{2}}) = 3 \qquad (14)$$

Now we define a new variable $\lambda = e^u$, thus $\lambda^{-\frac{1}{2}} = e^{-\frac{1}{2}u}$,
then eq. (14) becomes

$$f(u) + 2 f(-\frac{u}{2}) = 3 \qquad\qquad (15)$$

If one expands $f(u)$ in power series, i.e., $f(u) = \sum_{m=o}^{\infty} f_m u_m$ and substitutes into eq. (15). One obtains

$$\sum_{m=o}^{\infty} \left[1 + 2(-\tfrac{1}{2})^m \right] f_m u^m = 3$$

By equating the coefficient of $f_m u^m$ at the left hand side of equation to the right hand side, we have

$$f_o = 1, f_1 = ? \text{ and } f_2 = f_3 = \cdots f_n = 0 \qquad (16)$$

hence, we have

$$f(\lambda) = 1 + f_1 u = 1 + f_1 \ln\lambda \qquad\qquad (17)$$

where f_1 is an arbitrary constant to be determined from the experiment. For clarity, we set $f_1 = a$, where a is a material constant to be determined by experiment. Hence, one has linear expansivity in the direction of stretch

$$\frac{\beta||}{\beta_o} = 1 + a \ln\lambda_1 = \lambda_1 g^{-1}{}'(\lambda_1)/\beta_o \qquad (18)$$

where β_o is the linear expansivity before deformation. By integration of eq. (18), we have

$$g(\lambda_1) = \beta_o \left[\frac{\lambda_1^2}{2} + a\lambda_1^2 (\frac{\ln\lambda_1}{2} - \tfrac{1}{4}) \right] \qquad (19)$$

Then, from eq. (4), we have

$$G(\lambda_1, \lambda_2, \lambda_3) = \beta_o \sum_{i=1}^{3} \left[\frac{\lambda_i^2}{2} + a\lambda_i^2 (\frac{\ln\lambda_i}{2} - \tfrac{1}{4}) \right] \qquad (20)$$

Substituting eq. (20) into eq. (16), or from the eq. (10), one has the linear expansivity perpendicular to the stretched direction, i.e.,

$$\beta\bot = \beta_o (1 + a \ln\lambda_2) = \beta_o (1 - \frac{a}{2} \ln\lambda_1) \qquad (21)$$

Now, what we need here is to check whether eqs. (18) and (21) can predict the anisotropic behavior induced by simple extension. From the experimental data of Hellwege, Hennig and Knappe[3], we found that by using the following values for different materials, i.e., $a = 1.25$ for PC, $a = -0.66$ for PVC, $a = -0.29$ for PMMA and $a = -0.031$ for PS, eqs. (18) and (21) predict quite well both the parallel direction and perpendicular direction up to the extension ratio $\lambda_1 = 5$ (see figure 1). Thus, it verifies the basic assumption of eq. (4), even though eqs. (18) and (21) are resulted from the stringent condition imposed by eq. (11).

Remarks

It is important to stress that eqs. (18) and (21) are the special

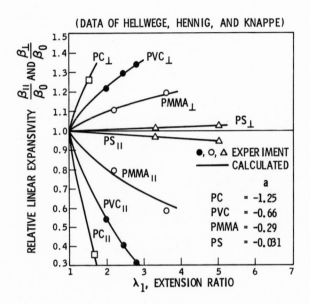

Figure 1. Relative linear expansivity of uniaxially stretched polymer
vs. stretch ratio, λ_1 [---calculated curves are based on
eqs. (18) and (21).]

case of underlying assumption, eq. (4), resulting from the phenomen observed experimentally in Ref. [2] and [3], i.e., the relation in eq. (11). For some other materials which do not obey eq. (11), we believe our basic assumption is still valid. Moreover, the assumption eq. (4) may hold at larger deformations, in which the relation in eq. (11) may break down. What we need here is a proper analytical form of $g(\lambda)$, which may be obtained from any stretched direction. Once we obtain $g(\lambda)$, then we can predict the linear thermal expansivity at any other direction through eq. (6). It may be useful to note that we may use an analytical expression for $g(\lambda)$ in terms of Seth's generalized measure of strain, i.e.,

$$g(\lambda) = \sum_{i=1}^{n} \lambda^{n_i}/n_i$$

where n_i is the exponent of stretch ration λ. This measure has recently been found[4-] to be a very powerful expression for describing the nonlinearity of elastic rubber.

Finally, it is important to note by employing our basic assumptions, the thermal expansion is due to the increased separation of neighborhood polymer chains from rising temperature and that the Gaussian and non-Gaussian chain network may be represented by the simplified three-chain model, one greatly simplifies the method needed to characterize the induced anisotropy. If the induced anisotropy is expressed in terms of invariants I_1 and I_2, then the equations needed to characterize it are complex, since one needs not only multiaxial deformation experiments, but also close control of many variables. Thus, our theory gives a very efficient and simple way to circumvent these difficulties.

REFERENCES

(1) L.R. G. Treloar, "The Physics of RubberElasticity," 2nd ed., Oxford University Press, Oxford, 1967, Chap. VI.

(2) T.C. Tjader and T.F. Protzman, J. Polymer Sci., Vol. 20, 591 (1956).

(3) K.H. Hellwege, J. Hennig, and W.Knappe, Kolloid-Z U.Z. Polymere, Vol. 188, 121 (1963).

(4) Ogden, R.W. Proc. R. Soc. Lond, A, 326, 565 (1972).

VISCOELASTIC BEHAVIOR OF POLY(METHYL METHACRYLATE):

PREDICTION OF EXTENSIONAL RESPONSE FROM TORSIONAL DATA

Gregory B. McKenna and Louis J. Zapas

Polymer Science and Standards Division
National Bureau of Standards
Washington, D.C.   20234

INTRODUCTION

Some years ago Rivlin[1] showed that for certain deformations, one can treat isochronal data from single step stress relaxation experiments on viscoelastic materials in the same fashion as if the data were obtained for an elastic material. We have conducted single step stress relaxation experiments on cylinders of poly (methyl methacrylate) (PMMA) where we measured torque and normal force responses as functions of time and angle of twist. By assuming that torsion is an isochoric motion and that volume effects are separable, we obtained[2] isochronal values for $\partial W/\partial I_1$ and $\partial W/\partial I_2$. Our results showed that $\partial W/\partial I_1$ is negative while $\partial W/\partial I_2$ is positive and greater in magnitude than $\partial W/\partial I_1$. These findings led to the possibility of explaining the phenomenon observed by Sternstein and Ho[3] that the single step stress relaxation response of PMMA is different in torsion than in simple extension. Specifically, it was found that the rate of decay of the stress with respect to time is significantly higher in torsion than in extension. This phenomenon was observed at small strains where the stress responses in both torsion and extension were linear in the appropriate strain measure. The difference could not be accounted for either in terms of a time dependent Poisson's ratio or the material compressibility.

Using a form for viscoelastic potential functions which is similar to the Valanis-Landel[4] form of the strain energy function for elastic materials, we calculated the response in simple extension from our torsion-normal force determined values of $\partial W/\partial I_1$ and $\partial W/\partial I_2$.

299

Although we obtained good agreement between our calculated and our observed behavior, we could not account fully for the observations of Sternstein and Ho[3].

EXPERIMENTAL

The poly(methyl methacrylate) used in this study was obtained from two sources. In one instance, tubes of PMMA made by a centrifugal casting technique were obtained from TFE Industries*. These tubes are known to be made from Rohm and Haas* PMMA monomer. The tubes were cut to length and turned on a lathe. They had a nominal inner diameter of 2.54 cm and a nominal outer diameter of 3.81 cm. These specimens were tested with no further thermal or mechanical treatment. We shall refer to them as unconditioned specimens[5]. The other source of PMMA was the NBS storeroom. We obtained solid rods of the material from the storeroom but have no knowledge of who the manufacturer was or who manufactured the monomer. These specimens were annealed at 100 °C for 24 hours, oven cooled, machined to a diameter of 2.52 cm and then tested.

Single step stress relaxation experiments on both the tubes and the rods were conducted using an Instron* servo-controlled tension-torsion hydraulic test machine. The machine is interfaced with a Hewlett-Packard* 2100 mini-computer for control and data acquisition. The tests on the tubes were conducted on a new tube for each individual strain level. Tests on the annealed rods were conducted using two specimens. The deformations were applied starting with the lowest and increasing to the highest. Between tests the samples were held at zero deformation for a minimum of ten times the duration of the previous test before beginning the test at the next higher deformation. Tests in tension were conducted after the series of torsional tests had been conducted--again starting at the lowest strain and increasing to the highest. Tests on the tubes were conducted at 24 ± 1 °C and those on the annealed cylinders were conducted at 23 ± 1 °C. Measurements on the diameter of the tubes did not show any change in dimension upon torsion to within 0.1%[5].

---

*Certain commercial materials and equipment are identified in this paper in order to specify adequately the experimental procedure. In no case does such identification imply recommendation or endorsement by the National Bureau of Standards, nor does it imply necessarily the best available for the purpose.

THEORETICAL CONSIDERATIONS

For simple viscoelastic materials Rivlin[1] has shown that for certain deformation histories, such as those obtained in single step stress relaxation experiments, one can treat isochronal data in the same same manner as for elastic materials. For incompressible materials, he showed that the stress in any deformation can be described using two material functions which are functions of time, t, and the in--variants in the principal stretches, I_1, and I_2. Though one need not assume the existence of a strain energy (or potential) function to describe single step stress relaxation histories, we shall use the BKZ[5]-type notation consistent with our prior work.[2,5,7]

In the BKZ theory of an elastic fluid[6], the existence of a time dependent strain potential function is postulated. If we consider single step stress relaxation deformations, then we can define iso-chronal values for the derivatives of the strain potential function as follows:

$$W_i(t) = \frac{\partial W}{\partial I_i}(I_1,I_2,t) = \int_{-\infty}^{t} \frac{\partial U}{\partial I_i}(I_1,I_2,t-\tau)d\tau \qquad (1)$$

where U is the potential function of the BKZ theory, and now the I_i's are the ith invariants of the left relative Cauchy deformation tensor, t is the present time, and τ is the past time.

For torsion of an elastic rod with fixed ends Nadai[8] and later Penn and Kearsley[9] showed that the derivatives of the strain energy function can be determined from torque and normal force measurements. By a similar analysis for a Viscoelastic (BKZ) rod we can find values for the $W_i(t)$ defined above:

$$W_1(t)+W_2(t) = \frac{1}{4\pi\Psi R^4}(3T+\Psi T_\Psi) \qquad (2)$$

$$W_1(t)+2W_2(t) = \frac{-1}{\pi\Psi^2 R^4}(N+\Psi^2 N_{\Psi^2}) \qquad (3)$$

where R is the radius of the rod, Ψ is the angle of twist per unit length, T is the applied torque, and N is the total normal force applied to the ends of the rod to keep the length constant. T_Ψ is the derivative of the torque with respect to Ψ and N_{Ψ^2} is the derivative of the normal force with respect to Ψ^2. Equations (2) and (3) can be solved simultaneously for $W_1(t)$ and $W_2(t)$ as functions of both twist and time.

In the case of torsion, the deformation state is one in which the strain invariants are equal. In general, one can not go from the values of W_1 and W_2 determined in shear to another state of deformation. This is possible only if one uses a restricted form of the strain energy function. In finite elasticity, such a special form was postulated by Valanis and Landel[4]. The V-L form of the strain energy function expresses the strain energy $W(I_1,I_2)$ in terms of only the principal stretches λ_1, λ_2, and λ_3 as:

$$W(I_1,I_2) = w(\lambda_1) + w(\lambda_2) + w(\lambda_3) \tag{4}$$

For a general strain energy function, the principal stress differences can be determined from the derivatives of the strain energy function with respect to I_1 and I_2. For example in simple extension:

$$\sigma_{11} - \sigma_{22} = \Delta\sigma = 2(\lambda_1^2 - \lambda_2^2)\ (W_1 + \frac{1}{\lambda_1^2 \lambda_2^2}\ W_2) \tag{5}$$

For the V-L form of the strain energy function this stress difference becomes:[10]

$$\sigma_{11} - \sigma_{22} = \lambda_1\ w'(\lambda_1) - \lambda_2\ w'(\lambda_2) \tag{6}$$

where $w'(\lambda)$ is the derivative of the V-L function with respect to λ.

Rivlin and Sawyers[11] have discussed the general relationships between $W(I_1,I_2)$ and $w(\lambda)$ and Kearsley and Zapas[10] have solved for the relationship between $w'(\lambda)$ and W_1 and W_2 in several deformations. From torsion the latter relationship is:

$$w'(\lambda) = \frac{2}{\lambda}(\lambda^2-1)(W_1 + \frac{1}{\lambda^2}\ W_2) \tag{7}$$

Following this approach we can use the values of W_1 and W_2 determined in torsion to evaluate $w'(\lambda)$ (knowing the principal stretches in torsion (see ref. 6 or 7)) from equation (7). Then, since w' is dependent only on λ, it is possible to use equation (6) to calculate the stress in a simple extension experiment.

Since isochronal values of $W_1(t)$ and $W_2(t)$ can be treated in a fashion analogous to the elastic strain energy function derivatives, we can also treat isochronal values of $w'(\lambda,t)$ as viscoelastic analogues to the V-L elastic strain energy function derivative to obtain $w'(\lambda,t)$.

In this work, we have assumed that torsion of a compressible material is an isochoric deformation and that $W_1(t)$ and $W_2(t)$ can be determined from equations (2) and (3) by assuming that the volume dependence of W_1 and W_2 is separable from the dependence on I_1 and I_2. In calculating the stresses for a simple extension deformation, we make the same assumption for $w'(\lambda)$ and make our calculations based only on the value of λ_1 imposed on the sample and values for $\lambda_2 = \lambda_3$ based on the assumption of a Poisson's ratio of 0.35.

The determination of W_1 and W_2 for tubular specimens is more difficult than for the solid rods due to the requirement that, in general, one needs to apply an internal pressure in order to prevent constriction of the tube. We made the assumption that not internally pressurizing the tube gives negligible error (we could not measure any constriction of the tube[5]) in our results and that the analysis used to obtain equations (2) and (3) for W_1+W_2 and W_1+2W_2 can be used to obtain similar equations with extra terms involved due to the geometry differences. The reader is referred to references 5,7,12 and 13 for discussion of various aspects of this problem. We also note that it was our concern about this problem and the effects of the tubular geometry on our measurements for determining W_1 and W_2 that led us to conduct the series of experiments on the solid rods.

RESULTS AND DISCUSSION

Table 1 shows some isochronal data for $W_1(t)$ and $W_2(t)$ for measurements made on PMMA in the form of both annealed rod and unconditioned tube. We remark on three things about these data. First, $W_1(t)$ is negative while $W_2(t)$ is positive and of greater magnitude than $W_1(t)$. This is different from the behavior usually observed in rubbers and polymer melts, where W_1 is usually positive and larger than W_2. Another observation we can make is that the time dependence of $W_1(t)$ and $W_2(t)$ is different. Also the time dependence of the modulus, $2(W_1+W_2)$, is different from that of either $W_1(t)$ or $W_2(t)$. (At the limit of small strains, i.e., in the linear range, the modulus $G(t) = 2[W_1(t)+W_2(t)]$). Finally, the magnitude of $W_1(t)+W_2(t)$ is different for the two different PMMA's. This is not attributable to the geometry difference but may be due to the differences in both source and thermal histories of the materials. (When the unconditioned PMMA is annealed at 100 °C for 24 hours significant shrinkage does take place even though the material is isotropic when viewed through crossed polarizers[5]).

Table I

Some Isochronal Values of $W_1(t) + W_2(t)$, $W_1(t)$ and $W_2(t)$ for PMMA

Annealed Rod

		W_1+W_2	W_1	W_2		W_1+W_2	W_1	W_2
	R	GPa	GPa	GPa		GPa	GPa	GPa
1.64s	0.00252	0.63	-2.19	2.82	1678s	0.488	-1.94	2.43
	0.00946	0.59	-1.98	2.57		0.456	-1.39	1.84
	0.0126	0.58	-1.82	2.40		0.442	-1.15	1.60
	0.0252	0.53	-1.10	1.63		0.379	-0.616	0.995

Unconditioned Tube

1.64s	0.0095	0.48	-1.24	1.72	1678s	0.36	-0.79	1.15
	0.012	0.46	-1.24	1.70		0.35	-0.76	1.11
	0.016	0.45	-1.14	1.59		0.33	-0.65	0.98
	0.0208	0.44	-0.96	1.40		0.31	-0.47	0.78

From these data we determined isochrones for the viscoelastic V-L function $w'(\lambda)$ for both the rods and tubes. From $w'(\lambda,t)$ and equation (6) we then calculated $\sigma_{11}(t)-\sigma_{22}(t)$ for simple extension. The agreement between the calculations and the experiments was within the uncertainty for our values of $w'(\lambda,t)$. (This uncertainty in the small strain region is approximately $\pm 15\%$ due to the large uncertainties in the normal stress measurements in this region.) Of greater interest is the comparison of the relaxation rates in extension and torsion.

Table 2 summarizes our results and those of Sternstein and Ho[3]. The relaxation rates given in Table 2 are calculated from the slopes of log(stress) vs. log (time) plots. There are several things about this data which need to be discussed. First, Sternstein and Ho[3] obtain greatly different rates of relaxation in extension and torsion (-.0195 vs -.0329) at strains of $\varepsilon = 0.005$ and $\gamma = 0.01$ respectively. Our data in torsion show a slightly higher relaxation rate than their's for $\gamma = 0.01$ which increases as γ increases. Also our data indicate that there is little difference in relaxation rates in extension and torsion at strains less than $\varepsilon = 0.02$ and $\gamma = 0.04$. But at these strains and higher we find that there is a more rapid torsional relaxation rate than extensional relaxation rate (-0.058 vs. -0.047). The fact that our data do not agree with those of Sternstein and Ho[3] requires some elaboration. First, their data were obtained on annealed PMMA made from a Rohm and Haas* resin. This is the same make of resin used in our unannealed samples. We do not know who the resin supplier was for the solid rod specimens of PMMA which we annealed. We also note the importance of thermal history (or supplier) which is apparent in the differences of W_1+W_2 values for the unconditioned tubes and the annealed rods. We are currently planning experiments to further elucidate these differences and perhaps account for the differences between our data and those of Sternstein and Ho[3].

Table 2

Relaxation Rates for PMMA for Different States of Deformation

Strain	Relaxation Rate[1]			
	Tension		Torsion[2]	Simple
	Predicted	Observed		Shear[3]
Data of Sternstein and Ho[1]:				
$\varepsilon = 0.005$; $\gamma = 0.01$	–	-0.0195	-0.0329	–
Unconditioned Tubes:				
$\varepsilon = 0.01$; $\gamma = 0.02$	-0.0397	-0.0384	-0.0377	-0.0450
$\varepsilon = 0.02$; $\gamma = 0.04$	-0.051	-0.047	-0.0581	-0.0634
Annealed Rods:				
$\varepsilon = 0.005$; $\gamma = 0.01$	-0.0370	-0.0319	-0.0345	-0.0395
$\varepsilon = 0.0075$; $\gamma = 0.015$	-0.0371	-0.0335	-0.0360	-0.0420

[1] Calculated from the slope of log(stress) vs. log(strain) plots. Rate is per decade of time.

[2] Observed experimentally. γ is the maximum strain in the tube or rod.

[3] Calculated from values of W_1 and W_2 obtained from torsion-normal force measurements. γ is the shear strain.

Finally we note that in Table 2 we have also given the values of the relaxation rates which are predicted for simple shear experiments, i.e., those calculated from $2[W_1(t)+W_2(t)]$. We can see that the relaxation rate in simple shear, even at small deformation, is predicted to be somewhat more rapid than the rate in simple extension. The reason for this can be seen if we consider equation (5) which gives the stress response in simple extension as:

$$\Delta\sigma = 2(\lambda_1^2 - \lambda_2^2)(W_1 + \frac{1}{\lambda_1^2\lambda_2^2} W_2) \tag{5}$$

for small strains this can be rewritten as:

$$\Delta\sigma = 2(\lambda_1^1 - \lambda_2^2)(W_1 + W_2 - \delta W_2) \tag{8}$$

where δ is a small number of the order of the strain. If we recall that the **shear** stress response to a given strain, γ, is given by:

$$\sigma_{12} = 2\gamma(W_1 + W_2) \tag{9}$$

we can see that if $W_1(t)+W_2(t)$ relaxes at a different rate than does $W_2(t)$, the shear stress will relax at a different rate than does the extensional stress. For the PMMA cylinders studies here, we found that $W_2(t)$ relaxes faster than does $W_1(t)+W_2(t)$. Therefore, equations (8) and (9) show why the extensional stress would relax more slowly than the shear stress. This would be true even though both the shear and extensional stresses were linear in their appropriate strain measures. In torsion, of course, we have an inhomogeneous state of strain. Since the relaxation rate increases with increasing strain, the rate of torsional relaxation will be slower than that for simple shear.

Though we show that the difference in relaxation rates for shear and extensions may be explained by the difference in relaxation rates for the $W_1(t)$ and $W_2(t)$, we recognize that there may be a dependence of $W_1(t)$ and $W_2(t)$ due to the material compressibility which would not be accounted for in our treatment. The possibility and importance of such effects need further study.

ACKNOWLEDGEMENTS

We would like to thank E. A. Kearsley of NBS for many fruitful discussions during the course of this work.

REFERENCES

1. R. S. Rivlin, "Stress Relaxation in Incompressible Elastic Materials at Constant Deformation", Quart. Appl. Math., 14 83 (1956).

2. G. B. McKenna and L. J. Zapas, "Determination of the Time Dependent Strain Potential Function in Poly(methyl methacrylate)", Paper presented at the 49th Annual Meeting of the Society of Rheology, Houston, Texas, Oct., 1978.

3. S. S. Sternstein and T. C. Ho., "Biaxial Stress Relaxation in Glassy Polymers: Poly(methyl methacrylate)", J. Appl. Phys. 43, 4370 (1972).

4. K. C. Valanis and R. F. Landel, "The Strain Energy Function of a Hyperelastic Material in Terms of Extension Ratios", J. Appl. Phys., 38, 2997 (1967).

5. G. B. McKenna and L. J. Zapas, "Nonlinear Viscoelastic Behavior. of Poly(methyl methacrylate) in Torsion", J. Rheology, 23 151 (1979).

6. B. Bernstein, E. A. Kearsley and L. J. Zapas, "A Study of Stress Relaxation with Finite Strain", Trans. Soc. Rheology, VII, 391 (1963).

7. G. B. McKenna and L. J. Zapas, "The Normal Force Response in Nonlinear Viscoelastic Materials: Some Experimental Findings", J. Rheology, in press.

8. A. Nadai, "Plasticity: A Mecnanics of tne Plastic State of Matter McGraw-Hill, New York, 1931

9. R. W. Penn and E. A. Kearsley, "The Scaling Law for Finite Torsion of Elastic Cylinders", Trans. Soc. Rheology, 20, 227 (1976).

10. E. A. Kearsley and L. J. Zapas, "Some Methods of Measurement of an Elastic Strain Energy of the Valanis-Landel Type", J. Rheology, in press.

11. R. S. Rivlin and K. N. Sawyers, "The Strain Energy Function of Elastomers", Trans. Soc. Rheology, 20, 545 (1976).

12. R. S. Rivlin and D. W. Saunders, "Large Elastic Deformations of Isotropic Materials. VII. Experiments on the Deformation of Rubber", Phil. Trans. Roy. Soc. London, A, 243, 251 (1951).

13. A. E. Green and J. E. Adkins, Large Elastic Deformations, 2nd Edition, Clarendon Press, Oxford, 1970.

DYNAMIC MECHANICAL BEHAVIOUR OF PLASTICISED AND

FILLED POLYMER SYSTEMS

R. STENSON

Ministry of Defence, PERME
Waltham Abbey, Essex
England

INTRODUCTION

Dynamic testing provides a sensitive fingerprint of the visco-elastic behaviour of solid polymers. Shear moduli have been determined in the oscillatory torsion mode for a series of plasticised nitrocellulose matrices and for two highly filled binders, viscous fluid and elastomer. Measurements were carried out using a Rheometrics Mechanical Spectrometer.

PLASTICISED NITROCELLULOSE

Nitrocellulose shows few features until plasticised when clearly defined glassy and transition regions are observed. A peak in the loss modulus G"/temperature curve corresponds at a frequency of 1.0 Hz with a change in base line (change in specific heat) determined by differential scanning calorimetry (DSC). The peak occurs between -30°C and -60°C depending on the plasticiser content and is referred to as the glass transition temperature (T_g) for a system consisting of a partially crystalline polymer and a plasticiser which shows a marked relaxation at -65°C. The peak in G" usually occurs at a temperature about 10°C lower than a peak in the internal friction G"/G' (= tan φ)/temperature curve. T_g is also marked by a loss of extensibility measured in the uniaxial tensile mode at a constant rate of strain.

The dominating effect of plasticiser on mechanical behaviour is summarised in Table 1 and examples are shown in Fig 1. For low plasticiser contents (P/NC $\ngtr$ 0.56) the peak in G" is broad and of low magnitude, the slope n in the transition region is low and only one small peak in tan φ occurs (curve B, Fig 1). With increase

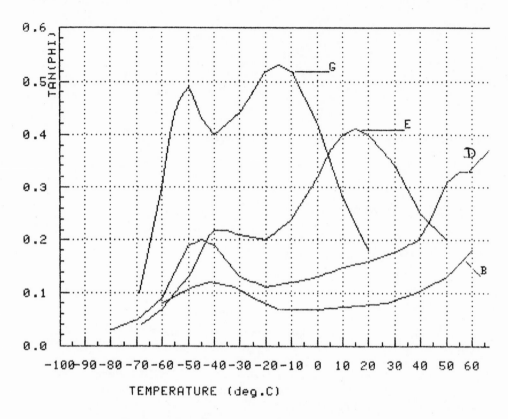

FIG.1 LOSS ANGLE OF PLASTICISED NC

Table 1. Effect of Plasticiser (P) on the Dynamic Mechanical
Behaviour of Nitrocellulose (NC)

Matrix	P/NC v/v	T_g °C	$n^{[†]}$	tan φ	
				Lower Peak	Upper Peak
A	0.26	-30	0.006	0.07 (-20°C)	>0.18 (>80°C)
B	0.56	-43	0.012	0.12 (-40°C)	>0.20 (>70°C)
C	0.83	-52	0.020	0.16 (-46°C)	- (>70°C)
D	1.04	-52	0.025	0.20 (-43°C)	0.33 (60°C)
E	1.20	-52	0.025	0.21 (-40°C)	0.41 (17°C)
F (filled)	2.30	-58	0.035	0.30 (-46°C)	0.56 (-10°C)
G	3.00	-60	0.053	0.49 (-52°C)	0.52 (-15°C)

[†] n = slope in the transition region of the log G*/temperature curve.

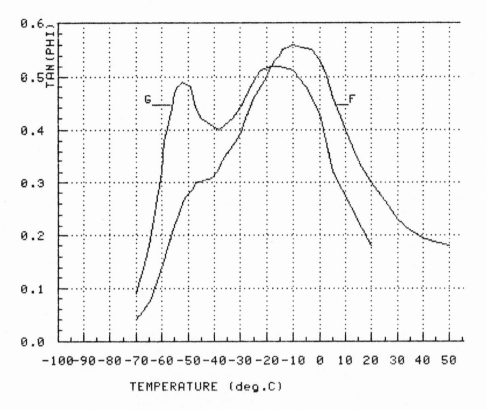

FIG.2 EFFECT OF FILLER ON LOSS ANGLE

in plasticiser content T_g is lowered, the peak in G" becomes sharper, n increases and two peaks in tan φ appear (curves D and E). Eventually for the highest plasticiser level examined, the matrix is very soft and a limiting T_g of -60°C is found, the complex modulus G* is very sensitive to temperature in the transition region and two large well defined peaks in tan φ are present (curve G). The temperature at which the upper peak in tan φ occurs is sensitive to plasticiser content and moves by about 90°C over the range of matrix formulations. The lower peak which may be related to relaxations in the plasticiser is less sensitive and moves by 30°C. For the less plasticised systems in Table 1 an upper peak is not present to swamp any fine structure and at intermediate temperatures (0°C to 30°C) small inflections occur in tan φ (not shown in Fig 1) which suggest additional relaxations in the material.

EFFECT OF FILLER

The addition of small particles of filler reinforces the matrix and increases G* above the T_g usually at the expense of extensibility. Fig 2 shows that for highly plasticised systems the filler

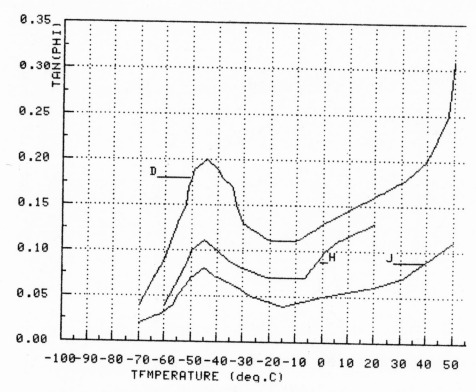

FIG.3 EFFECT OF INCREASING FILLER ON LOSS ANGLE

has broadened the upper peak in tan φ to such an extent that the
lower peak no longer appears and is replaced by a point of inflec-
tion (curve F).   The relaxation in G" at the T_g however is still
clearly defined.

     The effects of fillers for intermediate plasticiser levels
(P/NC = 1.0) are shown in Table 2 and Fig 3.   Curve D represents
the matrix to which increasing amounts of fillers are added (40
and 55 per cent w/w respectively in matrices H and J).   The temperature of the peak in tan φ is not displaced from-43°C but the peak
is broadened and its magnitude is lowered.   Matrix H represents a
similar composition to matrix I except that the former contains
particles of a greater aspect ratio such that the greater inter-
locking of the filler produces a lower tan φ over the whole
temperature range and a lower slope n.

Table 2. Effect of Fillers on Dynamic Mechanical Behaviour

Matrix	P/NC v/v	T_g °C	n	tan φ	
				Lower Peak	Upper Peak
D	1.0	−52	0.025	0.20 (−43°C)	0.33 (60°C)
H	1.0	−51	0.014	0.11 (−43°C)	−
I	1.0	−50	0.017	0.12 (−43°C)	−
J	1.0	−49	0.013	0.08 (−43°C)	0.12 (60°C)

HIGHLY FILLED BINDERS

Higher filler levels up to 88 per cent w/w (77 per cent v/v) have been examined dispersed in essentially a polyisobutene (PIB) binder to produce a stiff paste. The viscosity of the matrix at 25°C is about 5×10^4 Pa s and the resultant filler reinforced material is similar to modelling clay or plasticine. The plasto-viscosity of this system was described in the Proceedings of the VIIth Congress. Dynamic measurements show very large damping factors without the presence of any fine structure. Tan φ shows a sharp relaxation at −50°C with very little filler broadening, a minimum at 0°C and a progressive increase in damping up to 70°C (Table 3). Glassy, transition and flow regions are clearly defined and a sharp peak in G" occurs at −65°C.

Table 3. Dynamic Mechanical Behaviour of PIB and Elastomer
          (77 per cent v/v filled)

Matrix	Matrix Type	T_g °C	n	tan φ	
				Lower Peak	Upper Peak
K	PIB	−65	0.045	1.18 (−50°C)	>1.4 (>70°C)
L	Elastomer	−83	0.063	0.8 (−75°C)	0.5 (0°C)
M	Elastomeric Binder	−81	0.133	1.3 (−73°C)	0.3 (20°C)

The addition of a similar amount of filler to a prepolymer which can be crosslinked to an elastomeric binder produces a more rigid material than the PIB type, particularly in resistance to deformation at the higher temperatures. The T_g is low (−83°C) and relaxations in the transition region are very sensitive to tempera-ture. The appearance of the tan φ/temperature curve resembles that of the crosslinked binder and the sharp peak at −73°C is not broadened by the presence of filler; the upper peak is broad in both cases. For both PIB and elastomer systems the T_g correlates with a sharp change in extensibility.

PROPERTIES OF COMPATIBLE BLENDS OF

POLYCARBONATE AND A COPOLYESTER

P. Masi, D. R. Paul, and J. W. Barlow

Dept. of Chemical Engineering,
The University of Texas at Austin
Austin, TX 78712

INTRODUCTION

Blending of existing polymers to obtain a miscible mixture is widely being considered as a low capital cost route to new materials. Miscible mixtures of amorphous polymers are single phase and optically transparent, and they usually show improved mechanical properties of the pure components (1,2). Above all, miscible blends are characterized by a single, composition dependent, glass transition temperature, Tg.

Elementary thermodynamics indicates that miscible or partially miscible mixtures result if the free energy of mixing, ΔGm, is negative. Because the molecular weights of the blend components are high, the entropic contribution to ΔGm is very small relative to the enthalpic contribution, ΔHm. Consequently, miscible blends result most often for polymer components which mix exothermically (3). Since direct calorimetric measurement of ΔHm is impossible for polymer mixtures, indirect methods must be employed.

This paper presents transition behavior of a recently discovered miscible blend (4) of General Electric Polycarbonate, PC, and Eastman Kodar[R] A150, a copolyester of 1,4 cyclohexane dimethanol with a mixture of 20% isophthalic and 80% terephthalic acids. These data show that complete amorphous phase miscibility exists throughout the composition range. Equilibrium sorption data for CO_2 gas in PC/Kodar blends are also presented from which ΔHm is found to be -0.78 cal/cc for the 50:50 polymer mixture at 35°C by analysis of the Henry's Law portion of the isotherm.

315

Blend Transition Behavior

Melt blends of the carefully dried components were prepared in a Brabender Plasticorder at 280°C. Films, 5 mils thick, were prepared from blend stock by compression molding. These films were annealed for 4 hours at 200°C to completely develop the Kodar crystallinity prior to testing in the Rheovibron dynamic mechanical analyzer at 110 Hz.

As indicated in Figure 1, each blend shows a single α relaxation temperature which corresponds to its Tg and which is determined equivalently from the peak location of the phase angle curve, tan δ, or that of the imaginary component of the modulus, E". The α temperature varies uniformly with composition in support of the argument that PC/Kodar mixtures show miscible amorphous phase behavior. Figure 1 shows that these blends also have a single β relaxation whose temperature location varies smoothly with blend composition.

Blend Tg's observed by Differential Thermal Analysis for a 10°C/min heating rate are shown in Figure 2 to vary smoothly and uniquely with composition. While a comparison of Figure 1 and 2 shows some interesting differences, both results nevertheless confirm the miscible amorphous phase behavior of these blends.

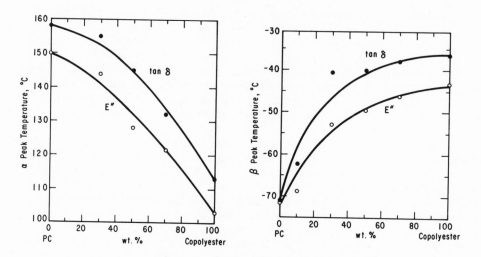

Fig. 1.   Effect of blend composition on the
α and β transition temperatures.

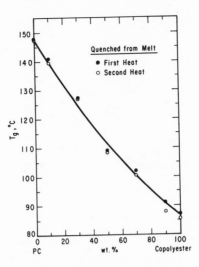

Fig. 2. Effect of blend composition on the glass transition
temperature as observed by DTA.

Gas Sorption Behavior

Equilibrium sorption of CO_2 in blends at 35°C was measured by
the pressure decay method (5). By this method the moles of gas
sorbed into the polymer are determined from the observed difference
between initial and equilibrium gas pressures, the gas volume in
contact with the polymer, and the gas law $p\tilde{V} = ZRT$. Amorphous film
was prepared for these measurements by extrusion at 20°C above the
melting point for each composition into an ice bath.

Previous investigations (6-9) have shown that the equilibrium
gas concentration in glassy amorphous polymers, C, is comprised of
Henry's Law and Langmuir contributions,

$$C = C_D + C_H = k_D p + C_H'bp/(1 + bp) \qquad (1)$$

where k_D is the Henry's Law coefficient, C_H' is the Langmuir capacity,
b is an affinity constant, and p is the gas pressure. The sorption
isotherms for PC/Kodar blends, Figure 3, quantitatively follow the
trends predicted by equation (1), and this equation has been fit to
the isotherms by non-linear regression analysis to compute the
parameters at the various blend compositions.

Figure 4 shows that the affinity parameter, b, is independent
of blend composition. This parameter is probably a function of the
gas, an hypothesis that is currently being investigated.

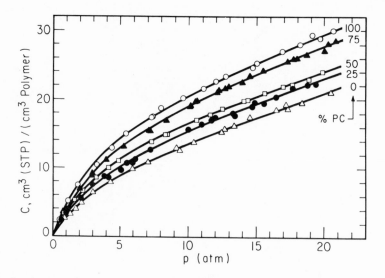

Fig. 3.   Sorption isotherms of CO_2 in PC/Kodar Blends at 35°C.

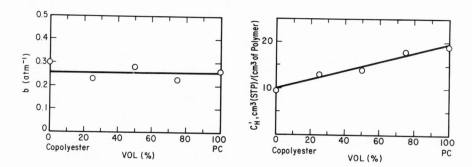

Fig. 4.   Effect of blend composition on the affinity coefficient,
          b and Langmuir capacity, C_H'.

        The Langmuir capacity, C_H', varies linearly with blend composi-
tion, Figure 4. Previous work with pure PC (9) has shown that C_H'
is related to the difference between Tg and sorption temperature,
T, or more precisely to the difference in free volume between the
actual non-equilibrium state of the glass and that which would
exist if equilibrium cooling from the melt were maintained. Since
blend Tg varies nearly linearly with composition, the observed

variation in C_H' is reasonable.

As shown in Figure 5, Henry's Law sorption is less in the blend than would be expected by averaging said behavior of the pure components. This result suggests that the heat of mixing of PC with Kodar is negative. Flory (10) has shown that the heat of mixing per unit volume of solution of polymers 2 and 3, ΔHm, may be written as

$$\Delta Hm = RT \, X_{23} \, \phi_2 \phi_3 / \tilde{V}_2 = B \phi_2 \phi_3 \qquad (2)$$

and that the corresponding activity of gas 1 sorbed in the blend in the Henry's Law region ($\phi_1 << 1$) is given by

$$\ln a_1 = \ln \phi_1 + (1 + X_{12} \phi_2 + X_{13} \phi_3) - \Delta Hm \, \tilde{V}_1 / RT \qquad (3)$$

where X_{ij} is the interaction parameter between components i and j, and ϕ_i and $\tilde{V}_i$ are the volume fraction and molar volume of i, respectively. Likewise, the activity of gas 1 sorbed in pure polymer j is just

$$\ln a_{1j} = \ln \phi_{1j} + (1 + X_{1j}) \qquad (4)$$

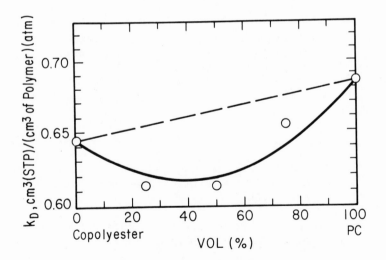

Fig. 5. Effect of blend composition on the Henry's law constant, k_D.

Noting that sorbant concentration $C_D = \phi_1/\tilde{V}_1$ and that $a = p/p_o$, where p and p_o are the pressure and sorbant vapor pressure at T, equation (4) can be rearranged to read

$$k_{1j} = C_{1j}/p = \exp(-X_{1j})/(p_o \tilde{V}_1 e) \tag{5}$$

where k_{1j} is the Henry's Law parameter. Equation (3) can similarly be rewritten in terms of the Henry's Law parameter for the blend, k_D, and equation (5) can be substituted to yield

$$\ln k_D = \phi_2 \ln k_{12} + \phi_3 \ln k_{13} + \Delta Hm \tilde{V}_1/RT \tag{6}$$

Equation (6) suggests that the deviation in k_D from the tie line average observed in Figure 5 is the result of the exothermic heat of mixing of PC with Kodar.

ΔHm can be evaluated from Figure 5 provided $\tilde{V}_1$, the molar volume of the "condensed" gas sorbant can be specified. Prausnitz and Shair (11) have estimated the hypothetical molar volume of CO_2 liquid at 25°C and 1 atm to be 55 cc/gmole by use of corresponding states arguments. Using this value, the maximum ΔHm is -0.78 Cal/cc at 35°C and $\phi_2 = \phi_3 = 0.5$, and the parameter, B, equation 2 is -3.1 cal/cc. This B value is comparable to those obtained for miscible blends of poly(vinylidene fluoride) with various oxygen containing polymers by analysis of melting point depression (12).

Summary and Conclusions

Blends of PC and Kodar are definitely miscible on the basis of their glass transition behavior. The sorption behavior of this miscible system seems to conform quite well to the dual sorption model, equation (1), and the thermodynamic information obtained by analysis of the Henry's Law portion of the sorption isotherm seems reasonable. This work demonstrates that the heat of mixing is negative for miscible PC/Kodar blends and that Henry's Law sorption data may prove useful for quantifying the intensity of interaction between the polymer components of miscible systems.

REFERENCES

1. W. J. MacKnight and F. E. Karasz, Modeling of Tensile Properties of Polymer Blends: PPO/poly(styrene-co-p- chlorostyrene), J. Appl. Phys. 50:6052 (1979).

2. A. F. Lee, Mechanical Properties of Mixtures of two Compatible Polymers, Polym. Eng. Sci. 17:213 (1977).

3. C. A. Cruz, J. W. Barlow and D. R. Paul, The Basis for Miscibility in Polyester-Polycarbonate Blends, Macromolecules, 12:726 (1979).

4. R. N. Mohn, D. R. Paul, J. W. Barlow and C. A. Cruz, Polyester-Polycarbonate Blends. III. Polyesters Based on 1,4-Cyclohexanedimethanol/terephthalic Acid/Isophthalic Acid, J. Appl. Polym. Sci., 23:575 (1979).

5. W. T. Koros, D. R. Paul and A. A. Rocha, Carbon Dioxide Sorption and Transport in Polycarbonate, J. Polym. Sci., 14:687 (1976).

6. W. R. Vieth, H. H. Alcalay and A. J. Frabetti, Solution of Gases in Oriented Poly(ethylene Terephthalate), J. Appl. Polym. Sci., 8:2125 (1964).

7. H. B. Hopfenberg and V. Stannett, in "The Physics of Glassy Polymers", R. N. Haward, ed., Applied Sciences Publishers, London (1973).

8. P. J. Fenelon, Theoretical Prediction of Pressure loss in Pressurized Plastic Container, Polym. Eng. Sci., 13:440 (1973).

9. W. J. Koros, A. H. Chan and D. R. Paul, Sorption and Transport of Various Gases in Polycarbonate, J. Membrane Sci., 2:165 (1977).

10. P. J. Flory, "Principles of Polymer Chemistry," Cornell University Press, Ithaca, New York (1953).

11. J. M. Prausnitz and F. H. Shair, A Thermodynamic Correlation of Gas Solubilities, A.I.Ch.E. Journal, 6:682 (1961).

12. D. R. Paul, J. W. Barlow, R. E. Bernstein and D. C. Wahrmund, Polymer Blends Containing Poly(Vinylidene Fluoride). Part IV: Thermodynamic Interpretations, Polym. Eng. Sci., 18: 1225 (1978).

RELAXATION PROCESSES IN GLASSY POLYMERS AND THE STRAIN AND TIME DEPENDENCE OF GAS PERMEATION

T.L.Smith, G.Levita°

IBM Research Laboratory, San Jose, CA, U.S.A.
°   University of Pisa, Italy

(Abstract)

The viscoelastic properties of glassy polymers are known to be affected by volumetric relaxation whose rate, as reflected in certain properties, is accelerated by tensile deformations. To obtain additional information on molecular processes in such polymers, a study was made of the effect of ostensibly homogeneous deformations in simple tension on the effective permeability and diffusion coefficients (P and D) of gases in a biaxially oriented polystyrene film and, to some extent, in a polycarbonate film.

Stepwise application of a tensile strain to the polystyrene film markedly increased P and D of CO_2, N_2, A, Kr, and Xe at 1 atm until the yield strain was reached, undoubtedly because the free volume was increased. (Increases in P up to 77% were observed). For each gas except CO_2, independent of molecular diameter, P and D increased 31% and 20%, respectively (25% and 15% for CO_2) per 1% strain; presumably, the size distribution of free-volume holes is not distorted initially by strain. At constant strain, P and D decrease with time, attributable to volumetric relaxation or to a size redistribution of free-volume holes. The observed decay rates, which are smaller than the stress-relaxation time, increase with strain magnitude and temperature. At 1.8% strain at 50°C, P and D for Xe decreased 13.8% and 11.8%, respectively, per decade of time, which are two to threefold larger than for CO_2. Apparently, the larger free-volume holes decrease in size more rapidly than the smaller holes.

For the polycarbonate film at small strains, ε, $(1/D)dD/d\varepsilon$ was found to be 14 at 50°C for CO_2 and between 5 and 8 at somewhat higher temperatures for N_2. If the effect (unknown) of volumetric relaxation on these data are overloocked, they can be accounted for semiquantitatively in terms of the derived equation :

$$(1/D)dD/d\varepsilon = (B/f^2)(\beta_f/\beta)(1 - 2\nu)$$

Where B depends on penetrant-polymer characteristics, f is the fractional free volume, β_f and β are the compressibilities of the free volume and the volume, respectively, and ν is Poisson's ratio. Estimation of these quantities from literature data leads to $(1/D)dD/d\varepsilon = 13.7$ for CO_2 and 12.3 for N_2 at 50°C.

APPLICATION OF A THERMODYNAMICALLY BASED SINGLE INTEGRAL

CONSTITUTIVE EQUATION TO STRESS RELAXATION AND YIELD IN ABS

C. J. Aloisio   R. A. Schapery

Bell Laboratories   Department of Civil Eng.
2000 Northeast Expressway   Texas A&M University
Norcross, Georgia 30071   College Station, Texas 77843

The main variation in linear viscoelastic time-dependent properties at different temperatures can be incorporated into the time scale.[1] When properties are a function of stress and strain – termed non-linear viscoelastic - it is often possible to use modulus and time scaling with respect to these variables.[2,3] The analytical framework supplied by Schapery's single integral constitutive equation based on a thermodynamic theory,[3,4,5] provides a rational approach to the characterization of linear and non-linear viscoelastic behavior. Initially developed in a generalized three-dimensional form,[5] simplified uniaxial forms have been frequently used in polymer characterization.[2,6]

In the thermodynamic development,[4] the rate of entropy production due to the irreversibility of the process is first derived. The entropy production is shown to be proportional to the difference between the actual force on the system and the "reversible" part of the actual force. Then the equations of motion of the system are obtained. It is assumed[5,7] that the entropy of a system sufficiently close to equilibrium exists and can be defined by thermodynamic state variables.

A simplified uniaxial, isothermal form is used here to analyze ABS data. In this case the constitutive equation becomes

$$\sigma = A_F \int_o^t E(\psi-\psi') \frac{d\varepsilon}{d\tau} \, d\tau \qquad (1)$$

The function A_F (σ, ε, τ) is the nonlinear contribution to the structural stiffness of the polymer.[5,6] In the linear viscoelastic case A_F may be the familiar rubber elasticity function from Rouse's theory.[1,6,8] Variations in A_F with temperature (T), stress (σ) or strain (ε) variations during testing, produce translations on the log modulus or stress coordinate. The function E(t) is the usual linear viscoelastic stress relaxation modulus. The reduced time difference is

$$\psi - \psi' = \int_\tau^t \frac{A_F}{A_D} (\sigma, \varepsilon, T) \, dt' \tag{2}$$

The function A_F/A_D accounts for time scaling in both the linear and non-linear viscoelastic range of behavior. Only if A_F is small, or removed, can such horizontal translation be treated as the contribution to polymer viscosity.

The ABS polymer studied here exhibited linear viscoelastic behavior (LVE) up to 0.5% strain. The LVE master curve of Figure 1 was synthesised from isotherms obtained over the range 82°F(27.8°C) to 140°F(60°C).[6] The stress relaxation curves at constant strains to 10%* did not appear to be superposable. Therefore, the simple application of a horizontal and vertical translation was not possible. The slope of the log modulus-log time data at high strains was about 0.1, considerably less than the 0.37 of the LVE curve at long times. Using a stress dependent form of A_σ

$$A_\sigma = \left(\frac{\sigma}{\sigma_o}\right)^{-m} \tag{3}$$

a nonlinear relaxation modulus of the form

$$E_N(t) = Kt^{-N} \tag{4}$$

and equation (1) and (2) the following is obtained

$$E_N(t) = A_F \, E(\psi) \tag{5}$$

$$\psi = A_F \, K' \, \frac{t^{(1-mN)}}{1-mN} \tag{6}$$

---

*A simple engineering strain measure was utilized, i.e. deformation over original length, and was measured with an extensometer attached to the sample.

Differentiation of the log of equation (5) with respect to log experimental time, t, one obtains

$$\frac{d \log E_N}{d \log t} = (1-mN) \frac{d \log E}{d \log \psi} = -N \tag{7}$$

The affect of a stress dependent non-linearity on stress relaxation data predicted by equations (6) and (7) is a horizontal translation corresponding to $\log A_F K'$ and a rotation to the LVE modulus long time slope (0.37).

Using an A_D of the following form

$$A_D = A_\varepsilon \left(\frac{\sigma}{\sigma_O}\right)^{-m(\varepsilon)}, \tag{8}$$

constant crosshead speed (CHS) data for 0.5 inch/min. CHS and the LVE modulus of Figure 1 the strain dependent functions of Figures 2 and 3 were obtained. As expected the strain-time relationship in the tensile bar gage length was not constant but of the form

$$\varepsilon = \left(\frac{t}{t_o}\right)^{1.6} \tag{9}$$

for constant crosshead speeds.

The comparison with experiment is shown in Figure 4 for 82°F. In an analogous approach the curves for 140°F in Figure 5 were obtained. The calculated curves predict many of the experimentally observed features of the stress-strain curve of ABS such as

1. gradual stress maximum without an abrupt stress decrease beyond yield,
2. increased yield stress with increased loading rate, and
3. increased yield strain with increased loading rate.

The yield stress values obtained are in quantitative agreement with experimental values over at least two decades of loading rate at 82°F and 140°F.

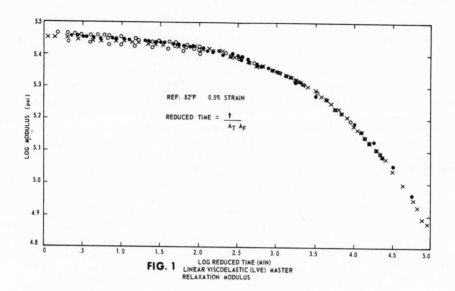

REF: 82°F   0.5% STRAIN

REDUCED TIME = $\dfrac{t}{A_T\,A_F}$

FIG. 1 LINEAR VISCOELASTIC (LVE) MASTER
RELAXATION MODULUS

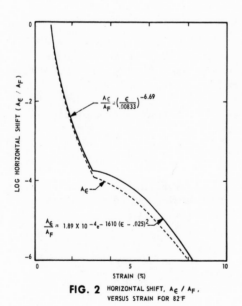

$\dfrac{A_\epsilon}{A_F} = \left(\dfrac{\epsilon}{.00833}\right)^{-6.69}$

A_ϵ

$\dfrac{A_\epsilon}{A_F} = 1.89 \times 10^{-4}\,e^{-1610\,(\epsilon - .025)^2}$

FIG. 2 HORIZONTAL SHIFT, A_ϵ / A_F,
VERSUS STRAIN FOR 82°F

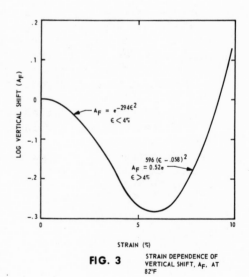

$A_F = e^{-294\epsilon^2}$
$\epsilon < 4\%$

$A_F = 0.52\,e^{596\,(\epsilon - .058)^2}$
$\epsilon > 4\%$

FIG. 3 STRAIN DEPENDENCE OF
VERTICAL SHIFT, A_F, AT
82°F

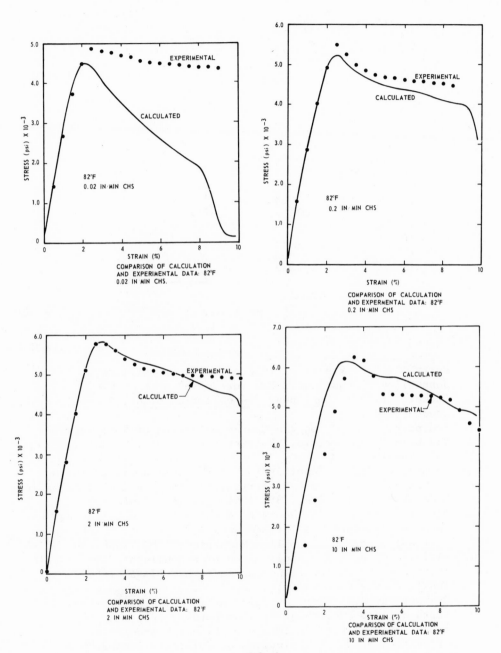

FIG. 4

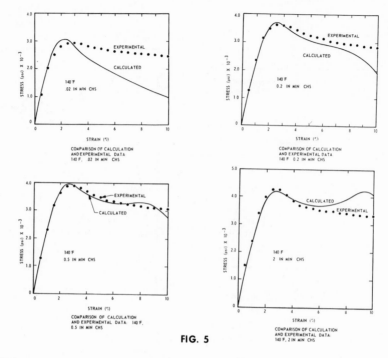

FIG. 5

REFERENCES

1. Ferry, J. D.: Viscoelastic Properties of Polymers,
    John Wiley & Sons, Inc., New York, 1969.
2. Lockett, F. J.: Nonlinear Viscoelastic Solids, Academic Press,
    London, 1972.
3. Schapery, R. A.: "On the Characterization of Nonlinear
    Viscoelastic Materials," Polymer Eng. & Sci., 9, July, 1979,
    pp. 295-310.
4. Schapery, R. A.: "Application of Thermodynamics to Thermo-
    mechanical, Fracture & Birefringent Phenomena in Viscoelastic
    Media," J. Appl. Phys., 35, 1964, pp. 1451-1465.
5. Schapery, R. A.: "A Theory of Nonlinear Thermoviscoelasticity
    Based on Irreversible Thermodynamics," Proc. of the Fifth U.S.
    National Congress of Applied Mechanics, ASME, 1966, pp. 511-530.
6. Aloisio, C. J.: The Application of a Nonlinear Viscoelastic
    Theory to the Characterization of an ABS Plastic. Ph.D.
    Thesis Purdue University, June, 1970.
7. DeGroot, S. R.: Thermodynamics of Irreversible Processes,
    North-Holland Publishing Co., Amsterdam, 1951.
8. Rouse, Jr., P. E.: "A Theory of Linear Viscoelastic
    Properties of Dilute Solutions of Coiling Polymers,"
    J. Chem. Phys., 21, pp. 1272-1280, 1953.

IRREVERSIBLE THERMODYNAMICS OF

GLASSY POLYMERS

K. C. Valanis, University of Cincinnati
Cincinnati, Ohio, U.S.A.
S. T. J. Peng, Jet Propulsion Laboratory
California Institute of Technology
Pasadena, California, 91003, U.S.A.

INTRODUCTION

In a previous paper[1] we laid the physical foundations for the atomic and molecular interpretations of the "internal variables" in theories of irreversible thermodynamics. This we did using the concepts of "deformation kinetics" as a point of departure. The results were two-fold:

(i)   Non-linear equations of evolution of internal variables were established.

(ii) Phenomenological parameters were given an atomic or molecular identity.

In this paper we utilize the above ideas for the understanding of certain aspects of the transient behavior of glassy polymers. Before we proceed with this task we review some of the results of the theory. With particular reference to pressure–volume behavior, let ϕ be the Gibbs free energy density of a system, where

$$\phi = \phi(P, T, q_\alpha) \tag{1}$$

where P is the pressure, T the absolute temperature and q_α are internal variables necessary to describe the internal molecular state of the system.

If we denote the volume of the system by V and entrophy by η, then it can be shown[2] that

$$V = \left.\frac{\partial \phi}{\partial P}\right|_{T,q} \quad (2), \qquad \eta = \left.\frac{\partial \phi}{\partial T}\right|_{P,q} \tag{3}$$

and

$$-\left(\frac{\partial \phi}{\partial q_\alpha}\right)_{T,P} \dot{q}_\alpha > 0 \quad (\alpha \text{ not summed}) \tag{4}$$

for all $\alpha = 1,2 \cdots N$.   Inequality (4) is the result of the Clausius-Duhem inequality regarding the non-negative nature of the rate of irreversible entropy production.

Inequality (4) is of extreme importance in that, it asserts that a relation must exist between $\frac{\partial \phi}{\partial q_\alpha}$ on one hand, and $\dot{q}_\alpha$ on the other, because otherwise it would be possible to ascribe values to these independently and in such a manner to violate inequality (4).   Also worthy of note is the fact eq.'s (2) and (3) apply to all materials irrespective of their constitution and it is the "internal constitutive equation" that relates $\frac{\partial \phi}{\partial q_\alpha}$ and $\dot{q}_\alpha$ that determines the materials behavior.   Such an equation was established in Ref. 1 from consideration of reaction kinetics.   These ideas will be reviewed here in brief leaving the reader to consult Ref. 1 for details.

For the purpose of the study of mechanical motion, we assume that in the case of a material which is initially in a state of equilibrium, the motion of a particle is governed by an initially symmetric potential near field.   (See Fig. 1)

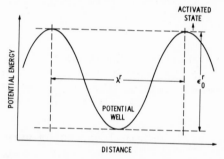

Fig. 1 Symmetrical potential energy barriers under no external stress.

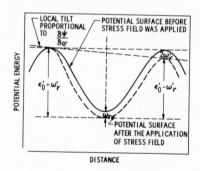

Fig. 2 Potential energy barriers under externally applied stress field.

Particles impeded by potential barriers of height ϵ_o^r belong to the group r whose mean motion is described by the internal variable q_r.   Obviously, the mean displacement is zero, if initially the barriers are symmetric.   Application of a force field results in coordinated motion which must be caused by a distortion of the potential barrier.   The local mean "internal force" Q_r experienced by the group r is given, by eq. (5)

$$Q_r = \frac{\partial \phi}{\partial q_r} \qquad (5)$$

The fundamental assumption of Ref. 1 is that the local distortion is a function of the local internal force.   With reference to Fig. 2, the distortion W_r, which is antisymmetric, is assumed to be a function of Q_r, i.e.

$$W_r = F(Q_r) \qquad (6)$$

In particular, it was assumed in Ref. 1 that the above relationship is linear, i.e.,

$$W_r = C_r Q_r \qquad (7)$$

where C_r is a material constant.

It was further shown that the number of particles N^r that will partake in a coordinated motion is given by the equation:

$$N^r = 2n^r \sinh (\beta W_r) \sum_{\varepsilon_i^r > \varepsilon_o^r}^{\infty} \alpha e^{-\beta \varepsilon_i^r} \qquad (6)$$

where n^r is the number of particles in "group r", $\beta = \dfrac{1}{kT}$, ε_i^r are the energy states of group r and

$$\alpha = \left\{ \sum_{\varepsilon_i} e^{-\beta \varepsilon_i^{(r)}} \right\} - 1 \qquad (7)$$

More precisely, since the distribution of energy states above ε_o^r is continuous, eq. (6) becomes

$$N^r = 2 n^r A_r \sinh (\beta W_r) e^{-\beta \varepsilon_o^r} \qquad (8)$$

where

$$A_r = \alpha \int_0^{\infty} A(x + \varepsilon_o^r) e^{-\beta x} dx \qquad (9)$$

in other words, A_r is a function of ε_o^r and the absolute temperature. Again, with reference to Fig. 2, if τ^r is the average time of the group r to traverse the distance λ^r across the barrier, then the average velocity of the group is $\dot{q}_r$ where

$$\dot{q}_r = \frac{\lambda^r}{\tau^r} \frac{N^r}{n^r} \qquad (10)$$

Equation (10) in conjunction with eq.'s (7) and (8) yields the equation of motion eq. (11), of the group r which is the equation of evolution of the internal variable q_r,

$$\dot{q}^r + K_1^r \sinh \left\{ K_2^r (\frac{\partial \phi}{\partial q^r}) \right\} = 0 \qquad (11)$$

where

$$K_1^r = \frac{2\lambda^r}{\tau^r} e^{-\beta \varepsilon_o^r} \cdot A(\varepsilon_o^r, T) \quad (12), \quad K_2^r = c^r \beta \lambda^r \quad (13)$$

Equations (1), (2), (3) and (11) completely describe the thermomechanical response.

Applications: Volumetric Thermal Relaxation or Creep of Glassy Polymers

I. T-jump experiment (sudden heating or cooling) from the same temperature. Let the specimen be initially in equilibrium at pressure p_o and temperature T_o. The temperature T_o is changed suddenly to a temperature T_1 at $t = t_1$ (i.e., $T_1 = T_o \pm \Delta T$). The volume V is then measured at time $t \geq t_1$. It is observed that V changes with time. For instance if $T_1 < T_o$ then V decreases with time until it reaches a new constant value. This is a case of thermal relaxation. If, on the other hand, $T_1 > T_o$ then the volume increases with time to a new constant value. This is the case of thermal creep.

Theoretical Description of the Phenomenon

Let $s = t - t_1$. The conditions of the experiment are the following;   At s = o: $p = p_0$, $T = T_1$, $q_r = q_r^o$ (13)

where q_r^o is an unknown function of the previous thermal history but
has a <u>fixed</u> value q_r^o at s = 0.

At s>o: $p = p_0$, $T = T_1$, $q_r = q_r(s)$ (14)

The purpose of the analysis is to determine the functions $q_r(s)$, r =
1,2····N.   Here we shall limit ourselves to the case where N = 1.
Though the resulting model is a very simple one indeed, we shall show
that it predicts the salient features of the experimental observations.

If we knew the precise form of the Gibbs free energy function ϕ,
and for r = 1, then application of eq. (11) would yield a first order
differential equation in q_1, which we simply denote by q shown below

$$\dot{q} + K_1 \sinh \left(K_2 \frac{\partial \phi}{\partial q}\right) = 0   (15)$$

where $\phi = \phi(p_0, T_1, q)$ in the time interval $(0 \le s > \infty)$.   If we limit
ourselves to situations where changes of temperature are small (of
the order of $10^{\circ}K$ or so) and changes of volume are also small (of the
order of one to two percent) then we can regard experiments leading
to such changes as perturbations from an initial state, thereby indicating changes in q are likely to be small.   Thus, we may write ϕ
in the form

$$\phi = \phi_o + \phi_1 q + \tfrac{1}{2} \phi_2 q^2   (16)$$

where ϕ_r (r = 0,1,2) are functions of P_o, T_1 and q_o.   Equation (16)
in conjunction with eq. (14) now leads to the relation

$$\dot{q} + K_1 \sinh K_2 (\phi_1 + \phi_2 q) = 0   (17)$$

Also, the volume V is obtained by using eq.'s (2) and (16) i.e.,
to a first order in q:

$$V = \frac{\partial \phi_o}{\partial P} + \frac{\partial \phi_1}{\partial P} q   (18)$$

where $\frac{\partial \phi_o}{\partial P}$ and $\frac{\partial \phi_1}{\partial P}$ depend only on P_o, T_1 and q_o, i.e. they are constant scalars independent of s.   To solve eq. (17) we introduce the
transformation

$$\mu = \phi_1 + \phi_2 q   (19)$$

in which case eq. (17) becomes:

$$\dot{\mu} + K_1 \phi_2 \sinh K_2 \mu = 0   (20)$$

with the initial condition:

$$\mu(0) = \phi_1 + \phi_2 q_o   (21)$$

The solution of eq. (20) is the following:

$$\mu = \frac{2}{K_2} \tanh^{-1} \left\{\mu_o \exp \left(- \, _2 K_1 K_2 t\right)\right\}   (22)$$

where $K_2 = K_2(T_1)$, $K_1 = K_1(T_1)$ and ϕ_1 are constant.

It follows from eq.'s (18), (19) and (22) that

$$V = (\frac{\partial \phi_0}{\partial P} - \frac{\phi_1}{\phi_2}\frac{\partial \phi_1}{\partial P}) + \frac{\partial \phi_1}{\partial P}\frac{2}{K_2 \phi_2} \tanh^{-1}\left[\mu_0 \exp (-\phi_2 K_1 K_2 t)\right]$$

$$(23)$$

Evidently, as $t \rightarrow \infty$ the second term on the right hand side of eq. (23) tends to be zero, so that

$$\frac{\partial \phi_0}{\partial P} - \frac{\phi_1}{\phi_2}\frac{\partial \phi_1}{\partial P} = V_\infty \qquad (24)$$

Thus, if we introduce the simplified notation

$$aV_\infty = \frac{\partial \phi_1}{\partial P}\frac{2}{K_2 \phi_2} \quad (25), \quad b = K_1 K_2 \phi_2 \qquad (26)$$

We then have an expression for the volumetric "strain" $\frac{V - V_\infty}{V_\infty}$ in the form:

$$\frac{V - V_\infty}{V_\infty} = a \tanh^{-1}\left[\mu_0 \exp (-bt)\right] \qquad (27)$$

It is clear that the constants a and b will be different from cooling and heating, i.e., $a(T_0 + \Delta T) \neq a(T_0 - \Delta T)$, $b(T_0 + \Delta T) \neq b(T_0 - \Delta T)$. Hence, the thermal creep and relaxation described by eq. (27) will be asymmetric.

II. T-jump experiment (sudden cooling or heating) from different temperatures T_1 and T_2 to $T_0(T_2 < T_0 < T_1)$. In this experiment, the sample will be suddenly cooled down from T_1 to T_0, or suddenly heated up from T_2 to T_0. The solutions of the experiments are still described by eq. (27), however, because of the different initial conditions μ_0 [i.e., eq. (21)] will be different. Thus, the thermal creep and relaxation described by eq. (27) will be asymmetric which have been observed by many authors[3].

Conclusion

We have shown by our very simple model, that the salient features of the experimental observations of non-equilibrium behavior of glassy polymer can be described, at least, qualitatively. Detailed description of thermal creep and relaxation with more complicated thermal history will be reported in later papers. Furthermore, attempts will be made by employing the concept of internal variables and deformation kinetics, to bring together in an unified fashion the coupling effects of temperature and stress or strain history manifested in the rate of physical aging.

References

1. K. C. Valanis and S. J. Lalwani, J. Chem. Phys. 67, 3980 (1977).

2. K. C. Valanis, Arch. Mech. Stosowanej, 23, 535, (1975).

3. A. J. Kovacs, Fortsch. Hochpolym. Forsch. (Ad. in Polymer Sci.), 3, 394 (1963).

CHARACTERIZING RHEOLOGICAL CURE BEHAVIOR OF EPOXY COMPOSITE

MATERIALS

R. J. Hinrichs and J. M. Thuen

Senior Analytical Chemists
Narmco Materials Inc.
Celanese Plastics and Specialties Company

ABSTRACT

The rheological properties of epoxy formulated polymers are directly related to the reaction cure dynamics, flow, gas-liquid transport, and laminant consolidation. However, epoxy formulations are not simple newtonian fluids. They require proper evaluation of instrumental test parameters to define an appropriate linear response conditions. This paper presents how the test conditions are derrived and illustrates the application of rheological measurements to evaluate composite manufacturing processes.

INTRODUCTION

Composite structures are now being fabricated into primary aircraft components. These laminantes are critical to the structural integrity of the aircraft and represent significant material-labor costs to produce. Unfortunetly, most fabrication facilities experience random, appearently unexplained, part rejections due to variations in cure ply thickness, porosity and voids. This study deals with developing the application of rheological characterization techniques to elucidate the unpredictable process behavior of composite structures. The rheological properties of epoxy polymers are directly related to the reaction cure dynamics, flow, gas-liquid transport and consolidation of the laminant.

RHEOLOGICAL TECHNIQUE

This study relied upon a Rheometric Visco-Elastic test unit modified for polymer cure behavior research. Specificaly, a 2000 gm-cm torque transducer was installed for sensitivity. An environmental chamber was modified for computer controled linear heat rate programming from 1-C/min to 99-C/min.   The instrument is also equipped with frequency and strain sweep options to automatically scan through appropriate ranges of operation.   Fifty millimeter parallel plate fixtures are used to measure the visco-elastic properties of the material during cure.   Sinusoidal oscillating rheometry was used to minimize shear stress affects.   It is recognized that parallel plate geometry represents a gradient of shear rates during oscillation but this results in a condition more realistic to laminant flow conditions.   This technique results in an average effective shear stress for a given frequency and strain parameter.   We then test the material for linearity of its dynamic viscosity over the temperature, frequency and strain conditions used to evaluate process behavior.   This results in an effective instrument operating region.

Narmco 5208 resin system was chosen to evaluate this technique. Figures 1 and 2 illustrates the frequency and strain sweep response at 50-C, 100-C, and 130-C.   From these tests, a frequency of 10 radians per second and 50% strain were chosen as the standard test parameters.

Sample handling techniques are also critical.   Epoxy materials are sensitive to temperature, aging, and moisture.   This is demonstrated under the applications section of this paper.   It is mentioned here to point out that control samples should be maintained under strict environmental conditions in order to determine if observed rheological effects are real or an induced artifact.

APPLICATION TO PROCESS BEHAVIOR RESEARCH

Three basic factors affecting the process behavior of composites were evaluated.   The first was resin advancement of staging during the impregnation process. (figure 3) Three advancement levels were studied representing the range over which the product could be supplied.   Several significant results were obtained.   First, the kinetic rate of reaction is much higher as the resin system is advanced.   This is evident by the slope and viscosity profile during the hold cycle.   The viscosity minimums are very much altered.   This viscosity profile determines how much flow will occur for a given pressure application cycle.   The resin of low advancement re-softens as the material continues to 177-C and gels approximately within five minutes after reaching 177-C.   The higher advancement system does not appreciably re-soften and is already very high in viscosity by end of hold cycle.   It also gels 10 minutes before reaching the cure temperature.   The intermediate staged sample bisects these two conditions.   This would indicate a dramatic difference in behavior. They would therefore not exhibit the same processing characteristics for a given fixed cure cycle.   Pressure application points would need to be changed to account for this effect.

The second factor investigated was the effect of room temperature aging during fabrication.   The material was maintained in a desicated control chamber at 75 ± 5F for 45 days.   Samples were evaluated and are summarized in figure 4.   Just as in staging, this

material changes its cure profile significantly with time. The
initial and minimum viscosities are increased. The important
region where pressure-laminant consolidation would occur, half
way through the isothermal hold, is so altered that under a fixed
cure cycle the aged material would not flow or consolidate.
This would result in poor laminant quality. This also illustrates
the importance of fabrication control during manufacturing.

The third important factor evaluated was the effect of
humidity in a fabrication facility causing moisture to absorb
into the epoxy material. The resin was exposed to 98% relative
humidity for several hours and the viscosity profile changed
dramatically. (Figure 5) There are three significant differences
in the profile. First, the initial viscosity measurements are
lower. This indicates a water induced plasticizing effect. The
presence of water increases the fluidity of the system and lowers
the apparent viscosity. Secondly, during the heat cycle the
minimum viscosity increases as the water volatilizes and expands.
It is interesting to note that the peak area observed in this
region corresponds to the water content within the resin. Also,
at 110-C the storage (or elastic) modulus suddenly becomes
significant. This would indicate some type of reaction or
polymer network formation is occuring as a result of the presence
of water. The viscosity remains higher during the remainder of
the cure cycle. Gel occurred before reaching cure temperature.

In short, the water appears to alter the resin behavioral
properties during volatilization similar to that of increasing
resin advancement. Another interesting effect was noticed on the
instrument plates. As the water concentration increased, so did
the number of void areas on the plates (photo 1, 2, 3). In fact,
the void area on the plates could almost be used to predict the
actual water concentration. Thus, along with air entrapment and
layup techniques, water acts to produce porosity areas and alter
the cure cycle behavior of the resin system.

Reversability of moisture effects was also investigated.
Exposed material was re-dessicated and dried back to pre-exposure
levels. The rheological curves are shown in Figure 7. Here the
sample prior to desorption clearly shows the moisture effects.
The dried sample returned to normal. Thus, at room temperatures
water does not tend to damage or alter the resin characteristics,
but must be controlled before going into an elevated temperature
cure cycle.

CONCLUSIONS

Rheological measurements of epoxy composite materials can
elucidate many manufacturing abnormalities. The test parameters
and sample handling techniques must be carefully controlled to
prevent aging and moisture effects from misleading experimental
results. The application to process behavior modeling and
investigation is significant. The effects of advancement, aging

and moisture on laminant process behavior is clearly evident.
This technique can therefore be used to model cure cycle parameters
and determine optimum time, pressure, temperature profiles. It
further illustrates the need for control of fabrication process
environment.

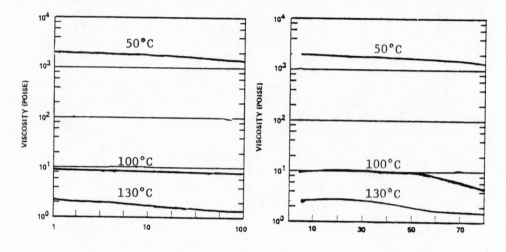

Fig. 1   Frequency Sweep (rad/sec)   Fig. 2   Strain Sweep (%)

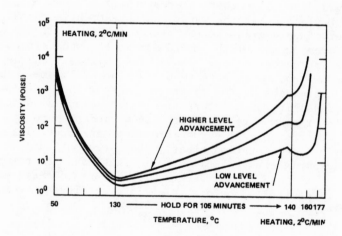

Fig. 3   Staging-Advancement

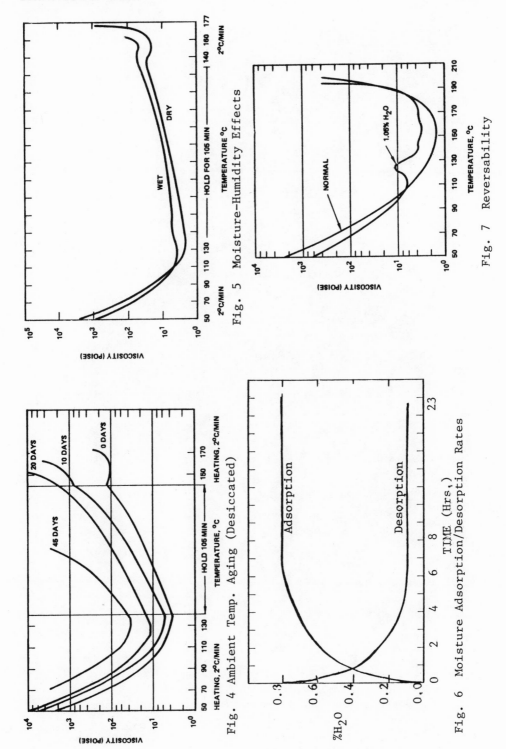

Fig. 5 Moisture-Humidity Effects

Fig. 7 Reversability

Fig. 4 Ambient Temp. Aging (Desiccated)

Fig. 6 Moisture Adsorption/Desorption Rates

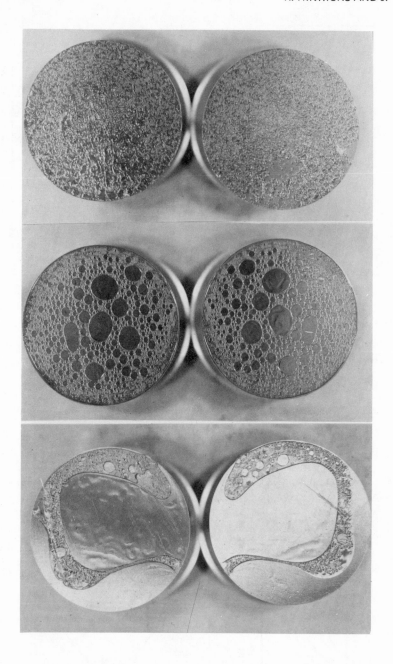

Photos. 1.   Top- 50mm rheological plates with 0.1% water
        2.   Middle- 50mm rheological plates with 0.34% water
        3.   Bottom- 50mm rheological plates with 1.0% water

CONFORMATIONAL ENERGIES OF POLYMERS

James M. O'Reilly

Webster Research Center
Xerox Corporation
Webster, New York   14580

INTRODUCTION

The difference in energy between different conformational states of macromolecules is a fundamental characteristic, in addition to molecular geometry, which determines the physical properties of chain molecules. Chain dimensions and the temperature dependence of chain dimensions[1] in solution and in the bulk are dependent upon the energy states of the chain backbone. For simplicity, only two energy states of the carbon to carbon backbone bonds of vinyl chains will be considered and the gauche state is assumed to be higher energy than the stable trans state.

Moreover, the conformational energy difference is important in glass formation[2] and in the changes in properties at T_g, such as, specific heat[3] and mechanical loss.[4] In spite of the importance of conformational energies, a proportionate amount of research has not been directed toward the determination of accurate values. A primary source of conformational energies is from rotational isomeric state calculations of chain dimensions and temperature coefficients of dimensions.[1] Derived energies are often compared with values for model compounds which are obtained by classical spectroscopic and structural methods.[5] Conformational energies have been measured for polyvinyl chloride by infrared spectroscopy. In this report, the infrared method has been applied to poly(methylmethacrylate) in different stereo-forms. The effect of stereo-form on the conformational energy can be measured directly on different stereo-polymers. In the Gibbs DiMarzio theory of glass formation, the conformational energy (flex energy) should scale with the glass transition temperature, that is,

ϵ/kT_g = constant. From the conformational energies measured by IR spectroscopy, the correlation of glass temperature with conformational energy could be tested in the most favorable conditions where only the stereo- configuration of the chain unit changes.

RESULTS

Three samples of poly(methylmethacrylate) of different stereoregularity denoted as isotactic (95%), syndiotactic (80%) and atactic (60%) were used for this study. More detailed characterization is reported[7] elsewhere. Infrared absorbance measurements were made on thin films cast upon KBr crystals using a Digilab FTS-15D spectrometer at 1cm^{-1} resolution and 200 scans. Absorbance was measured as a function of temperature using a Beckman RIIC cell at temperatures up to 180°C. Two or three temperature runs were made on each sample to establish reproducibility.

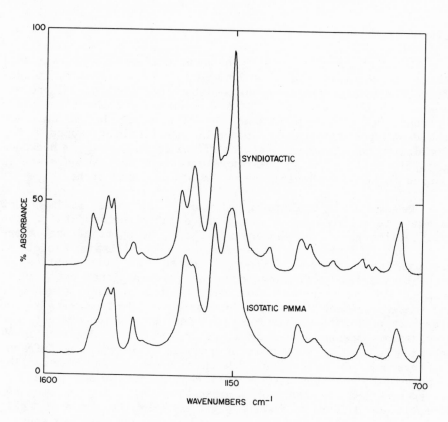

Fig. 1 - Absorbance spectra of PMMA

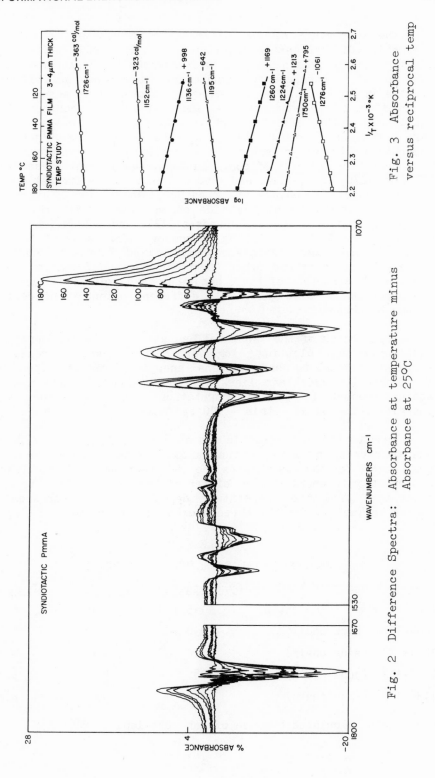

Fig. 3 Absorbance versus reciprocal temp

Fig. 2 Difference Spectra: Absorbance at temperature minus Absorbance at 25°C

The abosrbance of the different PMMA samples is shown in Figure 1. Differences in absorbance at $1485_{8,9}$ 1265, 1150cm^{-1} are apparent and have been previously discussed. Measurements as a function of temperature for syndiotactic are shown as difference spectra in Figure 2. The increases and decreases in absorbance as a function of temperature are indicative of a conformational equilibria. Other effects, such as intermolecular interactions or Fermi resonance could lead to the temperature dependent absorbance. The clear isobiestic points in the spectra support the interpretation in terms of conformational equilibria. Absorbances at 1730, 1270, 1240, 1190, and 1150cm^{-1} exhibit the largest changes in absorbance. The change in the carbonyl absorption is attributed to two conformations of the side chain ($COOCH_2$ with the methyl group cis to the carbonyl). Bands at 1240 and 1270cm^{-1} have been assigned^{9} the C-C-O stretching frequency. The bands at 1150 and 1190cm^{-1} have been associated with mixed vibration modes of skeletal stretching and other carbon hydrogen deformation modes. The doublet structure of these bands has been explained[10] as due to conformations of side group with different orientations relative to the backbone.

Application of the van't Hoff equation to the conformational equilibria by plotting log absorbance versus reciprocal temperature leads to conformational energies from the slopes shown in Figure 3. Excellent linearity is observed for all the data adding support to the interpretation. Duplicate runs yield energies which agree within $\pm$ 100 cal/mole.

Results for the isotactic and atactic polymer are similar but due to overlap of some of the bands certain energies cannot be evaluated. Yet the absorbances are different and lead to different conformational energies for different stereopolymers. Energies calculated from the temperature dependence of absorbance are listed in Table I for the three stereoregular polymers.

TABLE I

CONFORMATIONAL ENERGIES   (CAL/MOLE)

	Syndiotactic	Atactic	Isotactic
1725-1750cm^{-1} (side chain)	1145	970	678
1260-1276 (backbone)	2090	1426	718
1224-1244 (side chain)	1150	920	n.c.
1167-1195 (side chain)	1080	836	n.c.
1136-1151 (side chain)	1283	893	n.c.

n.c. - not calculated because of band overlap.

The energy determined from the 1730cm^{-1} band is attributed to the two conformations of the carbo-methoxy side group. A general trend, energy decreasing with increasing isotacticity, is found although differences are not large. The band at 1270cm^{-1} leads to the largest energy which would be assigned to the backbone conformations. This assignment is not inconsistent with the vibrational assignment to C-C-O stretching although one would have expected larger conformational energies for the skeletal stretching bands. The skeletal stretching frequency at 1150 and 1170cm^{-1} leads to energies of 1170 to 910 cal/mole as a function of tacticity.

The trends of the conformational energies with tacticity are reasonable and are self-consistent. Because many of the temperature dependent vibrational frequencies observed are associated with vibrational motion of side groups rather than the backbone, we must interpret these results with caution. More confidence will be gained in the interpretation as comparisons with other data agree and as equivalent measurements on other related polymers are obtained.

TABLE II

CONFORMATIONAL ENERGIES AND GLASS TEMPERATURES OF PMMA

	ΔH conf.		T_g	$\Delta H/kT_g = \epsilon/kT_g$
	FTIR cal/mole	R.I.S. cal/mole		
Isotactic	700 ± 200	$1330\ (^{550}_{2200})$	$45^\circ C$	1.1
Atactic	1400 ± 200		$105^\circ C$	1.85
Syndio-Tactic	2100 ± 200	$1930\ (^{1700}_{2300})$	$115^\circ C$	2.71

These results are compared with the rotational isomeric state (R.I.S.) calculations on PMMA dimensions in theta solutions in Table II. The agreement for syndiotactic PMMA is excellent and within the uncertainty of either result. Comparison of the results for isotactic PMMA is not as good, but there can be a large variation in the R.I.S. result depending on the cut-off distance beyond which the potential energy of interaction of different conformers with solvent is insensitive to distance. A lower value of the conformational energy could be accomodated by the R.I.S. calculations but the value from the infrared analysis represents an upper limit. In the R.I.S. calculation, the side group conformation is fixed and the influence of this on the calculation is unclear. Another factor is that the IR measurements apply to the bulk polymer and the dimensions are for theta solvent conditions. The results for the atactic PMMA are intermediate between the isotactic and syndiotactic polymers as would be expected.

Although, theta conditions and the bulk liquid are assumed to be equivalent, specific interactions could occur and influence the results. Thus, the infrared method provides an additional method of testing the conclusions of the R.I.S. calculations.

Encouraged by these comparisons, the results of Gibbs DiMarzio reduced energy parameter, ϵ/kT_g, tabulated in Table I can be discussed. As is immediately obvious from the results in Table I, ϵ/kT_g is not constant. The values are outside limits from results obtained from other methods. Early estimates of ϵ/kT are 1.8-2.3[2] and 1.8-2.4.[3] In spite of the uncertainty, we would conclude that the glass temperature is not controlled by conformational energy alone in PMMA. If this is true, then other factors, such as packing of chains and interactions of chains depend upon tacticity and affect the glass transition. The effect of packing of chains at T_g is probably the more important. These conclusions suggest the two parameters, flexed bands and empty lattice sites (holes), of lattice model theories (Gibbs DiMarzio) are not sufficient to explain the effects of polymer microstructure upon the glass transition temperature. Further tests of these conclusions are in progress.

REFERENCES

1. P. J. Flory, "Statistical Mechanics of Chain Molecules", Interscience, New York, (1969).
2. J. H. Gibbs, E. A. DiMarzio, J. Chem. Phys., 28, 807 (1958).
3. J. M. O'Reilly, J. Appl. Phys., 48, 4043 (1977).
4. J. M. O'Reilly, J. Poly. Sci., Poly. Sym., 60, 165 (1978).
5. W. Orville-Thomas, "Internal Rotation in Molecules", Interscience, N. Y., Chap. 5 & 6, (1973).
6. J. L. Koenig and M. K. Antoon, J. Poly. Sci., A-2, 15, 1379, (1977).
7. J. M. O'Reilly, Macromolecules (to be published).
8. U. Baumann, H. Schrieber, and K. Tessman, Makromol. Chem., 36, 81, (1960).
9. H. Nagai, J. Appl. Poly. Sci., 7, 1697, (1963).
10. S. Havriliak and N. Roman, Polymer 7, 387, (1966).
11. P. Sundararajan and P. J. FLory, J. Amer. Chem. Soc., 96, 5025 (1974).
12. P. Sundararajan, Polymer Letters, 15, 699, (1977).

THE RATE-DEPENDENT FRACTURE BEHAVIOR OF HIGH PERFORMANCE SULFONE POLYMERS

R. Y. Ting and R. L. Cottington

Polymeric Materials Branch, Chemistry Division
Naval Research Laboratory
Washington, D. C. 20375

Many fiber-reinforced composite systems have been given considerable attention for potential aerospace and advanced ship applications. This is primarily because of the superior specific modulus and tensile strength the composite materials exhibit. For many applications conventional polymer matrix materials such as polyesters and epoxies are not suitable since they have maximum use temperatures of only 90°C to 120°C. Recently, many "high performance polymers", which offer 170°C to 260°C temperature capability, have become available. These include tetrafunctional epoxies, thermosetting polyimides and some thermoplastic polymers such as polysulfone. One of the important mechanical properties that need to be evaluated for these polymers is their resistance to crack propagation. Bascom, Bitner and Cottington[1] determined the fracture energy of many high performance polymers and concluded that the thermoplastics are much "tougher" than the thermosetting materials. While this would certainly be translated into superior impact properties, the thermoplastics are also attractive because they can be formed easily and possess good storage and handling capability at room temperatures. In this paper, the rate-dependent fracture behavior of two high performance sulfone polymers is reported.

EXPERIMENTAL

Materials

Polysulfone (Udel, Union Carbide, abbreviated as PSF) and polyethersulfone polymers (Victrex, ICI America, abbreviated as PESF) were studied in this work. The chemical structures of these polymers are:

polysulfone   poly(ether sulfone)

Plates of various thicknesses produced by the manufacturers were
used as received for preparing test specimens. Prior to testing,
all specimens were treated according to manufacturer recommended
annealing cycles to remove the "thermal skin" effect resulting
from the molding or extrusion process.

Fracture Toughness

Polymer fracture toughness was determined by using standard
one-inch compact tension specimens[2]. A precrack was introduced
with a razor blade at the end of the saw-cut. Specimens were
then fractured in an INSTRON testing machine at various crosshead
speeds in order to determine polymer fracture toughness at
different loading rates. By measuring the critical failure load,
P_c, one may calculate polymer fracture energy by

$$G_c = Y^2 P_c^2 \, a/EW^2 b^2 \tag{1}$$

where Y is a geometrical factor given by

$$Y = 29.6 - 186 \, (a/W) + 656 \, (a/W)^2 - 1017 \, (a/W)^3 + 639 \, (a/W)^4$$

and a is the crack length, E the Young's modulus, W the specimen
width in the direction of the crack and b is the thickness[3].
The applicability of this equation is normally limited to the
range of $0.3 \leq a/W \leq 0.7$.

Impact Test

The standard Izod impact test was also carried out by using
a Tinius-Olsen impact tester for plastics. The impact load and
energy were recorded as a function of impact time. Based on the
result of a linear elastic fracture mechanics analysis[4], the
impact strength, ε, is shown to be related to the fracture energy
by

$$\varepsilon = G_c \phi bW \tag{2}$$

The dimensionless factor ϕ is related to the specimen compliance, C,
and its variation with respect to the crack length a:

$$\phi = \frac{C_{(a)}}{\left[\dfrac{d^{C}(a)}{d(a/W)}\right]}$$

This factor has been calculated and given in a tabulated form by Plati and Williams[5] for a standard Izod impact specimen. Therefore, once the impact energy ε is measured for specimens of various initial crack lengths, one may determine the fracture toughness G_c by using Eq. (2).

Torsion Pendulum Analysis (TPA)

Dynamic mechanical properties for both sulfone polymers were also determined by using a freely oscillating torsion pendulum[6] operating at ca. 1 Hz. The experiments were carried out in a nitrogen atmosphere in accordance with the recommended ASTM procedure, D-2236-70. The sample size was 10 cm x 1.25 cm x 0.075 cm. The frequency of the freely damped wave and the logarithmic decrement $\Delta = \ln (A_i/A_{i+1})$, where A_i was the amplitude of the ith oscillation of the wave, were directly measured as the sample was heated at a rate of 1°C/min. These parameters led to the determination of the dynamic shear modulus and the loss factor of the sample as a function of temperature.

RESULTS AND DISCUSSION

Figure 1 shows the TPA results for PSF and PESF. High dynamic shear moduli, ca. $G' \sim 10^{10}$ N/m^2, were observed for both polymers. In the glassy state, PESF had a slightly higher modulus than PSF; the difference was nevertheless very small. In terms of the loss factor Δ, both polymers had a common secondary relaxation peak at -90°C with a value of ca. 0.1. However, two major differences exist in the response of these two polymers to small amplitude oscillatory deformation. First, the glass-to-rubber transition temperature, T_g, where a rapid decrease in modulus and a sharp increase in Δ took place, was higher for PESF (203°C) than that for PSF (175°C). Second, in addition to the loss peak at -90°C, PSF exhibited another relaxation peak at ca. -20°C, suggesting possibly an additional loss mechanism for energy absorption. These differences may be related to the presence of bisphenol-A units in the PSF backbone. The additional aliphatic groups could introduce new rotational mobility but become restraints leading to a decrease in molecular flexibility. The result would be a different packing configuration in PSF than in PESF, which has only the bisphenol-S groups as the repeating backbone units.

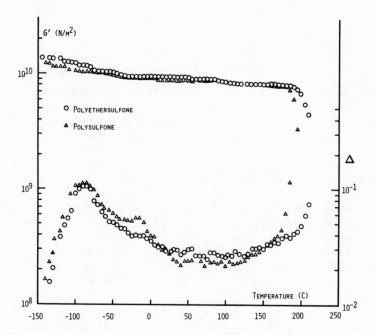

Fig. 1: Dynamic mechanical properties of sulfone polymers.

The fracture energies of both PSF and PESF were determined using compact tension specimens with thicknesses ranging from 0.3 cm to 2.5 cm. Figure 2 shows this specimen thickness

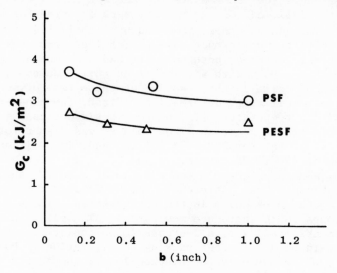

Fig. 2: Effect of specimen thickness on polymer fracture energy.

dependence of polymer fracture energy. For very thin specimens
the measured fracture energy was slightly higher than for thicker
specimens, indicating some plane-stress contribution. However, as
the specimen thickness was increased, the fracture energy quickly
approached a constant value. Based on this result, it is therefore felt that by using 1.3 cm thick specimens the plane-stress
contribution to fracture energy may be negligible, and the true
toughness value of polymers may be obtained.

Figure 3 shows polymer fracture energy as a function of loading rate at room temperatures. Since the actual strain rate based
on the local deformation pattern or stress distribution at the
crack tip is not known, the inverse of fracture time is used for
the abscissa of Figure 3. Both sulfone polymers exhibited very

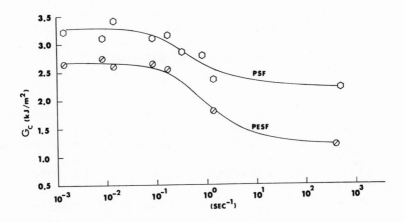

Fig. 3:   Loading-rate dependence of the fracture energy of sulfone
          polymers.

high fracture energy: at lower loading rates the G_c values are of
the order of 3 kJ/m^2 in contrast with the values of ca. 0.1 kJ/m^2
for many epoxies[1]. This high fracture energy may be attributed to
the large free volume in amorphous glassy polymers as opposed to
that in highly crosslinked epoxy systems. PSF fracture energy
is approximately 25% higher than that of PESF at these low loading
rates (10^{-3} sec^{-1}). This difference may be related to the
differences in their dynamic mechanical properties. More free
volume may be available to PSF than PESF for resisting crack
initiation and propagation in their amosphous glassy state because
of the differences in polymer backbone structure and molecular
packing.

The higher fracture energy PSF exhibited over PESF is amplified at higher loading rates. With increasing loading rate, both polymers show very rapid decrease in fracture energy, which eventually reaches a low, asymptotic value, typically represented by the impact test result. The transition from high to low fracture toughness with increasing loading rate is centered around 0.5 sec^{-1}. In PESF this decrease in fracture energy is more pronounced than in PSF. At the highest rate tested, corresponding to the impact condition, PSF is shown to be approximately twice as tough as PESF.

In conclusion, thermoplastic sulfone polymers are shown to be much more crack-resisting than thermosetting epoxy polymers. Polysulfone is found to have a higher fracture energy than polyethersulfone; the difference is attributed to the effects of molecular structure and packing configuration. It is also shown that the fracture energy of both polymers decreases rapidly with increasing loading rate. This suggests that the materials, if used for high-rate applications such as impact, would not appear to be as tough as the fracture energy determined in a low-rate experiment indicates.

REFERENCES

1. W. D. Bascom, J. L. Bitner and R. L. Cottington, Amer. Chem. Soc., Organic Coatings and Plastics Chemistry Preprint, 38, 477 (1978).
2. J. F. Knott, Fundamentals of Fracture Mechanics, Butterworth, London, 1973, p. 131.
3. W. Schultz, in Fracture Mechanics of Aircraft Structures, ed. M. Liebowitz, AGARDograph, AGARD Rept. AG-176, NTIS, Springfield, VA, 1974, p. 370.
4. P. E. Reed, in Developments in Polymer Fracture, Vol. 1, ed. E. H. Andrews, Appl. Sci. Pub., London, 1979, p. 121.
5. E. Plati and J. G. Williams, Polym. Eng. Sci., 15, 470 (1975).
6. L. E. Nelson, Mechanical Properties of Polymers and Composites, Vol. 1, Dekker, NY, 1974, p. 15.

THE INFLUENCE OF CRYSTALLINE STRUCTURE ON THE NECKING-FRACTURE

BEHAVIOUR OF POLYETHYLENE

U. Gedde, B. Terselius and J.F. Jansson

Department of Polymer Technology
The Royal Institute of Technology
S-100 44 Stockholm, Sweden

INTRODUCTION

An anomaly in the necking and fracture behaviour of polyethylene under constant uniaxial tensile loading has been reported in several papers[1-3]. At a certain stress level, the almost
instantaneous fracture of the neck formed at high loads was replaced by the formation of a neck that resisted fracture for a
considerable time. This marked transition was observed for high
density polyethylenes of comparatively high molecular weight. A
hypothesis has been proposed to explain the appearance of the
marked transition:

1. At high strain rates, large stresses concentrate to the taut
tie chains between the crystallites, favouring fracture processes.
At lower strain rates, the molecules rearrange in a way that distributes the local stresses more evenly over the structure which
then deforms further rather than fracturing.

2. Owing to the high concentration of the chains in the high density, high molecular weight polyethylenes, the transfer of stresses
and deformations between adjacent crystallites is very efficient.
Therefore the structure has a good resistance towards fracture and
the deformation of different lamellar regions occurs cooperatively.

3. The marked transition is a result of both (a) the rapid formation of high strength fibrillar structure, which is partly due to
reasons given in item 2 of this hypothesis, and (b) the influence
on the strength of the strain rate in accordance with item 1 of
the hypothesis. At the marked transition, the maximum in the local
strain rate is about the highest tolerable for the actual structure
without macroscopic fracture. When an element of the specimen

passes the maximum in local strain rate the probability for fracture of this element decreases due to the decreasing strain rate
and to the continuous formation of a high strength fibrillar structure. In addition, e.g. the rise in local temperature in the neck
caused by dissipation of deformation energy,may affect the necking
and fracture behaviour.

In this presentation the effect of annealing of a high density
polyethylene on the necking and fracture behaviour (especially on
the marked transition) is shown and discussed.

EXPERIMENTAL

Materials

Sheets of a high density polyethylene (MI_2=0.05, MI_5=0.2),
were produced by compression moulding of granules as described
earlier [1] . Some of the sheets were annealed in a nitrogene
atmosphere at different temperatures and times. The materials are
characterized in Table 1.

The supermolecular structure, i.e. the size of the spherulites,
was not affected by the heat treatments (spherulite radius is about
4.5 µm). The annealing influences the lamellar structure in several
ways:

• The crystallite size increases at moderate annealing temperatures. This is in agreement with the literature[6].

• A population of thin crystallites (4-6 nm) is observed in some
samples. These samples exhibit double peaks in DSC. Similar observations have been made by Wunderlich[7] and also by Dlugosz and
coworkers[8]. They showed that the thin, defect crystallites consisted of segregated low molecular weight material. DSC-measurements
further support this conclusion. For the sample annealed at
405.2 K only one melting peak is observed, indicating an almost
complete melting of the system without molecular fractionation.

From the sheets, dumb-bell specimens were prepared as previously described.

Mechanical measurements

The times to necking and fracture under constant uniaxial tensile loading in air at 313 K were measured as described earlier[2].
The local strain rates were also obtained according to the method
described in a previous paper.

RESULTS

The necking and fracture data are shown in Figure 1 and Table 2.

Table 1. Polyethylenes investigated

Materials	Original (not annealed)	393.2K 2h	396.7K 24h	398.2K 2h	401.2K 2h	401.2K 24h	403.2K 2h	405.2K 2h
Density[a] (kg/m³)	953.0	955.7	957.7	956.0	956.1	957.7	955.5	954.8
Crystallinity[b]	0.61	0.632	0.665	0.635	0.636	0.665	0.631	0.625
T_{peak} (K)[c]	406.3	407.2	407.8	407.8	408.3	408.3	407.8	406.5
T_{lp} (K)[d]	-	-	392	-[e]	404.0	403.6	405.3	-
Thickness of lamellae arithmetic mean (nm)[f]	12.0	-	13.1	12.8	13.0	13.5	12.7	12.5
Secondary lamellae (nm)[f]	-	-	-	4-6	4-6	4-6	-	-
DSC melting endotherm	one peak	one peak	double peak	peak with shoulder	double peak	double peak	double peak	one peak

a) using a density gradient column
b) from volumetric data according to Swan[4]
c) Melting peak temperature obtained by a Perkin Elmer DSC-Z (dT/dt=10K/min)
d) Melting temperature of low temperature peak (DSC, dT/dt=10K/min)
e) Marked low temperature peak absent although a shoulder at low temperature is present
f) From transmission electron microscopy using a technique developed by Kanig[5].

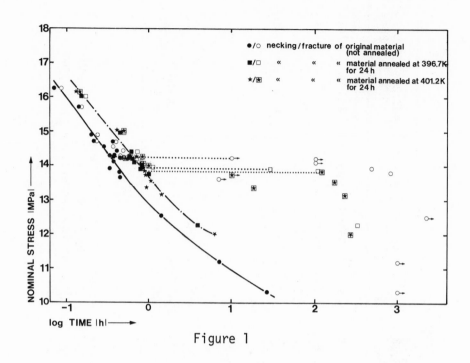

Figure 1

Table 2. Necking and fracture data of annealed polyethylenes

Materials: Annealing conditions Temp/time	a_0 [1] (log h)	a_1 [1] (log h/MPa)	r^2	τ_{trans} [2] (MPa)	t_{trans} [3] (h)
393.2K/2h	3.347	-0.2585	0.86	13.93	0.56
398.2K/2h	5.233	-0.3860	0.96	13.83	0.78
401.2K/2h	5.612	-0.4129	0.96	13.81	0.81
403.2K/2h	4.762	-0.3517	0.96	13.93	0.70
405.2K/2h	3.822	-0.2958	0.80	14.34	0.40

[1] Obtained by linear regressions analysis in accordance with
    $\log t_{necking} = a_1 \tau + a_0$ ← →r^2 is the coefficient of determination

[2] τ_{trans} and t_{trans} are respectively the nominal stress and
    the time for marked transition in the necking and fracture
    behaviour.

The following comments may be made with regard to these data:
   All materials exhibit marked transitions.
   The graph of the nominal stress of transition as a function of
annealing temperature shows a minimum close to 401 K.
   The necking time at 14 MPa shows simple positive correlations
with density and lamellar thickness of the non-oriented material.
   The fracture curves below transition were shifted to shorter
times with increasing annealing temperature (see Figure 1).
The results of the strain rate/time measurements are shown in
Figure 2 as the maximum strain rate plotted against necking time.
From fig. 2 it is obvious that under identical macroscopical conditions there is a variance in the maximum strain rate at different locations. However it seems reasonable to assume that different materials can be represented by a single function, i.e. by an
upper and lower limit, as indicated in the Figure.

DISCUSSION
   The tendency for necking decreases with increasing crystallinity and increasing lamellar thickness. It is not possible from
these experiments to distinguish the influence of crystallinity
from the influence of lamellar structure. However, since the
necking involves disintegration of lamellae, the dimensions of the
lamellae must be of major importance for the necking process.

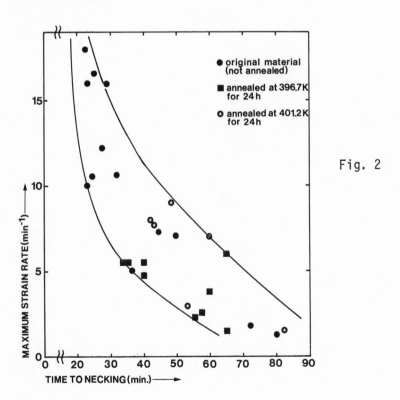

Fig. 2

In order to determine the maximum strain rate at the transition, data for the time to transition and the upper limit of the maximum strain rate/necking time function were used, see Fig 3. The upper limit was used since it seems reasonable to assume that the macroscopic fracture follows a path of points subjected to strain rates according to the upper limit.

A minimum at about 401K is observed in Fig. 3. The materials annealed at temperatures close to 401K showed two melting peaks and a thin lamellar population (of about 5 nm). As stated earlier, this indicates a molecular fractionation of the material generating a separate population of crystallites composed of segregated low molecular weight and branched material. The segregated parts exhibit a lower content of interlamellar tie chains than the rest of the structure, a fact which should increase the probability of fracture initiation in these segregated parts under the action of a stress field. The anticipated embrittlement of the material as a consequence was thus verified by the results depicted in Fig. 3.

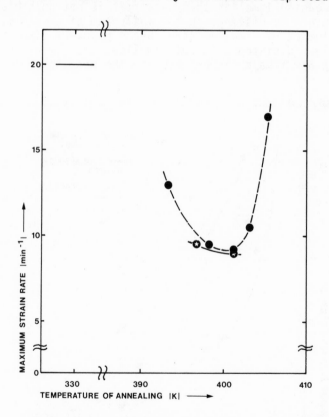

Fig. 3. Filled circles indicate materials annealed for 2 h., circles with asterisks indicate materials annealed for 24 h.

McCready and coworkers[9] explained the decrease in upper yield stress with increasing crystallization time by a decrease in interlamellar tie chain concentration caused by a special "cannibalistic" crystal growth mechanism[10]. It is not possible, as a result of experiments here presented, to separate the influence of molecular fractionation from the influence of this overall decrease in tie-chain concentration.

REFERENCES

1. U.W. Gedde and J-F. Jansson, Polym.Eng.Sci., 19:77 (1979)
2. U.W. Gedde and J-F. Jansson, to be published in Polym.Eng.Sci.
3. U.W. Gedde and J-F. Jansson, to be published in Polym.Eng.Sci.
4. P.R. Swan, J.Polym.Sci., 54:525 (1960).
5. G. Kanig, Coll.Polym.Sci., 255:1005 (1977)
6. B. Wunderlich, "Macromolecular Physics, Part 2", Academic Press, New York (1976).
7. B. Wunderlich, J.Polym.Sci, Polym.Symp., 43:29 (1973)
8. J. Dlugosz, G.V. Fraser, D. Grubb, A. Keller, J.A. Odell and P.L. Goggin, Polymer, 17:471 (1976)
9. M.J. McCready, J.M. Schultz, J.S. Lin and R.W. Hendricks, J.Polym.Sci., Polym.Phys.Ed., 17:725 (1979)
10. J. Peterman, M. Miles and H. Gleiter, J.Macromol.Sci.-Phys., B 12:393 (1976).

STRUCTURE AND FRACTURE OF THERMALLY OXIDIZED PIPES OF HIGH-
DENSITY POLYETHYLENE

B.Terselius, U.Gedde, J.F.Jansson

The Royal Institute of Technology, Stockolm, Sweden

(Abstract)

The structure and fracture properties of high-density poly
ethylene pipes, thermally oxidized during processing, were studied.

The inner wall of non-oxidized pipes has a rough, satin-mat
surface. Oxidized pipes develop a fine patterned, glossy surface,
which at 200 times magnification in SEM turns out to be covered by
craterlike, 30-50 μm large structures. At more heavily oxidized
pipes these craters tend to "dissolve" in the surface. As shown by
higher SEM magnifications the reason for the increased surface gloss
of the oxidized inner walls is that the smoothness of the micro
structure of the surface is increased in proportion to the
oxidation during processing.

Polarized microscopy of crossections of a glossy surface
clearly demonstrates a low crystalline top layer of about 15 μm.
Differential scanning calorimetry of the top layer material reveals
a very defect melting behavior with a peak value of 102°C and a
crystallinity of about 7 percent. The carbonyl content of this layer
as shown by infrared spectroscopy is very high decreasing success
ively to zero at about 150 μm from the inner wall surface. The
presence of ether crosslinks could also be shown in the oxidized
surface region. The gel content of the top layer material was as
high as 80 percent.

Thus the formation of a glossy almost amorphous, crosslinked oxidation skin could be demonstrated at highly oxidized inner walls of high-density polyethylene pipes.

Creep failure tests have shown that oxidized pipes exhibit considerably shorter life times than non-oxidized pipes.

At present work is carried out to clearify the fracture mechanisms by which the weak and brittle oxidation skin is reducing the strength of the oxidized pipes.

FIBRE FATIGUE IN VARIOUS ENVIRONMENTS

I.E. Clark   and J.W.S. Hearle

Department of Textile Technology
University of Manchester Institute of Science and
Technology, Manchester M60 1QD, U.K.

INTRODUCTION

Fibres in processing and use, generally in the form of fabrics, are inevitably subjected to repetitive straining. The result is that they undergo a fatigue action, often in the presence of a chemical environment, which can seriously reduce their useful life. The search for a laboratory test with a form of failure similar to that commonly found in worn materials, namely multiple splitting of the fibres, has led to a new form of fatigue test (figure 1). This form of test, known as "biaxial rotation over a pin" and which involves the tension-compression of fibres which are rotated whilst bent over a pin, was first suggested by Goswami and Hearle[1], and developed in the work of Hearle with Wong[2,3,4],   Calil[5,6], and Hasnain[7].

DEVELOPMENT OF THE NEW APPARATUS

There were limitations of cost, convenience and conditions in these earlier testers which were special pieces of research apparatus. We have therefore worked on the production of an improved and more versatile form of apparatus. As a prototype, a new single-station tester was designed and constructed at UMIST and has been reported by Clark and Hearle[8,9,10]. This apparatus (figure 2) works on a similar principle to previous designs but the jaw shafts are aligned parallel to facilitate gearing and allow a number of shafts to be driven off one another. In addition a new tensioning system has been devised in which the pin is mounted in a beam which is free to move vertically on bearings along shafts.

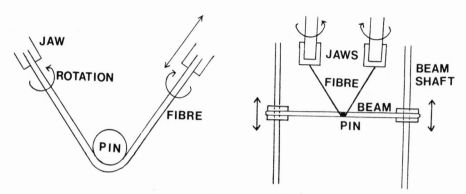

Fig. 1. Biaxial rotation over a pin. Fig.2. Modified arrangement for
                                         biaxial rotation over a pin.

A fixed tension is thus imparted to the fibre by the weight of
the beam itself, and also the tension may be increased by simply
adding weights. This system is ideal for adaption to a multi-
station design which is necessary if a large scale comparative
survey of the fatigue resistance of fibres is to be performed.

        An electronically controlled multi-station tester (figure 3)
based on the above has recently been completed at UMIST and is
described in a paper by Clark et al[11]. There are ten stations in
all, five on each side of the tester. Each station is similar to
the prototype shown in figure 2, except that the number of
revolutions, or cycles, of the jaw shafts required to break a
fibre (the fatigue life) is counted electronically as opposed to
the mechanical system used on the prototype.

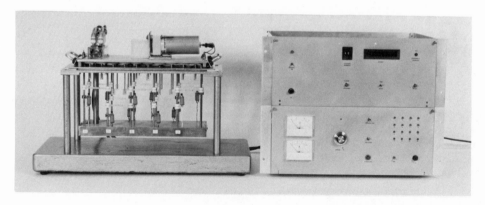

Fig. 3. Multi-station apparatus and control unit.

The tester has been designed to fit into a small (45 litre) environmental chamber which allows testing in gases other than air and also enables a wide range of temperatures and humidities to be investigated. Testing in liquid environments is performed by a simple modification to the beams. The new beam has two struts holding the pin below it. Hence the fibre and pin can be immersed in a liquid without interfering with the free movement of the beam.

EXPERIMENTAL METHOD

Two monofilament fibres have been tested initially:

(i) 1.7 tex medium tenacity polyester,
(ii) 1.7 tex medium tenacity nylon 6.

The polyester was produced on the pilot spinning unit of the Department of Polymer and Fibre Science, UMIST, and the nylon 6 was supplied by Courtaulds Ltd.

All the experiments were performed with an angle of wrap of the fibre around the pin, θ, of approximately 60°.

The tension in the fibre is given by:

$$T = (\frac{W + w}{2}) \sec (90-\theta/2)$$

where W is the beam weight, and w is any extra weight added to the beam. This was maintained at 98mN(10gf) throughout the experiments.

The apparent maximum bending strain, ε, experienced by the surface of the fibre is given to a close approximation by

$$\varepsilon = Rf/(Rf + Rp)$$

where Rf and Rp are the radii of the fibre and pin respectively. A strain amplitude of 10% was used in these experiments. The pins were short lengths of 0.36 mm diameter "Nichrome V" wire - an 80% Nickel/20% Chromium alloy, and the jaws were rotated at 2.5 revolutions per second.

EXPERIMENTAL RESULTS

The results presented as graphs in figures 4 and 5 are examples of our initial series of experiments. Each point represents the average of 10 individual tests, and the fractional error on each point is around 20%.

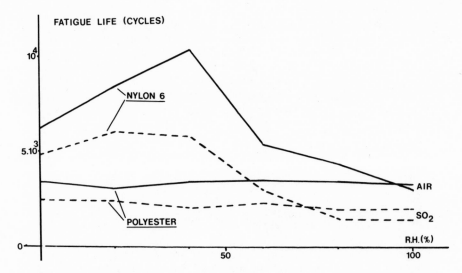

Fig.4. Fatigue life against relative humidity for nylon 6 and
       polyester in air and a mixture of air and SO₂ (10% by volume).

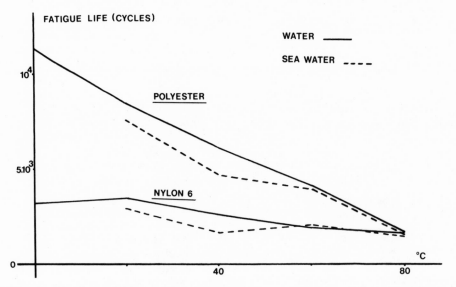

Fig. 5. Fatigue life against temperature for nylon 6 and polyester
        in tap water and simulated sea water.

(i) Gaseous Environments

     Figure 4 shows the effect of air, and of a mixture of air and
SO₂ (10% by volume), at different humidities, on the fatigue life
of polyester and nylon 6.

In air the polyester shows little or no variation in fatigue
life, which stays constant at around 3,500 cycles, regardless of
relative humidity. This is not unexpected as polyester has a small
moisture regain ($\sim$0.4%) which varies only slightly with changes in
humidity, and its physical properties are relatively unaffected by
the presence of moisture. The same is true for the polyester when
SO_2 is present, except that the average life is somewhat lower at
around 2,500 cycles, a decrease of $\sim$30%.

The nylon 6 in air behaves quite differently, having a fatigue
life which rises from about 6,000 cycles at 0% r.h., to a maximum at
about 40% r.h. of over 10,000 cycles, and then falls to below 5,000
cycles at the higher humidities. In general nylon has a higher
resistance to fatigue than polyester, and also a far higher moisture
regain (0$\rightarrow$9% depending on humidity) which will account, to some extent,
for its rather erratic behaviour. The curve takes a similar form
when SO_2 is present except that, as with the polyester, the average
life has dropped from around 7,000 cycles to 4,000 cycles, a decrease
of $\sim$40%.

(ii) Liquid Environments

Figure 5 shows the effect of tap water and simulated sea water
at different temperatures on the fatigue life of polyester and nylon 6.

There is little difference between the effects of the tap water
and sea water, although a slightly lower life was recorded for both
fibres with the sea water, particularly at lower temperatures. The
average nylon 6 fatigue life is less than half that in air and drops
only slightly as the temperature rises. The presence of water at
lower temperatures however, appears to enhance the fatigue life of
the polyester, although this decreases rapidly with increasing
temperature. This apparent increase in fatigue life is possibly
due to the water acting as a lubricant between the fibre and pin
and reducing damage by abrasion, an effect which is masked in the
case of the nylon 6 due to its weakness in the presence of water.

CONCLUSION

It can be seen from the graphs in figures 4 and 5 that fatigue
life is a property which is very sensitive to external conditions,
and as such, is a very useful indicator of the effect on a fibre
of its environment.

Although fatigue life is sensitive to environment, the fatigue
fracture morphology is relatively unaffected, and figures 6a and 6b
show typical fatigue breaks in polyester and nylon 6, broken in air,
respectively. Calil et al[12] have shown that the fibre is subject to
a combination of cyclic bending, twist, and tension, which act on

Fig.6.Fatigue break in (a) 1.7tex polyester and (b) 1.7tex nylon 6

the part of the fibre in contact with the pin during a fatigue
test to produce cracking and fibrillation. When a chemical
environment is present it can penetrate the fibre structure by means
of these cracks and accelerate the failure process. The weakened
fibre eventually fails under the applied tension and leaves the
characteristic fibrillated fracture ends shown in figure 6. In
general, fibrillation is less extensive in nylon which tends to split
into fewer, larger fibrils than polyester.

REFERENCES

1. B.C.Goswami and J.W.S.Hearle, A comparative Study of Nylon Fiber
   Fracture. Text.Res.J.,46,1,55(1976).
2. J.W.S. Hearle and B.S.Wong, Comparative Study of the Fatigue
   Failure of Nylon 6.6, PET Polyester, and Polypropylene Fibres.
   J.Text.Inst.,68,89(1977).
3. J.W.S.Hearle and B.S.Wong, Effects of Air,Water,Hydrochloric Acid,
   and other Environments on the Fatigue of Nylon 6.6 Fibres, J.Text.
   Inst.,68,127 (1977).
4. J.W.S.Hearle and B.S.Wong, Statistics of Fatigue Failure, J.Text.
   Inst.,4,155 (1977).
5. S.F.Calil and J.W.S.Hearle,Fatigue of Fibres by Biaxial Rotation
   over a Pin, ICF4 Conference, Canada, 1267-271 (1977).
6. S.F.Calil,Ph.D.thesis,University of Manchester (1977).
7. J.W.S.Hearle and N.Hasnain,Fatigue of Cotton Fibres and Yarns,
   Ann.Conf.of the Text.Inst., New Delhi, 163 (1979).
8. I.E.Clark and J.W.S.Hearle,Development of the Biaxial Rotation
   Test for Fatiguing Fibres,J.Phys.E.,12,1109-112(1979).
9. I.E.Clark and J.W.S.Hearle,Anomalous Breaks in the Biaxial
   Rotation Fatigue Testing of Fibres, J.Text.Inst.,to be published.
10. I.E.Clark and J.W.S.Hearle,The Effect of Test Conditions on the
    Fatigue of Fibres by Rotation over a Pin,J.Text.Inst., to be pub.
11. I.E.Clark, J.W.S. Hearle and A.R.Taylor,A Multi-station Apparatus
    for Fatiguing Fibres in Various Environments,J.Phys.E. to be pub.
12. S.F.Calil,B.C.Goswami and J.W.S.Hearle, The Development of Torque
    in Biaxial Rotation Fatigue Testing of Fibres, J.Phys.D., to be
    published.

CREEP-FATIGUE RUPTURE PREDICTION MODELS APPLIED ON 2 1.4 Cr -1 Mo

STEEL AT 550°C

A. Benallal

ENSET, Cachan, France

(Abstract)

Four methods for the prediction of creep fatigue interaction
are applied to 2 1/4 Cr 1 Mo steel at 550°C. These are :
i)   COFFIN's frequency modified relation ship
ii) MANSON's strain range partitionning
iii) MAYA and MAJURDUM's model
iv) Damage concepts, introduced recently by J.LEMAITRE and J.L.
     CHABOCHE.

These models are first established using strain controlled
fatigue tests over a range of frequencies for i) and iii), strain-
controlled fatigue - relaxation tests for ii), monotonic creep
tests and stress-and strain-controlled pure fatigue tests
(frequency 10 Hz) for iv).

These models are then applied to more complex tests in
respect to time extrapolation problems and compared with experiments.

The domains of validity of these methods are finally
established.

ULTIMATE STRENGTH OF POLY(ETHYLENE TEREPHTHALATE) FIBRES AND ITS RELATION TO THERMAL AND MECHANICAL HISTORY

Jiří Vaníček[+],Jiří Militký and Jaroslav Jansa

Research Institute for Textile Finishing
544 28 Dvůr Králové nad Labem,Czechoslovakia
+ Silon Combine,Planá n.Luž.,Czechoslovakia

INTRODUCTION

It is well-known that ultimate strength of poly-mers has a statistical nature and can be characterized by a suitable distribution function (e.g.Kausch (1971)). However,little attention has been paid to the relation between the first statistical moment of strength dis-tribution function (arithmetic mean value in the case of Gaussian distribution),structure and/or conditions of polymer preparation.

Peterlin (1978) described a fracture model of fibrous semicrystalline polymers composed of regularly alternating long elements with high elastic modulus and the short "deffect areas" having the low elastic modulus.The strong elements may be the microfibrils alone composed of the folded chain crystal blocks which, again,regularly alternate with amorphous layers bridged by a great number of taut tie molecules.The deffect areas contain few taut tie molecules and are practically identical with pure amorphous material.The break will occur in deffect areas.In this oversimplified model the stress to break σ_B is constant and dependent only on fracture properties of deffect areas.

In present work the dependence of mean ultimate strength on thermal and mechanical history of the modified poly(ethylene terephthalate) fibres is demon-strated.Simultaneously,validity of the Peterlin's model of fracture is verified.

EXPERIMENTAL

Copolyester composed from 90.5 molar per cent of poly(ethylene terephthalate),8% of isophthalic acid and 1.5% of 5-sulphoisophthalic acid was used for preparation of all samples.Undrawn fibres were spinned on a pilot plant device dosing 225 g/min melt,spinning speed 675 m/min.Fineness of the fibre was T_{o0} = 16.95 dtex (which corresponds to fibre diameter 40.16 μm) and birefringence Δn_0 = 2.76 10^{-3}.Limiting viscosity number (LVN) determined by a solvent dilution method in 1:3 phenol-tetrachlrethane was 57.2 ml/g.

From this starting fibre eleven various fibre types (Tables 1 and 2) were prepared under different conditions of drawing and annealing.Drawing was performed in an aqueous bath to two stages.Temperatures of water,T_{KI} in the first and T_{KII} in the second stage together with corresponding draw ratios are given in Table 1.After drawing the samples were dried and annealed in relaxed state at temperature T.

Structural Parameters

Volumetric crystallinity was calculated from the densities ρ measured in a density gradient column which contained the mixture of n-heptane and carbon tetra-chloride at 30^oC.Birefringence Δn was determined on a polarization microscope with the Berek compensator. WAXD analysis was used to assess orientation of crystal-line phase f_c (from α(101) azimuthal reflection). Orientation factor of amorphous phase f_a was computed from X,f_c and Δn according to Samuels (1974).All these parameters are shown in Table 1.

Ultimate Strength and Shrinkage

Stress-strain curves at 25^oC and deformation rate $d\varepsilon$/dt = 1.66 $10^{-2}s^{-1}$ were determined on the Instron testing machine TM-SM connected with a Schlumberger data logging unit Solartron 1420.2.For each sample ultimate strength f_B and corresponding deformation ε_B were assessed by arithmetic averaging from 100 indi-vidual stress-strain curves.Resulting values are pre-sented in Table 2.Shrinkages S_W (see Table 2) were computed from the length changes of 15 cm long samples after 10 minutes' treatment in boiling water.

Table 1. Conditions of Preparation and Structural
          Parameters of the Samples

No	λ	T_a $[^{\circ}C]$	T_d $[dtex]$	X $[\%]$	f_c	f_a	T_{KI} $[^{\circ}C]$	T_{KII} $[^{\circ}C]$
1	3.85	35[+]	4.90	17.8	0.713	0.611	50	50
2	3.96	35[+]	4.67	16.5	0.714	0.598	60	59
3	4.62	35[+]	4.07	17.0	0.716	0.671	70	64
4	4.18	35[+]	4.13	21.2	0.715	0.677	70	65
5	4.51	120	4.63	28.1	0.887	0.557	66	96
6	4.51	120	4.53	27.6	0.748	0.611	70	95
7	3.63	120	5.70	33.9	0.875	0.477	69	90
8	3.63	135	5.73	35.6	0.879	0.464	69	90
9	3.63	150	5.93	36.0	0.795	0.506	69	90
10	3.63	140	6.20	38.8	0.891	0.434	69	90
11	3.64	140	6.07	30.8	0.828	0.499	70	90

+/only dried

Strength Distribution

From particular strength values f_i (i = 1,...100)
the parameters of Weibull distribution function were
calculated for all samples by using the least squares
criterion.The Weibull distribution function may be
expressed

$$F(f) = 1 - exp\left[-(\frac{f-c}{d})^b\right] \qquad (1)$$

where F(f) is the number of all ultimate strength values
for which $f_i \leqslant f$ (cumulative frequency),c is the characteristics of location,d is the characteristics of
scale and b is the shape factor.Mean relative deviation
between model relationship from Eqn 1 and experimental
data was less than 5% in all cases considered.From the
values d,b,c the first statistical moments E(f) were
obtained with the aid of equation

$$E(f) = d\Gamma(1 + 1/b) + c \qquad (2)$$

where $\Gamma(x)$ is the Gamma function tabulated e.g. by
Janke et al. (1960).Resulting E(f) are given in Table2.

RESULTS

A detailed discussion of the results obtained
with respect to mutual relations between structure,
properties and conditions of preparation will be summarized in another paper (Militký at al. (1980)).Here,

the conclusions are restricted to the statement that
the conditions of drawing and those of annealing
strongly affect the structure of the fibres.To compare
the effect upon mechanical parameters as well as
shrinkage properties the mean values $\overline{X}$ and variation
coefficients $v(X)$ were computed from all eleven samples
(Table 2).As can be seen from these quantities the
conditions of drawing and annealing affect in particular
shrinkage S_W and deformation $\mathcal{E}_B$.The values of mean
strength f_B are affected very little ($v(f_B) \sim 5\%$).From
the comparison of first statistical moments $E(f)$ and
f_B it follows that these quantities are practically
identical.Therefore,they are further being used to
characterize strength distribution of arithmetic avera-
ges f_B (Gaussian distribution).

On the basis of these facts we may allow for the
hypothesis that fB is (for given deformation rate)
proportional only to the diameter of undrawn fibre P_0,
the mean molecular weight $\overline{M}$ and its orientation
(expressed by Δn_o).

According to Peterlin´s model of fracture it stands
to reason that the properties of the "deffect areas"
depend only upon the parameters of undrawn fibre preparation and not on those of drawing and annealing.

Table 2.   Ultimate Parameters,Shrinkage and the Para-
            meters of Weibull Distribution of the Samples
            and Their Statistical Characteristics

No	f_B [N]	$\mathcal{E}_B$ [%]	S_W [%]	S_T [%]	$E(f)$ [N]	σ_B [GPa]
1	0.1483	68.1	68.4	10.2	0.1486	0.63
2	0.1488	63.2	43.2	8.37	0.1481	0.61
3	0.1479	36.1	30.0	9.88	0.1488	0.61
4	0.1478	35.9	24.1	1.84	0.1475	0.60
5	0.1560	46.1	3.5	18.9	0.1551	0.60
6	0.1530	49.7	3.6	17.1	0.1519	0.62
7	0.1386	98;4	0.9	18.1	0.1376	0.67
8	0.1339	84.4	1.4	18.5	0.1331	0.57
9	0.1372	86.3	1.3	21.3	0.1371	0.57
10	0.1392	85.9	0.1	24.7	0.1381	0.56
11	0.1378	93.0	1.0	25.3	0.1375	0.60
$\overline{X}$	0.1444	67.9	16.13	17.19	0.144	0.61
$v(\overline{X})$ [%]	5.05	34.11	141.3	44.8	5.12	5.72

Relation between Strength,Deformation and the
Conditions of Preparation

Strength of the statement that f_B is independent
of thermal and mechanical history of the fibres,one
can expect stress at break σ_B to be the function of
$\overline{M},\Delta n_o$ and $P_{o\_}$(or the corresponding titre T_{do}),i.e.
$\sigma_B = \sigma_B(P_o,\overline{M},\Delta n)$.From the definition of strength it
can be deduced that $\sigma_B = f_B/P_B$,where P_B is the fibre
cross-section at break.To determine σ_B the relations
between both P_B versus ε_B and T_{do} versus T_d are to be
defined.

In deriving these relationships an assumption was
made for the fibre deformation to be homogenous up to
the break point (however,this is valid at the second
stage of drawing and with stress-strain experiments).
From definition of fineness T_d,draw ratio λ and relative shrinkage S_T it follows

$$P_B = T_d \cdot \rho^{-1}(1 - 2 r \varepsilon_B) \qquad\qquad (3)$$

$$T_d = T_{do} / \lambda (1 - S_T) \qquad\qquad (4)$$

where r is the Poisson´s ratio at break.By means of
Eqn 3 and Eqn 4 it is possible to express σ_B,viz.

$$\sigma_B(\overline{M},\Delta n_o) = f_B \cdot \rho \cdot \lambda(1-S_T) [T_{do} (1-2r \varepsilon_B)]^{-1} \qquad (5)$$

Provided that the Peterlin´s model holds,strength σ_B
is a constant of a given undrawn fibre (with fineness
T_{do}),regardless of deformation and thermal history
before breakage.To verify this presumption experimentally
it is necessary to assess shrinkage S_T and the Poisson´s
ratio r (for homogenous deformation where no change of
ρ takes place it results r = 0.292).The values S_T were
obtained from Eqn 4 (see Table 2).The parameter r was
calculated by one-dimensional optimization technique
with requirement of the minimum variation coefficient
$v(\sigma_B)$,where σ_B were computed from equation 5 for
all samples at various r.Hence,the following values
were obtained r = 0.255,$v_B(\sigma_B)$ = 5.7%.Optimal σ_B are
illustrated in Table 2.It is obvious that within the
framework of experimental errors an assumption may
be adopted for σ_B to be a constant having the value
0.61 GPa.

SUMMARY

   From the above results it may be concluded that:
1 - Structural parameters of the fibres,their shrinkage in boiling water and deformation at break are
strongly affected by thermal and mechanical history
2 - Stress at break σ_B is fully determined only by
original structure of undrawn fibre.

   These conclusions indicate that Peterlin´s simple
model of fracture of semicrystalline polymers with
fibrous structure is,in principal,valid for poly(ethylene terephthalate) fibres with restrictions due to
its extension to three dimensions - Peterlin (1978).

REFERENCES

Janke,E.,Emde,F.,Lösch,F.,1960,"Tafeln höheren
       Funktionen,B.G.Teulner,Stuttgart.
Kausch,H. H.,1971,The role of network orientation and
       microstructure in fracture initiation,Polym. Sci.
       Symp. No 32,1.
Militký,J.,Vaníček,J.,Jansa,J.,Čáp,J.,1978,Influence
       of a modifying component on stress-strain be-
       haviour of PET fibres,Proc. V-th Conf. on Modified
       Polymers,Bratislava,Contribution No M - 7.
Militký,J.,Jansa,J.,Vaníček,J.,1980,Work prepared for
       publication.
Peterlin,A.,1978,Fracture of fibrous polymers,Polym.
       Engn. Sci.,18: 1062.
Samuels,R.J.,1974,"Structured Polymer Properties,"
       J. Wiley and Sons,New York.

BUILDING OF STRESS-STRAIN CURVES FOR GLASSY POLYMERS ON THE BASIS OF THEIR DYNAMIC MECHANICAL CHARACTERISTICS

A.B.Sinani, V.A.Stepanov

A.F.Ioffe Physical-Technical
Institute,
Leningrad, 194021, USSR

It is known[1,2,3] that at small stresses and strains (in a linear viscoelastic range) it is necessary to take into account a wide spectrum of relaxation processes for the description of the regularities of polymeric glasses deformation.

The analysis of the glassy polymer unelastic deformation kinetics is made usually [4,5,6] with the aid of the following relation:

$$\dot{\varepsilon} = \dot{\varepsilon}_o \, exp\left(-\frac{Q_o - \alpha_o \sigma_b}{kT}\right) \, , \qquad (1)$$

where $\dot{\varepsilon}$ – the deformation rate, Q_o –the activation energy, α_o – the activation volume, $\dot{\varepsilon}_o$ – the frequency factor, σ_b – the yield stress, k –Boltzman's constant, T – absolute temperature. The molecular-kinetic models for polymers, with the aid of which the equation (1) was obtained, consider a single type of the molecular motion – that of the segments activated in the glassy condition by rather large stresses. However, there are many experimental data available [7,8,9], which cannot be regarded within the framework of this model.

In this connection, the aim of the present paper is to study the glassy polymer unelastic behaviour in a wide range of temperatures and strain rates. Taking into account the previous investigations[7,8,9], indicating the possible influence of the relaxation spectrum of polymeric glasses on their flow, the study of the relaxation characteristics was performed on the same materials.

Experimental

For the mechanical tests of polymers an experimental set up was made, which allowed to vary the test temperature from 120 to 370 K and the strain rates by 8 orders. The stress-strain curves were recorded by an automatic chart-recorder or an oscillograph. The test were made in uniaxial compression to suppress the fracture process, and to widen the range of unelastic deformation.

The shear modulus and the loss factor were obtained in the torsion pendulum within the temperature range from 90 K to Tg at the frequency ~1 Hz.

Four linear amorphous polymers were tested: polymethylmethacrylate (PMMA), polycyclohexylmethacrylate (PCHMA), polystyrene (PS), polyvinilchloride (PVC). The materials were chosen in accordance with the differences in their relaxation spectra below Tg.

Results and Discussion

Figures 1,2 illustrate the test temperature and strain rate dependences of the yield stress σ_b. As is seen, the dependencies within the wide range of the test conditions appear to be much more complicated then it would expected from equation (1). So, the model, considering a single kinetic process, does not describe the mechanical behaviour of the glassy polymers within a wide range of test temperatures and strain rates.

This anomalies could be explained due to the experiments on polymers with essentially different relaxation spectrum characteristics. Fig.3 shows the dependencies of the yield stress, the shear modulus and the factor on the test temperature. It is seen that the change in the yield stress is, to a large extent, similar to the shear modulus change[10]: the relaxation transition regions corresponding to the maximum values of the loss factor manifest themselves on the curves $\sigma_b(T)$ and $G'(T)$; it should be emphasized here that σ_b corresponds to the large stresses ~100 MPa, and to the strains 5 ÷ 10%, while the shear modulus has been found at very small stresses ~ 0.1 MPa, and strains ~0.01%. The results obtained have shown that the local molecular rearrangements occurring in the polymers below Tg contribute to the temperature and the time dependence of the unelastic deformation resistance. From this follows

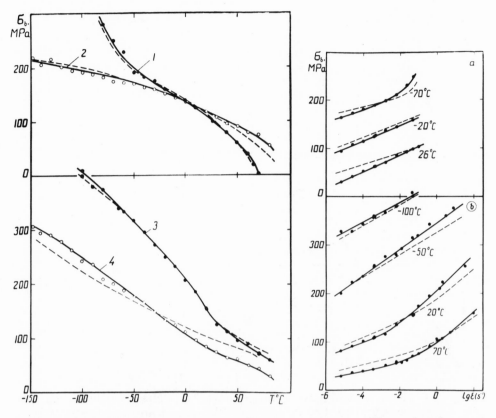

Fig.1. The temperature dependence of the yield point. 1 – PCHMA, 2 – PS, 3– PMMA, 4 – PVC; $\dot{\mathcal{E}} = 5 \cdot 10^{-2} s^{-1}$

Fig.2. The strain rate dependence of the yield point. a) PCHMA, b) PMMA The theoretical curves are shown by dotted lines.

that the relaxation spectrum consideration is absolutely necessary for building a realistic model of the unelastic deformation of polymeric glasses.

The present paper deals with analyzing the molecular-kinetic model, which regards the polymeric glasses as heterogeneous systems with the areas of different molecular mobility. In separate areas the displacement of different portion of macromolecules occurs by Frenkel's mechanism "of stochastic molecular re-arrangements at length on the distribution of processes by

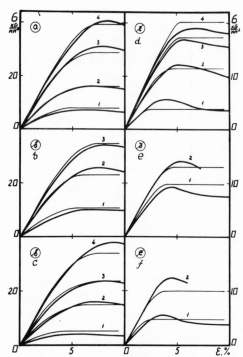

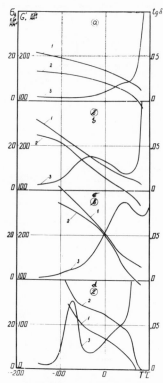

Fig.3. The comparison between the tempera-
ture dependence of
the yield point and
the mechanical dy-
namic characteris-
tics: a) PS,b) PVC,
c) PMMA, d) PCHMA;
1 - $6_b(T)$, 2 - $G'(T)$,
3 - $tg\,\delta(T)$.

Fig.4. The comparison of
the theoretical
(the thin lines)
and experimental
stress-strain
curves of PMMA:
a) $\dot{\varepsilon}=5\cdot10^{-2}s^{-1}$,T^{o}C:
70 (1), 20(2),
-50(3), -100(4);
b) T=20°C, $\dot{\varepsilon}$(s^{-1}):
10^{-4}(1), 36(2);
T= - 100°C,
$\dot{\varepsilon}$(s-1):10^{-4}(3);
c)T=70°C, $\dot{\varepsilon}$(s^{-1}):
10^{-4}(1), 100(2);
T= - 50°C, $\dot{\varepsilon}$(s^{-1}):
10^{-4}(3), 8(4) and
PS: d) $\dot{\varepsilon}=5\cdot10^{-2}s^{-1}$,
T^{o}C:80(1), 20(2),
-50(3),-100(4);
e)T=20°C, $\dot{\varepsilon}$(s^{-1}):
10^{-4}(1):10^{-4}(1),
5(2); f)T=60°C,
$\dot{\varepsilon}$(s-1)10^{-4}(1),100(2)

their activation energies, so the geometry of the kinetic units and their motion are considered in a simplest version of the liquid viscous flow model[11,12]. Summarizing of the processes at the transition from the local actions to the macroscopic behaviour has been made in an assumption of equality of viscoelastic shears in separate microvolumes of the total specimen shear deformation. The model suggested with respect to is kinetics and the summerizing procedure is close to the viscous flow model for heterogeneous systems of Ree-Eyring[13] but unlike that this model describes viscoelastic behaviour characteristic for polymer glasses both at small and large stresses and strains.

The model theoretical consideration results in the following kinetic equation for each of the processes:

$$\dot{\gamma} = \dot{\gamma}_i = \frac{f_i}{\left(\frac{\partial^2 u_i}{\partial x_i^2}\right)_0 \cdot \lambda_{3i}} + 2\nu_i \frac{\lambda_{1i}}{\lambda_{3i}} \exp\left(-\frac{U_{oi}}{kT}\right) \cdot sh \frac{f_i \cdot \lambda_{1i}}{2kT} \qquad (2)$$

Here f_i - is the projection on the reaction co-ordinate x_i of a mean (for $\gg 10^{-12}$s) force, acting on the i-th kinetic unit, $u_i(x_i)$ - the energy relief along the reaction co-ordinate, U_{oi} - the energy barrier height, ν_i -the complex oscillation frequency along x_i, λ_{1i} - the neighbouring energy minimum spacing, λ_{3i}- the mean intermolecular distance, $\dot{\gamma}$ - the shear rate. The equation (2), similar to that[14] obtained for a single process, in this paper is the main equation for the problems of the polymer glass viscoelastic behaviour. By integrating (2), the acting shear stress is obtained:

$$\tau = \sum_{i=1}^{N} n_i f_i \quad , \qquad (3)$$

N is the number of different kinetic processes, the first being that of glassing, n_i - the number of the kinetic units of the i-th group in a unit area.

Simulating different regimes of loading the model, relationships between the loading time (the strain rate), test temperature, deformation and applied stress can be obtained, which will take into account the relaxation spectra both at small and at large stresses. So, for the shear modulus and the loss modulus at cyclic loading by small stresses with the frequency ω, the customary formulars of the linear viscoelasticity can be used: $G' = \sum_{i=1}^{N} n_i G_i \frac{\omega^2 \tau_i^2}{1 + \omega^2 \tau_i^2}$, $G'' = \sum_{i=1}^{N} n_i G_i \frac{\omega \tau_i}{1 + \omega^2 \tau_i^2}$,

where $G_i = \left(\frac{\partial^2 u_i}{\partial x_i^2}\right)_0 \cdot \lambda_{3i}$, $\tau_i = \frac{kT}{\nu_i \lambda_{1i} \left(\frac{\partial^2 u_i}{\partial x_i^2}\right)_0} \exp\left(\frac{U_{oi}}{kT}\right) = \tau_o \exp\left(\frac{U_{oi}}{kT}\right)$. As in all the loading regimes the same kinetic processes take part, we succeeded in performing a certain comparison between the influence on polymer viscoelastic properties of the test temperature and time (frequency) and the change in stress-strain curve parameters at shear $\tau(\gamma)$ with deformation. The calculations give:

I. within the linear viscoelastic range $0 < \gamma \leq \gamma_{bm}$:

$$\left(\frac{\partial \tau}{\partial \gamma}\right)_{\dot{\gamma}_1, T_1} \approx G'(\omega, T_1) \qquad \text{at} \qquad \omega = \frac{\dot{\gamma}_1}{\gamma} \tag{3}$$

(the strain-frequency correlation)

$$\left(\frac{\partial \tau}{\partial \gamma}\right)_{\dot{\gamma}_1, T_1} \approx G'(\omega_1, T) \qquad \text{at} \qquad \left(\frac{T}{T_1}\right)^{\xi} = \frac{a_0(\omega_1) - \ln \frac{\omega_1 \gamma}{\dot{\gamma}_1}}{a_0(\omega_1)} \tag{4}$$

(the strain-temperature correlation)

II. within the non-linear viscoelastic range

$$\left(\frac{\partial \tau}{\partial \gamma}\right)_{\dot{\gamma}_1, T_1} \approx G'(\omega, T_1) \qquad \text{at} \qquad \ln \frac{\omega_m}{\omega} = \frac{\alpha/\beta \, a_m \gamma - 1}{1 - \alpha/\beta \gamma} \tag{5}$$

$$\left(\frac{\partial \tau}{\partial \gamma}\right)_{\dot{\gamma}_1, T_1} \approx G'(\omega_1, T) \qquad \text{at} \qquad \left(\frac{T}{T_1}\right)^{\xi} = \frac{a_m - 1}{a_0(\omega_1)(1 - \alpha/\beta \gamma)} \tag{6}$$

where $\dot{\gamma}_1$, T_1 - the shear rate and the test temperature, respectively, $\alpha = \frac{\lambda_{1i}}{2 U_{0i}} \left(\frac{\partial^2 U_i}{\partial x_i^2}\right)_0$ - the parameter determined by the shape of the energy relief (for the sinusoidal profile $\alpha = \pi^2$), $\beta = \frac{\lambda_{1i}}{\lambda_{2i}}$ - a geometrical factor, $\gamma_{bm} = \frac{\beta}{\alpha \ln(2 \nu_0 \cdot \beta \cdot \dot{\gamma}_1^{-1})}$ - the deformation, separating the linear and non-linear ranges of viscoelasticity at constant strain rate, $\nu_0 \approx 10^{11} s^{-1}$ - the frequency factor, $\omega_m = \dot{\gamma}_1/\gamma_{bm}$, $a_m = \frac{\beta}{\alpha \cdot \gamma_{bm}}$, $a_0(\omega_1) = 1/\ln(\omega_1 \cdot \tau_0) \approx 30$ at $\frac{\omega_1}{2\pi} = 1 Hz$; $\xi = \frac{\ln G'(T_1)/G'(T)}{\frac{2 a_0(\omega_1)}{\pi} \int_{T_1}^{T} tg \delta \, d\ln T}$

- the parameter connected with the temperature dependence of G_i, as well as with the deviation from a simple exponential dependence of the kinetic process relaxation times near Tg. The physical significance of (3--6) is based on the fact that at deformation of solids with a set of molecular re-arrangements the applied stress increment stimulate the mobility of the kinetic units with greater activation energies; the same process of "re-freezing" can be realized in a linear viscouselastic range (at very small stresses) be increasing the test temperature or time.

So, knowing, for instance, $G'(T)$ and integrating eqs (4,6) the equation of the stress-strain curve can be formulated:

$$\tau(\gamma)_{\dot{\gamma}_1, T_1} \approx \int_0^{\gamma} G' d\gamma \qquad 0 < \gamma \leq \gamma_{bm} \quad G'(\gamma) \text{ by } (4) \tag{7}$$

$$\tau(\gamma)_{\dot\gamma_1,T_1} \approx \int_{\gamma_{bm}}^{\gamma} G' d\gamma \qquad \gamma > \gamma_{bm} \qquad\qquad G'(\gamma) \text{ by (6)}$$

For the yield stress we have:

$$\tau_b(\dot\gamma_1,T_1) \approx \int_{0}^{\gamma_{b_1}} G' d\gamma \ , \tag{8}$$

where γ_{b_1}, the deformation responsible for the inception of the yielding, can be estimated from (6), assuming that T=Tg.

The only fitting parameter in the theory is the magnitude of α/β , which can be evaluated from simple assumptions: α is the shape coefficient of the energy relief ($\alpha = \pi^2$ for the sinusoidal barrier), β characterizes here the geometry of the molecular re-arrangements and depends on the local stacking of the different portions of macromolecules. If we consider the simplest cases of the most feasible macromolecule stacking pattern in amorphous polymers, then β =1 for a tetragonal lattice, $\beta=\frac{2}{\sqrt{3}}$ for the close packed hexagonal structures, and so the relation α/β should be of the order of $\sim 10 \div 8$.

With the aid of the eqs.(4-6) for the materials studied at different test temperature and strain rates the stress-strain diagrams were calculated; when passing to the uniaxial compression the maximum shear stress criterion was applied. Figure 4 shows both the calculated and experimental stress-strain curves for PMMA, PS. It follows from the comparison between the theoretical and experimental curves that at small strains (2-4%) and the test temperatures appreciably far from Tg there is good agreement between the theory and the experiment. However, as we approach the yield point, especially near Tg, the discrepancy, as a rule, increases. There may be several reasons responsible for this fact: the violation of the strain - temperature correlation (6), the violation of the maximum shear stress criterion, and so on.

In conclusion, it should be emphasized that the temperature and rate dependencies of the yield point, Figs 2,3, calculated by the formula (8) contain all the peculiarities of the experimental curves, the discrepancy being an average as low as 10%. The values of the parameters α/β for different polymers found by the technique of the least mean squares were also close to the estimations made above: α/β =9.3 for PMMA, 14.0-for PS, 13.0 - PVC, 8.7 - PCHMA.

Thus, the model considered enables us to analyse
the strain-stress diagrams of the glassy polymers at
different stages of the deformation process, it also
predicts by the linear viscoelasticity spectrum characteristics the polymer behaviour at large stresses within a wide range of the test conditions. So, it is
shown that the main physical idea inherent in the model suggested and promoting the role of the wide spectrum of relaxation processes in the unelastic deforma -
tion of polymeric glasses has been proved experimentally.

References

1. J.Ferry, Viscoelastic properties of polymers,
    N.Y.-London (1961)
2. G.M.Bartenev, U.V.Zelenev, Processi mekhani-
    cheskoi relaksacii v polymerakh., Mekhanika
    polymerov, 1:30, (1969).
3. I.I.Perepechko, Akusticheskie metodi issledo-
    vaniya polymerov, M. (1973).
4. U.S.Lasurkin, R.L.Fogelson, O prirode bolschikh
    deformacii viscomolecularnich veshestv v
    stecloobraznom sostojani, Zhwan.Tekn.Fiz.
    v.21:267,(1951).
5. T.E.Bradey, G.S.Y.Yeh, Mechanism of yielding
    and cold flow in glassy polymers. - J.Macro-
    mol. Sci.- Phys. B9:659 (1974)
6. N.N.Peschanskaya, V.A.Stepanov, Activacionie
    parametri polzuchesti lineinikh polymerov,
    Fiz.Tv.Tela, 20:2005,(1978).
7. J.A.Roettling, Yield stress behaviour of poly-
    methylmethacrylate Polymer, v.6:311,(1965).
8. A.H.Sinani, V.A.Stepanov, Issledovanie sopro-
    tivleniya polymerov deformirovaniu metodom
    izmerenija tverdosti, in: Issledovanija v
    oblasti izmerenija tverdosti, ed.B.I.Pilip-
    chuk, M.-L., izd.Standartov, 91: 180,(1967).
9. C.Bauwens-Crowet, J.C.Bauwens, G.Homes, Ten-
    sile yield-stress behaviour of glassy poly-
    mers, J.Polym.Sci: pt A-2:Polym.Phys.,7:735,
    (1969).
10. A.S.Argon, R.D.Andrews, J.A.Godrick, W.Whitney,
    Plastic deformation bands in glassy polysty-
    rene, J.Appl.Phys., 39:1899 (1968).
11. J.I.Frenkel, Kineticheskaja teorija zhidkos-
    tei, izd.AN SSSR, M-L. (1945).

12. S.Glasstone, K.J.Laidler, H.Eyring, The Theo-
    ry of Rate Processes, Mc Craw-Hill, N.-Y.
    (1941).
13. T.Ree, H.Eyring, Theory of non-neutonian
    flow, J.Appl.Phys., 26:793,(1955).
14. A.Tobolsky, R.Powell, H.Eyring, Elastic-
    Viscous Properties of Matter, in: Frontiers
    in chemistry, v.1., The chemistry of large
    molecules, Princeton, New Jersey, 125(1944).

KINETICS OF DEFORMATION AND

INTERMOLECULAR FORCES IN POLYMERS

W.A.Stepanov, W.A.Bershtein,
N.N.Peschanskaya

A.F.Ioffe Physical-Technical
Institute
Leningrad, USSR

Study of the nature of deformation as a kinetic process implies the following determination of the kinetic parameters in the equation for the strain rate[1-4]. The enthalpy of activation $Q(t_m)$ as well as the activation volume α are usually found experimentally. Knowing $Q(t_m)$ and α, the activation energy for the process, Q_o, can be found from the relation $Q_o = Q(t_m) + \alpha t_m$ (1). Experimental data available and the present understanding of the nature of the activation barrier of deformation are controversial[5-7].

In the present paper an attempt was made to suggest a correct definition of Q_o and α and to make clear the physical significance of the activation barrier, determining the kinetics of deformation, for these question are fundamental in building the concrete model of deformation process. This study deals with a systematic consideration of creep parameters of the glassy polymers in a wide range of temperatures and stresses.

Cylindrical specimens of 6 mm length and 3 mm in diameter were tested in creep (t_m=const). Compression deformation was chosen to make the process of deformation preferential as to that of fracture having another nature [8]. Tests were made at 20° on polymethylmethacrylate (PMMA): polystyrene (PS), copolymers of styrene and methacrylic acid (PS+n% MAA), high pressure polyethylene (PEHP), polyvinyl-chloride (PVC), polycarbo-

nate (PC), polyvinyl-butyrate (PVB), polyacrylonitrile
(PAN), polyethylene terephthalate (PET), and on PMMA,
PS, PVC, PC and PE- within a wide temperature range
below Tg.

A differential method, i.e. measuring the creep
rate (ε_1) before and (ε_2) after a small "step" in the
test temperature or stress was used. A sensitive meth-
od of measuring of the creep rate with the aid of the
laser interferometer was applied[9]. This method enables
us to take $\dot\varepsilon$ in any "point" of the creep curve with
the 1% exactness on the basis of the strain increment
as small as ~0.005% which is important because the
constancy of structure should be provided during the
measurement of   and [10]. At small stepwise variations
of temperatures and stresses, 3-5°C and ≤ 10%, respectively, and small strain increments (0.01-0.005%), the
creep parameters $Q(t_m)$ and α were assumed to remain
constant; however, generally speaking, we presume that
they may have the temperature and stress dependencies
of their own. Restoration of creep rate in returning
to the initial conditions (the "back step") was taken
as a criterion of the structure constancy.

Results and Discussion. Figure 1 shows the typical
dependencies of activation parameters Q_0 and α on the
strain ε at constant stress and temperature for PS +
+16% MAAc. The curves $Q_0(\varepsilon)$ and $\alpha(\varepsilon)$ have a maximum at $\varepsilon = \varepsilon_\ell$ (ε_ℓ corresponds to the minimum value of
the creep rate and the yield point on the $\sigma - \varepsilon$ curve[11],
and show then a sharp decrease passing into a "plateau".
For different polymers the parameters vary slightly as
deformation grows from 10 to 40%.

In the works by Kobeko, Eyring, Alexandrov[2-4,12]
as well as in those developing modern models[13,14] the
deformation is considered in terms of overcoming the
intermolecular energy barrier (IME). Experimental data available prove these suggestions. In particular,
a correlation was found between the resistance to deformation ability and IME[15,16]. However, the kinetic
parameters obtained in [5,6] as well as by the present
authors are inconsistent with such a consideration,
just due to the fact that Q_0 exceeds greatly the energy
of a single intermolecular bond (1-5 kcal/mole). So,
an assumption can be made that the inconsistency mentioned might be caused by the cooperative nature of
an elementary deformation event.

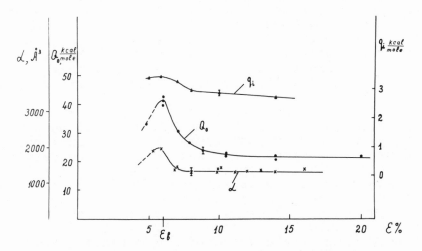

Fig.1.   Dependences of the activation energy Q_o, the activation volume α and the energy barrier $q_i = Q_o V/\alpha$ (V = 157 Å³) on strain in creep for PS+16% MAAc; T = 20°C, t_m=35 MPa, $\dot{\varepsilon}$ =10⁻⁵s⁻¹.

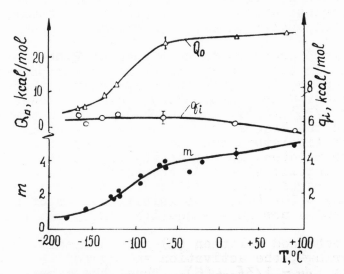

Fig.2.   The temperature dependence of the activation energy Q_o, the activation volume $m = \alpha/V$ and the energy barrier. Polycarbonate[21].

Assumed that an elementary deformation event is a shear transportation of a certain molecular mass having the volume $\alpha = m \cdot V$ over the potential barrier $Q_o = \Sigma q_i$ [17]. Here V is a volume of a periodic unit of the structure (a monomeric unit), and m is a number of the monomeric units in the kinetic unit. The barrier q_i being related to one monomeric unit, the activation energy is $Q_o = m q_i$

If our understanding of the nature of the barrier Q_o is right, the magnitude $q_i = Q_o / m$ should be determined by the chemical structure of the monomer unit, while by its physical significance it is the barrier of intermolecular energy related to a monomer mole. The value of q_i characterizing the IME for a given polymer should change insignificantly with the experiment conditions (ε, T)[15,18]. Indeed, the results given in Figure 2 show that Q_o and α or m strongly varying with temperature, the value of $q_i = Q_o / m$ remains nearly constant [18]. Alongside with this, as at follows from Figure 1, the deformation ($\varepsilon > \varepsilon_\ell$) results in reduction of q_i as well as of the intermolecular energy barrier[15].

As it assumed that q_i is the energy for the molecular interaction related to a monomer unit, it can be compared with a well-known characteristic of the molecular interaction for the corresponding monomer, i.e. the cohesion energy ($E_{coh.}$), for example.

Figure 3 shows that the values of q_i obtained in this work for eleven polymers of different chemical composition and the cohesion energy values taken from literature[19,20] are governed by a single linear dependence $q_i = 1/3 E_{coh.} (2)$. The natched band takes into account both the discrepancy of the values of $E_{coh.}$ estimated from literature and the variations of q_i. The top border relates, mainly to the region close to ε_ℓ (when the initial system of the intermolecular bonds is not distroyed by deformation), the bottom one to the strains $\sim 20\%$ (when the initial system of bonds is violated and a new quasi- equilibrium system is formed).

The obtained relation (2) is remarkable. According to Eyring[2], the activation energy for viscous flow of liquids $E_{visc.} = 1/3 E_{coh.} (6)$. Thus, the barrier q_i in the region of developed plastic deformation corresponds by its physical significance and magnitude to the activation energy of the viscous flow of monomer liquids. From this follows that the barrier $Q_o = m \cdot E_{coh} / 3$, characterizing an elementary deformation event, is determined by the energy of the molecular interaction and the

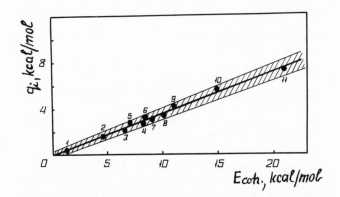

Fig.3. The connection between the deformation energy
        barrier per one monomeric unit (q_i) and the
        cohesion energy of the corresponding monomer
        ($E_{coh.}$);
        1 - PE; 2 - PVC; 3 - PVB; 4 - PAN; 5 - PMMA;
        6 - PS; 7 - PS + 16%MAAc; 8 - PS+33% MAAc;
        9 - PS+60% MAAc; 10 - PET; 11 - PC.[21]

number of monomer units in the activation volume α.
The volues of q_i were obtained in compression. In the
case of extensive breaking of chemical bonds and formation of cracks and cavities (in tensile tests, for
example) the creep rate is also influenced by the kinetics of fracture [8]. Here the parameters and their in-
terrelationship might change and q_i no longer has such
a clear physical significance.

So, the results obtained convincingly show that
irrespectively of a concrete model the process of deformation of glassy polymers is controlled, mainly, by
the surmounting of intermolecular energy barriers; the
cold flow phenomenon in polymers is similar to the viscous flow of an olygomer, the molecular volume of which
is equal to the volume of a kinetic deformation unit α.

References

    1. Y.I.Frenkel, "Kineticheskaya teoriya zhidko-
        stei", AN SSSR, M.L.(1945).
    2. S.Glasstone, K.I.Laidler, H.Eyring, "The Theo-
        ry of Rate Processes", Mc Graw-Hill,New-York
        (1941).

3. A.P.Alexandrov, Morozostoikost vysokomoleku-
    lyarnych soedinenii, Trudy 1 i 2 conferen-
    ziy po vysokomolekulyarnym soedineniyam,
    AN SSSR, M.-L. (1945).
4. A.S.Krausz, H.Eyring,"Deformation Kinetics",
    Wiley, New York (1975).
5. S.B.Patner, U.I.Brokhin, Temperaturno-vremen-
    naya zavisimost predela vynuzhdennoi elas-
    tichnosti, Dokl.AN SSSR, 188; 807 (1969).
6. T.E.Bradey, G.S.Y.Yeh, Mechanism of Yielding
    and cold low in Glassy Polymers, J.Macromol.
    Sci.Phys., B9: 658 (1974).
7. P.M.Pakhomov, M.L.Shermatov, W.E.Korsukov,et al
    Svyaz konformacionnykh perekhodov s defor-
    maciei bolshogo perioda v polietilene, Vyso-
    komolek.soyedineniya, A18:132.
8. W.A.Stepanov, Deformaciya i razrushenie poli-
    merov, Mekhanika polimerov, 1:95 (1975)
9. N.N.Peschanskaya, W.A.Stepanov, aklivacionnye
    parametry polzuchesti lineinykh amorphnykh
    polimerov, Fizika tverdogo tela, 20: 2005
    (1978).
10. W.L.Indenbom, A.N.Orlov, Vstupitelnaya statya,
    in: Sb."Novosti Fiziki Tverdogo Tela", Mir,
    M. (1973).
11. W.N.Borsenko, A.B.Sinani, W.A.Stepanov, Svyaz
    krivykh polzuchesti polimerov s diagrammami
    deformirovaniya, Mekhanika polimerov, 5:787,
    (1968).
12. P.P.Kobeko, Amorfnye veschestva, AN SSSR,
    M.-L.(1952).
13. P.B.Bowden, S.Raha, A molecular model for
    yield and flow in amorphous glassy polymers
    making use of a dislocation analogy, Philos.
    Magaz., 29: 149 (1974)
14. A.S.Argon, M.I.Bessonov,Plastic deformation in
    polyimides with new implication on the theo-
    ry of plastic deformation of glassy polymers
    Philos.Magaz, 35: 917 (1977).
15. W.A.Bershtein, Z.G.Razgulyaeva, A.B.Sinami,etal.
    Mezhmolekulyarnoe vzaimodeistvie i neupruga-
    ya deformaciya amorphnykh polimerov, Fizika
    tverdogo tela, 18: 3017 (1976).
16. W.A.Bershtein, N.N.Peschanskaya, W.A.Stepanov,
    Mezhmolekulyarnoe vzaimodeistvie i razrushe-
    nie amorfnykh polimerov, Mekhanika polimerov,
    6: 963 (1977).
17. N.N.Peschanskaya, W.A.Bershtein, W.A.Stepanov,
    Svyaz energii aktivacii polzuchesti lineinykh
    polimerov s energiei kogezii,Fizika tverdogo
    tela, 20:3371 (1978)

18. W.A.Bershtein, M.Z.Petkevich, D.G.Razgulyae-
     va *et al.* Temperaturnaya zavisimost mezhmoleku-
     lyarnykh vzaimodeistvii v stekloobraznykh
     polimerakh, Vysokomol.Soyedineniya, A20:
     2681 (1978).
19. D.W.Van Krevelen, Properties of polymers Cor-
     relations with Chemical Structure, Amster-
     dam - London, New York (1972).
20. A.A.Askadskii, L.K.Kolmakova, A.A.Tager *et al.*
     Ob ocenke plotnosti energii kogezii nizko-
     molekulyarnykh zhidkostei i polimerov,
     Vysokomol. Soyedineniya, A19:1004 (1977).
21. W.A.Bershtein, N.N.Peschanskaya, A.B.Sinani
     et al. Fizika Tverdogo tela, 22:767 (1980).

MEASUREMENT OF DYNAMIC MECHANICAL PROPERTIES

Takayuki Murayama

Monsanto Triangle Park Development Center, Inc.
P. O. Box 12274
Research Triangle Park, N. C.   27709

ABSTRACT

Measurements of dynamic mechanical properties of viscoelastic materials have been reviewed. The technology for investigating the dynamic mechanical properties of materials has advanced greatly due to new electronic circuits, improved transducers, signal receiving systems, and computerized controlling systems. Some representative instruments and techniques are discussed. Molecular interpretation and effects of structure geometry are also reviewed. In addition, methods of measurement of dynamic properties in gas and liquid media are presented.

INTRODUCTION

The dynamic mechanical behavior of viscoelastic materials has been discussed as a part of mechanical properties such as stress-strain behavior, strength of material, creep, and stress relaxation. However, in recent years the technology of investigating the dynamic mechanical properties of materials has advanced greatly. The interpretations and applications of dynamic data have stimulated interest from both the practical and scientific standpoint. Consequently, widespread research work on these topics has established diverse results and principles.[1,2]

The commonly used dynamic mechanical instruments measure the deformation of a material in response to vibrational forces. The dynamic modulus, the loss modulus, and a mechanical damping or internal friction are determined from these measurements. The modulus indicates stiffness of material, and it may be a shear, a tensile, or a flexile modulus, depending upon the experimental equipment. The mechanical damping (internal friction) gives the

amount of energy dissipated as heat during the deformation. The
internal friction of materials is important, not only as a property
index, but also for environmental and industrial applications. Since
noise is radiated by the vibration of an object--(0.001-0.004)--the
application of damping materials to the vibrating surface will
convert the energy into heat, which is dissipated within the damping
materials rather than being radiated as airborne noise.

Amorphous viscoelastic polymers are good damping materials,
having high internal friction (0.1-0.3). High damping or internal
friction is essential in decreasing the effect of undesirable vibration, in reducing the amplitude of resonance vibration to safe limits,
and in all kinds of structures from airplanes to buildings.[3]

The investigation of the dynamic modulus and internal friction
over a wide range of temperatures and frequencies has proven to be
very useful in studying the structure of high polymers and the
variations of properties in relation to end-use performance. These
dynamic parameters have been used to determine the glass transition
region, relaxation spectra, degree of crystallinity, molecular orientation, crosslinking, phase separation, structural or morphological
changes resulting from processing, and chemical composition in poly-
blends, graft polymers, and copolymers.

In addition to these structure-property relationships, dynamic
mechanical studies have been extended to material composites and
structural systems. Considerable attention has been given to the
dynamic behavior of structures of composite materials. This is
primarily because most aircraft and aerospace structures are subjected to dynamic loadings. Knowledge of the dynamic moduli and
properties of composite materials is indispensable to intelligent
design with these materials.

Instrumentation and Technique

The development of characterization techniques for measuring
dynamic mechanical properties has been an important activity in the
area of dynamic mechanical analysis. An accurate measurement of the
response of a polymeric material to vibrational forces requires many
instrumentational considerations. A very large number of techniques
has been developed by using different vibrational principles at
various frequencies. The methods are free vibration, resonance
forced vibration, non-resonance forced vibration, and the wave or
pulse propagation technique. The frequency for each method has a
limited range. For instance, the frequencies of forced vibration
are about 10^{-3} Hz to 10^2 Hz. In the case of wave propagation, the
frequency range used is from 10^3 Hz to 10^7 Hz.

Since polymeric materials exhibit a wide variety of shapes and
moduli (polymer melts, rubbers, plastics, and fibers), the

instrument is often specialized for one type of material. It is
very difficult to develop one instrument which has the capability
of characterizing all types of polymers over a wide range of frequency and temperature. Many instruments are described in the
literature.[1,2,4-6]

During the past decade, however, the technology for investigating the dynamic mechanical properties of materials has advanced
due to new electronic circuits, improved transducers, signal
receiving systems, and computerized controlling systems. A great
quantity and variety of research has been carried out by these
means.[4,7-9]

One of the most used instruments for measurement of dynamic
mechanical properties of polymers is the standard Rheovibron visco-
elastometer DDV-II. This is designed to measure the temperature
dependence of the complex modulus (E^*), dynamic storage modulus
(E'), dynamic loss modulus (E''), and loss tangent (tan Delta) of
viscoelastic materials at specific selected frequencies (0.01 to 1 Hz,
3.5, 11, 35, 110 Hz) of strain input. This instrument was developed
by Takayanagi.[7] It was unique because it was designed to read
tan Delta directly, greatly simplifying and speeding the characterization of a material. The Rheovibron was commercialized by Toyo
Measuring Instruments Co. A number of improvements in this instrument have been reported by users. A mechanical model and modified
drive expression for the moduli to take into account system compliance, sample yielding within the tensile grips, and system inertia
were proposed.[10] These effects require a correction to the raw data
for glassy polymer samples in the tensile mode. In addition to the
standard tensile grips, a new sample holder has been introduced
which adapts the Rheovibron to the flexural and shear mode for
investigation of the rheological properties of styrene-ethylene
oxide block copolymers and polyurethane elastomer.[11,12] These
modifications of grips have been extended to the compression mode[13]
and the bending mode[14] for measurement of anisotropic dynamic mechanical properties. In addition, the Model DDV-III-B has been introduced, with larger stress capability (5 kg) and a hydraulic driving
system. Using this model, a new type of instrument has been developed for measurement of the complex piezoelectric stress constant e*
and the complex tensile modulus E* with devices installed in the
same equipment. The principle of measurement combines the merits
of the Null-detection method and the direct reading method.[15] The
polarization charge on the sample surface generated by the sinusoidal
strain is detected with a charge amplifier. From the strain, stress,
and polarization signals, the complex tensile modulus and the complex piezoelectric stress constant under the same conditions of
temperature and frequency are evaluated. Detailed studies of piezoelectric properties of oriented films of poly (Gamma-methyl
Phi-glutamate) using this instrument have been reported.[16] The
piezoelectric properties of this PMDG film are related to molecular

motion and the Alpha-helix. The piezoelectric properties of some
biopolymers, such as silk and collagen, and some synthetic polymers,
such as polyvinylidine fluoride, have been found to be comparable to
those of inorganic piezoelectric crystal, such as quartz and Rochelle
salt.[17]

The automated Rheovibron was originally developed by Monsanto
Company,[18] and the automation system for the Rheovibron has been
commercialized by IMASS, Inc. This automatic instrument involved
primarily a measurement system and an automatic tension system. The
key component in the measurement system is a modified Princeton
Applied Research Model 129A two-phase vector lock-in amplifier. This
amplifier is used in the vector-voltmeter mode to measure the amplitude and phase of both the strain and the stress gauge outputs. The
information of the vector voltmeter is connected to the data logging
system. A digital thermocouple thermometer and a sample length
measurement circult are used for length and temperature change information for logging by the data system.

A torsion pendulum is the simplest instrument for measuring the
dynamic shear modulus and mechanical damping of polymer. The frequency range of operation is from 0.01 Hz to 50 Hz. This is a free
vibration instrument. The specimen is a cylindrical rod or a rectangular block; one end is rigidly clamped and the other is attached
to an inertia disk which is free to oscillate. The oscillations
twist and untwist the specimen. The time required for one complete
oscillation is the period. The successive amplitudes will decrease
with time because of the damping, which gradually converts the elastic energy of the system into heat. Instruments based on the torsion
pendulum have been described and used by many investigators.[19-23]
A number of commercial variants are now on the market, and ASTM
standard D2236-64T describes the procedure to be used.

Torsional braid analysis (TBA) is an extension of the torsion-
pendulum method for examining the dynamic mechanical properties of
materials. It was developed by J. K. Gilham,[24] and his extensive
studies on this subject can be found in more than 30 publications.
This method is the only one available for measuring properties of
solids or melts with less than 100 mg of polymer per experiment
using the torsion pendulum. The technique differs from the simple
torsion-pendulum one mainly in that the sample specimen is prepared
by impregnating a multifilament glass braid substrate with a solution
of the material which is to be investigated. Changes in the mechanical properties of the composite are attributed to changes in the
polymer. An advantage of using a supported polymer technique is
that specimen fabrication is generally simplified to deposition of
the material from a melt or from solvent onto the substrate.
Further, by the use of a support, a material can be investigated
through various regions of the mechanical spectrum, including those
in which the material may not be capable of supporting even its own

weight. This provides an opportunity for studying resin-forming reactions and high temperature reactions of thermoplastic polymers, both of which often start in the liquid state and proceed through gel, rubbery, and glassy states.

In addition, Boyer's[25] comprehensive review on mechanical motion in amorphous and semi-crystalline polymers describes how TBA provides information for the oligomeric range of molecular weights as well as the region above Tg, or even the liquid region above Tm, by using pendulum with different moments of inertia.

The vibrating reed,[26-29] or cantilever, is a simple and widely used technique for measuring Young's modulus and internal friction or damping. A small specimen in the form of a plastic strip or reed is clamped at one end in an electromagnetic vibrator which is driven by a variable frequency oscillator. As the frequency of the vibrations is changed, the natural resonance of the reed will be reached, and the amplitude of the free end of the reed will go through a maximum. The amplitude of fibration of the specimen is measured as a function of frequency at a fixed temperature.

The mechanical spectrometer is a rotational rheometer for measurement of the basic rheological properties, such as shear viscosity, normal stresses, dynamic moduli, and melt elasticity in fluids and solids. The Weissenberg Rheogoniometer[30] by Sangamo Controls Ltd., England, and the Rheometrics Mechanical Spectrometer by Rheometric, Inc.[31] U. S. A., are examples of the commercial instruments.

This apparatus has been used extensively to study polymer melts[32-36] by using various geometrics such as eccentric rotating disks, cone and plate, parallel plates, and concentric cylinders[37,38]. Sample specimens of polymer melts, rubbers, and elastomers (about 1.5 grams) are placed between the two disks. The instrument deforms the specimen rotationally and measures the stress responses with a unique transducer, which is capable of simultaneous detection of three orthogonal forces, and the torque about the z axis. This solid-state unit continuously monitors three forces: the normal force and the torque in the cone-and-plate flow, the biaxial stress relaxation in a solid rod, or the three orthogonal forces generated by the flow between eccentric rotating disks.

A newer commercial mechanical spectroscope is the duPont 980 Dynamic Mechanical Analyzer (DMA). This instrument measures the resonant frequency and energy dissipation without tension on a wide range of samples as a function of temperature (or time).[39] The resonant frequency is related to the modulus by the sample geometry. Energy dissipation relates to such properties as impact resistance, brittleness, and noise abatement.

The analyzer contains the mechanical oscillator, driver, and digital display portions of the DMA system. The complex modulus apparatus has been developed to determine the complex modulus of elasticity and the loss factor of solid materials. In the development of this apparatus, there was special regard for the measurement of plastic materials in the vital frequency range of 10 to 10,000 Hz.

This instrument consists of four units: the beat-frequency oscillator (Type 1022), the audio-frequency spectrometer (Type 2112), the complex modulus apparatus (Type 3930), and the level recorder (Type 2305). These units are made by the Bruel and Kjaev Company.[40]

The sonic technique has been employed to determine the dynamic Young's modulus of a wide range of materials (e.g., metal rods,[41] concrete beams,[42] paper,[43] and fibers.[44] The dynamic modulus E of a solid material is calculated from the acoustic velocity and density. This velocity is independent of sample geometry.[41] In practice, a sonic pulse is transmitted along a length of solid, and the flight time between transmitting and receiving transducers is measured. From a knowledge of the instrument delay time and sample length, the sonic velocity can be computed.

The PPM-5 (pulse propagation meter) dynamic modulus tester developed by Henry H. Morgan Company has been commonly used in the fiber and paper field. The recent developments in technique for investigating structure/property relationships of polymers have been the combination of several methods: dynamic mechanical measurements of polymers carried out with x-ray diffraction measurements, polarized light scattering, and small angle x-ray or neutron scattering.

Rheo-optical studies of semicrystalline polymers[45-47] have utilized electromagnetic radiation in studying the deformation and flow of polymers. In this study, optical as well as mechanical responses of the material are investigated simultaneously against mechanical excitation. Multi-channel narrow sector technique for dynamic x-ray diffraction measurements was developed.[46] This technique can be applied in principle to oscillatory measurement of every rheo-optical response, not only to the dynamic wide-angle x-ray diffraction but also to dynamic birefringence, dynamic infrared absorption, dynamic wide- and small-angle light scattering, and dynamic small-angle x-ray scattering.

These techniques will be powerful instruments for a better understanding of dynamic properties of materials.

Effect of Molecular Parameters

Dynamic mechanical properties can be determined by a number of previously discussed instruments. Some instruments are designed for rubbers and polymer melts, whereas others are for rigid polymers or fibers. The dynamic data are obtained as a function of temperature and frequency.

The dynamic modulus and internal friction are the most basic of all mechanical properties, and their importance in any end-use application is well known. The dynamic loss modulus, or damping, is sensitive not only to many kinds of molecular motion but also to various transitions, relaxation processes, structural heterogeneities, and the morphology of multiphase systems (crystalline polymers, polymer blends, and copolymers). Therefore, interpretations of the dynamic mechanical properties at the molecular level are of great scientific and practical importance in understanding the mechanical behavior of polymers.

Much research has been done on the relationships between molecular parameters and the dynamic mechanical behavior of polymers. The book by McCrum, Read, and Williams[48] gives a comprehensive review of dynamic mechanical spectroscopy in relation to the molecular structure of polymers. A number of useful generalizations and interpretations on this subject are found in texts by Nielsen[5] and Ferry.[4] In Recent years, Boyer also has reviewed the molecular motions in amorphous and semicrystalline polymers with respect to the anelastic spectra.[8] This has been the most active area in the dynamic mechanical analysis of polymers.

Motions of long chain segments in the polymer structure have a profound effect on the loss factors (tan Delta and loss modulus) of the dynamic mechanical properties. The tan Delta value at the Alpha peak (the glass transition) is of greater magnitude than at the dispersion peaks for lower temperatures, and is accompanied by the greatest decrease in dynamic modulus with increasing temperature.

The loss factors plotted against temperature often are called relaxation or anelastic spectra. These contain several loss peaks. Analyses of these spectra are useful for determining molecular motions in combination with dielectric nuclear magnetic resonance (NMR) measurements. Numerous relaxation spectra on all major polymers are discussed by McCrum et al.[48] Interpretations of the spectra are found in texts[4-7] and review articles.[8,25]

The loss factors are most sensitive to the molecular motions. For example, in the glassy region the dynamic modulus shows small changes, yet the loss factors exhibit damping peaks.

The largest loss peak is generally the Alpha peak, which is associated with the glass transition temperature. This transition occurs in the amorphous regions of the polymer with the initiation of the micro-Brownian motion of the molecular chains. The magnitude of Alpha peak in the amorphous polymer is much higher than in the semicrystalline polymer because the chain segments of the amorphous polymer are freer from restraints imposed by crystalline polymers in the glass transition region.

The glass transition temperature is affected by a number of chemical and molecular structures. Flexibility of the molecular chain, bulkiness of the side groups attached to the backbone, and molecular polarity affect the value of T_g. The peak height and the temperature at which the Alpha peak occurs are closely related to the glass transition phenomenon. Therefore, similar interpretations were used for both the Alpha peak and T_g.

The magnitude of the Beta peak is small in comparison with the Alpha peak. The temperature at the Beta peak is about $0.75\ T_g$. This dispersion is associated with an amorphous phase relaxation in most polymers. It is resolved into a double or triple peak and appears to involve local motion in the chain. In the case of amorphous polymers, the Beta peak is very broad and may appear as a shoulder of the Alpha peak. It is associated with motion about the chain backbone of a relatively small number of monomer units or with motion of side groups.[8]

The Gamma peak temperature is below the Beta peak temperature. In this region, the chain segments are frozen in, while side group motion is made possible by defects in packing or configuration in the glassy and crystalline states. This Gamma peak of polymers can be related to the following general classifications: side group motions in the amorphous and crystalline phases, end group rotation, crystal-line defects, backbone-chain motions of short segments or groups, and phase separation of impurities or diluents.[5,8]

The Gamma relaxation in both amorphous and crystalline polymers could in many cases be attributed to a restricted motion of the main chain which required at least four $-CH_2$ groups in succession on a linear part of the chain.[49] This is the crankshaft mechanism of Shatzki[50] and Boyer.[51] It has been proposed that this mechanism is relevant to the Gamma peak in polyethylene, polyamids, polyoxy-methylene, and polypropylene oxide.

The Alpha' (Alpha$_c$) peak occurs at temperatures between T_g and the melting temperature. This process relates rotation about the chain axis of chain-folded polymers to a second order expansion of lattice parameters perpendicular to the chain direction in the crystalline phase.

In the case of amorphous polymers the Alpha' dispersion is often designated the T_{11} relaxation, and involves a premelting process. The T_{11} relaxation is exhibited in polystyrene,[52] polyvinyl chloride,[53] and polymethylmethacrylate.[54] It has been extensively studied in synthetic rubber blends of the polybutadiene type.[55] In this study it was found that T_{11} dispersion is related to the motion

of the low molecular weight components of a polymer with a broad
molecular weight distribution.

The effects of molecular parameters such as degree of crystallinity, crystal size, molecular orientation, molecular weight, and
polymer composition on dynamic mechanical properties were also
reported.(56-61)

Effects of Structure Geometry

Many polymeric materials are combined as structural units.
Examples are the laminated plastics, the fiber-reinforced composites,
tires, and fabrics. Dynamic mechanical analysis is useful in determining the properties of these structural units and effects of the
structural geometry on the dynamic properties. The end use performance of these structural assemblies is influenced by the inherent
properties of the components and the external structural factors of
the systems.

The damping characteristics of structural systems have been the
sybject of investigation for many years. In the past several decades,
engineers have expressed considerable interest in the damping (internal friction) properties of structures at stress levels encountered
in engineering design. They have contended that resonant structural
vibrations could be controlled by the effective use of this property.

In addition to the internal damping properties of structural
materials, attention has been paid to the use of structural fabrica-
tions that involved Coulomb-slip and viscoelastic-shear damping
mechanisms. With the advent of high speed space vehicles, there was
great need for degrees of damping in structures that exceeded what
was available from the structural material itself. Investigators
thus turned to such structural damping techniques as interfacial
Coulomb-slip damping, viscoelastic laminate damping, and Coulomb or
viscoelastic junction damping. Detailed studies of structural
damping can be found in books from the ASME[62] and ASTM.[63] In this
section the effect of the structural geometry on dynamic mechanical
properties will be discussed in examples of interfaces in assemblies[64]
and adhesion.[65]

The defining characteristic of a structural assembly is its
joints. These are interfaces or mating surfaces which are maintained
in contact. Thus, in analyzing the dynamic properties of a structural assembly, it is important to consider not only the component
materials but also the dynamic response and the energy dissipation
caused by interface effects. Three types of interfaces are important
in the dynamic analysis in structural mechanics: dry interface
surfaces, lubricated interface surfaces, and adhesive bonded interfaces.

Unjoined interfaces are generally subjected to a large variety of loads and displacements in service. For the case of dry interfaces (metal-to-metal contact), Coulomb friction provides an important mechanism for dissipating energy under cyclic shear displacement.

Textile assemblies (yarns, woven fabrics, knitted fabrics, and nonwoven structures) and fiber-reinforced plastics have extremely largy interfaces between fibers and matrix. This is because the fiber has a large surface area compared to its volume. The characteristics of the interface have a strong effect on the dynamic mechanical properties of these polymer assemblies. Thus, dynamic analyses of twisted yarn and nonwoven fabrics are good examples for exhibiting the effects of the interfaces on mechanical properties.

REFERENCES

1. T. Murayama, Dynamic Mechanical Analysis of Polymeric Material, Elsevier Scientific Publishing Co., Amsterdam-Oxford, New York, 1978.
2. B. E. Read and G. D. Dean, The Determination of Dynamic Properties of Polymers and Composites, Wiley, New York, 1978.
3. M. P. Blake and W. S. Mitchell, Vibration and Acoustic Measuring Handbook, Spartan, New York, 1972.
4. J. D. Ferry, Viscoelastic Properties of Polymers, 2nd. ed., Wiley, New York, 1969.
5. L. E. Nielsen, Mechanical Properties of Polymers and Composites, Marcel Dekker, New York, 1974.
6. I. M. Ward, Mechanical Properties of Solid Polymers, Wiley, New York, 1971.
7. M. Takayanagi, Viscoelastic Properties of Crystalline Polymers, Mem. of the Fac. of Eng., Kyushu Univ., Vol. 23 (1963), No. 1.
8. R. F. Boyer, in Polymeric Materials: Relationship Between Structure and Mechanical Behavior (E. Baer and S. V. Radcliffe, Ed.), Am. Soc. Metals, Ohio, 1974, Chap. 6, pp. 277-368.
9. J. K. Gillham, S. J. Stadnicki, and Y. Hazony, J. Appl. Polym. Sci., 21, 401 (1977).
10. D. J. Massa, J. Appl. Phys. 44, 2595 (1973).
11. P. F. Erhardt, J. J. O'Malley, and R. G. Crystal, in Block Copolymers (S. L. Aggarwal, Ed.), Plenum, New York, 1970, pp. 195-213.
12. T. Murayama, J. Appl. Polym. Sci., 19, 3221 (1975).
13. T. Murayama, J. Appl. Polym. Sci, 20, 2593 (1976).
14. T. Murayama, J. Appl. Polym. Sci., 23, 1647 (1979).
15. K. Koga, T. Kajiyama, and M. Takayanagi, J. Phys. E: Sci. Instrum., 8, 299 (1975).
16. K. Koga, T. Kajiyama, and M. Takayanagi, J. Polym. Sci., Phys. Ed., 14, 401 (1976).
17. E. Fukada, Progr. Polym. Sco., 2, 329 (1971).
18. A. S. Kenyon, W. A. Grote, D. A. Wallace, and M. C. Rayford, ACS Polymer Preprints, 17, #2, 7-13 (Aug. 1976).

19. L. E. Nielsen, Rev. Sci. Instrum., 22, 690 (1951).
20. K. H. Illers and Jenckel, J. Polym. Sci., 41, 528 (1958).
21. J. Koppelmann, Kolloid-Z.Z-Polym., 144, 12 (1955).
22. D. J. Plazek, M. N. Vrancken, and J. W. Berge, Trans. Soc.
     Rheol., 2, 39 (1958).
23. N. Tokita, J. Polym. Sco., 20, 515 (1956).
24. J. K. Gillham, in Techniques and Methods of Polymer Evaluation,
     Vol. 2 (P. E. Slade, Jr., and L. T. Jenkins, Eds.), Marcel
     Dekker, New York, 1970, p. 225.
25. R. F. Boyer, Polymer, 17, 996 (1976).
26. K. Fujino, H. Kawai, and M. Horio, Textile Res. J. 25, 722
     (1955).
27. M. Horio, S. Onogi, C. Nakayama, and K. Yamamoto, J. Appl.
     Phys., 22, 966 (1951).
28. R. Buchdahl, R. J. Morgan, and L. E. Nielsen, Rev. Sci. Instrum.,
     41, 1342 (1970).
29. A. N. Gent, Br. J. Appl. Phys., 11, 165 (1960).
30. Sangamo Controls Ltd., Weissenberg Rheogoniometer Instruction
     Manual, Sussex, England.
31. Rheometrics, Inc., Mechanical Spectrometer, Product Bulletin,
     Union, N. J., U.S.A., 1976.
32. R. L. Cerro, C. W. Macosko, and L. E. Scriven, Nat. Phys. Sci.,
     241 (1973) 146.
33. F. G. Mussatti and C. W. Macosko, Polym. Eng. Sci., 13, 236 (1973).
34. J. H. Southern and D. R. Paul, Polym. Eng. Sci., 14, 560 (1974).
35. N. Nakajima, D. W. Ward, and E. A. Collins, J. Appl. Polym. Sci.,
     20, 1187 (1976).
36. C. W. Macosko and P. J. Morse, VII International Congress on
     Rheology, Gothenburg, Sweden, Proceeding Preprint 376 (Aug.
     1976).
37. E. Broyer and C. W. Macosko, SPE Tech. Papers, 21, 343 (1975).
38. W. M. Davis and C. W. Macosko, SPE Tech. Papers, 22, 408 (1976).
39. DuPont Instruments, 980 Dynamic Mechanical Analysis System
     (DMA), 1976.
40. Gruel and Kjaer Instruments, Instructions and Applications of
     Complex Modulus Apparatus Type 3930, Copenhagen, Denmark,
     1974.
41. R. C. McMaster, Nondestructive Testing Handbook, Ronald Press,
     New York, 1963, Sect. 43.9.
42. E. A. Whitehurst, Evaluation of Concrete Properties from Sonic
     Tests, ACI Monogr. No. 2, 1966, p. 3.
43. J. K. Craver and D. L. Taylor, Nondestructive Sonic Measurement
     of Paper Elasticity, Tappi Plastic Conference, 19-21,
     Octo. 1964.
44. C. F. Zorowski and T. Murayama, Textile Res. J., 37, 852 (1967).
45. R. S. Stein and T. Oda, J. Polym. Sci., B, 9, 543 (1971).
46. S. Suehiro, Dynamic X-ray Diffraction Technique for Measuring
     Rheo-Optical Properties of Crystalline Polymeric Materials,
     Dissertation, Kyoto University, Kyoto, Japan (1978).

47. S. Suehiro, T. Yamada, H. Inagaki and H. Kawai, Polymer Journal, 10, 3, pp. 315 (1978).

48. N. G. McCrum, B. E. Read, and G. William, Anelastic and Dielectric Effect in Polymeric Solids, Wiley, New York (1967).

49. A. H. Willbourn, Trans. Faraday Soc. 54, 717 (1958).

50. T. F. Shatzki, J. Polym. Sci., 57, 496 (1962).

51. R. F. Boyer, Rubber Rev. 34, 1303 (1963).

52. R. F. Boyer, J. Polym. Sci. C. 14, 267 (1966).

53. L. Utracki, J. Macromol. Sci., Phys., B10 (3) 477 (1974).

54. K. Ueberreiter and J. Naghagaden, Kolloid-Z, Z-Polym., 250, 927 (1972).

55. E. A. Sidorovitch, A. I. Marei, and N. S. Gashtol'd, Rubber Chem. Technol., 44, 166 (1971).

56. J. P. Bell and T. Murayama, J. Polym. Sci., Part A-2, 7, 1059 (1969).

57. P. R. Pinnock and I. M. Ward, Polymer, 7, 255 (1963).

58. K. H. Illers and H. Breuer, J. Colloid Sci., 18, 1 (1963).

59. J. H. Dumbleton and T. Murayama, Kolloid-Z., Z-Polym., 220, 41 (1967).

60. T. Murayama and B. Silverman, J. Polym. Sci., Polym. Phys. Ed., 11, 1873 (1973).

61. E. L. Lawton, T. Murayama, V. F. Holland, and D. C. Felty, J. Appl. Polym. Sci. (in press).

62. J. E. Ruzicka, Structural Damping, Am. Soc. of Mech. Eng., New York (1959).

63. ASTM Special Technical Publication No. 378, Internal Friction, Damping and Cyclic Plasticity (1964).

64. T. Murayama, J. Appl. Polym. Sci., 24, 1413 (1979).

65. C. F. Zorowski and T. Murayama, Proc. First Int. Conf. on Mech. Behav. of Mater., Vol. 5, Soc. of Mater. Sci., Kyoto, p. 28 (1972).

RHEO-OPTICAL STUDIES ON THE NATURE OF ALPHA AND BETA MECHANICAL DISPERSIONS OF POLYETHYLENE IN RELATION TO THE DEFORMATION MECHANISMS OF SPHERULITIC CRYSTALLINE TEXTURE

Hiromichi Kawa, Takeji Hashimoto,
and Thein Kyu

Department of Polymer Chemistry, Faculty of Eng.
Kyoto University, Kyoto 606, Japan

INTRODUCTION

The previous dynamic X-ray diffraction studies(1) of a row-nucleated HDPE revealed that the intralamellar crystal reorientation corresponding to the $\alpha1$ mechanical dispersion occurs in the following two preferential fashions. A mechanism of lamellar detwisting involving crystal rotation around the crystal b-axis is dominant in the machine direction of fabrication when applied stress is imposed perpendicularly to lamellar growth direction or lamellar axis, while the other mechanism of intralamellar shear involving crystal rotation around the crystal a-axis is prominant in the transverse direction when the applied stress is parallel to the lamellar axes. These two preferential crystal reorientation mechanisms are believed to serve as fundamental retardation processes in spherulitic polyethylene in which the former mechanism must be dominant in the equatorial zone while the latter mechanism must be accentuated in the polar zone of uniaxially deformed spherulite. The apparent crystal lattice compliance revealed a frequency dispersion at elevated temperatures which was believed to be a manifestation of the $\alpha2$ mechanical dispersion. The primary objective of this study is to examine how the two fundamental mechanisms contribute to the $\alpha1$ process and whether or not the frequency dispersion of the apparent lattice compliance is associated with the $\alpha2$ process. Furthermore, dynamic birefringence will be conducted to elucidate the nature of β process. Four kinds of bulk-crystallized polyethylenes, i.e. two HDPEs (Sholex 6009 and MPE 200), one MDPE (KP 119), and one LDPE (G 201), which possess spherulitic crystal texture are used as test samples to account for the variances in the crystal reorientation retardation behavior in terms of the relative contribution of the two fundamental deformation processes.

RESULTS AND DISCUSSION

    <u>Mechanical Dispersion</u>. Figure 1 demonstrates the unresolved
and resolved master curves of mechanical loss compliance functions
of the four kinds of polyethylenes all reduced to a reference temperature of 50 °C. The unsolved master curves are very broad for
all of the samples as typical for semicrystalline polymers. The
resolved master curves obtained from a resolution procedure proposed by Tajiri et al.(2) indicate that the broadness of the curve is
due to multiple retardation processes. It is seen that there appear
three mechanical dispersions in the loss compliance curves of HDPEs
and also in the Arrhenius plots of the shift factor corresponding
to the β, α1, and α2 processes. However, the α2 process becomes not
obvious in MDPE and its activation energy is found to be relatively
low. In LDPE, the α2 process is diminished and only a single α process is observed even at elevated temperatures. It is likely that
the α dispersion of LDPE seems to decompose into α1 and α2 components with progress of the degree of crystallinity of the samples.
The activation energy of the α2 process accordingly increases with
the progress of the degree of crystallinity, while that of the α1
process is almost identical for all of the samples. At low tempear-
tures, the β mechanical dispersion can be seen for all of the samples, appearing at almost the same reduced frequency range. It can
be noted that the activation energy of the β process has a tendency
to become somewhat smaller with the progress of the degree of crystallinity of the samples.

Figure 1.
Unresolved and
resolved master
curves of loss
compliance fun-
ctions of four
kinds of PEs
reduced to 50 °C.

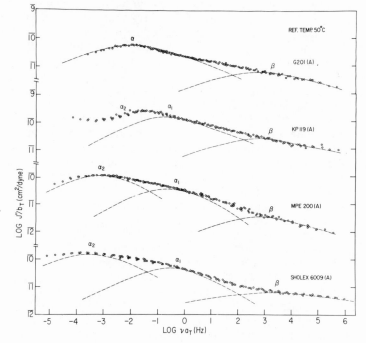

Dynamic Crystal Orientation Dispersion. Figure 2 illustrates the comparison of the dynamic crystal orientation dispersion for the four kinds of polyethylenes in terms of the dynamic strain-orientation coefficient function(3), $C_k^* = C_k' + iC_k''$, for the k-th crystal plane or crystallographic axis. It is seen that the crystal orientation dispersion behavior is significantly different between the HDPE and LDPE, although the activation energy required to bring about the α crystal orientation dispersion is the same for all of the samples and well corresponds to the α1 mechanical dispersion.

For a HDPE(Sholex 6009), the C_k' for the crystal a, b, and c-axes are almost zero at low temperatures. The C_c' at first becomes negative then crossover to positive value while the negative C_a' continuously decreases both with decreasing frequency at a rate with increasing temperature. Similarly, the C_b' at first becomes positive then reduces to zero. At low temperatures, the crystals are frozen in and would be difficult to change their orientation. When temperature increases, the crystal lamellae will align their axes towards the stretching direction concomitant with the spherulite deformation, leading to the positive b-axis orientation. Consequently, the c-axis orientation becomes negative. Further increasing of temperature will enhance the reorientation of crystal grains within the lamellae, resulting in the positive c-axis orientation. The change of C_b' with respect to reduced frequency is relatively gradual as compared with those of C_a' and C_c' and, accordingly, the retardation intensity of C_b'' is relatively small. The predominant mechanism must be rotation of the crystal grains around the crystal b-axis and is believed to be accentuated at the equatorial zone of spherulites.

For LDPE, on the other hand, the C_b' is slightly positive at high frequencies while the C_c' is positive and C_a' is negative. The C_c' increases and C_a' decreases both with decreasing frequency or increasing temperature, whereas C_b' crossover to negative values and continues to decrease. The C_k'' for all axes exhibit remarkable frequency dispersion in this range of reduced frequency. The retardation intensity of C_a'' is considerably smaller than those of C_b'' and C_c'', suggesting that the rotation of the crystal grain around the crystal a-axis must be more pronounced than that around the crystal b-axis. The crystal orientation behavior of the intermediate crystallinity samples lie in between those of the two extremes. It is therefore believed that these orientation behavior should be possible to explain in terms of the relative contribution of the two fundamental processes of crystal reorientation. Depending upon the perfectness of spherulitic texture and relative stiffness of constituent structural units, the relative contribution of the two processes may differ, in which the crystal orientation in HDPE depends more on the deformation of spherulite than those of lower crystallinity samples. It should be emphasized that the crystal orientation retardation time is longer in HDPE than in LDPE.

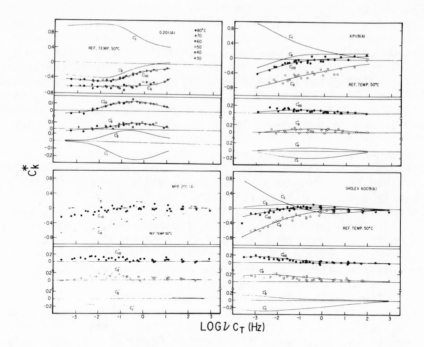

Fig. 2. Master curves of dynamic strain-orientation coefficient functions, $C_k^* = C_k' + iC_k''$, of the k-th crystallographic
axes of the four PEs reduced to 50 °C.

Dynamic Crystal Lattice Dispersion. The dynamic crystal lattice dispersion is analyzed in terms of complex dynamic apparent
lattice compliance, $J_k^* = J_k' - iJ_k''$ defined as the ratio of complex amplitude of dynamic lattice strain to that of bulk stress.
In figure 3, the J_{110}^* for the (110) crystal plane is plotted
against logarithmic frequency at various temperatures. The J_{110}'
in LDPE shows no frequency dependence with no appreciable values
of J_{110}'' fluctuating near zero. This indicates that the crystal
lattice deformation responds in-phase with the bulk stress. This
result is also consistent with the absence of the $\alpha2$ process in
mechanical data of LDPE. With the progress of the degree of crystallinity, however, the J_{110}' varies with frequency and J_{110}'' shows
distinct dispersions, suggesting that the crystals in HDPEs are no
longer elastic but viscoelastic in their responses. The difference
in the behavior of J_k^* of LDPE and HDPE may be inherent in the
difference in stress level in which the stress concentration would
be greater and the interaction between the crystals would be more
severe in HDPE than in LDPE due to the presence of large fraction
of crystalline phase in HDPE.

The frequency-temperature superposition of J_k^* for the (110) and (200) crystal planes of HDPEs are conducted, in which two crystal lattice dispersions corresponding to the α1 and α2 mechanical dispersions are observed in the Arrhenius plots. The observation of α1 process in the J_k^* is unexpected but one probable account may be that it must be driven by the α1 crystal orientation process. In other words, the frictional force which is produced between crystals must be relaxing as a result of the α1 crystal orientation process. If the relaxing frictional force transmits through the crystal contacts into interior chains, it may be possible to detect the α1 process in J_k^* as lattice response, depending on the degree of interaction between the crystals. The observation of α2 process is as expected and believed to be the inherent one arising from the onset of rotational vibration of interior chains associated with crystal disordering transition.

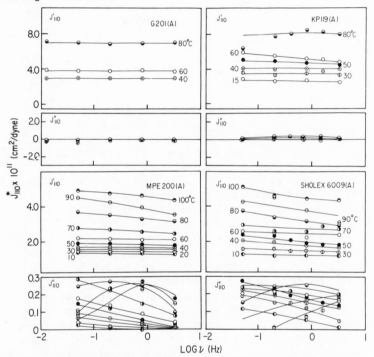

Fig. 3. Frequency dispersion of crystal lattice compliance J_k^* of four PEs at various temperatures.

Dynamic Birefringence Dispersion. The dynamic birefringence dispersion is analyzed in terms of complex dynamic strain-optical coefficient function, $K^* = K' + iK''$, which is defined as the ratio of complex amplitude of dynamic birefringence to that of bulk strain. The frequency-temperature superposition of the K^* indicates that the birefringence dispersions are composed of β and α optical processes for all of the samples. The activation energies of these processes correspond to those of β and α1 mechanical dispersions,

respectively. The activation energy and corresponding temperature
region of α birefringence dispersion also correspond to those of
αl crystal orientation dispersion, suggesting that the α optical
dispersion is mainly associated with the crystal orientation.

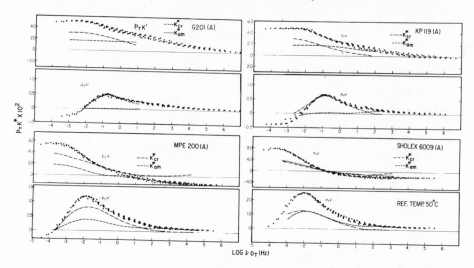

Fig. 4. Master curves of dynamic strain-optical coefficient
function, $K* = K' + iK''$, of the four PEs reduced to 50 °C.

The superposed master curves of K' and K'' are further resolved
into crystalline and amorphous contributions on the basis of two-
phase hypothesis as depicted in Figure 4. In LDPE and MDPE, it is
seen that the α optical dispersion is mainly attributable to the
crystal orientation process whereas amorphous phase responds in-
phase with the bulk strain. In HDPEs, a substantial amount of amorphous contribution to the optical dispersion can be visualized. This
may be due to the fact that the amorphous chains in HDPEs may be
appreciably restrained in conformation due to the presence of large
fraction of crystalline phase which would require to reorient or
rotate to reach a more stable conformation. On the contrary, the
amorphous chains in LDPE may be less restrained and reach at their
limiting orientation in the vicinity of α process, thus responding
in phase with the bulk strain. At low temperatures, the K'_{cr} in
Sholex HDPE is negative as a consequence of lamellar orientation
associated with spherulite deformation while those of lower crystallinity samples are positive. The K''_{cr} in MDPE and HDPEs approach
zero at high reduced frequencies. The major contribution to the β
birefringence dispersion must be arisen from the amorphous phase
and probably due to amorphous orientation process.

(1) S. Suehiro et al., Polym. Eng. & Sci., 19, 929 (1979)
(2) K. Tajiri et al., J. Macromol. Sci.-Phys., B4, 1 (1970).
(3) H. Kawai, Rheol. Acta, 14, 27 (1975).

ON THE CREEP-BEHAVIOUR OF MATERIALS UNDER COMPRESSION BETWEEN PARALLEL PLATES

J. Betten

Technische Hochschule Aachen, F.R.Germany

(Abstract)

The paper will discuss the creep behaviour of materials under compression between parallel rough plates.

The theory will be developed from both an Eulerian and a Lagrangian point of view.

The material is considered in the "secondary" creep stage. Therefore, Norton-Bailey's power law is adopted, to describe the creep behaviour.

Creep deformations of the "secondary" stage are large and of a similar character as "pure" plastic deformations. For instance, creep deformations of metals will usually be uninfluenced if a hydrostatic pressure is superimposed.

The investigation is based upon the usual assumption of incompressibility, and the case of plain creep strain will be discussed. The creep formations are considered to be of such magnitude that the use of finite-strain theory is necessary. Therefore, Hencky's strain tensor is used.

Some results of the numerical evaluation will be represented.

PROCESSING INDUCED SUPERSTRUCTURES IN MOULDED AMORPHOUS POLYMERS

K. P. Großkurth

Institut für Baustoffe, Massivbau und Brandschutz
Technische Universität Braunschweig,Beethovenstraße 52
D-33 Braunschweig, Federal Republic of Germany

INTRODUCTION

Although HOUWINK [1] suggested already in 1936 the existence
of colloidal superstructures in amorphous polymers newly developed
methods of electron microscopic preparation techniques allow in the
recent years for the first time to obtain direct evidence of well
defined morphologies in many high polymers. So for example it was
possible to show in the case of atactic polystyrene [2], styrene/
acrylonitrile copolymer [3], polyethylene terephthalate [4], polycarbonate [4] and polymethyl methacrylate [5] that the isotropic
and slightly oriented states are characterized by globular network
structures; the highly oriented materials posses a line by line
structure. Geometrical shape and regularity of the morphology only
depend on the degree of molecular orientations, but not on the way
producing themselves by varying the hotstretching parameters.
Therefore similar relationships can be assumed between morphological changes and processing induced molecular orientations in injection moulded polymers.

EXPERIMENTAL

The material used for the experiments was commercial bulk
polystyrene with a molecular weight of $\bar{M}_n$ = 111 000. In order to
mould specimens with significant different molecular orientations
in the cross section a low injection temperature of only 180 $^\circ$C
and a mould cavity wall temperature of 40 $^\circ$C have been chosen. The
dimensions of the longitudinally moulded prismatic specimen were
120 x 15 x 4 mm^3.

To get informations about the processing induced internal
structure due to the wanted distance to the sample surface thin
layers have been cut by a microtome knife. Now the morphology of the
remained uncovered internal surface could be identified by means
of electron microscopy and one-step-replica method after oxygen ion
etching. The selective etching process is based on different degradation rates of regions with locally differing densities and/or
differing packing types of the chain molecules.

Fig. 1 shows the scheme of the used etching apparatus type
GEA 004- S. The bulk specimen is treated on the watercooled sample
plate. Oxygen ions are produced in the high-frequency field (27.12
MHz) of the annular electrode at a partial pressure of about 10^{-4}...
10^{-3} mbar. Superimposed low dc voltages (max. 500 V) between the
electrode at the top of the recipient and the sample plate accelerate the oxygen ions onto the sample surface. In addition to the
selective chemical etching effect of the activated oxygen a mechanical microerosion process based on the kinetic energy of the accelerated ions is degradating the polymer surface. By means of mass

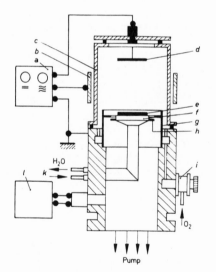

a	electronic control unit
b	Hf annular electrode
c	glasrecipient
d	dc electrode
e	sample
f	watercooled sample plate and opposite electrode
g	cylinder
h	deflector
i	proportionating valve
k	cooling water
l	vacuum gauge

Fig. 1 : Scheme of the etching apparatus GEA 004- S

spectroscopy it was possible to point out that the degradation products mostly are gaseous and removed over the diffusion or turbo-
molecular pump [6]. The etching time of normally some hours strongly
depends on the plastic deformation during microtome pretreatment in
the surface layers. Otherwise artefacts would be expected.

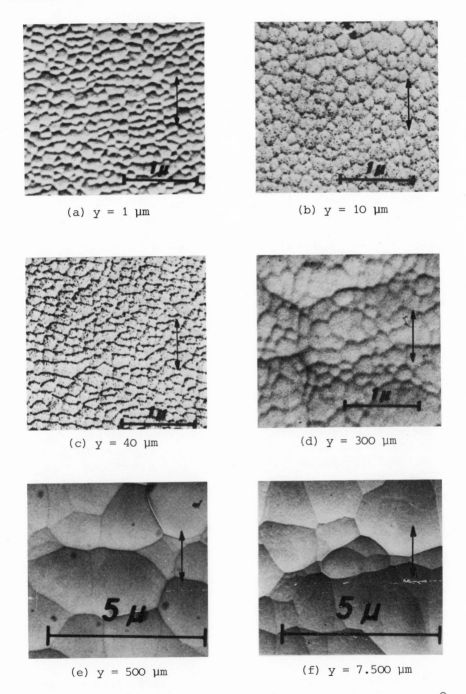

(a) y = 1 μm

(b) y = 10 μm

(c) y = 40 μm

(d) y = 300 μm

(e) y = 500 μm

(f) y = 7.500 μm

Fig. 2: Morphology of moulded PS. Injection temperature 180 °C.
Depth y. Arrow marks the direction of injection.

RESULTS

        Fig. 2 represents a characteristic series of processing indu-
ced supermolecular structures in moulded polystyrene at the sprue
distant end of the specimen. Qualitatively, however, the demonstrated morphological spectrum is also typical für the other sample
sections. The depth values include the microtome abrasion and
etching degradation.

        For the investigation of the direct surface structure a mi-
crotome cutting was not necessary. Thus, the original sample surface was etched for only 30 minutes. In a depth of nearly 1 µm a
superlattice structure ordered perpendicular to the injection direction can be observed (Fig. 2a). With increasing distance to the
surface the morphology at first becomes globular (Fig. 2b). An
alignment of the globules normal to the injection direction again is
clearly visible. The morphological regularity reaches its maximum
in a depth of 20 ... 40 µm, where a line by line structure is realized (Fig. 2c). Then the structure becomes more and more irregularly. At a distance of 300 µm to the surface the change from the
fine globular morphology into a coarse globular one is indicated
(Fig. 2d); in a depth of 500 µm the change has been finished already (Fig. 2e) and now the structure remains nearly constant
(Fig. 2f).

DISCUSSION

        It is well known, that the injection moulding process effects
molecular orientation distributions in the cross section. Therefore the observation of rather similar superstructures in polystyrene subjected to uniaxial hotstretching is a significant factor
for interpreting these phenomena.

        The morphology of the isotropic and oriented state also was
detected by means of electron microscopy and the same preparation
techniques [2]. Fig. 3a shows the typical structure of the isotropic
material. It consists of an irregularly formed network similar to
grain boundaries with a mesh size of some µm. After stretching of
only 25 % a globular morphology can be observed (Fig. 3b). An orientation of the globules in the transverse direction is indicated.
With further drawing the globules become smaller and change after
stretching of 50 % to a line by line structure (Fig. 3c). The lines
are ordered perpendicular to the direction of molecular orientations.
Their regularity increases with the stretching ratio (Fig. 3d).

        The observed superstructures are based on locally disconti-
nuities of deformation processes, which produce density differences
[3],[4].   Privious conclusions obtained from structural studies
on amorphous polystyrene [2] indicate that the line boundaries pos-

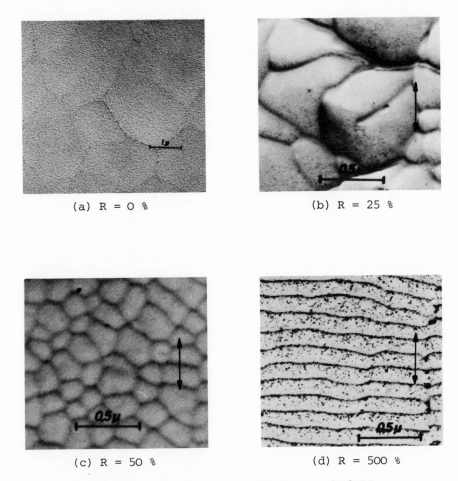

(a) R = 0 %   (b) R = 25 %

(c) R = 50 %   (d) R = 500 %

Fig. 3: Morphology of isotropic and hotstretched PS.
          Stretching parameters: time 15 s, temperature 115 °C,
          degree R. Arrow marks the direction of stretching.

sess a higher density than the intermediate material. So a nearly
parallel alignment of the macromolecules within the boundaries was
suggested.

     By quantitativ analysis of the electron micrographs mean
spacing of structure details - boundaries of globules or lines -
and corresponding standard deviation were determined. Both parameters characterize geometrical shape and regularity of the supermolecular structures. In Fig. 4 the standard deviation as a size for

the reciprocal regularity is plotted against the chosen stretching
conditions. The three-dimensional diagram illustrates that the
formation of structure becomes more regularly with increasing degree of orientation and decreasing temperature of stretching. The
most significant changes result from drawing up to about 200 % and
stretching temperatures just beyond the glass transition.

The relationships between stretching parameters and mean spacing of structure details are rather similar.   A minimum
line mean spacing of about 100 nm has been found. It is nearly independent of further drawing.

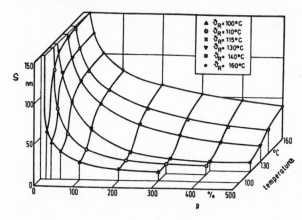

Fig. 4 :

Standard deviation s
of structure details
mean spacing as func-
tion of stretching de-
gree R and temperature.
Stretching time 15 s.

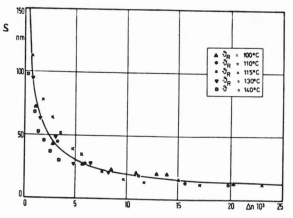

Fig. 5 :

Like Fig. 4, plotted
against birefringence
Δn.

Using the birefringence as a measure for molecular orientation
instead of the stretching parameters the standard deviation reduces
to a very simple function as shown in Fig. 5. Thus, it must be pointed out, that geometrical shape and regularity of the morphology
only depend on the degree of molecular orientations, but not on the
way producing themselves.

This is why the morphological results found at the moulded
material readily can be understood. Frozen-in molecular orientations
produced by melt shearing are the reason for changes in morphology.
Owing to the strongly oriented chain molecules the area near the
surface possesses a well-defined line by line structure; contrary
to this region internal a globular structure due to a low degree of
molecular orientation is evident.

A quantitative analysis comparison of the processing induced
and hotstretching effected morphology now is suitable to detect the
local degree of molecular orientation in microregions of mouldings.
The results plotted in Fig. 6 show extremely differing orientations
in the cross section of moulded polystyrene. These conclusons are
of high technical significance, because the local mechanical and
fracture behaviour mainly depends on the orientation distribution
function in the moulding section.

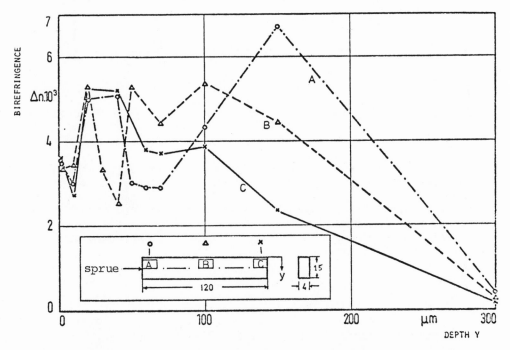

Fig. 6 : Birefringence calculated from the morphological data
as a function of depth. Moulded PS. Injection temperature 180 $^{\circ}$C.

The connections between morphology, mechanical properties and
molecular orientations, as measured by birefringence, quantitively
can be studied on hotstretched samples. As shown in Fig. 7 tensile
stress (σ_B) and strain at maximum load (ϵ_B) reach their greatest
values at the beginning of the line by line structure area and then

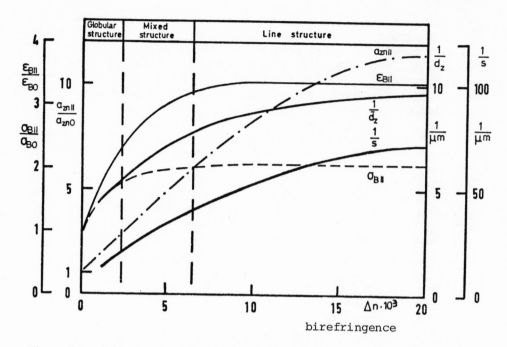

Fig. 7 : Effect of molecular orientations on morphology and
mechanical properties of hotstretched PS in the
longitudinal direction

remain approximately constant. Better connections exist between tensile impact strength (a_{zn}) and the structural parameters mean line
concentration d_z^{-1} and regularity s^{-1}. Obviously morphological defects are representing internal notches and therefore effect a decreasing of the expected tensile impact strength in the area of the
globular and mixed morphology as well as at the beginning of the
line by line structure.

REFERENCES

[1] Houwink, R., Trans. Farad. Soc. 32:122 (1974)
[2] Großkurth, K.P., Colloid & Polymer Sci. 255:120 (1977)
[3] Großkurth, K.P., Gummi Asbest Kunststoffe 25:1159 (1972)
[4] Kämpf, G. and H. Orth, J. Macromol. Sci. - Phys. B 11 (2):151
        (1975)
[5] Großkurth, K.P., Progr. Colloid & Polymer Sci. 66:281 (1979)
[6] Großkurth, K.P., Gummi Asbest Kunststoffe 30:848 (1977)

ACKNOWLEDGEMENT

I would like to thank Professor Dr. S. Wintergerst, Institut für
Kunststoffprüfung und Kunststoffkunde, Universität Stuttgart, for
the experimental possibilities and helpful discussions as well as
the Deutsche Forschungsgemeinschaft for their financial support.

EFFECT OF VARIABLE MOLECULAR ORIENTATIONS ON STRESS-CRAZING IN MOULDED AMORPHOUS POLYMERS

K.P. Grosskurth

Institut für Baustoffe, Massivbau und Brandschutz
Technische Universität Braunschweig, Beethovenstrasse 52
D-3300 Braunschweig, West Germany

(Abstract)

Initiation and propagation of crazes in injection moulded transparent polystyrene subjected to uniaxial tensile load at room temperature were investigated by means of conventional and high frequency cinematography. Different strain rates between 10^{-7} s^{-1} and 10^{1} s^{-1} were used.

In the course of processing internal stress and molecular orientation distributions exists in the specimen cross section. These effect that moulded tensile bars always craze internally. Low elongation speeds show a small craze concentration. Near the highly oriented surface the material remains still uncrazed. Higher strain rates cause a great number of crazes growing partially into the oriented region near the surface. A relation between strain rate, macroscopic sample deformation and craze concentration is clearly visible. The formation of shear bands and macroscopic necking only can be observed at very low strain rates verified in creeping tests.

Adiabatic heating above glass transition temperature occurs during craze propagation as evidenced by means of electron microscopy and one-step-replica method after oxygen ion etching. This result is given by comparing the quantitative micrograph analysis of craze morphology and superstructures in hotstretched polystyrene. The electron micrographs also demonstrate that the craze consists of stretched material. Significant differences

in the degradation rate of ion etching between crazes and intermediate areas directly are seen in the scanning electron microscope.
So it is indicated that craze propagation produces lower density
and a lot of microholes. Those morphological defects are representing internal notches. Therefore the stress distribution becomes
inhomogeneously and initiates cracking processes.

VISCOELASTIC BEHAVIOR OF COMPOSITE SYSTEMS COMPOSED OF POLYBUTADIENE

PARTICLES AND POLYSTYRENE AT ELEVATED TEMPERATURES

T. Masuda, M. Kitamura, S. Onogi

Department of Polymer Chemistry, Kyoto University
Japan

(Abstract)

Dynamic viscoelastic and extensional properties have been measured for composite systems composed of crosslinked polybutadiene particles and polystyrene at elevated temperatures. The experimental results have been discussed in terms of particle content, molecular weight of medium polystyrene and so on. Effect of temperature on viscoelastic functions can be expressed by the WLF equation by use of different reference temperatures according to particle content. The rubbery plateau modulus of the system can well be espressed by the Okano equation, in which the relative stiffness of particle to that of medium is taken into account. The characteristic time constant for slipage of entanglement couplings in medium polymer is proportional to the 3.5 power of molecular weight, and slightly increases with increasing particle content. The frequency dependence curve of dynamic modulus exhibits the second plateau at lower frequencies in spite of the isolation of particles in the material, suggesting a relaxation mechanism associated with motion of each particle. The second plateau modulus have been expressed by particle content, molecular weight of medium polystyrene and the mean distance between two particles in the system. From these results, a relaxation mechanism related to the second plateau in particle-filled system has been discussed.

The relation between viscoelastic properties and extensional behavior of the composite systems has also been discussed.

INTERACTIVE ENHANCEMENT OF PVC AND ABS TOUGHNESS IN THEIR BLENDS,

A FRACTURE MECHANICS INVESTIGATION

T. Riccò, M. Rink and A. Pavan

Istituto di Chimica Industriale del Politecnico di Milano
Sezione di Chimica Macromolecolare e Materiali
Piazza L. da Vinci, 32 - 20133 Milano Italy

INTRODUCTION

Blending of PVC with ABS shows the singular effect of an enhancement in strength and toughness over the values of the individual components, when conventional testing methods are used to measure these ultimate properties.

Previous impact experiments carried out both on unnotched and standard V-notched specimens and covering an entire range of compositions of the ternary PVC/BR/SAN system, have shown that (i) the effect is present in notched specimens even at constant butadiene rubber (BR) content,[1] but (ii) is absent when unnotched testpieces are examined.[2] These results prompted us to reexamine the fracture behaviour of these blends following a fracture mechanics approach, so as to encompass specimen geometry effects.

EXPERIMENTAL

A series of PVC/ABS blends prepared under constant compounding conditions were examined. The parent materials used were described previously.[1] The ABS sample used in the present work contained 32 wt-% of butadiene rubber. PVC and ABS powders were dry-blended in a lab-size ball mill at room temperature for 23 hours and subsequently melt-mixed on a two-roll mill at 160°C for 5 minutes. The same thermo-mechanical treatment was also given to the two parent materials. Each sample was compression moulded at 180°C into plates from which 63.5 x 12.7 x 6.35 mm rectangular bars were cut out. The bars were single-edge notched to various depths on their side of thickness B = 6.35 mm by means of a very sharp cutter. Impact testing was conducted Izod-wise by means of a pendulum instrumented to

429

record the whole force–deflection curve generated during the pendulum stroke.[3] All measurements were carried out at room temperature.

The force–deflection curves obtained invariably show a rise
of the force up to a maximum, after which the force drops following
different paths depending on sample and notch depth. As the peak
of the curve appears to be a threshold beyond which the material
fails, special significance can be attached to the values of the
force F_m, and the energy, W_m at the peak.[3,4] In Fig. 1(a) and (b)
these two quantities are shown as a function of blend composition
for each notch depth examined. These results reproduce the general
features of the data obtained on standard V–notched specimens[1] with
a marked enhancement both in F_m and W_m at some intermediate composition. Further, as the notch depth, a, is increased, the force F_m
regularly decreases while the peak energy, W_m, shows a more complex
pattern.

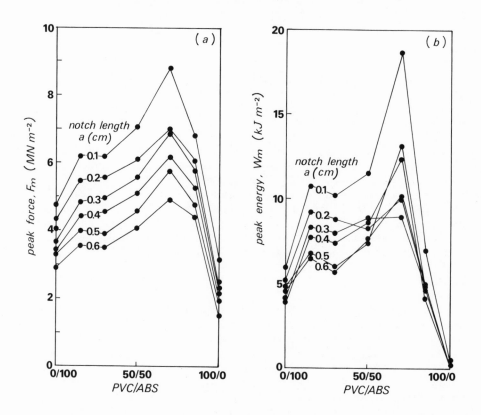

Fig. 1. Force, F_m, (a) and energy, W_m, (b) at the peak of the force-
        -deflection curve as a function of PVC/ABS ratio for differ-
        ent notch lengths, a. Both F_m and W_m are conventionally
        referred to unit initial ligament area .

FRACTURE MECHANICS ANALYSIS

In view of the ductility of most of the blends studied, in the present investigation Rice's contour integral, J, was used as an appropriate method to analyze force-deflection data up to incipient fracture. To evaluate J-integral Begley and Landes' procedure [5] was adopted. Accordingly, J is obtained as

$$J = -\frac{\partial U'}{\partial a}$$

where $U' = U/B$ is the mechanical energy input normalized per unit thickness of the test specimen and a is the crack length. U' can be determined as a function of crack length, a, and displacement of the loading point, u, from a family of force-deflection records experimentally obtained from test specimens of various initial crack (notch) length. U' can then be plotted as a function of a for different values of deflection u. An example of such a plot is shown in Fig. 2(a) for the blend PVC/ABS = 15/85 . From plots such as this, J can be measured as the slope of each curve at any crack length: we have calculated J at the experimental values of notch length, a, by averaging the slopes of the two straight lines converging on each point. A plot of J versus deflection, u, can then be constructed for each notch length, a. An example is given in Fig. 2(b). Since the present testing configuration produces a tensile opening mode of fracture (mode I), values of J_I are so ob-

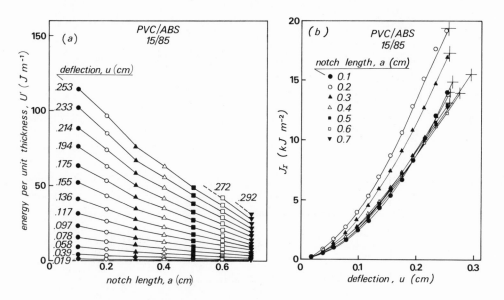

Fig. 2. Example of dependence of energy per unit thickness, U' (a) and J-integral (b) on deflection and notch length .

tained. To determine the critical value of J_I for fracture initiation, J_{IC}, a critical value of deflection must be determined. It
is assumed that fracture initiation occurs at the point of maximum
load in the force-deflection curve. Therefore, J_{IC} is obtained
from curves such as those in Fig. 2(b) by taking the value of J_I
corresponding to the critical deflection. In this way a value of
J_{IC} is determined for each notch length a. The values so obtained
for each material appear to vary randomly with notch length a. The
average values over all the notch lengths tested for each material
were thus calculated. In Fig. 3 the distribution of the percent
deviation of J_{IC} from its mean (points), together with its standard
deviation (shaded range), is shown for each PVC/ABS ratio.

From these results we conclude that Begley and Landes' procedure here adopted to evaluate J-integral yields critical values
J_{IC} reasonably constant over the range of notch depths explored
for each material. This result pays a tribute to the validity of

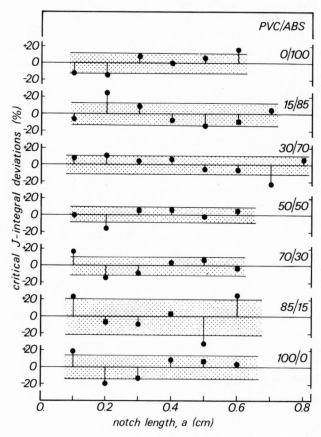

Fig. 3. Analysis of J_{IC} variation with notch length (see text).

J_{IC} as a fracture criterion for these materials, at least under our testing conditions.

CONCLUSION

The dependence of J_{IC} on blend composition is represented in Fig. 4. It clearly shows that blending of PVC with ABS has the effect of enhancing J_{IC} over the values of the individual components. Thus, the effect previously observed on strength and toughness determined by conventional testing methods, is still present in this fracture mechanics parameter. In view of the significance attached to J_{IC} as a "material property" (within limits of specimen and notch dimensions), the present result definitely indicates that the effect under study stems from the intrinsic deformation mechanism of the material. The hypothesis of a "synergistic" interaction between distinct but simultaneously operating deformation mechanisms appears now supported.

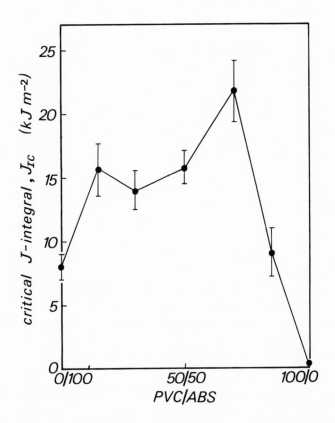

Fig. 4. J_{IC} versus PVC/ABS ratio. Bars denote standard deviations.

REFERENCES

1. A. Pavan, T. Riccò and M. Rink, High performance polymer blends.
   I. Impact behaviour of PVC/ABS blends as a function of PVC/
   /BR/SAN composition, Mater. Eng. Sci. (in press).
2. A. Pavan, M. Rink and T. Riccò, High performance polymer blends.
   II., (in preparation).
3. W. Lubert, M. Rink and A. Pavan, Force displacement evaluation
   of macromolecular materials in flexural impact tests. I. Apparatus and data handling., J. Appl. Polymer Sci., 20:1107
   (1976).
4. M. Rink, T. Riccò, W. Lubert and A. Pavan, Force displacement
   evaluation of macromolecular materials in flexural impact
   tests. II. Influence of rubber content, degree of grafting
   and temperature on the impact behaviour of ABS resins, J.
   Appl. Polymer Sci., 22:429 (1978).
5. J. A. Begley and J. D. Landes, The J integral as a fracture
   criterion, in: "Fracture Toughness, Proceedings of the 1971
   National Symposium on Fracture Mechanics", Part II, ASTM STP
   514, American Society for Testing and Materials, Philadelphia
   (1972).

ENVIRONMENTAL CRAZING OF POLYMETHYLMETHACRYLATE AND POLYCARBONATE

A.T.Di Benedetto[+] P.Bellusci, M.Iannone, L.Nicolais

University of Naples, Italy
[+] University of Connecticut, U.S.A.

INTRODUCTION

Polymethylmethacrylate and polycarbonate are finding many uses as industrial materials, such as in automobiles and other consumer goods. Of particular interest to the manufacturer and consumer alike, is their resistance to mechanical degradation under stress in certain unfavorable environments, such as organic solvents. Under certain conditions these materials are subject to the phenomenon of crazing and other forms of microcavitation. The formation of crazes is the precursor to cracking and subsequent failure of the material under stress. In the present work, the effects of stress and temperature on the kinetic of crazing of materials in an environment of n-butyl alcohol are studied.

EXPERIMENTAL

The materials chosen for study were polymethylmethacrylate (PMMA, Vedril Montedison) and polycarbonate (PC, Lexan General Electric). Compact tensile specimens, of geometry specified by ASTM Standard E 399-74 were machined from 1 meter by 3 meter x 12mm extruded sheets of these materials.

A mechanical operated wedge, was used to propagate a natural crack from the machined notch. The length of the natural crack could be precisely controlled by putting a light compression at the end of the desired growth zone, perpendicular to the direction of growth. Surfaces appeared silvery and flat and the leading edge

435

of the crack was nearly rectilinear. The same technique was applied successfully to polycarbonate at T = 80°K in liquid nitrogen.

All craze propagation measurements were carried out at constant temperature and load while the specimen was immersed in n-butyl alcohol. Since the material was transparent, it was possible to use a transparent vessel to contain the alcohol so that the propagation could be observed optically.

The glass tank was approximately two litres in volume. The top was open and the bottom possessed a gasketed hole to accept the two grips and connecting rods on the Instron 1112 Testing Machine. The vessel was fixed to the lower grip and rod through the gasketed hole. The sample was fixed between the two grips by loading pins and the upper grip and rod were connecting directly to the load cell of the Instron machine.

Upon one surface of the specimen a grid of uniform lines was scratched   so as to measure the rate of craze propagation directly. The specimen was back-lighted to accent the craze and the craze growth was observed using a cathetometer.

The rate of craze growth was correlated to the stress intensity factor, which can be defined by the following equation :

$$K = (P/Bw^{\frac{1}{2}}) \left[29.6(a/w)^{\frac{1}{2}} - 185.5(a/w)^{3/2} + 655.7(a/w)^{5/2} - 1017.0 \right.$$
$$\left. (a/w)^{7/2} + 638.9(a/w)^{9/2} \right] \qquad (1)$$

where K is the stress intensity factor, P is the load, B is the thickness of the specimen, w is the width of the specimen and a is the initial length of the natural crack from which the craze propagates. The above relation is valid when the ratio a/w is between 0.45 and 0.55, a ratio which was maintained for all specimens tested[1].

RESULTS AND DISCUSSION

A. Polymethylmethacrylate

Typical data of craze length versus time at a constant value of stress intensity factor K_0 are shown in Figure 1, plotted in logarithmic coordinates. A straight line with a slope of 0.5 indicates a dependence of craze length on the square root of time, in accordance with the end-flow model of Marshall.

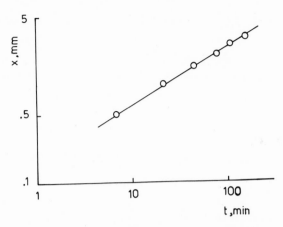

Fig. 1.   Typical data of craze length, x, versus time, t, at a
          constant value of stress intensity factor for PMMA.

In fact using a simplified form of the Navier-Stokes equation
for convective flow in channels, Marshall et al.[2] developed theor
etical equations for craze length as a function of time for the
case of a rate limiting flow through the end of the craze, along
its length, and also for the case of a rate limiting flow through
the sides of the specimens, normal to the direction of craze growth.

For end-flow conditions, tha craze length, x, is proportional
to the square root of time :

$$x = (1_o \frac{\bar{P}}{10 \; \sigma_{yE} \eta})^{\frac{1}{2}} (K_o - K_m) t^{\frac{1}{2}} \tag{2}$$

where 1_o is the average distance between sites of dishomogeneity
in the plastic zone, K_o is the stress intensity factor calculated
using the initial length of the crack, K_m is a lower limiting value
of K_o, σ_y is the yield strength of the material at the crack tip,
E is the modulus of elasticity, $\bar{P}$ is the pressure of the environ
ment at the edge of the notched specimen and η is the viscosity
of the solvent at the temperature of the test.

For side-flow conditions, the craze length, x, is proportional
to time :

$$x = \frac{1}{2} \Delta H (1_o \frac{\bar{P}}{10 \; \sigma_{yE} \eta}) \left[(K_o - K_m)^2 - (K_n - K_m)^2 \right] t \tag{3}$$

where Δ is a constant, H is the specimen thickness, K_n is an upper
limiting value of K_o and other quantities are described above.

It was found that below a certain value of the stress intensity factor, $K_o = K_m$, there was no visible craze propagation. For values of stress intensity factor between K_m and a higher limiting value of K_n, the rate of craze propagation was described by the end-flow model. For values of K_o greater than K_n, up to the critical stress intensity factor, the craze length increased with an exponent of time between $\frac{1}{2}$ and 1, indicating that a combination of end-flow and side-flow controlled the rate of growth.

In a more recent study, Kramer and Bubeck[3] hypothesize a model similar to that of Marshall et al., but using the driving force of capillary action to characterize the flow of solvent through the craze. The Kramer-Bubeck model reduces to the same form as the Marshall model in the limit of end-flow, but can take on exponential values both greater and less than $\frac{1}{2}$ in other cases.

Equation 2 may be rewritten in the form :

$$x/t^{\frac{1}{2}} = m(T)K_o - m(T)K_m \qquad (4)$$

where

$$m(T) = (\frac{l_o \, \overline{P}}{10 \, \sigma y \, E \, \eta})^{\frac{1}{2}} \qquad (5)$$

is a measure of the rate constant and K_m is the lower limiting value of the stress intensity factor, below which craze propagation does not occur. The quantity $m(T)$ should be temperature dependent, primarily because of the temperature dependence of the fluid viscosity.

Figure 2 is a plot of the data for various values of K_o and temperature. Each mean value of $x/t^{\frac{1}{2}}$ was obtained from linear regression analysis of the raw data. At a value of $K_o = 50$ N/mm$^{3/2}$ the material fractures spontaneously, signifying that the material is at the critical stress intensity factor K_{oc}. Both the critical stress intensity factor K_{oc} and the minimum value K_m appear to be relatively independent of temperature.

In order to estimate values for $m(T)$ and K_m as a function of temperature, the values of $x/t^{\frac{1}{2}}$ below K_o at the maxima (i.e. 35 N/mm$^{3/2}$) were fit to a straight line using a linear regression analysis. The results are tabulated in Table 1 for the three temperatures. The value of $m(T)$ increases monotonically with

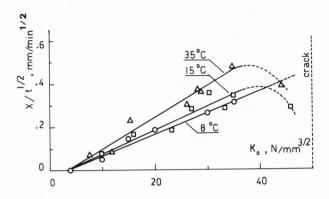

Fig. 2. $(x/t^{\frac{1}{2}})$ data versus K_o at different temperatures as
indicated for PMMA.

Table 1.   Minimum stress intensity factor, K_m, and $m(T)$
values as a function of temperature for PMMA
and PC; r^2 represents the correlation coefficient.

Polymer	$T(°C)$	$m(T) \times 10^2$	$K_m(N/mm^{3/2})$	r^2
PMMA	8	0.991	3.02	0.98
	15	1.085	2.89	0.90
	35	1.450	3.23	0.96
PC	8	0.393	23.69	0.81
	15	0.463	19.23	0.97
	25	0.699	16.06	0.99

temperature, while K_m is probably independent of temperature. The
correlation coefficient r^2 indicates a reasonably good correlation
to a straight line.

From equation 5, one can see that the slope $m(T)$ is related
to the properties of the polymer and solvent. The ratio $(1_o \bar{P}/ \sigma_{yE})$

should be relatively independent of temperature over the range studied, so that the ratio of m(T) at two temperatures ought to be approximately equal to the inverse ratio of the square roots of the viscosities of the n-butyl alcohol. This, in fact, is precisely the case:

$$\frac{m(8)}{m(15)} = 0.91 \quad \text{and} \quad (\frac{\eta_{15}}{\eta_8})^{\frac{1}{2}} = 0.92$$

and

$$\frac{m(8)}{m(35)} = 0.68 \quad \text{and} \quad (\frac{\eta_{35}}{\eta_8})^{\frac{1}{2}} = 0.73$$

Thus it appears that, at least for stress intensity factors less than about 35 $N/mm^{3/2}$ (about 2/3 of the critical stress intensity), convective flow of solvent through the end of the craze controls the rate of craze propagation in polymethylmethacrylate.

B. Polycarbonate

The kinetic of craze propagation of polycarbonate in n-butyl alcohol was studied at 8°, 15° and 25°C.

As with the polymethylmethacrylate data, the craze length varied with the square root of time for every temperature and value of K_o studied. Values of $x/t^{\frac{1}{2}}$ vs K_o are shown in Figure 3 and the general trends are similar to those previously obtained.

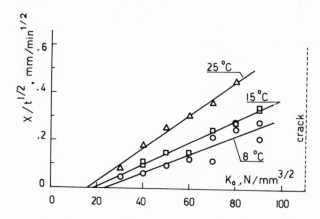

Fig. 3.   ($x/t^{\frac{1}{2}}$) data versus K_o at different temperatures as indicated for PC.

A linear regression analysis was used to correlate the data to a straight line. The values of m(T) and K_m from a best fit of the data are shown in Table 1. In this case it appears that both the slope, m(T), and the minimum value of the stress intensity factor K_m vary monotonically with temperature. The inverse ratios of slopes at different temperatures may be compared to the ratios of the square roots of viscosities :

$$\frac{m\ (8)}{m(25)} = 0.56 \quad \text{and} \quad (\frac{\eta_{25}}{\eta_8})^{\frac{1}{2}} = 0.80$$

$$\frac{m(15)}{m(25)} = 0.66 \quad \text{and} \quad (\frac{\eta_{25}}{\eta_{15}})^{\frac{1}{2}} = 0.82$$

One does not see a very close agreement, indicating that the temperature dependence of the other factors in m(T) (i.e. $l_o\bar{P}/\sigma_{yE}$) may be important also. There is not enough information to separate the various factors, but it is likely that the local yielding properties and the density of microvoid formation are more sensitive to temperature than was in the case of the PMMA. The minimum stress intensity factor, K_m, decreases by 32 percent over the temperature range 8°C to 25°C, indicating that craze initiation is easier (i.e. requires less stress).

ACKNOWLEDGEMENT

The authors wish to express their gratitude to the National Science Foundation (U.S.A.) for providing support to one of them under the Science Faculty Professional Development Program.

REFERENCES

1. ASTM Standard E 399-74
2. G. P. Marshall, L. E. Culver, J.G.Williams, Proc.Roy.Soc. London, A,319, 165 (1970)
3. E. J. Kramer, R. A. Bubeck, Journal of Polymer Science, Polymer Physic Edition, 16, 1195, (1978).

RECOVERY OF UNIAXIALLY ORIENTED POLYPROPYLENE

S. Piccarolo

University of Naples, Italy

INTRODUCTION.

Thermal treatments strongly influence physical properties of isotropic polymers. This influence is very dramatic in crystalline polymers were the process deeply modifies the morphology. In amorphous polymers, annealing has the effect of changing the free volume modifying particularly long term physical properties; phenomenon which has been given the name of physical aging. Moreover if the polymer is oriented, the recovery will follow a different mechanism depending on whether the polymer is crystalline or not and whether the annealing is made on samples with free or fixed ends. For an amorphous oriented polymer a small degree of recovery takes place when the samples are exposed at temperatures below the equilibrium glass transition temperature, Tg, and the process reaches asymptotic values which are function of the annealing temperature[1]. At temperatures above Tg the recovery phenomenon is much faster and the equilibrium recovery is not a function of temperature. As a matter of fact the process can be interpreted as a relaxation phenomenon which takes place in the sample and the behavior can be predicted by using W.L.F. equation[2].

In oriented crystalline polymers recrystallization takes place during annealing and the mechanism of recovery changes hardly depending on whether the annealing is made on specimens with free or fixed ends[3]. This has very much to do with the mechanism of plastic deformation as outlined by Peterlin[4]. This author has also

443

pointed out the key role that on the recovery properties have the
intra- and interfibrillar taut tie molecules (TTM) which bridge
crystalline blocks to each other. As a matter of fact during cold
drawing the density of TTM's increases and moreover their physical
state is the amorphous one[3].

In this paper an attempt has been made to study the recovery
kinetics of oriented crystalline polypropylene cold drawn at different temperatures. Symilar studies on polypropylene have already
been accomplished[5,6]. In both cases the asymptotic recoveries are
considered and are related to the previous mechanism of plastic
deformation[5] or to the orientation parameters in the two phases[6].

Here particular attention has been devoted to the kinetics of
the recovery process as influenced by the previous conditions of
cold drawing. The approach is phenomenological trying to schematize
the recovery of oriented crystalline polymers as a process with two
distinct contributions, one due to the recrystallization and the
other to the relaxation phenomena.

EXPERIMENTAL

The polymer used is a commercial polypropylene (Montedison
Moplen MOF41). The samples for cold drawing were prepared by
extrusion with a Negri & Bossi single screw extruder. The extruder
was used in order to avoid surface irregularities arising from
molding and machining the samples. A temperature of 200 °C was
maintained along all the extruder in order to obtain a high density
of heterogeneous nuclei. The molten strip of polymer was hot drawn
and rapidly quenched to room temperature in a water bath. Quenching
was adopted in order to obtain a high density of tie molecules[7].
In fact occurrence of the phenomenon of neck formation and plastic
flow is strongly dependent on density of tie molecules[4].

The recovery of the as extruded samples annealed with free ends
at temperatures close to the melting point was insignificantly low.
The randomizing processes in a polymer melt of a crystalline polymer
being very rapid with respect to the crystallization rate due to the
low viscosity of the melt[8], therefore inducing an almost nonexistent
orientation.

Table 1. Natural Draw Ratios (NDR) at the drawing temperatures
(T_S) used in the experiments

T_S	25	60	90	120	150
NDR	5.5	6.5	7	7.2	7.5

The extruded samples were cold drawn with an Instron mechanical
testing machine model TT/DM at an initial deformation rate of 0.16
min^{-1} and at temperatures ranging from 25 to 150 °C. When the cold
drawn sample necks, the draw ratio (defined as the ratio between
the distances of two material points after and before the deformation) is only dependent on temperature and deformation rate[9],
(i.e. Natural Draw Ratio, NDR). The values of the NDR's at the
different drawing temperatures used in our experiments are listed
in table 1.

The annealing experiments were performed in a thermostated oil
bath at temperatures ranging from 120 to 165 °C on samples obtained
from the cold drawn strips. The data were recorded as ratios between
the initial length, L_o, and the length at time t, L_t. The length
L_t was measured after various immersion times and after quenching
the samples in ice water.

Figure 1 shows a typical plot of the recovery behavior for the
samples drawn at 25 °C. It can be noticed that a large amount of the
total recovery takes place in very short times not experimentally
observable according to the results already obtained in reference 5.

RESULTS AND DISCUSSION

Annealing with free ends an oriented crystalline polymer tends
to mobilize the amorphous intercrystalline links with a force that
is proportional to the temperature of annealing. The magnitude of
this force is relevant due to their high extension and it is of
entropic nature. The randomization of the TTM's is opposed by the
cohesive forces between the adjacent chains in the crystalline
blocks. These resistant forces against the axial chain displacement
decrease rapidly with increasing temperature. Furthermore, depending
on the thermal level, the TTM's tend to crystallize under the stress
applied by the crystal bloks in such a way as to limit the probability of further recoil of the links themselves[3].

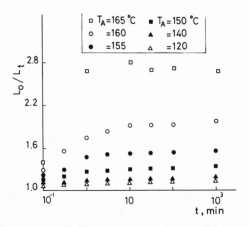

Fig. 1.   Recovery behavior versus time for samples cold drawn at
          25 °C for various annealing temperatures.

Therefore during the recovery process the crystallinity increases.
Consequently the L_o/L_t versus time plots (Fig.1) obtained at different annealing temperatures could be better compared if referred
to the common asymptotic deformation L_o/L_{inf} corresponding to the
highest annealing temperature used. Here L_o/L_{inf} stands for the
asymptotic deformation that can be achieved at a given temperature
if the time of observation is sufficiently large. These values, for
every annealing and drawing temperatures, have been evaluated
assuming that the data can be correlated by an exponential function
and then extrapolating the limiting value with a numerical
technique[10].

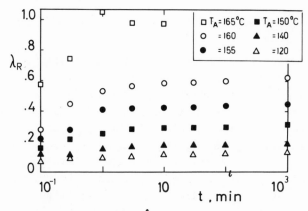

Fig. 2.   Reduced deformation, λ_R, versus time for samples cold
          drawn at 25 °C for various annealing temperatures.

In figure 2 the reduced deformation, λ_R, as a function of time is reported. It is defined in the following way:

$$\lambda_R = \frac{L_o - L_t}{L_o - L_{infR}} = \frac{1 - 1/(L_o/L_t)}{1 - 1/(L_o/L_{infR})}$$

were L_o/L_{infR} is the highest asymptotic deformation which is identical to the asymptotic deformation at the highest temperature of annealing. The whole set of curves is referred to the same drawing temperature and in particular figure 2 refers to 25 °C. This reduced deformation takes into account the amount of recovery taking place at each annealing temperature with respect to the maximum allowable deformation.

The data reported in figure 2 can be superimposed in a master curve if also a vertical shift is adopted. The vertical shift can be explained according to the model already mentioned as due to a different level of crystallinity reached by the TTM's inducing a different force of cohesion in the crystalline blocks.
The horizontal shift is made according to the time-temperature superposition principle. The master curves referred to different drawing temperatures are shown in figure 3.

It can be noticed that the higher is the drawing temperature the longer are the times covered with the superimposition technique. In fact under these conditions the mobility of the links in the subsequent annealing process is decreased due to the simultaneous crystallization of the links while drawing[3].

Finally figure 4 shows plots of the horizontal shift factor, a_T, versus reciprocal annealing temperature. Here a_T is defined as the ratio between the time scale relative to the data which are being shifted and the time scale relative to the reference temperature. Therefore the master curve time scale is $t'=t/a_T$.

The plots are nonlinear at all the drawing temperatures tested, the overall recovery process should therefore be a double activated one. At higher temperatures crystallization phenomena should prevail justifying the high values obtained for the activation energy. At lower temperatures, on the other hand, the lower activation energy obtained can suggest that relaxation phenomena are dominant.

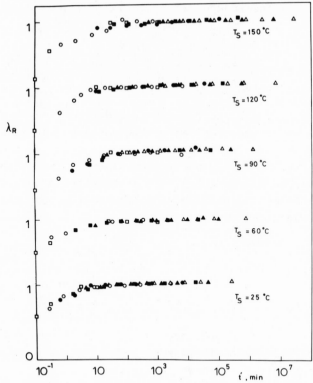

Fig. 3. Master curve of the data reported in figure 2. Symbols
         as in previous figures.

    In table 2 the maximum and minimum activation energies referred
to each drawing temperature, E_{aM} and E_{am} respectively, are reported.
The increasing value of E_{aM} with drawing temperature can be justified
with the higher difficulty for the partly crystallized links to
bring toghether the crystalline blocks.

Table 2. Maximum and minimum activation energies, E_{aM} and E_{am},
         obtained from figure 4 at the different annealing
         temperatures listed

T_S	25	60	90	120	150
E_{aM}	68.2	83	104.6	130.4	171.3
E_{am}	28.1	32	27.8	37	24

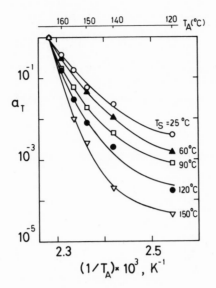

Fig. 4.   Shift factor, a_T, versus reciprocal annealing temperature,
T_A, at the drawing temperatures indicated

In conclusion, the kinetics of the recovery behavior of cold
drawn polypropylene has been experimentally evaluated. The data
obtained at different annealing temperatures can all be superimposed
in a unique master curve at each drawing temperature. The non-
linearity of the horizontal shift factors with respect to reciprocal
annealing temperature can be explained with the model proposed by
Peterlin[3].

REFERENCES

1. Booij H. C. and J. H. M. Palmen, Polym. Eng. Sci., 18:781 (1978)
2. Apicella A., L. Nicodemo and L. Nicolais, "Recoil Kinetics of
        uniaxially oriented Polystyrene", Rheol. Acta, in press
3. Peterlin A., Polym. Eng. Sci., 18:488 (1978)
4. Peterlin A., Jour. Mater.Sci., 6:470 (1971)
5. Balta Calleja F. J. and A. Peterlin, Die Makromol. Chem.,
        141:91 (1971)
6. Samuels R. J., Jour. Macrom. Sci., Phys., B8:41 (1973)
7. Keith H. D., F. J. Padden and R. G. Vadimsky, Jour. Pol. Sci.,
        A-2,4:267 (1966)

8.   Marrucci G., Polymer Eng.Sci., 15:229 (1975)

9.   Balta Calleja F. J. and A. Peterlin, J.Materials Sci., 4:722
     (1969)

10. Kittrel J.R., Adv.in Chem.Eng., 8:97 (1970).

DIMENSIONAL STABILITY OF UNIAXIALLY ORIENTED POLYSTYRENE

COMPOSITES

A.Apicella and L.Nicodemo

University of Naples, Italy

INTRODUCTION

Particulate fillers are often used to improve the dimensional stability of polymeric items obtained through processing
techniques such as injection molding,deep drawing, hot stamping,
etc. In these operations the material undergoes large multiaxial
deformations which result in molecular orientation and remain
frozen-in stresses during cooling of the material. Once the
formed objects are exposed to sufficiently high temperatures,
various degrees of spring-back take place and subsequent changes
in shape and dimensions occur[1].

The effect of draw ratio and glass filler on the recoil
kinetics of polystyrene composite sheets is here analyzed.

EXPERIMENTAL

Drawn composite samples have been obtained with the same
processing procedures described in previous papers[2-4]. The constituent materials used were Montedison's Deistir FA, general
purpose polystyrene, (Mw=200.000, T_g=83°C) and Silenka 8041
E-glass fibers with 95% of the fibers having a nominal length of
3 mm and 5% having a nominal length of 6 mm. The fiber diameter
was 10 microns. Composites obtained with polystyrene and glass
beads have been also studied. The glass beads (Ballotini Europe
3000 CP/01) have a diameter between 40 and 50 microns.

The recoil experiments were performed by immersing the

drawn samples into a thermostatic bath filled with glycerol,
which is not absorbed into polystyrene. Specimens were removed
after various residence times at temperatures ranging from 108°C
to 165°C and for residence times as long as 200 min. The initial
length of the samples (l_0) cut from the extruded strips was 9 mm.
Longer strips gave spiral-like shrinkage due to inhomogeneous
deformation[3]. As previously discussed[2], although the deformation
during extrusion, drawing and recoil is inhomogeneous, the specimens
approach homogeneous uniaxial extension in the region around
their center line. For these reasons the recoil kinetics of drawn
composite samples have been studied by measuring the length at
the central line using a micrometer with a sensitivity of .01 mm.

       Tensile tests have been performed on rectangular specimens
by means of an Instron 1112 tensile machine.

RESULTS AND DISCUSSION

       The recoil kinetics for the polystyrene samples filled
with =0.017% by volume of glass fiber, at DR=6.1 and different
temperatures are reported in fig.1. The data are plotted as
l_0/l_t vs t and display the same behaviour as the unfilled drawn
polystyrene previously studied[2]. At high temperatures an asymptotic equilibrium value l_f is reached. The data have been superimposed by horizontal shifts to give the master curves shown in
figure 2 and relative to a reference temperature of 135°C.

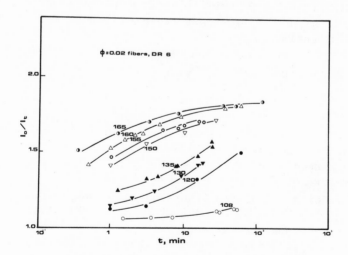

Figure 1 : Recoil of an oriented filled ribbon at
            different temperatures (°C) above T_g, =0.017%.

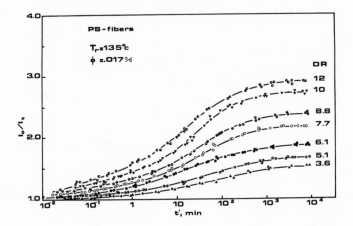

Figure 2 : Recoil master curves referred to 135°C for
short glass fiber polystyrene composites

The master curves for materials drawn to different DR are
also reported in figure 2. The variable t' is related to the
real time by t' = t/ a_T, where a_T is the shift factor given by[5]:

$$\log a_T = \frac{-C_1(T-T_g)}{C_2+(T-T_g)} - \log a_{To} \qquad (1)$$

with T_g = 83°C, C_1= 13.3, C_2= 47.5°C and a_{To}=1.12 x 10^{-7}.
The values of C_1 and C_2 are those reported for polystyrene[5] and
used in reference 3.
        From these curves it is possible to determine the values of
l_f. The quantity (l_o-l_f) represents the deformation due to orientation and consequentely the maximum value of the shrinkage that
can be obtained for a sample with a given processing history.
The rate of recoil at a time can be defined as a function of the
fractional distance from the equilibrium length, $\lambda =(l_o-l_t)/(l_o-l_f)$.
        It is expected that in a λ vs t' plot a single curve can be
obtained for different draw ratios at constant composition
The data for differently filled and oriented samples have been
replotted in this form in figure 3 and are well fitted by a
single master curve. This procedure is similar to the one
followed previously[3] to correlate the data obtained for unfilled
PS at different DR. The presence of the filler does not influence
the recoil kinetics of the polymeric matrix, but has a strong
effect on the equilibrium value of the reversion ratio l_o/l_f.

        Lower equilibrium values of the length reversion ratios(LRR)
have been indeed observed for glass bead and glass fiber compos

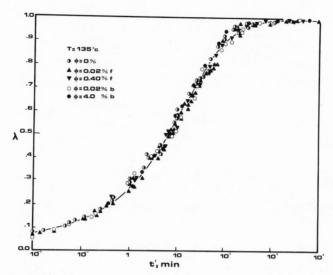

Fig. 3.   Normalized master curve for differently filled and
          oriented Polystyrene.

ites of the same DR at increasing filler volume fraction.   A
possible reasons for this difference could be a different degree
of induced or retained molecular orientation for filled and
unfilled samples subjected to the same thermal and mechanical
history. However the data reported in Figure 4 indicate that this

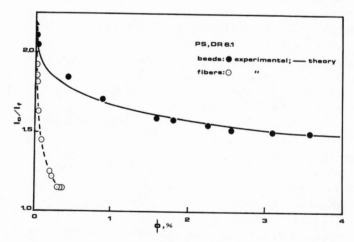

Fig. 4.   Equilibrium reversion ration vs.volume fraction for glass
          beads and glass fiber-polystyrene composites of nominal
          drawing ratio 6.1.

is not the case, at least for glass bead composites. In fact in Figure 4 the experimental values of l_o/l_f relative to bead composites are plotted as a function of volumetric filler content $\emptyset$, as full dots together with a full line which represents the relation :

$$\varepsilon_{rf} = \varepsilon_{rp} \left[1 - (\emptyset/\emptyset_{max})^{1/3} \right] \tag{2}$$

In this equation ε_r is defined as :

$$\varepsilon_r = \frac{l_o - l_f}{l_o} = 1 - (\frac{l_o}{l_f})^{-1} \tag{3}$$

and the subscripts f and p refer to the bead filled composites and the unfilled polymer respectively; $\emptyset_{max}$ represents the maximum volume fraction of beads for random packing of large spherical particles and is expected to be equal to 0.63 (6, 7). Equation 2 is based on the hypothesis that ε_{rp} represents, under fixed thermal and mechanical history, a property of the drawn polymer both when unfilled or bead filled. Therefore, using the Smith's analysis[8] relative to the strain at break of glass bead composites equation 2 follows. The good agreement between the experimental data and the predicted values (calculated with $\emptyset_{max} = 0.63$, $(l_o/l_f)_p = 2.3$ in eq.2) indicates that the molecular orientation in the glass bead filled system and in the unfilled polymer is the same for a fixed DR.

In Figure 4 the equilibrium value of the LRR for fiber filled composites are also reported. In this system the filler effect on the dimensional stability is much stronger than for bead composites. In fact at very small fiber concentration ($\emptyset = 0.35\%$) l_o/l_f is reduced from a value of 2.3 (i.e. 130% of the deformation of unfilled PS) to a value of 1.15 (i.e. 15% of deformation). Also in this case the decrease of the equilibrium values of LRR as a consequence of filler content can be attributed to the effect of fibers on the recovery process more than on the amount of molecular orientation of the polymeric matrix at a fixed DR.

The stronger influence on the equilibrium value of the LRR of the glass fibers with respect to the glass beads is probably due to a different recovery mechanism. In fact if one puts an unfilled drawn specimen with fixed ends at $T > T_g$, after some time the internal stresses relax without modifying the shape of the

sample, while if the ends are free to move, the specimen creeps
until an equilibrium value of LRR is reached. The fibers oriented
in the drawing direction act as a constraint for a certain amount
of polymer surrounding them and consequently the fiber filled
specimens partly creep and partly stress relax. The stress relaxation contribution on the overall process increases at increasing
fiber content.

In Figure 5, the elastic moduli of unfilled PS and composites
containing 0.87% of glass fibers and glass beads respectively are
reported as a function of DR. The modulus increases rapidly at
increasing DR, up to DR = 3.5 and then increases slowly. This
effect of DR on the elastic modulus is similar to that on the
tensile strength[9] or on birifrangence[10] reported in literature for
similar systems. However the effect of filler on the modulus is
very small as expected for this very low filler content. The data
show that the molecular orientation of the matrix in both fiber
and bead composites is the same of that of the unfilled polymeric
matrix at the same DR. This, once more, indicates that the strong
reduction of the equilibrium values of LRR should be attributed
to the relaxation process and not a low molecular orientation
induced during the drawing process.

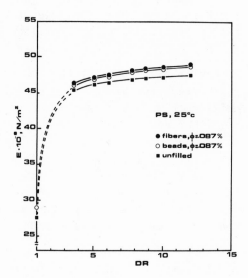

Fig. 5.   Elastic moduli for unfilled (■), glass bead (○) and
         glass fiber (●) polystyrene composites at different
         drawing ratios, T = 25°C.

In conclusion, recovery experiments have shown that small amount of short glass fibers can enhance the dimensional stability of oriented polystyrene sheets. This effect has been demonstrated not to be due to a reduction in molecular orientation of the polymeric matrix. A model to predict the equilibrium values of the length reversion ratio of glass bead composites is developed and well compared with experimental data.

ACKNOWLEDGEMENTS

The financial support of CNR grant n° 77.01354 is gratefully acknowledged.

REFERENCES

1. K. M. Kulkarni, Polymer Eng.Sci., 19:474 (1979)
2. L. Nicolais, L. Nicodemo, P. Masi, A. T. Di Benedetto, Polymer Eng.Sci., 19 :1046(1979)
3. A. Apicella, L. Nicodemo, L. Nicolais, "Recoil kinetics of uniaxially oriented polystyrene", Rheol.Acta, in press
4. L. Nicolais, A. Apicella, L. Nicodemo, "Filler effect on the recoil kinetics of drawn polystyrene", Polymer Eng.Sci. in press
5. J. D. Ferry, "Viscoelastic Properties of Polymer", J.Wiley, New York (1970)
6. R. F. Fedors, J.Colloid Interf.Sci., 46:546 (1974)
7. G. D. Scott, Nature (London), 188:908 (1960)
8. T. L. Smith, Trans.Soc.Rheol., 3:113 (1959)
9. S. S. Sternstein, J. Rosenthal, Adv. in Chemistry Series N°154 "Thoughness and Brittleness of Plastics", R.D.Deanin and A.M.Crugnola Eds., (1976)
10. T. T. Jones, "The Effect of Molecular Orientation on the Mechanical Properties of Polystyrene" Macromolecular Div., IUPAC, Aberdeen (September 10-14, 1973).

MECHANICAL PROPERTIES OF HIGH DENSITY POLYETHYLENE-POLYPROPYLENE BLENDS

R.Greco, G.Ragosta, E.Martuscelli, G.Mucciariello

Istituto di Ricerche su Tecnologia dei Polimeri e
Reologia - CNR - Arco Felice - Naples, Italy

(Abstract)

High density polyethylene (HDPE) - isotactic polypropylene (iPP) blend specimens, directly obtained by extrusion, have been characterized by the following techniques:
• Wide angle X-ray diffraction (WAXD), to get informations on the orientation of the two homopolymers.
• Differential scanning calorimetry (DSC), to detect melting points and crystallinity contents of the polyolefins.
• Equilibrium swelling in n-hexane, to know about the overall anisotropy of the samples.

On the same specimens, so characterized, mechanical tensile and impact tests have been performed.

From the experimental results the following considerations can be drawn:
• No cocrystallization phenomena have been detected both in unoriented as well as in oriented specimens.
• The melting points of both HDPE and iPP decrease slightly with composition. This is probably due to kinetic limitation to crystal growth during the sample crystallization.
• The oriented sample melting points appear to be slightly higher than the unoriented sample ones, showing probably thickening or improved perfection of the crystals of both the polyolefins.
• The crystallinity content of HDPE drops drastically and that of iPP increase abruptly with enhancing the HDPE and the iPP

percentage respectively, at low iPP concentrations. Elsewhere they remain constant. Such effects can be attributed to reciprocal kinetic interactions.

- The Young modulus changes slightly with composition with a trend proportional to the overall crystallinity content.
- The yield strngth and the relative elongation show a synergistic effect with a maximum at about a concentration of 75% in weight of iPP.
- The strength and the elongation at break both show a marked drop for the blends with respect to the homopolymer values. The Izod impact strength also shows a very small decrease for blend specimens in regard to the homopolymer values.

In the present work preliminary results of the addition of a suitable "compatibilizer" on the mechanical properties of HDPE-iPP blends are also reported.

The experimental data show a strong improvement of the ultimate properties and of the Izod impact strengths for the ternary blends due to the addition of a random ethylene-propylene copolymer (Dutral produced by Montedison with a 47% of C_2 in weight).

Work is in progress in our Institute to investigate the morphology of such systems by optical and electron microscopy.

PROPERTIES OF POLYSTYRENE-POLYOLEFIN ALLOYS

E.Martuscelli, C.Silvestre, R.Greco, G.Ragosta

Istituto di Ricerche su Tecnologia dei Polimeri e
Reologia - CNR - Arco Felice - Naples - Italy

(Abstract)

In the area of new plastic materials, polymer blends are
increasingly replacing homopolymers. The morphological studies of
these blends are indispensable to establish relationship between
morphology and properties as well as to point out how the process
ing conditions may govern the morphological features.

The formation of ribbons, rods or droplets of the disperse
phase in a polymer binary blend can be predicted on the basis of
Van Oene's theory which relates interfacial tensions, molecular
weight, molecular weight distribution and shear stress.

According to the theory, the modes of dispersion depend only
on the deformability, molecular weight of the disperse phase and
initial particle sizes of individual components.

The main goal of the present paper is to review the most
important morphological features of extrudate samples of atactic
polystyrene, high density polyethylene, and atactic polystyrene/
isotactic polypropylene blends, and to compare the results found
in the literature on such blends processed by usual apparatus
(screw extruder, capillary rheometer) with those obtained by us
on extrudate samples made using a special laboratory screw-less
mini extruder.

STRAIN AND STRENGTH PROPERTIES OF LINEAR POLYMERS AT UNIAXIAL

EXTENSION ABOVE THEIR GLASS TRANSITION TEMPERATURE

V.Ye. Dreval, Ye.K. Borisenkova, G.V. Vinogradov

Institute of Petrochemical Synthesis of the USSR
Academy of Sciences, Moscow, USSR

(Abstract)

This work involves studies on the general regularities of straining and rupture of linear polymers under the effect of high strain rates and stresses with particular reference to uniaxial extension of incured rubbers with a narrow molecular weight distribution. The results indicate that when incured rubbers are extensioned in a broad range of strain rates ($\dot{\varepsilon}$) covering more than five orders of magnitude, as the value of $\dot{\varepsilon}$ increases, they lose fluidity and pass successively to a high elastic and a glassy state.

In the fluid state linear polymers are capable of accumulating unlimited strains. In the other two states their strain is limited and in the case of continuous straining they rupture. The strain rate corresponding to the transition from the fluid to highly elastic state is unambiguously related to the viscosity of the polymer and varies inversely with it. Despite the limited extent of high elastic strains of linear polymers, their values may be as high as two thousand per cent, which is much in excess of the recoverable strains of cured rubbers.

Transition to the glassy state with increasing $\dot{\varepsilon}$ is accompanied by a sharp decrease in the strain of linear polymers. A result of relaxational transition from one physical state to another is the correlation between the behaviors of linear polymers at continuous and periodic low-amplitude straining, established in this work.

Therewith, corresponding to the maximum on the loss modulus-versus
frequency curve $G''(\omega)$ is the minimum on the continuous strain rate-
versus-strain curve $\varepsilon(\dot{\varepsilon})$, while corresponding to the minimum on
the $G''(\omega)$ curve is the maximum on the $\varepsilon(\varepsilon)$ curve. Studies into
the strength of linear polymers have shown that its value is
unambiguously determined, irrespective of the way in which straining
takes place, by the high elastic strain accumulated at rupture.
At the same time such polymers feature minimum values of high
elastic strain and stress, below which the polymer does not rupture
but passes into the fluid state. As is shown in this work, similarly
to cured rubbers, the long term durability of linear polymers in a
broad range of stresses covering six orders of magnitude is a
diminishing power function of the latter. The general results
indicate that the strain-strength behavior of linear polymers
features a number of common traits with that of cured rubbers.
However, unlike cured rubbers, their rupture is due to the rupture
of the fluctuation entanglement rather than chemical network.

BIORHEOLOGY

MOLECULAR RHEOLOGY OF HUMAN BLOOD: ITS ROLE IN HEALTH

AND DISEASE (TO DAY AND TO MORROW ?)

Leopold Dintenfass

Haemorheology & Biorheology Department
Medical Research KMI, Sydney Hospital
and Department of Medicine, University
of Sydney, Sydney, NSW, Australia

ABSTRACT

Hyperviscosity syndrome (or 'hyperviscosaemia'), in which an elevation of one or more of the blood viscosity factors (such as viscosity of plasma, viscosity of whole blood, aggregation of red cells, rigidity/deformability of blood cells, aggregation of platelets, thrombus formation, etc.) might take place, can lead to ischaemia, infarction and necrosis of the tissue; in the case of increased aggregation of red cells it might play a role in the cancer metastasis; in the case of rigidity of red cells, it might play a role in hypertension and diabetes.

The theoretical studies of blood cells are of great interest, as the red cells represent a nearly ideal emulsion. The liquid -
- crystalline components of the red cell membrane and of the cell interior supply an array of catalytic and mechano-chemical opportunities.

The practical applications of rheological findings will find a place in diagnosis and therapy of heart diseases, hypertension, diabetes and cancer; and might become of particular importance in screening for the silent (not detectable by the usual means) cardio-vascular diseases.

A scope opens slowly for a study of rheology of blood under conditions of near-zero gravity. This might be not only of interest to the future space-travellers, but might answer certain basic questions, an answer to which cannot be obtained under 1 gravity.

467

INTRODUCTION

As is well known,microrheology sets itself two following tasks:
it either aims at getting a picture of the molecular and colloidal
structure of the material studied from the observed flow curves, or
it aims to explain the rheological behaviour of a complex material
from the known rheological behaviour of its components. The first task
is called "structural analysis", and the second, "structural theory".
The latter task is complicated by interactions between phases or subphases of the material studied. with a consequent change in the
rheological properties of these subphases. As will be seen, a study
of rheology of blood brought some new understanding of flow properties
of the concentrated suspensions of fluid drops and, thus, contributed
to the general field of molecular physics and molecular rheology.

This study deals with microrheology of blood in health and disease,
and covers work carried out in my laboratories at Sydney Hospital and
at the Department of Medicine, University of Sydney. Some of this work
was carried out at the University of Glasgow Department of Medicine,
Royal Infirmary, and in the Department of Medicine I, Hadassah-Hebrew
University Hospital, Jerusalem. A long-distance collaboration existed
also with the University Central Hospital, Helsinki, Finland, and with
the University of Padova, Italy. As rather early it became apparent
that rheology of blood sheds new light on the fundamental aspects of
circulation, and on the specific pathologies of dianetes, ischaemic
heart disease, renal failure, various forms of cancer, etc., a series
of collaborative studies was set up in association with cardiologists,
renal physicians, haematologists, vascular surgeons, psychiatrists,
etc.

INSTRUMENTATION

No adequate instrumentation for study of blood existed in 1961,
and at the time, and during subsequent years, various viscometers
have been constructed.

Rotational viscometers

The first viscometer used was the cone-in-cone rotational viscometer[1,2,3] in which the inner cone was made of Teflon, and the outer
cone of brass. Guard rings[4,5] were found to be of no advantage when
freshly shed blood, in absence of anticoagulants, was used; furthermore,
results depended on the shape and position of the guard rings when
such were used. In order to avoid an air-blood interface, another
version was constructed: a rhombospheroid viscometer in which the
inner rhombospheroid (double cone) was made of Teflon, and the outer
rhombospheroid (double hollow cone) was made of brass and Perspex.
The cones were centered by means of an adapted microscopic stage[6,7];
furthermore, the torsion strip was equipped with an universal joint.

The variable-frequency thromboviscometer (VFTV)

This was an oscillatory version of the cone-in-cone viscometer, working at high amplitude (arc of 110° of the circle) and a range of frequencies of oscillation from 10 to 250 cycles per minute [8,9,10]. Coagulation of blood in the VFTV resulted in formation of a spectrum of artificial thrombi, of morphologies ranging from red clots, through red and white thrombi, to 'suspension' coagula [11,12].

Slit-capillary viscometer

The slit-capillary viscometer[13,14] was formed of two optically flat glass plates, with gaps ranging from 4 to 55 micrometers. This type of viscometer allowed not only determination of blood viscosity but also photography of red cells and aggregates of red cells. This instrument is now being adapted for a study under near-zero gravity.

MICRORHEOLOGY OF BLOOD

Blood viscosity varies greatly over the range of shear rates. At near zero shear rates the viscosity of whole blood may be 100 to 1000 times higher than that of water; however, at high shear rates it will be only 3 to 6 times higher than water. When blood viscosity is measured in capillary viscometers, blood viscosity at high shear rates can be about twice that of water. Obviously, the viscosity of blood depends on the shear rate and on the geometry of the instrument. Basically, the viscosity of blood depends on the concentration of red cells, on the degree of aggregation (and morphology of aggregation) of red cells, on the rigidity (inverse deformability) of red cells, and on the viscosity of plasma. White cells and platelets can play a very important role in the capillary flow, but have relatively small effect on blood viscosity measured in a rotational viscometer. Fig. 1.

Viscosity of blood decreases as the shear rate increases, and vice versa. For many years a controversy raged whether blood is shear-thinning or thixotropic. Although I reported thixotropy of blood in 1962 [15,16], only during last few years a concensus was obtained that blood is thixotropic. It is easy to demonstrate a hysteresis loop. Additionally, when observing formation of cell aggregates by microscopy, one can note that a time needed for an aggregate to form is up to 10 minutes. The relatively high viscosity of blood at low shear rates is due mainly to aggregation of red cells. This is affected by proteins and lipids of plasma, and also by the surface characteristics of red cell membrane. A fever, or an increase in temperature, leads to an increase of aggregation of red cells [17]. Aggregation-disaggregation is a dynamic process depending on the attractive forces between the red cells and on the shear rates. In general, the degree of aggregation of red cells is elevated in patients with myocardial infarction, lung or intestinal cancer, multiple myeloma, etc. Cigarette smoking increases aggregation of red cells, especially in blood group O.

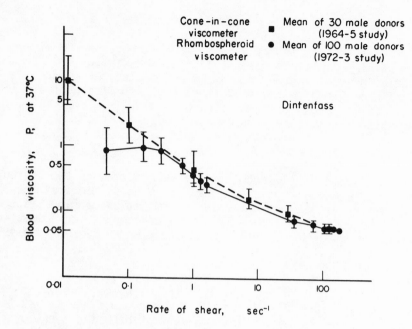

Fig. 1.   Blood viscosity of healthy men measured by means of the
cone-in-cone viscometer or rhombospheroid viscometer.

Rheology of the red cell

While suspensions of rigid particles exhibit a consistency of
concrete at concentrations not much above 65%, the viscosity of blood
remains quite low even at highest concentrations of an about 95%.
Thus, it was not too difficult to conclude that the red cell must be
a fluid drop[18, 19], [20, 21]. Utilizing a theory of ideal emulsions of Taylor
and Oldroyd, and Roscoe's equation for viscosity of suspension
of concentrated spheres[22], I suggested in 1968 a blood viscosity
equation of the following type:

$$\eta_r = (1 - TkC)^{-2.5}$$

in which η_r is the relative viscosity of blood (that is, the
viscosity of blood, particularly when measured at higher shear rates
of about 200 rec. sec., divided by the viscosity of plasma); C is the
volume concentration of red cells; k is a coefficient of plasma
trapping; T is Taylor's factor by which the volume of red cells is
decreased due to free transmission of tangential and normal forces
and internal circulation within the red cell. The numerical value
of Taylor's factor is

$$T = (p + 0.4)/(p + 1)$$

in which p is the ratio of the apparent internal viscosity of the
red cell (or fluid drop) and the viscosity of plasma. Experimentally
one can obtain values for T using the following transformation:

$$TkC = (\eta_r^{0.4} - 1)/\eta_r^{0.4}$$

Note that terms T and k are susceptible to shear rates and decrease
with an increase of shear rates.

There are many problems associated with this approach. One is the
structure of the red cell membrane. This membrane must behave as a
fluid film, otherwise the red cell would behave as a rigid particle.
I suggested [23,24] that the membrane is a two-phase system, composed
of liquid crystals restrained by a protein network. Some studies in
the n.m.r. appear to confirm this approach [25,26]. Another problem
concerns viscosity of plasma. It does appear that Taylor's vortices
might be formed in plasma when studied in rotational viscometers,
and it might be safer to use for this purpose capillary viscometers.

Viscosity of plasma

Plasma viscosity is supposed to be Newtonian, at least in normal
donors. However, there is an evidence that, for instance, in leukaemia
plasma is shear-thinning [27,28]. One cannot exclude a possibility of
some plasmas being dilatant or shear-thickening.

Viscosity of blood in microcapillaries

Viscosity of blood decreases as the radius of capillary decreases.
This phenomenon, known as Fahraeus-Lindqvist phenomenon, and reported
originally for tubes of diameters well above 200 micrometers, has been
extended in our laboratories down to 4 micrometers. In this study,
using slit-capillary viscometers, I noted a sudden and dramatic
'inversion' of this phenomenon at certain critical capillary radii,
the value of the latter depending on the rheology of blood, and
especially on the rigidity of the red cell. This phenomenon was named
"inversion phenomenon" [28, 29, 30]; and is illustrated by Fig. 2.

The critical capillary radius of the 'inversion phenomenon' is
affected by blood pH, the type of haemoglobin, osmolarity, presence
of some abnormalities in the red cell membrane (i.a., Heinz bodies),
presence of platelet aggregates or white cells. Both platelets and
white cells are much more rigid than the red cells. A localized suspension
of rigid blood cells can act as a suspension of rigid particles and
exhibit extremely high viscosity and even dilatancy. Toxins, hormones
and drugs may affect rigidity of blood cells and thus the critical
radius of the inversion phenomenon.

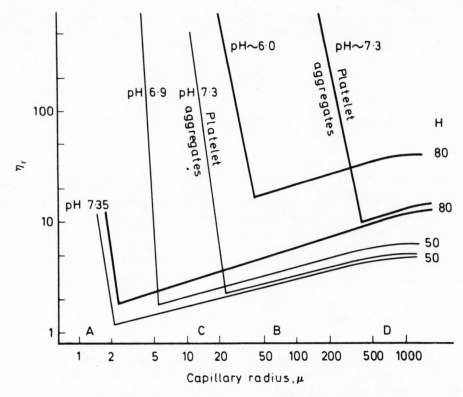

Fig. 2. Fahraeus-Lindqvist phenomenon and the inversion phenomenon.
Note that blood viscosity decreases when capillary radii
decrease; but at a certain critical capillary radius there
is a sudden increase in the viscosity (that is, an increase
in the resistance to flow. This is a most crucial factor in
many ischaemic diseases.

Thrombus formation

Practically every aspect of blood coagulation is affected by the
flow velocity of blood and the shear rate existing during the clotting
process. Clotting times, thrombus formation times, and apparent
viscosity (or consistency) of blood clots decrease when the 'casting'
shear rate is increased. The morphology of blood clots changes in
parallel, showing characteristics of a glass-clot at near-zero shear
rates (a three-dimensional network of fibrin), of a red/white or white
thrombus at medium shear rates, 20-80 sec^{-1} (an orientated fibrin
network with participation of platelet aggregates), and of 'suspension
coagula' at high shear rates (globular particles of fibrin and platelets
The 'casting' shear rate determines the final properties of clots and
thrombi. Coagula formed at very low shear rates exhibit gel-like and
viscous and thixotropic properties, reaching viscosities of thousand

poises; red and white thrombi exhibit also thixotropic properties,
but their viscosity range is nearer to 1 poise. Suspension coagula
show viscosity of the same order as viscosity of the original blood
sample.

Platelet aggregation is greatly enhanced by high flow velocity
and/or high shear rates. There exists a quantitative relationship
between the rate of shear and the degree of platelet aggregation; it
depends on the intrinsic characteristics of blood of individual donor.
In some donors the effect of shear rate on blood clotting and
platelet aggregation is rather small, in other dramatic. A progressive
process of platelet aggregation at increasing shear rates suggests an
irreversible phenomenon due to biochemical and/or mechano-chemical
transformations in the membrane of the platelets.

Thrombus viscosity is not necessarily related to the concentration
of fibrinogen or concentration of platelets. It appears that the
'rheological activity' of fibrinogen, and of platelets, depends on
many factors, including also ABO blood groups. In some series it was
possible to correlate thrombus viscosity with the degree of aggregation
of red cells.

BLOOD RHEOLOGY IN CLINICAL SCIENCE

Rheological factors such as internal viscosity or rigidity of
red cell, aggregation of red cells, concentration of red cells, plasma
viscosity, aggregation of platelets, etc., play a profound role in the
stability of circulation and, especially, the microcirculation. An
increase in the rigidity of red cell, due either to the genetic factors,
tonicity, hydrogen ion concentration, antigen-antibody reaction, drug
action, age of the cell, or any other acquired or hereditary factor,
may be expected to impair a proper tissue perfusion. The latter will
be affected also by increased aggregation of red cells, and by other
viscosity factors.

Blood viscosity and viscosity factors (subphases of blood) were
studied in patients with myocardial infarction, hypertension, diabetes,
peripheral vascular disease, chronic anxiety and psychosomatic pain,
malignant melanoma, multiple myeloma, Waldenström's macroglobulinaemia,
deep vein thrombosis, lung cancer, visceral cancer, leukaemia, etc.
Within each series we worked out means and standard deviations, using
either normal or log distribution. All these values were compared with
the values obtained from 'normal' population. With time it became
apparent that the so called 'normal' and 'healthy' population is not
exactly healthy. By observing elevation of blood viscosity factors we
were able to detect 'silent' cases of heart disease, cancer, and diabetes.
A subsequent study of athletes indicated that by comparison with
athletes, most of the 'normal' population is not normal rheology-wise.
Thus, blood rheology will be of importance in the preventive medicine.

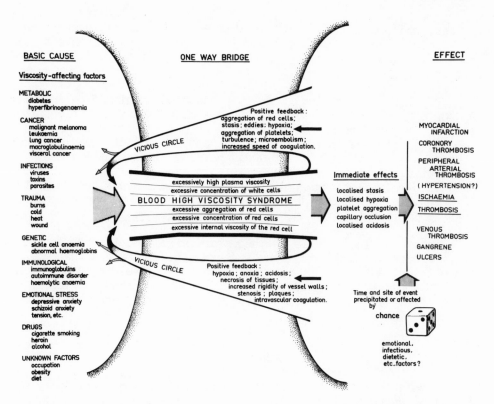

Fig. 3. A schematic outline of the blood hyperviscosity role in
        different diseases. A positive feedback exists. Increase
        of blood viscosity, whatever the cause, will lead to
        abnormalities in tissue perfusion, thus leading to a range
        of cardio-renal and vascular diseases.

Viscosity factors were related to concentrations of fibrinogen,
albumin, protein, various enzymes and hormones. Subsequently, blood
viscosity factors were related to the clinical indices: such as
cardiac ouput and submaximal work ouput, arterial blood pressure,

All the aforementioned aspects are illustrated by Figs. 3, 4,
and 5. The actual pathway of the effect of elevated blood viscosity
on tissue perfusion and cardiovascular disorders is illustrated by
Fig. 6.

These studies led to development of the concept of blood hyperviscosity (hyperviscosaemia), and its latest definition is quoted
from my book on "Rheology of Blood in Diagnostic and Preventive
Medicine" :"Blood hyperviscosity can be due to elevation of any one
of the blood viscosity factors: elevation of plasma viscosity, or
elevation of haematocrit, elevation of the degree of aggregation of

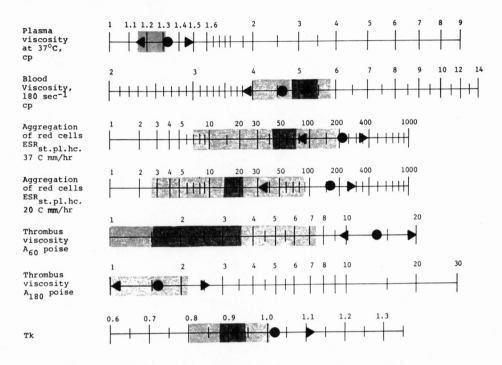

Fig. 4. A rheological 'fingerprint' for a series on diabetes.
Shaded areas: normals; left-hand shaded area: athletes;
right-hand shaded area: 'normals; means given as dots, plus
minus one standard deviation given as an arrow. Tk is the
measure of rigidity of red cells. Aggregation of red cells
has been estimated from erythrocyte sedimentation rates,
corrected for plasma viscosity and adjusted to haematocrit
of 30%.

of red cells, increase of the internal viscosity and rigidity of red
cells, etc. Blood hyperviscosity can be accompanied by elevation of
the viscosity of whole blood, but can be present in spite of normal
or even decreased viscosity of whole blood. The crucial role of blood
hyperviscosity is apparent in the microcirculation. The effect of
increased rigidity of the blood cells or of the presence of micro-
thrombi, microemboli or other products of blood coagulation, is
amplified by the inversion phenomenon in the microcapillary flow."
See references [30,31,32,33,34].

We studied effects of drugs [35], correlations between blood
viscosity and ECG [36]: in the latter we found elevation of blood
viscosity at abnomalities (ST-segment depression) of ECG. We were
able to relate physical fitness to the viscosity of blood. We noted
that abnormal elevation of blood viscosity is a better predictor of
future health than the submaximal work ouput or cholesterol level.

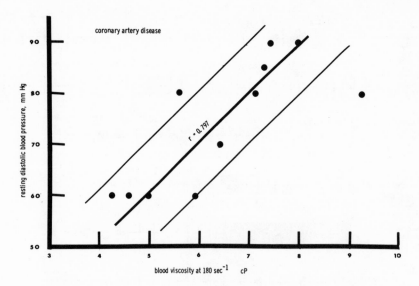

Fig. 5. Resting diastolic blood pressure plotted against the blood
viscosity determined at shear rate of 180 rec. sec. in
patients with severe coronary occlusive disease.

We observed that patients with chronic anxiety[37,38] showed increase
in some of the blood viscosity factors. Patients complaining of the
so called 'psychosomatic pains', and who were considered clinically
healthy by usual examinations, did show a pronounced elevation of
the blood viscosity factors. Patients undergoing kidney transplantation
showed a pronounced decrease in their blood viscosity factors when
transplant was successful; and a pronounced increase when transplant
was rejected or otherwise unsuccessful[39].

Blood viscosity factors have been found elevated in many types
of cancer; furthermore, it is likely that increased aggregation of
red cells plays a role in metastases[40].

CONCLUSIONS AND COMMENTS

The evidence obtained during our studies, as well as data
presented by many new overseas centres, indicate that blood viscosity
is of importance in health and disease. The problems in application
of the blood viscosity factors (and blood rheology) to the clinical
medicine are due in part to inertia: it is difficult to introduce
new methods and new way of thinking. Due to historical reasons,
clinical medicine is based on biochemistry; physics, physical
chemistry and rheology are usurpers.

From the fundamental point of view there is also a great scope

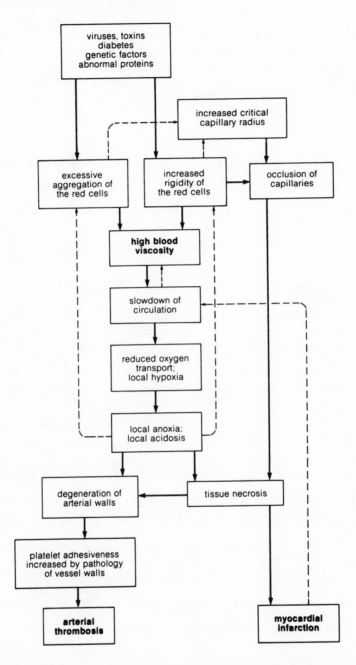

Fig. 6. A schema of the effect of high blood viscosity on tissue
perfusion and consequent arterial thrombosis or myocardial
infarction. Note that arterial thrombosis can be a result
of occlusion of 'vasa vasorum'.

in definition of the autoregulatory mechanism of blood viscosity[41];
exploration of the role of liquid crystals not only as structural
units of the red cell membrane[24] or blood clots and thrombi[42], but
also as units of the regulatory system and receptors; and finally, the
great opportunity in the outer Space for study of blood rheology[43] as
well as of metabolism of blood cells.

REFERENCES

1. L. Dintenfass, Thixotropy of blood and proneness to thrombus
   formation. Circ. Res. 11:233 (1962)
2. L. Dintenfass, An application of a cone-in-cone viscometer to the
   study of viscosity, thixotropy and clotting of blood.
   Biorheology, 1: 91 (1963)
3. L. Dintenfass, A trolley viscometer for estimating viscosity and
   clotting of blood in hospital wards. Lancet 2: 567 (1964)
4. L. Dintenfass, Guard rings, surface films and artefacts in the
   viscometry of human blood. Nature, Lond., 213: 179
5. L. Dintenfass, Some observations on the role of guard-rings in the
   cone-in-cone viscometer. Observations on the surface-film in
   water, oil and human blood. Rheol. Acta, 7: 197 (1968)
6. L. Dintenfass, A coaxial rhombo-spheroid viscometer: a further
   development of the cone-in-cone viscometer. Biorheology, 6: 33
   (1969)
7. L. Dintenfass, Microrheology of blood in health and disease. Effect
   of blood subphases (i.e., the internal viscosity and aggregation
   of the blood cells) on blood viscosity and microcapillary flow,
   occlusion, and infarction. In: "Proc. Fifths Int. Congress on
   Rheology," S. Onogi, ed., vol. 2, p. 27. University of Tokyo
   Press and University Park Press, Tokyo and Baltimore (1970)
8. L. Dintenfass, Effect of velocity gradient on the clotting time of
   blood and on the consistency of clots formed in vitro. Circ. Res.,
   18: 349 (1966)
9. L. Dintenfass, Dynamics of blood coagulation. Introducing a new
   coagulation factor: 'velocity gradient'. Haematologia, 1: 387
   (1967)
10. L. Dintenfass and J. H. Stewart, Formation, consistency and
    degradation of artificial thrombi in severe renal failure. Effect
    of ABO blood groups. Thromb. diath. Haemorrh., 20: 267 (1968)
11. L. Dintenfass and M. Rozenberg, The influence of the velocity
    gradient on in vitro blood coagulation and artificial thrombosis.
    J. Atheroscl. Res., 5: 276 (1965)
12. M. Rozenberg and L. Dintenfass, Platelet aggregation in the variable
    -frequency thromboviscometer. Nature, Lond., 211: 525 (1966)
13. L. Dintenfass, An inversion of the Fahraeus-Lindqvist phenomenon
    in blood flow through capillaries of diminishing radius. Nature,
    Lond., 215: 1099 (1967)
14. L. Dintenfass, Viscosity of blood at high haematocrit measured in
    microcapillary (parallel-plate) viscometers of r = 3-30 microns.
    In: "Haemorheology, Proc. First Conf. Reykjavik", A.L. Copley ed.
    p. 197. Pergamon Press, Oxford (1968)

15. L. Dintenfass, Thixotropy of blood at very low rates of shear. Kolloid Ztschr., 180: 160 (1962)

16. L. Dintenfass, A study in flow, viscosity and clotting of human blood. Med. J. Aust., 1: 575 (1963)

17. L. Dintenfass and C. D. Forbes, About increase of aggregation of red cells with an increase of temperature in normal and abnormal blood (i.e., cancer). Effect of ABO blood groups and proteins. Biorheology, 10: 383 (1973)

18. L. Dintenfass, Considerations of the internal viscosity of red cell and its effects on the viscosity of whole blood. Angiology, 13: 333 (1962)

19. E. Hatschek, Die Viskosität von Blutkörperchen-Suspensionen. Kolloid Ztschr., 27: 163 (1920)

20. G. I. Taylor, The viscosity of a fluid containing small drops of another fluid. Proc. Roy. Soc., 138A: 41 (1932)

21. J. C. Oldroyd, The effect of interfacial stabilizing films on the elastic and viscous properties of emulsions. Proc. Roy. Soc., 232A: 567 (1955)

22. L. Dintenfass, Internal viscosity of the red cell and a blood viscosity equation. Nature, Lond., 219: 956 (1968)

23. L. Dintenfass, Molecular and rheological considerations of the red cell membrane in view of the internal fluidity of the red cell. Acta Haemat., 32: 299 (1964)

24. L. Dintenfass, The internal viscosity of the red cell and the structure of the red cell membrane. Considerations of the liquid crystalline structure of the red cell interior and membrane from rheological data. Mol. Cryst., 8: 101 (1969)

25. A. G. Lee, N. J. M. Birdsall, and J. C. Metcalfe, Nmr studies of biological membranes. Chemistry in Britain, 9: 116 (1973)

26. S. J. Singer and G. L. Nicolson, The fluid mosaic model of the structure of red cell membrane. Science, 175: 720 (1972)

27. L. Dintenfass, Fluidity (internal viscosity) of the erythrocyte and its role in physiology and pathology of circulation. Haematologia, 2: 19 (1968)

28. L. Dintenfass, "Blood Microrheology, Viscosity Factors in Blood Flow, Ischaemia and Thrombosis.", Butterworths, London (1971)

29. L. Dintenfass and J. Read, Pathogenesis of heart failure in acute-on-chronic respiratory failure. Lancet, 1: 570 (1968)

30. L. Dintenfass, "Rheology of Blood in Diagnostic and Preventive Medicine.", Butterworths, London and Boston (1976)

31. L. Dintenfass, A preliminary outline of the blood high viscosity syndromes. Arch. Intern. Med., 118: 427 (1966)

32. L. Dintenfass, Clinical applications of blood viscosity factors and functions: especially in the cardiovascular disorders. Biorheology, 16: 29 (1979)

33. L. Dintenfass, The role of blood viscosity in occlusive arterial disease. Practical Cardiology, 5: 77 (1979)

34. L. Dintenfass, Viscosity factors in hypertensive and cardiovascular diseases. Cardiov. Med., 2: 337 (1977)

35. L. Dintenfass and B. Lake, Beta blockers and blood viscosity.
    Lancet, 1: 1026 (1976)
36. L. Dintenfass and B. Lake, Blood viscosity factors in evaluation
    of submaximal work output and cardiac activity in men.
    Angiology, 28: 788 (1977)
37. L. Dintenfass and I. Zador, Blood rheology in patients with
    depressive and schizoid anxiety. Biorheology, 13: 33 (1976)
38. L. Dintenfass and I. Zador, Hemorheology, chronic anxiety and
    psychosomatic pain: an apparent link. Lex et Scientia, 13: 154
    (1977)
39. L. Dintenfass and J. H. Stewart, Formation, viscosity and
    degradation of artificial thrombi after cadaveric-donor kidney
    transplantation. Thromb. diath. Haemorrh., 26: 24 (1971)
40. L. Dintenfass, The place of blood viscosity studies (rheology)
    in the diagnosis and prognosis of primary and advanced malignant
    melanoma and in cancer generally. In: "Advances in Medical
    Oncology, Research and Education. Vol. 10. Clinical Cancer
    Principal Sites 1.", S. Kumar ed. p. 131. Pergamon Press, New
    York and Oxford (1979)
41. L. Dintenfass, Malfunction of viscosity-receptors (viscoreceptors)
    as the cause of hypertension. Am. Heart J., 92: 260 (1976)
42. L. Dintenfass, On the possible liquid crystalline structures in
    artificial thrombi formed at arterial shear rates: effects of
    disease, protein concentrations, and ABO blood groups.
    Mol. Cryst., 20: 239 (1973)
43. L. Dintenfass, Aggregation of red cells and blood viscosity under
    near-zero gravity. Biorheology, 16: 29 (1979)

ANALYSIS OF DRAG REDUCTION PHENOMENON IN BLOOD FLOWS

P.N. Tandon, A.K. Kulshreshtha[+]

Department of Mathematics, H.B.T.I., Kanpur, India
[+]Department of Mathematics, D.B.S.P.G.College, Kanpur
University, Kanpur, India

(Abstract)

Recently, the phenomenon of drag reduction has been shown to
play a very important role in transporting blood in human arteries.
This paper presents a drag reduction problem in a circular as well
as a stenotic tube in reference to blood flows. Considering a thin
peripheral layer of drag reducing polymer near the wall, with blood
flowing in the central region steadily in both the cases, the
percentage of drag reduction has been estimated with actual drag
without this assumed peripheral layer. It has been observed that
although the amount of drag is two less in laminar flows as
compared to turbulent flows, the percentage of drag reduction is
of the same order. Various other important results have also been
brought out and presented graphically.

AUGMENTED RATES OF OXYGEN TRANSFER TO BLOOD UNDER THE INFLUENCE OF IMPOSED MAGNETIC FIELDS

V.G. Kubair

Department of Chemical Engineering
Indian Institute of Science
Bangalore - 560012, India

INTRODUCTION

The effect of magnetic field on the increase in the number of erythro-cytes and hemoglobin of blood and the concentration of potassium ion in the plasma is reported in the literature [1,2] and has been observed in experimental investigations. The transfer of oxygen to blood is very important in blood oxygenators. Several methods of augmenting the rates of oxygen transfer to blood and associated theories are available.[3] Since the experiments have shown the significant effect of magnetic field on the constituents of blood, it is very important to understand the effect on oxygen transfer to blood. The transfer of oxygen to blood involves diffusion and chemical reaction between oxygen and hemoglobin under the influence of magnetic field.

FORMULATION OF THE MATHEMATICAL MODEL

Consider oxygen absorbed by blood over a vertical plate. The natural convection currents are set up as a result of the solutal concentration gradients of oxygen at the wall and the bulk. Oxygen reacts with hemoglobin forming oxyhemoglobin, accompanied by first order chemical reaction.

Using the following assumptions

(i)   steady state laminar free convection currents prevail

483

(ii)   blood is characterised by Casson's model

(iii)   oxygen is paramagnetic

(iv)   the solutal concentration gradients of oxygen are affected by magnetic field, which is governed by Maxwell's equations

(v)   the operation is isothermal

(vi)   the thickness of concentration boundary layer is affected by diffusion, chemical kinetics of first order reaction and hydrodynamics

(vii)   the physical properties are constant

(viii)   viscous dissipation is neglected

(ix)   cartesian system of coordinates is used

(x)   as oxygen is absorbed, the concentration decreases from bulk to the wall as shown in Figure 1.

the equations of continuity, motion and diffusion are given by

$$\partial u / \partial x + \partial v / \partial y \quad = \quad 0 \tag{1}$$

$$u(\partial u / \partial x) + v(\partial u / \partial y) = (1/\rho)(\partial \tau_{yx} / \partial y) + g(C_w - C_\infty / C_\infty) \pm$$

$$(\sigma \mu^2 H^2 \cos^2 \phi) \, u / \rho \tag{2}$$

$$u(\partial C / \partial x) + v(\partial C / \partial y) = D(\partial^2 C / \partial y^2) - KC \tag{3}$$

With the boundary conditions

$$y = 0 , \quad u = v = 0 , \quad C = C_w ; \quad y \rightarrow \infty \quad u \longrightarrow 0 , \quad C \longrightarrow C_\infty$$

In the above equations, the shear stress τ_{yx} for blood is characterised by Casson's model, is given by

$$\sqrt{\tau_{yx}} = \sqrt{\tau_0} + \sqrt{\mu_0} \, (\partial u / \partial y)$$

The term $(\sigma \mu^2 H^2 \cos^2 \phi / \rho) \, u$ indicates the resultant velocity of the interaction of magnetic flux with free convection velocity. The positive or negative sign indicates the direction of magnetic field in the case of transverse magnetic field (north to north) respectively as shown in Figure 1.

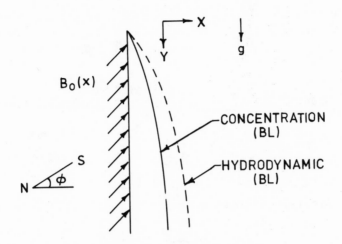

Figure 1. Physical Model

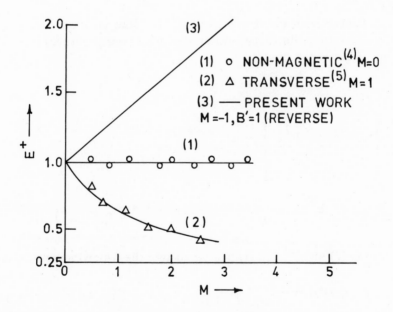

Figure 2. Plot of E^+ vs. M

With the aid following dimensionless groups and similarity transformation,

$$\psi = 4 E \mathcal{V} x^{3/4} f(\eta) \qquad \eta = E y x^{-1/4}$$

$$u = \partial \psi / \partial y \qquad v = -\partial \psi / \partial x \qquad \theta = \frac{C_\infty - C}{C_\infty - C_w}$$

$$N_{Gr} = \text{Grashof number} = \frac{x^3 g (C_\infty - C_w)}{\mathcal{V}^2 C_\infty}$$

$$N_{Sc} = \text{Schmidt Number} = \mathcal{V}/D$$

$$N_D = \text{Damkohler Number} = \frac{K x^2}{D}$$

$$M = \text{Magnetic parameter} = \frac{\sigma B_0^2 L^{1/2}}{E^2} Cos^2 \phi$$

$$B = B_0 (x/L)^{-1/4}$$

B' = Function of (Casson's parameters, like yield stress, Casson's viscosity, which depend on temperature) (dimensionless)

$$E = \left[\frac{g(C_\infty - C_w)}{4 \mathcal{V}^2 C_\infty} \right]^{1/4} = \frac{N_{Gr}^{1/4}}{\sqrt{2} x^{3/4}}$$

$$u = 4 x^{1/2} E^2 f' \quad , \qquad v = E x^{-1/4} (\eta f' - 3f)$$

the equations (2) and (3) are transformed into the following equations

$$f''' \quad 1 + B' (f'')^{-1/2} \quad + 3 f f'' - 2 (f')^2 + 0 \pm M f' = 0 \qquad (3)$$

$$0'' + 3 N_{Sc} 0' - 2 N_D N_{Gr}^{-1/2} 0 = 0 \qquad (4)$$

with the boundary conditions

$$f(0) = 0 = f'(0) \ , \ \ 0 \ (0) = 1.0 \ ; \ f'(\infty) = 0 = 0(\infty)$$

The boundary conditions imply that the solutal concentration of oxygen decreases as oxygen reacts with the hemoglobin of blood.

The parameters are estimated for blood-oxygen system from the literature.[3] The range of parameters used in the present investigation is

Magnetic parameter (M)	-1, 0, 1
Damkohler number (N_D)	1.7
Schmidt number (N_{Sc})	0.74
Grashof number (N_{Gr})	10^6
B'	0, 1

RESULTS AND DISCUSSION

The equations (3) and (4) reduce to the case of non-isothermal absorption of gases accompanied by chemical reaction in a solvent containing a reacting compound[4], when M = 0 and B' = 0 . The equations are solved by Runge Kutta Gill's method with a built-in iterative procedure and the results check very well within 0.1% for M = 0 and B' = 0, with the available results.[4]

From Figure 2, it can be observed that E^+ , enhancement factor, the ratio of dimensionless concentration of oxygen for magnetic case to that in nonmagnetic case increases with M for the case of reverse magnetic field (M = -1), whereas it decreases for transverse magnetic field (M=1). For M = 0 , the results agree with the corresponding heat transfer problem[5], when N_D = 0 , (transverse magnetic field). It can be concluded that rates of oxygen transfer to blood is augmented in the case of reverse magnetic field, which can be recommended as an alternative powerful tool in augmenting the rates of oxygen transfer to blood.

LITERATURE CITED

1. G.K.Gerasimova, Erythrocyte potassium ion transport in animals under the ;effect of continuous magnetic field. Kosm.Biol. Aviokosm Med. (3), 11, 63 (1977).

2. T.M.Guseinov, Effect of sodium selenite in the state of peripheral blood under the influence of a permanent magnetic field. Selen Biol. Mater Nauchn. Konf._1, 133 (1975).

3. D.O. Kooney, 'Biomedical Engineering Principles', Marcel Dekker Inc., New York (1976).

4. V.G.Kubair, 'Non-isothermal absorption of gases in liquids accompanied by reaction and Benard cells', personal communications.

5. R.S.Nanda and Y.K.Mohanty, 'Hydromagnetic free convection for high and low Prandtl numbers', Phy.Soc.Japan, 20, n6, p 1608(1970).

Nomenclature

B'	=	dimensionless constant
B	=	product of and H
C	=	concentration
D	=	diffusivity
E	=	constant
E^+	=	enhancement factor
Ψ	=	stream function
g	=	gravitational constant
H	=	magnetic field intensity
K	=	reaction rate constant
L	=	length
M	=	Magnetic parameter
N_D	=	Damkohler number
N_{Gr}	=	Grashof number
N_{Sc}	=	Schmidt number
u, v	=	component of velocity in x and y direction respectively
x, y	=	distance in system of Cartesian coordinates

Greek symbols

η	=	similarity variable
ϕ	=	angle
θ	=	dimensionless concentration
ρ	=	density
τ_{yx}	=	shear stress
μ	=	magnetic permeability
μ_0	=	Casson's viscosity
ν	=	kinematic coefficient of viscosity
τ_0	=	yield stress

Suffix

w and ∞ refer to wall and bulk conditions

MASS TRANSFER IN TIME DEPENDENT BLOOD FLOW

G. Akay

Polymer Research Institute, Department of Chemistry, Middle East Technical University, Ankara, Turkey

INTRODUCTION

The particulate nature of blood is usually ignored in the analysis of mass transfer in the cardiovascular system; see for example, reviews by Middleman (1972), Lightfoot (1974) and Fletcher (1978). However, both the flow field and mass transfer rates may be modified significantly by micro-rotation of the red blood cells responsible for the facilitated oxygen transport to the tissue. The exact analysis of mass transfer in the cardiovascular system, taking into account the particulate nature of the blood, is obviously very difficult because of the complex geometry of the red blood cells and the complexity of flow conditions. One promising approach in describing the particulate nature of blood is the microcontinuum approach in which blood is treated as a suspension containing spherical non-deformable particles (red blood cells) whose rotational motion is described through a dynamic kinematical variable, called spin vector. The exact solution of the time dependent uniaxial laminar flows of micropolar fluids are given in a series of papers by Arıman and his colleagues, see for example, Arıman et al.(1974). However, in these studies, a spin boundary condition has to be assumed in order to be able to solve the necessary equations. The axial velocity profile is strongly dependent on this spin boundary condition, but the lack of any definite spin boundary condition, places limitations on the application of polar fluids as a model for the flow of suspensions.

In this study, we assume a two-phase model (to describe blood flow) in which a micropolar fluid (core fluid) is surrounded by less viscous Newtonian plasma fluid (annular fluid). This model avoids the spin boundary condition at the wall and allows a more accurate description of the diffusion processes. In single-phase description of blood flow, a mean diffusion coefficient for the dissolved species in blood is assumed. Furthermore, the effect of the micro-rotation of the blood cells on the mass transfer rate of oxygen is not considered, despite the fact that mass transfer to rotating bodies increases with the angular velocity (see for example Ellison and Cornet, 1971). Micropolar fluids can be conven - iently used to include the effect of cell rotation on the mass transfer rate of gases by assuming that the diffusion coefficient in whole blood is also a function of the cell spin.

TWO-PHASE MODEL OF TIME-DEPENDENT BLOOD FLOW

Two-phase model of blood flow has in fact been proposed earlier and recently extended by Chaturani and Upadhya (1979) for fluids with couple stress in time independent flows. Chaturani and Upadhya (1979) also give an account of the previous studies of steady two-phase blood flow. As regards the time-dependent stratified multiphase flow of Newtonian and various non-Newtonian fluids, a numerical solution has been given recently by Akay (1979) and this method is modified to apply to the time-dependent stratified two-phase flow of Newtonian annular fluid and micropolar core fluid.

Let us consider the two-phase flow of two immiscible fluids of equal density in which a micropolar core fluid is surrounded by less viscous Newtonian plasma fluid in a circular capillary with radius R. The interface is ripple-free and forms a circle of radius R_1. The physical quantities are referred to the cylindrical polar coordinates (r, θ, z) in which z-axis coincides with the axis of the capillary. This circular cylinder is surrounded by the tissue cylinder (Krogh's tissue cylinder model) of radius R_2.

The flow is brought about by the application of a time depen - dent pressure gradient $p(t) = p_0 E(t)$ where p_0 and $E(t)$ are the time independent and time dependent parts of the pressure gradient and t is time. Hence the suitable velocity and micro-rotation distributions are

$$w_i = (0, w(t,r), 0) \; ; \; v_i = (0, 0, v(t,r)) \qquad 0 < r < R_1 , \qquad |1|$$

$$u_i = (0, 0, u(t,r)) \qquad R_1 < r < R , \qquad |2|$$

where w_i, v_i are the micro-rotation and velocity vectors for the micropolar fluid and u_i is the velocity vector for the Newtonian fluid in the annular region. $w(t,r)$, $v(t,r)$ and $u(t,r)$ are the only non-zero components of the vectors w_i, v_i and u_i .

The equations of motion are given by (Arıman et al., 1974)

$$\rho \frac{\partial v}{\partial t} = p_0 E(t) + (\eta + \mu) \left\{ \frac{\partial^2 v}{\partial r^2} + \frac{1}{r} \frac{\partial v}{\partial r} \right\} + 2 \mu \left\{ \frac{\partial w}{\partial r} + \frac{w}{r} \right\} , \qquad |3|$$

$$\rho \psi \frac{\partial w}{\partial t} = \gamma \left\{ \frac{\partial^2 w}{\partial r^2} + \frac{1}{r} \frac{\partial w}{\partial r} - \frac{w}{r^2} \right\} - 4 \mu w - 2 \mu \frac{\partial v}{\partial r} , \qquad |4|$$

for the region $0 < r < R_1$ and for the annular region $R_1 < r < R$,

$$\rho \frac{\partial u}{\partial t} = p_0 E(t) + \mu \left\{ \frac{\partial^2 u}{\partial r^2} + \frac{1}{r} \frac{\partial u}{\partial r} \right\} \qquad |5|$$

The equation of state for micropolar fluids is (Eringen, 1966)

$$q_{ij} = (-p' + \lambda v^n_{,n}) g_{ij} + (\eta - \mu)(v_{i,j} + v_{j,i}) + 2\mu (v_{j,i} - \varepsilon_{ijn} w^n) \qquad |6|$$

$$m_{ij} = \alpha w^n_{,n} g_{ij} + \beta w_{i,j} + \gamma w_{j,i} \qquad |7|$$

where as, for Newtonian fluid, the equation of state is

$$p_{ij} = -p' g_{ij} + \eta (v_{i,j} + v_{j,i}) \qquad |8|$$

Here, we use standard tensor notation throughout; covariant suffices are written below, contravariant suffices are written above and the usual summation convention for repeated suffices is assumed. Covariant differentiation of vectors and tensors is denoted by a comma. In Eqs. $|3| - |8|$, ε_{ijn} is the alternating tensor, q_{ij} and p_{ij} are the stress tensors, m_{ij} is the couple stress tensor, g_{ij} is the metric tensor, p' is the pressure, η is the viscosity of the plasma fluid, $\alpha, \beta, \gamma, \lambda, \mu$ and ψ are the material constants of the micropolar fluid. We note that $p_{ij} = p_{ji}$ and $q_{ij} \neq q_{ji}$ when $i \neq j$.

Boundary and Initial Conditions

     The following boundary and initial conditions apply;

$$u(t,R) = 0, \quad u(t,R_1) = v(t,R_1), \qquad |9|$$

$$\frac{\partial v}{\partial r} \{t,0\} = 0, \quad w(t,0) = 0, \qquad |10|$$

$$q_{13}(t,R_1) = q_{31}(t,R_1), \quad p_{13}(t,R_1) = q_{13}(t,R_1) \qquad |11|$$

Boundary conditions given by $|11|$ require that the shear stress should be continuous across the interface. Using the equation of state, it can be shown that Eq. $|11|$ is satisfied if

$$w(t,R_1) = -\frac{1}{2} \frac{\partial v}{\partial r} \{t,R_1\} \qquad \frac{\partial u}{\partial r} \{t,R_1\} = \frac{\partial v}{\partial r} \{t,R_1\} \cdot \qquad |12|$$

     In order to obtain the solution of the single-phase flow in a capillary, one has to assume a spin boundary condition on the wall. No-spin boundary condition is usually assumed but Arıman et al. (1974) assume that $\partial(rw)/\partial r = 0$ at the wall. As seen here, no specific spin boundary condition is required and it is provided by the continuity of the shear stress at the interface. This approach however, introduces a new parameter into the problem; the thickness of the cell free plasma layer. This parameter is quite well studied experimentally and it can be predicted theoretically from the solution of the stress induced diffusion equation.

The initial conditions are;

$$u(0,r) = u_s(r), \quad v(0,r) = v_s(r), \quad w(0,r) = w_s(r) \qquad |13|$$

For start-up flows, $u_s=v_s=w_s=0$.

Solution by Finite Difference Technique

The numerical solution of Eqs. $|3| - |5|$ subject to the boundary and initial conditions given by Eqs. $|9| - |13|$ can be obtained by using the finite difference technique given by Townsend (1973) and Akay (1977,1979). A single tri-diagonal finite difference equation is obtained for the axial velocity distributions $u(t+\Delta t,r)$ and $v(t+\Delta t,r)$ from Eqs. $|3|$ and $|5|$ after using the last equation in Eq. $|12|$. However, in this tri-diagonal equation, $w(t+\Delta t,r)$ is estimated from $w(t,r)$ and $w(t-\Delta t,r)$ by linear extrapolation. Here Δt is the step length in time. Once $u(t+\Delta t,r)$ and $v(t+\Delta t,r)$ are obtained, the correct distribution of $w(t+\Delta t,r)$ is obtained by using Eq. $|4|$. It involves the solution of another tri-diagonal equation. In principle, this technique can be applied to any number of phases in co-current flow.

MASS TRANSFER IN TIME DEPENDENT BLOOD FLOW: OXYGENATION OF TISSUE

Quantitative analysis of mass transfer in the cardiovascular system (such as oxygenation of tissue or renal function) or mass transfer in extracorporeal circulation, requires the solution of the diffusion equation. Here, we apply the numerical technique given by Akay (1980) to the oxygenation of tissue. The relevant equations to be solved are (Lightfoot, 1974, p.334; Eqs. 5.4.1 and 5.4.2 are modified to fit the case considered here)

$$\frac{\partial c}{\partial t} = - v \frac{\partial c}{\partial z} + \frac{1}{m+1} \left(\frac{1}{r} \frac{\partial}{\partial r} \left\{ D_1 r \frac{\partial c}{\partial r} \right\} + \frac{\partial}{\partial z} \left\{ D_2 \frac{\partial c}{\partial z} \right\} \right), \; 0 < r < R_1 \qquad |14|$$

$$\frac{\partial b}{\partial t} = - u \frac{\partial b}{\partial z} + D_3 \left\{ \frac{\partial^2 b}{\partial r^2} + \frac{1}{r} \frac{\partial b}{\partial r} + \frac{\partial^2 b}{\partial z^2} \right\} \qquad R_1 < r < R \qquad |15|$$

$$\frac{\partial a}{\partial t} = D_4 \left\{ \frac{\partial^2 a}{\partial r^2} + \frac{1}{r} \frac{\partial a}{\partial r} + \frac{\partial^2 a}{\partial z^2} \right\} \qquad R < r < R_2 \qquad |16|$$

where m is the oxygen-binding capacity of blood (Lightfoot,1974), $a(t,r,z)$, $b(t,r,z)$ and $c(t,r,z)$ are the oxygen concentration in tissue, cell free plasma layer and core fluid respectively, D_3 and D_4 are the diffusion coefficients of oxygen in plasma and in tissue, where as D_1 and D_2 are the diffusion coefficients of oxygen in blood in radial and axial directions. One would expect D_3 and D_4 to be independent of concentration and flow field, but D_1 and D_2 may be taken as a function of the invariant of the spin vector W_i. In particular, let us assume that

$$D_1 = D_2 = D_0 (1 + \kappa I^\delta) \equiv D (t,r) \qquad\qquad |17|$$

where I is the invariant of W_i ($I = W_i w^i$) and D_0, κ and δ are constants. The results of Ellison and Cornet (1971) indicate that $\delta \approx 1/4$. Note that in our case $I = w^2$.

Boundary and Initial Conditions and Numerical Solution

The following boundary and initial conditions apply,

$$c (t,R_1,z) = b (t,R_1,z) \qquad\qquad |18|$$

$$D (t,R_1) \frac{\partial c}{\partial r}\{ t,R_1,z \} = D_3 \frac{\partial b}{\partial r}\{ t,R_1,z \} \qquad\qquad |19|$$

$$D_3 \frac{\partial b}{\partial r} \{ t,R_1,z \} = D_4 \frac{\partial a}{\partial r} \{ t,R,z \} \qquad\qquad |20|$$

$$a (t,r,0) = a_0 (t,r); \quad b (t,r,0) = b_0 (t,r); \quad c(t,r,0)=c_0(t,r) \quad |21|$$

$$a(0,r,z) = a_1(t,r) ; \quad b(0,r,z) = b_1(t,r) ; \quad c(0,r,z) = c_1(t,r) \qquad |22|$$

$$\frac{\partial c}{\partial r} \{t,0,z\} = 0, \quad \frac{\partial a}{\partial r} \{t,R_2,z\} = 0 \qquad\qquad |23|$$

$$\frac{\partial a}{\partial z} \{t,r,z\} = \frac{\partial b}{\partial z} \{t,r,z\} = \frac{\partial c}{\partial z} \{t,r,z\} = 0 \quad \text{when} \quad z = 0 \text{ or } z = L \qquad |24|$$

Here, we assume that arterial blood enters the capillary at $z = 0$ and venous blood leaves at $z = L$.

Eqs. $|14|-|16|$ subjected to the boundary and initial conditions given by Eqs. $|18|-|24|$ are solved by using the numerical tech nique given by Akay (1980). The use of interface mass transfer conditions, Eqs. $|19|$ and $|20|$, results in a single tri-diagonal finite difference equation, for the concentration profile in the region $0 \leq r \leq R_2$.

Estimation of the Thickness of the Cell Free Plasma Layer

Stress induced diffusion of macromolecules results in concentration depletion in high shear rate regions due to the migration of macromolecules to the low shear rate regions. Diffusion of macromolecules is considered to be driven by gradients of entropic potential arising from distortion of molecular conformation by deformation, as well as by gradients of concentration of individual molecular weight species, Tirrell and Malone (1977). The same concept can also be applied to the migration of red blood cells with the aim of estimating the cell concentration distribution in flow.

If $s(t,r)$ is the concentration of the red blood cells (or all other enclusions; hematocrit) during flow, it can be shown that in a long capillary ($\partial/\partial z \to 0$), $s(t,r)$ is governed by, Akay (1980-b)

$$\frac{\partial s}{\partial t} = D_5 \left\{ \frac{\partial^2 s}{\partial r^2} + \frac{1}{r}\frac{\partial s}{\partial r} + g_0 \left(\frac{\partial s}{\partial r}\frac{\partial h}{\partial r} \right)^n + s \frac{\partial^2 h}{\partial r^2}^n + \frac{s}{r}\frac{\partial h}{\partial r}^n \right) \right\} \qquad |25|$$

where D_5 is the diffusion coefficints of the cells in plasma, n and g_0 are two constants associated with stress induced diffusion, h is the second invariant of the rate of deformation tensor. In one dimensional axial flows $h = (\partial v/\partial r)^2$.

The thickness of the cell free layer can be estimated from the solution of Eq. $|25|$ subject to the appropriate boundary conditions by assuming that in this region the concentration of the cells is below a certain value.

ACKNOWLEDGMENT:This research is supported by the Science Research Foundation of the Middle East Technical University, Project no:79-01-03-505.

REFERENCES

Akay,G.,1977, Numerical solutions of some unsteady laminar flows of viscoelastic fluids in concentric annuli with axially moving boundaries, Rheol. Acta, 16:589.
Akay,G.,1979, Non-steady two-phase stratified laminar flow of polymeric liquids in pipes, Rheol. Acta, 18:256.
Akay,G.,1980, Numerical solution of unsteady convective diffusion with chemical reaction in time dependent flows , Chem. Eng. Comm. 4(1).
Akay,G.,1980-b, Mechanochemical degradation of macromolecules during laminar flow, presented in the VIII th Int. Congress on Rheology.
Arıman,T., Turk,M.A., and Sylvester,N.D.,1974, On steady and pulsatile flow of blood, J. Appl. Mech., 41:1.
Chaturani,P., and Upadhya,V.S.,1979, A two-fluid model for blood flow through small diameter tubes, Biorheol., 16:109.
Ellison,B.T., and Cornet,I., 1971, Mass transfer to rotating disk, J. Electrochem. Soc., 118:68.
Eringen,A.C.,1966, Theory of micropolar fluids, J.Math.Mech., 16:1.
Fletcher,J.E.,1978, Mathematical modeling of the microcirculation, Math. Biosci., 38:159.
Lightfoot,E.N.,1974, "Transport Phenomena and Living Systems",Wiley Interscience, New York.
Middleman,S.,1972, "Transport Phenomena in the Cardiovascular System" Wiley Interscience, New York.
Tirrell,M., and Malone,M.F.,1977, Stress-induced diffusion in macromolecules, J. Polym. Sci., Polymer Physics Edition, 15:1569.
Townsend,P.,1973, Numerical solutions of some unsteady flows of elastico-viscous liquids, Rheol. Acta, 12:13.

ERYTHROCYTE ELASTICITY IN MUSUCLAR DYSTROPHIC MICE

Yannis Missirlis, Marianne Vanderwel,
Ludmila Weir and Michael Brain

Departments of Engineering Physics and Medicine
McMaster University
Hamilton, Ontario

INTRODUCTION

Various abnormalities of the erythrocyte membrance structure
and properties have been reported in patients with muscular dys-
trophy diseases as well as in animal models of these diseases,
such as chickens and mice. These biophysical and biochemical abnormalities of the erythrocyte membrane include morphologic changes[1,2]
altered lipid composition[3], enzyme activities[4,5], transport properties[6], deformability[7-9], etc. However, there is no consensus
regarding these findings[7,9] while the relationship between the
biochemical and biophysical alterations of these erythrocyte
membranes is unknown.

In this report the stress-strain relationships of normal and
dystrophic erythrocytes are examined by using two different methods:
the flow channel shear deformation technique, and the Nuclepore
filter partial aspiration of erythrocytes.

MATERIALS AND METHODS

Blood samples were obtained from the tail of C57 black strain
normal mice and D1/2J strain dystrophic mice. The blood was dil-
uted in Tris buffer containing glucose and bovine serum albumin as
well.

The detailed description of the flow chamber apparatus is found
elsewhere[10,11]. In summary, clean glass coverslips, coated with
polylysin, formed one of the two parallel plate chamber; the other
was a machined cavity on a rectangular piece of perspex. The dis-

tance between the plates was approximately 150 µm. Erythrocytes
were allowed to settle on the coverslips, where they partially
adhered. The buffer fluid flow was controlled with a syringe pump.
The pressure drop across the 2.352 cm long chamber was measured with
a sensitive differential pressure transducer (Validyne).

   The shear stress on the channel floor (coverslip), approximating the shear stress acting on the free surface of the cell (cell
thickness 2–3 µm << 150 µm), τ_s, is given by

$$\tau_s = \frac{\Delta P \cdot h}{2L}$$

with

$$h^3 = \frac{12 \cdot L \cdot n}{w} \cdot \frac{Q}{\Delta P}$$

where

   ΔP: pressure drop

   h: height of channel

   L: length of channel

   W: width of channel

   n: buffer fluid viscosity

   Q: volumetric flow rate

Projections of microphotographic slides were used to determine the
length and maximum width of all point–attached cells at various
flow rates. Therefore for each erythrocyte a series of τ_s values
was determined and corresponding values of the uniaxial extension
ratio, $\lambda_x = \ell/\ell_o$, where ℓ_o is the original length, were measured
and the uniaxial strains, $\varepsilon_x = \frac{1}{2}(\lambda_x^2 - 1)$, were obtained.

   The detailed description of the Nuclepore filter apparatus is
found elsewhere[12]. In summary, Nuclepore polycarbonate filters,
25 mm in diameter with pores 0.8 µm in diameter and desnity 2×10^6
pores/sq cm were sealed onto glass tubes. A very dilute suspension
of erythrocytes in Tris buffer was placed in the glass tube and a
controlled positive hydrostatic pressure was maintained across the
filter by a system of Pasteur pipettes connected to vacuum lines.
The partially aspirated cells were fixed in position with 1% glutar-
aldehyde, postfixed in O_sO_4, dehydrated and the filter was dissolved
with chloroform with the cells adhering to a glass coverslip, their

"tongues" facing upwards. The cells were then coated and examined by scanning electron microscopy at a magnification of 12,000 x.

By selecting erythrocytes with a single tongue protruding from or near the center of the cell and measuring the length and diameter of the tongue at two different angles the actual length-over-radius ratio, D_p/R_p of each tongue could be measured as:

$$\frac{D_p}{R_p} = \frac{H/R_p - 1}{\sin\Theta} + 1 - \cot\Theta$$

where H: length of tongue at angle $\Theta°$

R_p: $\frac{1}{2}$ diameter of tongue at angle $0°$

Therefore for each pressure (2,4,6 and 8 mm H_2O) an average value of D_p/R_p could be calculated for 20-30 erythrocytes from the same sample.

RESULTS

Flow Channel

The measurements and the analysis of the data were made without knowing a priori whether a particular sample was normal or dystrophic. Altogether 56 normal and 51 dystrophic mice erythrocytes were considered for analysis. The shear stress, τ_s, was plotted as a function of the shear strain, ε_x, in cartesian coordinates, with the resulting relationship being clearly nonlinear. A similarly nonlinear relationship has been shown also by others[10,13] for human erythrocytes.

If the elastic constant (shear modulus), $\mu = \tau_s/\varepsilon_x$, is considered not constant but a linear function of the strain product $\varepsilon_x\varepsilon_y$[14], i.e. $\mu = a + b\varepsilon_x\varepsilon_y$, and assuming that the membrane is incompressible, i.e. $\lambda_x\lambda_y = 1$, then

$$\tau_s = a\varepsilon_x + \frac{b\varepsilon_x^3}{2\varepsilon_x + 1}$$

The mean values of the two coefficients are a = 2.76 for the normal erythrocytes and a = 2.39 for the dystrophic erythrocytes whereas the mean values of b = 51.07 for the normal and b = 40.55 for the dystrophic erythrocytes with a correlation coefficient $r^2 = 0.956$ for the normal and $r^2 = 0.979$ for the dystrophic cells.

In order to compare these mechanical characteristics of the two populations of erythrocytes statistically one should have a strain-stress relationship, i.e. $\varepsilon_x = f(\tau_s)$, since τ_s is the independent variable. The equation that best-fitted our data was

$$\varepsilon_x = \alpha\tau_s + \beta\tau_s^2 + \gamma\tau_s^3$$

where α, β and γ are constants. The statistical analysis of variance indicated that a large percentage of the variance between the cells, the samples and the cell populations was due to the randomness of the data. However calculation of the 95% confidence levels about the difference of the strain-stress curves for the normal and dystrophic erythrocytes revealed that for the range of stresses $0 < \tau_s \leq 5.5$ dyn/cm^2 the dystrophic mice erythrocytes were slightly more deformable than the normal ones.

Nuclepore Filter Aspiration

A typical erythrocyte aspirated at 4 mm H$_2$0 pressure is shown in Figure 1. Whereas the fluid channel experiments provided stress-strain relationships on few individual erythrocytes, the Nuclepore filter aspiration method resulted in a pressure-deformation (or stress-strain) relationship on many erythrocytes considered together.

When the mean values of D_p/R_p are plotted as a function of pressure the resulting relationship is linear, if the pressure does not exceed 10 mm H$_2$0. If furthermore the dimensionless deformability parameter D_p/R_p is plotted against the dimensionless quantity $P.R_p/\mu$, where μ is considered constant over the range of extension ratios measured, i.e. $1 \leq D_p/R_p \leq 3$, this relationship is also linear.

This linearity has been established previously[12] for human erythrocytes and is in good agreement with the linear elastic membrane behaviour found in micropipet aspiration experiments of individual erythrocytes[7,14,15].

The elastic shear modulus, μ, calculated from the linear relationship that best-fitted the data was $\mu = 0.012$ dyn/cm for the dystrophic erythrocytes and $\mu = 0.010$ dyn/cm for the normal erythrocytes. These values, however, are not significantly different although the dystrophic mice erythrocytes appear to be slightly more deformable than the normal controls.

DISCUSSION

The flow channel technique is purported to estimate the

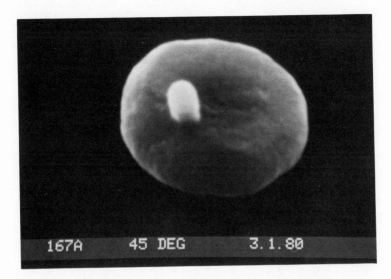

Fig. 1: A dystrophic mouse erythrocyte aspirated in a 0.8 μm
         Nuclepore filter pore, under 4 mm H_2O pressure, as seen
         by SEM at 12,000 x magnification.

elastic shear modulus of the adhered erythrocytes in uniaxial
loading[10]. The experimental data of this work as well as of other
investigations[13,16] clearly indicate a non-linear stress-strain
relationship, whereas it has been established that the elastic
shear modulus of the erythrocytes is constant over a wide range
of extension ratios. It has been suggested[14] that μ may be a
function of strain and therefore, not constant. However, it is
more likely that inherent limitations of the flow channel technique such as unknown and uncontrolled area of cell attachment,
may be responsible for the observed nonlinearity. Furthermore
is is possible that under these experimental conditions the cell
surface is under biaxial strains resulting in the observed non-
linear uniaxial stress-strain relationship.

     The Nuclepore filter partial aspriation technique confirmed
that a linear relationship exist between the deforming forces
and the resulting deformation, which has been shown to be the
case using this technique for human erythrocytes[12].

     Both techniques, however, failed to show a significant dif-
ference in the elastic properties between normal and dystrophic
mice erythrocytes. Two other studies[17,18] have reported membrane
differences between normal and dystrophic mice erythrocytes but
failed to identify the observed differences to particular parameters characteristic of the membrane. Therefore it is difficult
to compare our results with those studies. In view of the fact
that the findings on the mechanical properties of erythrocytes

from dystrophic patients contradict each other[7,9] it is important
that carefully designed experiments should be conducted to investigate well specified properties of the dystrophic erythrocyte membranes.

REFERENCES

1. D.W. Matheson, J.L. Howland, Erythrocyte Deformation in Human
    Muscular Dystrophy, Science 184:165,(1974).
2. S.E. Miller, A.D. Roses, S.H. Appel, Scanning Electron Micro-
    Scopy Studies in Muscular Dystrophy, Arch. Neurol. 33:172,(1976).
3. R.I. Sha'afi, S.B. Rodan, R.L. Hintz, S.M. Fernandes, G.A.
    Rodan, Abnormalities in Membrane Microviscosity and Iron Trans-
    port in Genetic Muscular Dystrophy, Nature 254:525,(1975).
4. A.D. Roses, M.H. Herbstreich, S.H. Appel, Membrane Protein
    Kinase Alteration in Duchenne Muscular Dystrophy, Nature 254:
    350,(1975).
5. A.C. Roses, M.J. Roses, S.E. Miller, K.J. Hull, S.H. Appel,
    Carrier Detection in Duchenne Muscular Dystrophy, New Eng. J.
    Med. 294:193 (1976).
6. J.L. Howland, Abnormal Potassium Conductance Associated with
    Genetic Muscular Dystrophy, Nature 251:724,(1974).
7. Y.F. Missirlis, I.L. Kohn, J.D. Vickers, M.P. Rathbone, D.H.K.
    Chui, A.J. McComas, M.C. Brain, Alterations in Erythrocyte
    Membrane Material Properties: A Marker of the Membrane Abnor-
    mality in Human and Chicken Muscular Dystrophy. In W.C.
    Kruckeberg, J.W. Eaton, G.J. Brewer (editors) Erythrocyte
    Membranes New York, A.R. Liss,(1978).
8. A.K. Percy, M.E. Miller, Reduced Deformability of Erythrocyte
    Membranes from Patients with Duchenne Muscular Dystrophy,
    Nature 258:147,(1975).
9. H. Somer, S. Chien, L.A. Sung, A. Thurn, Erythrocytes in
    Duchenne Dystrophy: Osmotic Fragility and Membrane Deforma-
    bility, Neurology 29:519,(1979).
10. R.M. Hochmuth, N. Mohandas, P.L. Blackshear, Measurement of the
    Elastic Modulus for Red Cell Membrane using a Fluid Mechanical
    Technique, Biophys. J. 13:747,(1973).
11. Y.F. Missirlis, M. Vanderwel, and M.C. Brain, Membrane Elas-
    ticity of Erythrocytes from Normal and Dystrophic Mice,
    (submitted for publication).
12. Y.F. Missirlis, M.C. Brain, An Improved Method for Studying
    the elastic Properties of Erythrocyte Membranes, Blood 54:
    1069,(1979).
13. S. Chien, Principle and Techniques for Assessing Erythrocyte
    Deformability, Blood Cells 3:71,(1977).
14. E.A. Evans, New Membrane Concept Applied to the Analysis of
    Fluid Shear and Micropipette-Deformed Red Blood Cells, Biophys.
    J., 13:941,(1973).
15. S. Chien, K.P. Sung, R. Skalak, S. Usami, A. Tozeren, Theoreti-

cal and Experimental Studies on Viscoelastic Properties of
Erythrocyte Membrane, Biophys. J. 24:463,(1978).

16. T.C. Hung, L.W. Shen, T. Akutsu, N.H.C. Hwang, Some Physical
    Characteristics of Normal Erythrocyte Membrane from Experi-
    mental Animals and Humans, Trans. A.S.A.I.O. XXIV:573,(1978).

17. P. Di Stefano, H.B. Bosmann, Erythrocyte Membrane Abnormality
    in Muscular Dystrophy, Cell Biol. Intern. Reports 1:375,(1977).

18. R.F. Morse, J.L. Howland, Erythrocytes from Animals with
    Genetic Muscular Dystrophy, Nature 245:156,(1973).

KINETICS AND MORPHOLOGY OF AGGREGATION OF RED CELLS:

PREPARATIONS FOR THE NASA SPACELAB

Leopold Dintenfass

Haemorheology & Biorheology Department
KMI Sydney Hospital, Sydney 2000, and
Department of Medicine, University of
Sydney, Sydney 2006, Australia

INTRODUCTION

Experiment MPS77F113 , scheduled for NASA Space Lab 3, has as
its objective to study aggregation and kinetic of aggregation of red
cells under near-zero gravity. In the years 1978 and 1978 it was
possible to construct the instrument required (parallel-plate slit-
-capillary viscometer, equipped with cameras and a microscope) for
such a study; furthermore, it was possible to carry out preliminary
experiments to to obtain a background information on aggregation of
red cells, their kinetics and morphology under 1 gravity condition.

INSTRUMENTATION

The 'heart' of the instrument is a cell formed of two parallel
plates made of optical quality glass ('Crown' glass) and polished
to tolerance of few light wave length only. The size of plates is
about 15 x 15 cm, and thickness of glass plates is about 1.3 cm.
The plates are spaced by plastic 'spacer' of 12.5 micrometer and 55
micrometer thickness. Each plate contains a semicircular channel of
0.2 cm diameter for distribution or collection of blood which flow
like a sheet between the plates. Plates are held together by means
of metal bars and also by a layer of epoxy resin. Plastic tubing of
external diameter of 0.2 cm connects the plates to hypodermic syringes
containing blood or saline. This is schematically indicated on Fig. 1.
One syringe is used at a time; high speed, low speed, and zero speed
are used; simultaneously measurements of pressure drop and of the
temperature within the plates are carried out; also simultaneously
photography of the blood is carried out using Olympus cameras; one
camera is equipped with bellows, and the other with a microscope.

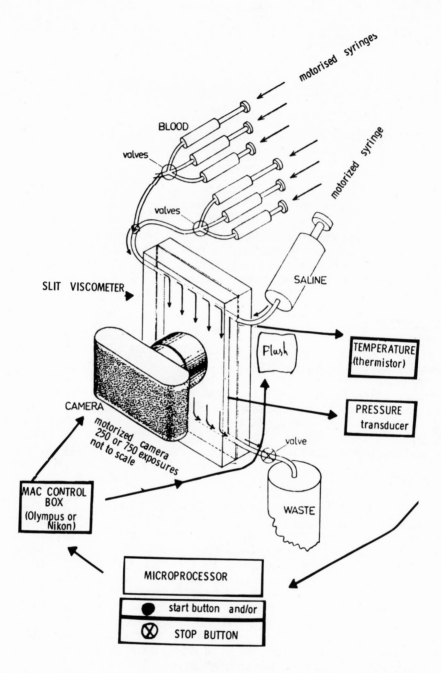

Fig. 1.   A schematic representation of the instrument intended for
study of aggregation of red cells and viscosity of blood
under zero and one gravities. Note that the microscope is
not included into the schema above.

Fig. 2   A photograph of the slit-capillary. Note a well in the lower
left hand corner; this well accommodates the objective of
the microscope.

OM2 camera, equipped with bellows, has a macro lens 20 mm, f 3.5, and
gives magnification x10; high magnification, of up to x300, is
supplied by a Zeiss microscope equipped with Leitz objective L32/0.40
and Reichert condenser 0.95. Two flash guns Auto Quick 310 (Olympus)
are employed. The objective of the microscope fits into a well of
one of the plates forming the slit-capillary. The 'window' at the
bottom of the well is 1.3 mm thick (Fig. 2).

Photography is carried out at preset time intervals using Mac
control box. During stasis (zero flow condition), photography is
carried out at 30 sec intervals for a period of 10 minutes, both at
low and at high magnifications.

Although at this stage (February 1980) the instrument operates
only in manual mode, it is expected to have it completely automated
by April 1980. A microprocessor will be used.

During the current laboratory experiments a mechanized syringe
(Braun) is being used; however, for the future experiments we will
use a multi-syringe revolving drum which will allow a sequential
study of eight blood samples.

Cameras, flash guns, motorized syringe, control boxes and the
microprocessor operate on DC current obtained from batteries.

At this stage of the project, all experiments are run at 21°C,
although we intend to include later a facility for 37°C experiments.

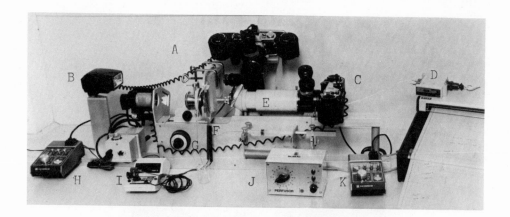

Fig. 3. A photograph of the slit-capillary viscometer in the
laboratory set up. A: Olympus camera with 250 back; B: flash
gun; C: second camera placed on the microscope phototube;
D: recorder (used only in the ground laboratory experiments -
will be replaced by the microprocessor memory for the space
experiments); E: Zeiss phototube; F: glass plates; note
immediately left the Reichert condenser; H and K: Mac control
boxes; I: voltmeter, temperature read-out; J: Braun motorized
syringe.

EXPERIMENTATION

     The freshly drawn blood (by clean venepuncture) is anticoagulated
with EDTA (sequestrene) and adjusted, using native plasma, to the
constant haematocrit of 30 per cent. The slit capillary is filled with
saline which is replaced by blood sample. Fig. 4 shows macro and micro
photographs of blood sample obtained from a lymphoma patients, at
stasis time of 1 and 5 minutes. Note that an increase of aggregation
of red cells is obvious at longer stasis time when observed at low
magnification; such increase in the degree of aggregation is much
less noticeable at micro level. Observe formation of long rouleaux
consisting of 10 to 20 red cells in face to face linear structures;
other red cells form clumps containing from from 50 to 100 cells.

     It does appear that aggregation at microstructure˜level is more
rapid than at macro structure level.

     Experiments, photographs of which are not included here, show
that micro structure is not visible when gaps of 55 micron (micrometer) are used; on the other hand, the macro structure obtained in
the gaps of 12.5 and 55 micron (micrometer) are nearly identical.
However, sedimentation of red cell aggregates is more noticeable in
slits of larger gap.

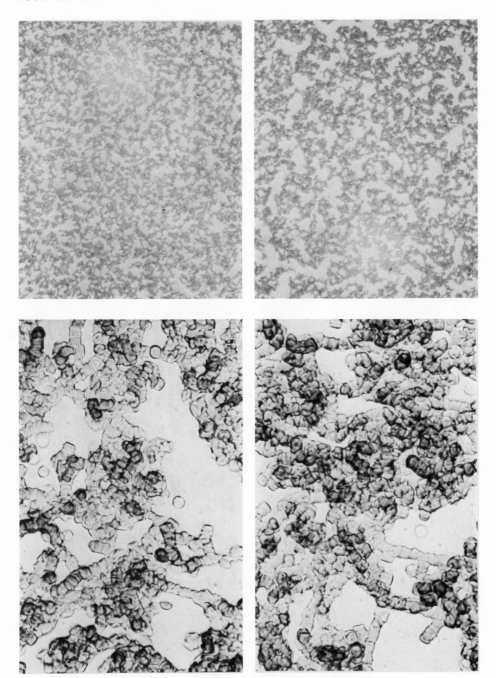

Fig. 4. Macro photography (top) and micro photography (bottom) of
a blood sample obtained from a lymphoma patient. Photographs
on the left were taken after 1 min stasis; at the right,
after 5 min stasis. Camera magnifications x10 and x150.

DISCUSSION

It has been tacitly assumed that aggregation of red cells can take place within few seconds; however, the present study shows that time of formation of an aggregate requires at least five minutes (as indicated by the photographs presented). Other current work suggests that time periods of uo to 10 or 12 minutes might be rquired for red cell aggregation.

The area of slit-capillary is rather large when compared with the field of usual microscopic or rheomicroscopic techniques. As consequence of larger area of the slit-capillary, the size of red cell aggregates is unimpeded, and on some occasions aggregates of up to few millimeters in diameter have been observed. The morphology of aggregates may differ. The case of lymphoma, presented in this paper, show structural units of aggregates formed either from rouleaux or from compact clumps of red cells. Other experiments indicated that the random and loose structure is more typical of normal blood, while compact clumps are encountered mainly in the patients with cancer.

This study forms a part of the background preparations for the Space Lab investigation. The detailed theoretical background for the study of kinetics of aggrégation have been reported (Dintenfass, 962), while the details/of the concept of space investigation have been supplied recently ,(Dintenfass, 1979a,b).

ACKNOWLEDGMENTS

Photography has been carried out by Mr. Henry Jedrzejczyk. Grants-in-aid have been supplied by the Clive and Vera Ramaciotti Foundations and the National Heart Foundation of Australia.

REFERENCES

Dintenfass, L., 1962, A study in thixotropy of concentrated
        pigment suspensions. Part II. Thixotropic recovery and non-
        -equilibria states. Rheol. Acta, 2: 187
Dintenfass, L., 1979a, Aggregation of red cells and blood viscosity
        under near-zero gravity. Biorheology, 16: 29
Dintenfass, L., 1979b, Aggregation of red cells and blood viscosity
        under zero and one gravities - instrumentation for Space
        Lab 3. Australasian Phys. Sci. Med., 2: 437

SOME RHEOLOGICAL ASPECTS OF ENZYME MANUFACTURE

J.J.Benbow

Imperial Chemical Industries Limited, U.K.

(Abstract)

The increasing interest in manifactured enzymes introduces a new area of importance to applied rheologists. Both in the preparation and use of these novel materials rheology plays a major role and an understanding of how it affects production processes and product quality is extremely important. This paper deals with the harvesting of Glucose Isomerase. This is an enzyme used to convert glucose to fructose. Attention is focused on the extrusion forming step which shapes the flocculated cells into particles suitable for use in packed beds.

A description will be given of the effects of material composition on rheological parameters. The flow behavior will be compared to that of molten polymers and ceramic pastes. The effects of altering screw geometry and die size on output rate will be discussed. Finally the use of laboratory methods of measuring flow behavior on small samples to predict the performance of production equipment will be demonstrated.

SYNOVIAL FLUID RHEOLOGY, HYALURONIC ACID

AND MACROMOLECULAR NETWORK STRUCTURE

Henning Zeidler, Siegfried Altmann

Div. of Rheumatic and Metabolic Diseases
Medical Department, MHH
3000 Hannover 31, FRG

INTRODUCTION

Former investigations [1,2,3,4,5,6] of the synovial fluid with modern rheological methods clearly showed a high reduction of the viscosity and elasticity in rheumatoroid arthritis and other arthropathies. The reduction of the concentration and the molecular weight of the hyaluronic acid (HA) is the obvious cause of pathorheological changes. For rheumatoid arthritis the dissociation of a macro-molecular complex of HA and proteins [7,8] is also taken into consideration.

The current investigations have following purpose:

1) Comparison of the rheological properties of the normal and pathological synovial fluid with HA isolated out of the last mentioned synovial fluid.
2) The proof of intra and/or intermolecular changes by rheological measurements.
3) The description of the solution structure of the HA with regard to the conception of a macromolecular network structure.

METHODS

The measurement of flow curves under shear rates of 9.13 s^{-1} up to 9 13 s^{-1} were done with a Weißenberg rheogoniometer, type R 18 (Messrs. Sangamo Controls) with a cone/plate system. Normal force measurements and prestationary measurements were done with the same apparatus. The rotation rheometer 'Low Shear 100' with a rheomate 30 as driving unit (Messrs. Contraves, Stuttgart) functioning according to the couette principle was applied for the determination of the viscosity value at the zero shear rate (η_o) applying very low shear rates from 4.6 s^{-1} to 2.9 x 10^{-3}s^{-1}. The measurement of the intrinsic viscosity was done with the same rheometer.

The concentration of the hyaluronic acid was determined colorimetrically by means of the hexuronic acid method according to Bitter and Muir [9]. The determination of the sedimentation coefficient (s) was done in a Beckmann-Spinco-E-ultra centrifuge with schlieren optic at 60.000 rpm and 21 °C.

RESULTS

We have investigated 193 pathological synovial fluids, with 95 inflammatory arthropathies, 69 non-inflammatory and 29 crystal-induced synovitis. Compared with a 'normal synovial fluid' post mortem the flow curves are evidently reduced, the high deviation and intersection of the different diseases and diagnosis groups, however, hardly allow a precise diagnosis for the individual case. This is clearly demonstrated by Fig. 1, comparing the flow curves of the rheumatoid arthritis with meniscus lesions. The same applies to the normal forces which can be measured more rarely with inflammatory arthropathies than with non-inflammatory arthropathies. For measurable normal forces there do result rather insignificant differences between rheumatoid arthritis and meniscus lesions.

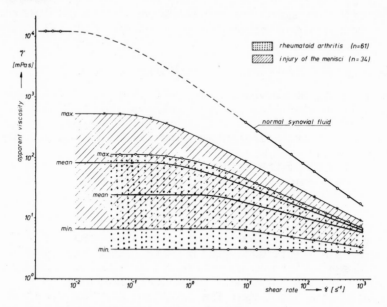

Fig.1: Flow curves in normal synovial fluid, rheumatoid arthritis and injury of menisci.

$(\vartheta = 37\,°C)$

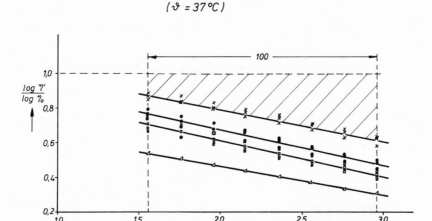

Fig.2 Master-curve with concentration variation.   □ Osteo-
arthritis (c = 79 to 158 mg/100 ml).   ● Rheumatoid arthritis
(c = 46 to 137 mg/100 ml).   ✕ Chondro calcinosis (c = 88 to
175 mg/100 ml). △ Normal synovial fluid (c = 277 mg/100 ml).

        The comparison of HA-concentrations of the normal synovial
fluid (277 mg ./100 ml) with the rheumatoid arthritis   (113 ∓
39 mg/100 ml), meniscus_lesions (137 ∓ 61 mg/100 ml) and crystal-
induced synovitis (113 ∓ 44 mg/100 ml) show as expected a clear
reduction of the pathological synovial fluids. The molecular
changes are a further cause for the reduced viscosity and normal
forces. This can be measured directly and quantatively by means of
a standardized procedure of the flow curves and the thus extracted master curves (Fig. 2). As to be able to compare the master
curves of the many samples statistically we calculated an area
(A) between master curves and ordinate section of 1.0 multiplying
the abscissa section simplified with the value 100.

        The surface measure of the master curves of the different
diagnoses are shown in table 1. The difference between inflammatory arthropathies and non-inflammatory arthropathies are not
significant. All values are reduced compared with the normal synovial fluid. There is a linear dependance between A and $[\eta]$
with a correlation coefficient of r = + 0.86. As for the calculation of the molecular weight of polymere in general $[\eta]$ is taken,
it can be first of all supposed that also the reduction of A in
the pathological articular effusions clearly shows a molecular
weight reduction of HA. In this sense the findings comply with the
generally known reduction of concentration and molecular weight of
the HA in pathological synovial fluids, but showing the difference

Table 1. Area of master-curve from normal synovial fluid,
rheumatoid arthritis and other joint diseases.

diagnosis		number of samples	area
normal synovial fluid		1	58
inflammatory joint diseases	rheumatoid arthritis	32	31 ± 7
	other arthritis	7	25 ± 6
	arthritis of unknown origin	6	32 ± 6
			30 ± 7
non-inflammatory joint diseases	injury of the menisci	18	34 ± 10
	trauma	4	34 ± 12
	osteoarthritis	3	42 ± 4
			35 ± 10
crystal-induced synovitis		8	28 ± 6

between inflammatory and non-inflammatory arthropathies
not as clear as it could have been expected.

An important additional aspect results from the investigation of two HA specimens isolated out of the synovial fluid of rheumatoid arthritis and meniscus lesions. Compared with the native punctate the flow curves
of the isolated HA are evidently lower and the normal forces noted
before are now no longer measurable (table 2). Also [?] has
dropped whilst the sedimentation coefficient has practically not
changed. This outstanding finding indicates that the rheological
reaction of the synovial fluid is not only caused by HA but also
by an interaction with other macromolecules of the synovial fluid.
This not only applies as already mentioned [7,8] to non-inflammatory, but also to inflammatory joint effusions. Recent tests of
the recombination of HA with the isolated protein fraction and individual protein components try to find out which polymeres of the
synovial fluid cause the formation of complexes.

For the comprehension of the disturbance of intermolecular
interactions a view on the network structure can be of great help.
The determination of the critical polymere concentration served to
find out whether there still is a network solution in the pathological synovia fluid or a particle solution can already be noted
(Fig. 3). Whilst the high value of 16.3 for the ratio $C/C_{crit.}$
is an evidence for a very narrow-mashed and dense network structure of the normal synovial fluid, the values between
1.4 and 6.5 of the pathological synovia make a reduced network
structure evident, whilst a particle solution cannot be precisely

Table 2. Rheological data and sedimentation coefficients of pathological synovial fluid and its hyaluronic acid

	synovial fluid (inflammatory)	isolated HA	pooled synovial fluid (non-infl.)	isolated HA
η_o (mPa s)	19.4	4.2	23.2	3.5
η_{100} (mPa s)	8.1	3.2	13.0	3.1
A	35	25	24	14
$[\eta] \cdot 10^{-3}$ (ml/g)	9.0	4.2	6.5	1.4
S_{20}^o (S)	5.0	4.6	3.2	3.9
$C/C_{crit.}$	3.1	1.4	3.6	0.8
$\nu \cdot 10^{-13}$ (cm^{-3})	9.5	2.6	15.9	2.7
$\tau_{11}-\tau_{22}$ (N/m^2)	25.6	not measurable	36.2	not measurable

determined. Independantly of the critical concentrations determinated by means of $[\eta]$, the prestationary measurements allow quantative statements on the macromolecular network structure of the synovial fluid. There is a high correlation (r = + 0.92) between the number of entanglements resulting from prestationary measurements and the HA-concentration ratio.

CONCLUSION

The HA isolated out of the pathological synovial fluid shows a highly reduced viscosity value as well as a lost elasticity. This leads to the conclusion that for the special rheological properties of the normal as well as of the pathological synovial fluid, a molecular interaction between HA and other macromolecules is of utmost importance. There is a more or less dense macromolecular

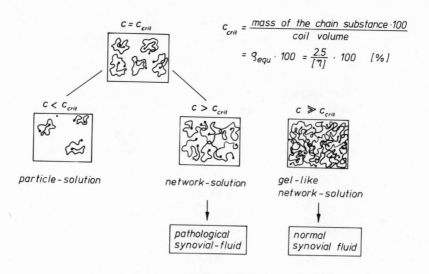

Fig.3: Types of polymer solution

network structure which can be determined quantitatively by means of prestationary measurements and the resulting number of entanglements.

Supported by: Deutsche Forschungsgemeinschaft, SFB 54

REFERENCES

1 R.R.Myers, S. Negami and R.K. White: Dynamic mechanical properties of synovial fluid. Biorheology 3: 197 (1966)
2 D.V. Davies and A.J. Palfrey: Some of the physical properties of normal and pathological synovial fluids. J. Biomechanics 1 : 79 (1968)
3 F. Ferguson, J.A. Boyle, R.N.M. McSween and M.K. Jasani: Observations on the flow properties of the synovial fluid from patient with rheumatoid arthritis. Biorheology 5 : 119 (1968)
4 G.B. Thursten and H. Greiling: Viscoelastic properties of pathological synovial fluids for a wide range of oscillatory shear rates and frequencies. Rheol. Acta 17: 433 (1978)
5 I. Anadere, H. Chmiel and W. Laschner: Viscoelasticity of normal and pathological synovial fluid. Biorheology 16: 179 (1979)
6 H. Zeidler, S. Altmann, B. John, R. Gaffga and W.-M. Kulicke: Rheologie pathologischer Gelenkflüssigkeiten. 1. Weitere Ergebnisse zur Viskoelastizität. Rheol. Acta 18: 151 (1979)
7 J. Ferguson, J.A. Boyle and G. Nuki: Rheological evidence for the existance of dissociated macromolecular complexes in rheumatoid synovial fluid. Clin.Sci. 37: 739 (1969)

8 G. Nuki and J. Ferguson: Studies on the nature and significance of macromolecular complexes in the rheology of synovial fluid
   from normal and diseased human joints. Rheol. Acta 10: 8 (1971)
9 T. Bitter and H.M. Muir: A modified uronic acid carbazole reaction. Anal. Biochem. 4: 330 (1962)

A MATHEMATICAL FORMULATION OF THE GROWTH OF BIOLOGICAL MATERIALS

S.L. Koh and P. Hore

Purdue University,
West Lafayette, Indiana, U.S.A.

(Abstract)

Several theories have been developed to characterize growth, more specifically that of tumors. Among these theories are those advanced by Laird, Burton, Greenspan and Deakin. While these theories consider the diffusion of blood, oxygen and other nutrients into regions surrounding the tumor, interactions between biochemical and mechanical processes have been ignored.

In the present paper, we focus on such interactions using the basic concepts of continuum physics. The kinematics of the continuum mixture is briefly reviewed. The basic field relations are then derived for a growing material composed of several species. A constitutive theory is developed. Within the context of recent theories of chemically reacting media (e.g.Eringen and Ingram), the general global balance equation is formulated. From this the basic governing equations and the associated boundary conditions are derived. For the treatment of a general physical problem, a complete set of ten differential equations in terms of ten thermemechanical field variables are explicitly determined. The corresponding initial and boundary conditions are likewise specified.

To simplify the theory, the constitutive equations are linearized retaining the feature of the general theory in that the equations characterize the biomaterial undergoing growth, mechanical deformation, chemical and diffusion processes. To demonstrate the

application of the theory, two special cases are considered, namely:
i) diffusion of mass of one species in a binary growing mixture;
and ii) a unidirectional growth process. In the first case, a
generalized Fick's Law is obtained and compared with the classical
Fick's Law. In the second case, the growth of a material A in one
direction due to the diffusion of material B is considered where
material A retains its cylindrical configuration .

It is conceded that the growth process is extremely complex
if one is to consider the innumerable genetical, physiological
and biological factors that influence the process. Nevertheless,
it is demonstrated in this study that it is possible and meaning—
ful to focus on some of the more pertinent factors, in the present
mathematical study on the thermomechanical and biochemical
influences.

EFFECTS OF MEMBRANE ANISOTROPY UPON THE

VISCOELASTICITY OF SPHERICAL CELL SUSPENSIONS

Akio Sakanishi[*] and Yo Takano

Department of Chemistry, University of Wisconsin, Madison,
Wisconsin 53706, U.S.A.[*] and Department of Physics,
Faculty of General Studies, Gunma University, Maebashi,
Japan (On leave from the Department of Physics, Faculty
of Science, University of Tokyo, Tokyo, Japan[*])

INTRODUCTION

In order to develop and to apply the method how the information
about the mechanical property of biological cells including red blood
cells is attained from measurements of cell suspensions, we have studied both theoretical and experimental.[1] Theoretically the homogeneous and isotropic system has been treated in calculation of bulk
modulus,[2] shear viscosity[3] and rigidity[4] in spherical shell structure
suspensions. However in biological cells the cellular membrane has a
quite peculiar anisotropy: the translational diffusion constant of
lipid molecule is in the order of 10^{-8} cm^2 sec^{-1} within the membrane
surface for instance while it is in the rough order of 10^{-15} cm^2
sec^{-1} in the direction perpendicular to the membrane. In this paper
the bulk modulus and the shear rigidity are calculated in spherical
cell suspension taking into consideration of membrane anisotropy.

MODEL AND EQUATIONS OF MOTIONS

As a model of biological cells we consider the spherical shell
structure shown in Fig. 1 that is simple but at least contains
effects of the membrane. In Fig. 1, we adapt the spherical polar
coordinate system (r, θ, φ); a and a' are the outer and inner radii
of the cell, so a-a' is the thickness of the membrane, κ and μ are
the bulk modulus and the rigidity in the medium, κ' the bulk modulus
in the internal region of the cell. Supposing that the cell suspension is dilute and the interaction between cells is neglected, we
can regard that there exists another homogeneous system equivalent to
the whole suspension with the bulk modulus κ* and the rigidity μ* to

521

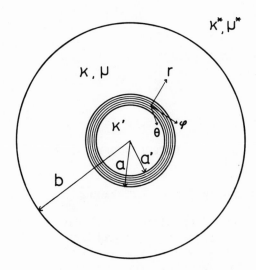

Fig. 1 Model of spherical cell suspensions

be solved in the outermost region $r \geq b$, where $b = a/\sqrt[3]{c}$ for c the volume concentration of cells.

Although the medium, the internal region of the cell and the equivalent system to the suspension are considered as isotropic, the membrane is assumed to be anisotropic: the elastic stiffness constants are different between those in the direction of r and within θ-ϕ plane as shown in Table 1, and the uniformity within the membrane of θ-ϕ gives us $C_{44} = (C_{22} - C_{23})/2$, where the components 1,2,3 correspond to normal stresses in r, θ, ϕ directions and 4, 5, 6 to shear stresses in θ-ϕ, ϕ-r, r-θ planes respectively.

Table 1. Elastic Stiffness Tensor of the Membrane
          with the Anisotropy of Tetragonal Symmetry

	C_{11}	C_{12}	C_{12}	0	0	0
	•	C_{22}	C_{23}	0	0	0
$C =$	•	•	C_{22}	0	0	0
	•	•	•	C_{44}	0	0
	•	•	•	•	C_{55}	0
	•	•	•	•	•	C_{55}

The differential equations of motion are written in spherical polar coordinates taking into account the symmetry of the solution as

$$\frac{\rho \partial^2 u_1}{\partial t^2} = \frac{\partial \sigma_1}{\partial r} + \frac{1}{r}\frac{\partial \sigma_6}{\partial \theta} + \frac{1}{r}[2\sigma_1 - \sigma_2 - \sigma_3 + \cot\theta \cdot \sigma_6] = 0, \qquad (1)$$

$$\frac{\rho \partial^2 u_2}{\partial t^2} = \frac{\partial \sigma_6}{\partial r} + \frac{1}{r}\frac{\partial \sigma_2}{\partial \theta} + \frac{1}{r}[3\sigma_6 + \cot\theta(\sigma_2 - \sigma_3)] = 0, \qquad (2)$$

where u_i is the displacement component in i direction (i = 1, 2, 3 for r, θ, ϕ respectively), and the stress component

$$\sigma_i = \sum_{j=1}^{6} C_{ij}\gamma_j . \qquad (3)$$

Here the strain components γ_j's are expressed by

$$\gamma_1 = \frac{\partial u_1}{\partial r} , \quad \gamma_2 = \frac{u_1}{r} + \frac{1}{r}\frac{\partial u_2}{\partial \theta} , \quad \gamma_3 = \frac{u_1}{r} + \frac{u_2}{r}\cot\theta ,$$

$$\gamma_6 = \frac{1}{2}\left[\frac{1}{r}\frac{\partial u_1}{\partial \theta} + r\frac{\partial}{\partial r}\left(\frac{u_2}{r}\right)\right] . \qquad (4)$$

The elastic stiffness constants C_{ij}'s are given by the bulk modulus and the rigidity in the isotropic system: for example in the medium

$$C_{11} = C_{22} = \kappa + \frac{4}{3}\mu , \quad C_{12} = C_{23} = \kappa - \frac{2}{3}\mu ,$$

$$C_{44} = C_{55} = \mu . \qquad (5)$$

BULK MODULUS

As an external boundary force we exert the following macroscopic stress

$$\mathbb{P} = \begin{pmatrix} -p & 0 & 0 \\ 0 & -p & 0 \\ 0 & 0 & -p \end{pmatrix} \qquad (6)$$

where p is generally a function of time. Corresponding displacements are given by

$$u_1 = -\frac{pr}{3\kappa^*}, \quad u_2 = u_3 = 0, \tag{7}$$

In the microscopic systems the equation (1) should be solved and rewritten as

$$\frac{\partial^2 u_1}{\partial r^2} + \frac{2}{r}\frac{\partial u_1}{\partial r} - 2\varepsilon\frac{u_1}{r^2} = 0, \tag{8}$$

where

$$\varepsilon = (C_{22} + C_{23} - C_{12})/C_{11}. \tag{9}$$

The equation (8) leads us to the general solution

$$u_1 = \alpha_m r^x + \beta_m r^y, \tag{10}$$

for example in the membrane with constants α_m and β_m, where

$$x = (-1 + \sqrt{1 + 8\varepsilon})/2, \quad y = (-1 - \sqrt{1 + 8\varepsilon})/2. \tag{11}$$

We have the displacement u_1 similar to the equation (10) with the different constants in the medium and the internal region in the isotropic systems of which $x = -1$ and $y = -2$.

We solve the simultaneous equations representing the continuity of u_1 and σ_1 obtained from the equations (3) and (4) on the boundaries at $r = a'$, $r = a$, and $r = b$. Finally the bulk modulus κ^* is

$$\kappa^* = (\kappa Q + \frac{4}{3}\mu Q'c)/(Q - Q'c), \tag{12}$$

where

$$Q = (K + \frac{4}{3}\mu)(\kappa' + \frac{4}{3}M) - \frac{4}{3}(M - \mu)(K - \kappa')A^{x-y}, \tag{13}$$

$$Q' = (K - \kappa)(\kappa' + \frac{4}{3}M) - (K - \kappa')(\kappa + \frac{4}{3}M)A^{x-y}, \tag{14}$$

$$K = (xC_{11} + 2C_{12})/3, \quad M = -(yC_{11} + 2C_{12})/4. \tag{15}$$

The parameter ε characterizes the effect of the anisotropy in bulk deformation. These results are not consistent with the volume additivity of the compressibility $1/\kappa^*$ but are rewritten to a consistent formula by using the modified thickness $(K/C_{11})(a - a')$, while they are consistent in the case of the isotropic membrane with the use of the real thickness $a - a'$.

SHEAR MODULUS

As another macroscopic force we take the shear stress

$$\begin{pmatrix} 2T & 0 & 0 \\ 0 & -T & 0 \\ 0 & 0 & -T \end{pmatrix} \rightarrow \begin{pmatrix} T(3\cos^2\theta - 1) & -3\sin\theta\cos\theta & 0 \\ \cdot & T(2 - 3\cos^2\theta) & 0 \\ \cdot & \cdot & -T \end{pmatrix} \quad (16)$$

where T is a function of time, and the arrow means the transformation of a matrix from cartesian to spherical polar coordinates. The components of displacement

$$u_1 = \frac{1}{2\mu^*} Tr(3\cos^2\theta - 1), \quad u_2 = -\frac{3}{2\mu^*} Tr\sin\theta\cos\theta \quad (17)$$

correspond to the stress (16).

Furthermore we assume material incompressibility as $C_{11} = C_{22}$ and $C_{12} = C_{23}$ tend to infinity. Then the equations (1) and (2) are described with only one parameter $\mu_r = C_{44}/C_{55}$ and we have the general solution

$$u_1 = \sum_{i=1}^{4} A_i r^{x_i}(3\cos^2\theta - 1),$$

$$u_2 = -\sum_{i=1}^{4} A_i(x_i + 2)r^{x_i}\sin\theta\cos\theta, \quad (18)$$

where A_i's are constants and x_i fulfills

$$x_i^4 + 2x_i^3 + (9 - 22\mu_r)x_i^2 + (8 - 22\mu_r)x_i + 8(2 + \mu_r) = 0. \quad (19)$$

Four roots of the equation (19) is given by

$$x_i = (\pm\sqrt{22\mu_r - 15/2 + \sqrt{54\mu_r + 225/4}} - 1$$

$$\pm\sqrt{22\mu_r - 15/2 - \sqrt{54\mu_r + 225/4}})/2. \quad (20)$$

Thus we have the two cases: First x_i's in the equation (20) are real numbers when $\mu_r \geq 96/121$ and second are complex ones when $\mu_r < 96/121$. Especially for an isotropic system it is known that x_i's are -4, -2, 1, and 3.[3]

From the equation (3) and Table 1, we have an expression of the stress component

$$\sigma_1 = 2C_{44}\gamma_1 + C_{12}\gamma \quad (21)$$

where $\gamma = \gamma_1 + \gamma_2 + \gamma_3$. Although the volume change $\gamma = 0$ by incompressibility, the pressure

$$p = -C_{12}\gamma = C_{55}\sum_{i=1}^{4}[2\mu_r - (5\mu_r - 2)(x_i + 2)/3$$

$$+ x_i(x_i + 1)(x_i + 2)/6]A_i r^{x_i-1}(3\cos^2\theta - 1) \tag{22}$$

is finite and derived from the equation (2). Finally we get

$$\sigma_1 = C_{55}\sum[2\mu_r(x_i - 1) + (5\mu_r - 2)(x_i + 2)/3$$

$$-x_i(x_i + 1)(x_i + 2)/6]A_i r^{x_i-1}(3\cos^2\theta - 1),$$

$$\sigma_6 = -C_{55}\sum(x_i^2 + x_i + 4)A_i r^{x_i-1}\sin\theta\cos\theta. \tag{23}$$

In the second case that $\mu_r < 96/121$ the equation (18) may be reformed as

$$u_1 = (B_1\sinh\xi + B_2\cosh\xi)(D_1\sin\zeta + D_2\cos\zeta)(3\cos^2\theta - 1),$$

$$u_2 = -\{[(B_1 + \alpha B_2)\sinh\xi + (B_2 + \alpha B_1)\cosh\xi]x$$

$$(D_1\sin\zeta + D_2\cos\zeta) + (B_1\sinh\xi + B_2\cosh\xi)[(D_1 - \beta D_2)x$$

$$\sin\zeta + (D_2 + \beta D_1)\cos\zeta]\}\sin\theta\cos\theta, \tag{24}$$

where B_1, B_2, D_1, and D_2 are constants, $\xi = \alpha\ell nr$, $\zeta = \beta\ell nr$; α and β are respectively the real and the imaginary parts of x_i in the equation (20).

Similarly but with more complexity the calculation can be made on the shear rigidity μ^* of the suspensions.

REFERENCES

1. A. Sakanishi, S. Mitaku and Y. Takano, Linear viscoelasticity of suspensions of spherical shell structures and erythrocyte membrane, Thrombosis Research Suppl. II 8:35 (1976).
2. A. Sakanishi and Y. Takano, Bulk modulus of the disperse system of spherical cells with compressible and elastic membrane, J. Phys. Soc. Japan 32:1160 (1972).
3. A. Sakanishi and Y. Takano, Complex viscosity of dispersions of spherical cells with elastic membrane, Japan J. Appl. Phys. 13:882 (1974).
4. Y. Takano and A. Sakanishi, Complex rigidity of dispersions of spherical shell structures with material compressibility, J. Appl. Phys. 44:4023 (1973); ibid. 45:2811 (1974).

WALKING, RUNNING, JUMPING - AN INTERACTION OF TWO RHEOLOGICAL SYSTEMS

Bertil Olofsson

Chalmers University of Technology

S-412 96 Gothenburg, Sweden

INTRODUCTION

At the Department of Textile Technology we have been working together with a group from the Orthopedic-Surgical Hospital Department at the Gothenburg University on relationships between injuries in sports as football and properties of ground as well as shoes, preventive tapes etc. The present paper presents some fundamental rheology of the ground and its context with the rheological activity of the human being (player, athlet).

GROUND OR FLOOR RHEOLOGY AS EVALUATED BY THE "SIMULATED ATHLET"

The "simulated athlet"[1] (Kunstlicher Sportler) is a commercially available equipment,[1] which imitates a step or jump by a weight falling on and rebounding from a testing foot via a piston. By a pressure cell the force P as well as displacement s are measured and drawn as functions of time t (fig. 1). We found that an appropriate model curve for s might be written

$$s_D(t) = S_M[1 - (1 - t/t_M)^2]; \quad 0 \le t \le t_M \tag{1a}$$

$$s_U(t) = s_m \{1 - [(t - t_M)/(t_E - t_M)]^2\}; \quad t_M \le t \le t_E \tag{1b}$$

where $s_D(t)$ means the phase of increasing displacement from $s_D(0) = 0$ to $s_D(t_M) = s_M$, while $s_U(t)$ means the phase of decreasing displacement from $s_U(t_M) = \check{s}_m$ to $s_U(t_E) = 0$. At $t = t_M$ there might be a frictional (plastic) displacement from s_M to s_m.

We now assume that the rheology of the ground might be described by a spring (constant E_0) working in parallell with a dashpot (constant V_0) and a frictional element (constant F_0). Also introducing a mass (constant M_0) in series with these mechanisms yields for the resulting force $P(t)$

$$P(t) = \pm F_0 + E_0 \ s + V_0 \ s + M_0 \ s \qquad\qquad (2)$$

where $+F_0$ means the s_D- and $-F_0$ the s_U-phase. In agreement with fig. 1 we might further write the conditions for $P(t)$

$$P(0) = P(t_F) = 0; \ \dot{P}(t_N) = 0 \text{ and } P(t_N) = P_N \text{ (or } P_n) \qquad (3)$$

The calculations are much simplified by putting $t_E = 2t_M$ and $t_F = 2t_N$. This makes (1a) valid for all the range ($0 \le t \le 2t_M$) and by substitution in (2) and fitting to conditions (3) the parameters are evaluated

$$E_0 = (P_N/s_M)(t_M/t_N)^2; \ V_0 = E_0(t_M-t_N); \ M_0 = V_0 t_M; \ F_0 = 0 \ (4)$$

E_0 and V_0 are most significant. As $t_M - t_N$ is generally small compared to t_M or t_N, the approximation $(t_N/t_M)^2 = 2t_N/t_M - 1$ is acceptable and so are also the substitutions of P_N for P_n and s_M for s_m. For $t_M = t_N$ we get $V_0 = 0$ as expected and $M_0 = 0$ which seems peculiar. But the model formalism is probably not very significant regarding either M_0 or F_0. Table 1 illustrates differences between materials (coefficients of variation in parenthesis).

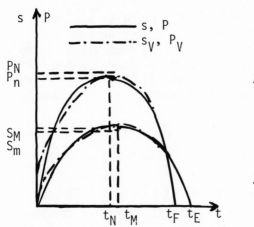

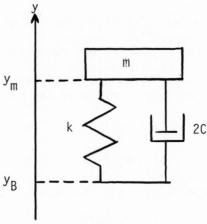

Fig. 1. s- and P-characteristics for the "simulated athlete".

Fig. 2. Rheological model of the "human athlete".

Table 1. Some results from "artificial athlet"

	E_0 (N/mm)	V_0 (N·sec/mm)	t_M-t_N (sec·10^{-3})
Gravels	5873 (0.24)	76.3 (0.53)	13
Grass	940 (0.08)	11.2 (0.42)	12
Art.turf I	608 (0.14)	7.6 (0.40)	12.5
Art.turf II	303 (0.04)	4.6 (0.14)	15

Evidently the time lag between P and s is $\approx 10^{-2}$ sec for all the materials yielding approximate proportionality between V_0 and E_0. As expected E_0 seems very significant and measurable with good accuracy.

In fig. 1 there are also indications of vibrational effects, which might be described by

$$s_V(t) = s_A \exp(-\alpha t) \sin \omega t \qquad (5a)$$

$$P_V(t) = P_A \exp(-\alpha t) \sin(\omega t + \delta) \qquad (5b)$$

However, these trends are not distinct enough for alternative calculations of the rheological parameters

$$(P_A/s_A)^2 = [E_0-V_0\alpha-M_0(\omega^2-\alpha^2)]^2 + (V_0\omega-2M_0\alpha\omega)^2 \qquad (6a)$$

$$\tan \delta = (V_0\omega-2M_0\alpha\omega)/[E_0-V_0\alpha-M_0(\omega^2-\alpha^2)] \qquad (6b)$$

Another complication is that the ground stiffness might change with the displacement. Also a "point elastic" floor should increase in stiffness with displacement as compared to a "surface elastic" floor[2]. In our model (2) it might be appropriate to add a term E_1s^3 and to settle E_1 from the s- and P-curves by the times t_{M1} and t_{N1} for half the maximum values ($s_M/2$ and $P_N/2$). But $E_1 \neq 0$ also requires that the parabolic approximation of s (1) is supplemented with higher power terms in t. However, a symmetry for s_D and s_U requires that $\ddot{s}(t) = -2s_M/t_M^2$ is constant and s thus independent of such high power terms. Thus a primary refinement of the model might be to keep $t_E \neq 2t_M$ in (1b) yielding $E_{0D} \neq E_{0U}$. Another possibility to account for such effects is to add a term E_1s^2 instead of E_1s^3 in equ. (2).

A RHEOLOGICAL MODEL OF THE "HUMAN ATHLETE" AND ITS SECURITY CRITERIA

There are several approaches on modelling significant parts of the human body. Most interesting approaches have been made in connection with NASA studies of space pilot landing, where elemental visco-elastic models are coupled to take a stiff "spinal" mode of high frequency as well as a soft "viscara" mode of low frequency in consideration [3]. For the walking-running-jumping activity considered here one single mode seems most appropriate [4] as drawn in fig. 2. There the mass m is balanced by a spring of stiffness k and a dashpot of viscosity 2C in parallell. This has also a definite anatomic background: a muscle-leg stiffness (almost independent of deformation) modified by a rate dependence of the muscle force together with the nerv reflexes. The equation for this model is written:

$$m\ddot{y}_m = k \cdot e + 2C \cdot \dot{e} - m\underline{g} \tag{7a}$$

where e = contraction = $(y_{m0} - y_{b0}) - (y_m - y_b)$ \hfill (7b)

and the coordinate y is positive upwards. Putting

$$w^2 = k/m; \quad c = C/m; \quad \lambda^2 = w^2 - c^2 \tag{8}$$

where certainly $\lambda^2 > 0$ as the damping is subcritical, we get (7a) in the form

$$\ddot{e} + 2c\dot{e} + w^2 e = \ddot{y}_b + \underline{g} \tag{9}$$

which at equilibrium yields $e_0 = \underline{g}/w^2$ \hfill (9a).

From the previous discussion and other experimental studies [2] we conclude that the reaction of the ground or floor can be divided in elemental steps of retardations r_0 with durations T_0, i.e. corresponding velocity changes ΔV_0

$$\Delta V_0 = r_0 T_0 \tag{10}$$

Thus there is a "stand"-phase in walking and running with $T_0 = 10^{-1} - 1$ sec and $r_0 = 0 - 1g$. In jumping, however, this is preceeded by a "peak"-phase with $T_0 = 1 - 2 \cdot 10^{-2}$ sec and $r_0 = 2 - 5g$. Thus we study the effect of a single retardation step according to (10) on the mechanics assuming that

$$\ddot{y}_b = r_0 \quad (0 < t < T_0) \quad \text{and}$$

$$\ddot{y}_b = 0 \quad (t > T_0) \tag{11}$$

Using the conventional Laplace transform method we get from (8), (9) and (11) the result

$$e = \exp(-ct)\{\sin \lambda t[\dot{e}_0 + ce_0 - c(\underline{g} + r_0)/w^2]/\lambda +$$

$$+ \cos \lambda t[e_0 - (\underline{g} + r_0)/w^2]\} + [\underline{g} + r_0 R(t)]/w^2 \tag{12a}$$

where $R(t) = 1$ for $0 < t < T_0$ $\hfill$ (12b)

and $R(t) = \exp[-c(t-T_0)][\sin \lambda(t-T_0) \cdot (c/\lambda) + \cos \lambda(t-T_0)]$

for $t > T_0$ $\hfill$ (12c)

(here e_0 and $\dot{e}_0$ means values of e and $\dot{e}$ at $t = 0$).
Further simplifications are obtained by putting

$$\dot{e}_0 = 0; \ e_0 = \underline{g}/w^2 \hfill (12d)$$

From (12) we calculate two alternatives for maximum stress also
introducing the nondimensional damping $\underline{c}$

$$\underline{c} = c/w, \text{ thus } \lambda = w\sqrt{1-\underline{c}^2} \hfill (13)$$

We first study the case $t < T_0$ where (12b) is valid and evaluate as
a first alternative maximum force on the spring mechanism (fig. 2)
yielding from (12)-(13)

$$F_{T1} = k \cdot e_{T1} = m\underline{g} + mr_0 \cdot f_1(\underline{c}) \hfill (14a)$$

where T_1 is the time for maximum occuring and $\lambda T_1 = \Pi$;

$$f_1(\underline{c}) = 1 + \exp(-\lambda T_1 \ \underline{c}/\sqrt{1-\underline{c}^2}) \hfill (14b)$$

As a second alternative the maximum force on the spring + dashpot
mechanism (fig. 2) is evaluated

$$Q_{T2} = k \cdot e_{T2} + 2C \cdot \dot{e}_{T2} = m\underline{g} + mr_0 \cdot f_2(\underline{c}) \hfill (15a)$$

where now T_2 is the time for maximum occuring

$$\lambda T_2 = \Pi - \sin^{-1} 2\underline{c} \sqrt{1-\underline{c}^2} \text{ for } \underline{c} \le 1/\sqrt{2} \text{ and } \lambda T_2 = -\sin^{-1} 2\underline{c}\sqrt{1-\underline{c}^2}$$

for $\underline{c} \ge 1/\sqrt{2} = 0.707$ $\hfill$ (15b)

and $f_2(\underline{c}) = 1 + \exp(-\lambda T_2 \ \underline{c}/\sqrt{1-\underline{c}^2})$ $\hfill$ (15c)

Analysing the case of $t > T_0$ from (12c) in the same way we find
that a non-controversial solution demends $T_0 \Rightarrow 0$ and to maintain a
finite value of ΔV_0 in (10) $r_0 \Rightarrow \infty$, yielding a "Dirac-δ" function
for the retardation step. Performing the further calulations as be-
fore the alternative of maximum spring force is written

$$F_{T3} = k \cdot e_{T3} = m\underline{g} + mw\Delta V_0 \ f_3(\underline{c}) \hfill (16a)$$

with $\lambda T_3 = \sin^{-1}\sqrt{1-\underline{c}^2}; \ f_3(\underline{c}) = \exp(-\lambda T_3 \ \underline{c}/\sqrt{1-\underline{c}^2})$ $\hfill$ (16b)

and the alternative of maximum total force means

$$Q_{T_4} = k \cdot e_{T_4} + 2C \cdot \dot{e}_{T_4} = m\underline{g} + mw\Delta V_0 f_4(\underline{c}) \tag{17a}$$

where the time of maximum T_4 is written

$$\lambda T_4 = \pm\sin^{-1}[(1-4\underline{c}^2)\sqrt{1-\underline{c}^2}] \quad (\pm \text{for } \underline{c} \lessgtr 0.5) \tag{17b}$$

and $\quad f_4(\underline{c}) = \exp(-\lambda T_4 \underline{c}/\sqrt{1-\underline{c}^2}) \tag{17c}$

Damping functions according to (14b), (15c), (16b) and (17c) are calculated in Table 2.

Table 2. Calculated damping functions.

$\underline{c}$	$f_1(\underline{c})$	$f_2(\underline{c})$	$f_3(\underline{c})$	$f_4(\underline{c})$	f_1/f_3	f_2/f_4
0	2.00	2.00	1.00	1.00	2.00	2.00
0.2	1.53	1.57	0.76	0.82	2.02	1.91
0.4	1.25	1.26	0.60	0.86	2.08	1.46
0.6	1.09	1.25	0.50	0.76	2.19	1.64
0.8	1.02	1.18	0.42	0.20	2.39	5.93
1.0	1.00	1.14	0.37	0.05	2.72	22.80

Introducing $\underline{g}$ = 10 N/kg and further the representative values[2-4] $\underline{c}$ = 0.2, m = 50 kg, k = 500 N/mm (thus w^2 = 10 N/kg·mm) we consider as an example the maximum spring force for an "impact" retardation peak of r_0 = 40 N/kg and T_0 = 10^{-2} sec, i.e. ΔV_0 = 0.4 N·sec/kg, which yields $w\Delta V_0$ = 40 N/kg to be used for F_{T3} and Q_{T4}. As a comparison we apply the long time peak criterion with r_0 = 40 N/kg to get F_{T1} and Q_{T2}. The results are: F_{T1}=3560, Q_{T2}=3640, F_{T3}=2020, Q_{T4}=2140 N and the differences are only due to the f($\underline{c}$) factors, which at this $\underline{c}$ are about twice as large for the low long time peak criterion as the high impact peak criterion. Of course the mathematical "Dirac-δ"-peak is practically not limited to $T_0 \Rightarrow 0$, $r_0 \Rightarrow \infty$ but an upper limit of T_0 is obtained by assuming that a critical limit for both r_0 and ΔV_0 must be attained, yielding the same critical limit for $F=F_B=F_{T1}=F_{T3}$ (or alternatively $Q_B=Q_{T2}=Q_{T4}$). From (10), (14a), (15a), (16a) and (17a) we get for this T_0

$$wT_0 = f_1/f_3 \quad (F_B \text{ considered}) \tag{18a}$$

$$wT_0 = f_2/f_4 \quad (Q_B \text{ considered}) \tag{18b}$$

REFERENCES
1. Otto-Graf-Institut (Stuttgart) and Bundesinstitut für Sport-
   wissenschaft (Köln): Sport- und Freizeit anlagen, Berichte B2(1973).
2. Laboratorium für Biomechanik der ETH Zürich (editor B.M.Nigg):
   Biomechanische Aspekte zu Sportplatzbelägen.(Zürich 1978).
3. P.R.Payne, The Dynamics of Human Restraint Systems. Frost report
   No 101-2.National Academy of Science Symposium, San Antonio,USA(1978)
4. T.A.Mc Mahon-P.R.Greene: Fast Running Tracks, Sci.Amer.112.Dec.(1978)

RHEOLOGICAL PROPERTIES OF PULLULAN BY TSCr

K. Nishinari, T. El Sayed, D. Chatain and C. Lacabanne

Laboratoire de Physique des Solides
Université Paul Sabatier
118, Route de Narbonne - 31077 Toulouse Cédex (France)

INTRODUCTION

Pullulan films have been used as food packaging materials and food materials. However, its rheological properties have not yet been studied in detail (1). So it is desirable to know the relationship between the chemical structure and rheological properties not only for scientific interest but also for developing further utilization. We have studied its rheological properties by thermally stimulated creep. Many investigations have been done on dielectric and mechanical properties of polysaccharides, and so comparison of the properties of pullulan with those of other polysaccharides may shed some light on the relaxational process.

EXPERIMENTAL

The pullulan film has been kindly supplied by Hayashibara Seikagaku Kenkyusho Co. Its chemical structure is as follows :

$$
\begin{array}{l}
\downarrow \\
o - o - o \\
\quad\quad \downarrow \\
\quad\quad o - o - o \\
\quad\quad\quad\quad \downarrow \\
\quad\quad\quad\quad o - o - o \\
\quad\quad\quad\quad\quad\quad \downarrow \\
\quad\quad\quad\quad\quad\quad o \ \ldots
\end{array}
$$

where o represents the glucose residue, o - o the α-1,4-glucosidic linkage and $\overset{o}{\underset{o}{\downarrow}}$ the -1,6-glucosidic linkage. The typical size of the sample was 1.0 cm × 7.0 cm × 0.06 cm.

The film was subjected to a shear stress at room temperature, allowing some orientation of the mobile units. Then the temperature

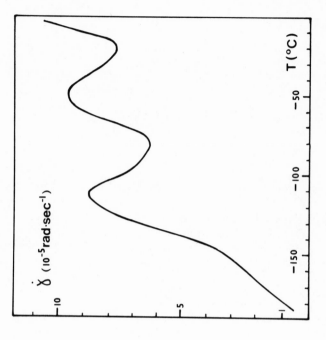

Figure 2. Low temperature TSCr spectrum of pullulan film left at 30% relative humidity for 2 days

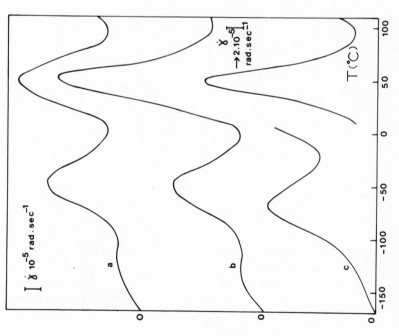

Figure 1. Overall TSCr spectrum of pullulan film for highly dehydrated sample (curve a) and for sample with increasing levels of moisture content (curve b to c)

was lowered to liquid nitrogen temperature (LNT) at which molecular
motions were frozen ; then the stress was removed. Next, the sample
was heated at a linear rate so that the mobile units can reorientate.
The strain γ, its time derivative $\dot{\gamma}$, and the temperature T were
recorded as a function of time t. For detailed explanations on the
method, see references 2 and 3.

RESULTS AND DISCUSSION

Typical TSCr spectra of pullulan film with different moisture
contents are shown in Fig.1. For highly dehydrated film (curve a),
two peaks were observed : one at 50°C and the other at -50°C.

The higher temperature peak

The peak at 50°C involves absorbed water. Indeed, the magnitude increases slightly and the maximum shifts to lower temperatures as the moisture level was increased. This behavior might be
explained by the presence of water as plasticizer and so facilita--
ting molecular movements. From viscoelastic relaxation, Bradley and
Carr (4) attributed a peak observed at the same temperature in
hydrated amylose and dextran to dehydration. The influence of water
on TSCr peak at 50°C is analogous to that on the peak observed at
the same temperature in thermally stimulated current (TSC) experiment performed on the same material (5). Moreover, a broad DSC
endotherm is exhibited by the hydrated sample over the temperature
range from 20°C to 50°C (6). From these facts, the TSCr peak at
50°C has been attributed to dehydration.

The lower temperature peaks

Increases in moisture content caused the magnitude of the peak
at about -50°C (curve a) to increase slightly (curbe b). Further
absorption of moisture made the peak shift to -70°C (curve c).

Hydrated film -the specimen was left at about 30% relative
humidity for two days- showed two main peaks at -50°C and at -120°C
(Fig.2).

The TSCr peak at -120°C seemed to correspond to the TSC peak
at -130°C (5) and to the peak of ε'' at about -100°C measured at
10 Hz (7) for the same material. The activation energy obtained
from the fractional depolarization technique in TSC was about 0.45eV
$\simeq$ 10 kcal/mol. This value is very close to that for methylol rotation in amylose (4, 8) and so this peak seemed to have the same
origin as in amylose.

Bradley and Carr (4) studied the viscoelastic properties of
cellulose, amylose and dextran and they ascribed the loss tangent
peak at about -75°C observed in cellulose and amylose to the

methylol rotation. They didn't find the corresponding peak in
dextran which lacks methylol groups attached to the C_6 atoms. Kimura,
Usuda and Kadoya (9) made the viscoelastic investigations on cellulose and hemicelluloses by torsion vibration, and found the same
process at about -100°C for 0.2 to 2 Hz.By comparing with the results
on triphenyl-methylcellulose, they also ascribed this process to
a rotation of methylol group.

Mikhailov, Artyukov and Shevelev (10) made the dielectric and
NMR studies on cellulose and its derivatives to make a closer asses-
sment of the molecular mechanism of the observed relaxation process.
Examining the change of spin-lattice relaxation times caused by deu-
teration, they found that the change was too small if the mobile
unit was only the primary radical. Thus, they concluded that the
mobile unit causing the relaxation process in cellulose must be the
methylol group as a whole.

Norimoto (11) studied in detail the dielectric properties of
cellulose, xylan and other polysaccharides by bridge measurement
in the frequency range from 300 Hz to 1 MHz and he also ascribed
the peak at about 0°C to the methylol rotation. He didn't find the
corresponding peak for xylan in which there is no methylol group.
Sawatari (12) studied the dielectric properties of cellulose and
its derivatives by TSC, and he found the similar peak at about -130°C.
He suggested to assign this process to methylol rotation because
the peak intensity decreased by tritylation and acetylation of
cellulose, and moreover, this peak was not found in xylan.

Zhbankov (13) calculated the dependence of potential function
for internal rotation of methylol group of α-glucose on the rotation angles ϕ_1 and ϕ_2. Here, ϕ_2 is the rotation angle about the

C_6 - O_6 bond and is measured clockwise from the cis position of C_6
and C_4 atoms, and ϕ_1 is the rotation angle about the C_6 - O_6 bond
and is measured from the trans position of the H_{O6} and C_5 atoms.
From their results we can see that the rotation of OH group does
not need a high energy, and we may need to consider only the ϕ_2
rotation. There are three stable sites of methylol groups at ϕ_2 =80°,
177° and 300°. Since the potential barrier between ϕ_2 = 177° and
300°(about 6 kcal/mol) is much larger than that between ϕ_2 = 80°
and 300° (about 1 kcal/mol), Nishinari and Fukada (8) smeared out
the potential curve as having only two minima (sites). Using these
results, they estimated the dielectric relaxation strength by two
sites transition theory. They compared this value with the experimental one and found the fairly good agreement in spite of the

roughness in the approximation used.

So the relaxation mode observed at $-50°C$ in TSCr experiments on highly dehydrated pullulan might involve the methylol group. Loosely bound water would act as plasticizer and induce a shift of the TSCr peak towards lower temperature. As for the peak observed at $-120°C$ in pullulan film at room humidity, it might be added another possibility : movements involving tightly bound water. Two analogous regimes of bound water have been successfully analyzed to explain the molecular motions in poly-L-proline (14, 15). The structural stabilization by hydrogen bonding in the presence of water has been observed also in amylose (8), cellulose (16), poly-α-amino acids (17) and poly-amides (18).

REFERENCES

1.   S. Yuen, Shokuhin to Kagaku, 73 (1977)
2.   J.C. Monpagens, Thèse (Université Paul Sabatier, 1977)
3.   J.C. Monpagens, D.G. Chatain, C. Lacabanne and P.G. Gautier, J. Polym. Sci. Polym. Phys. Ed. 15, 767 (1977)
4.   S.A. Bradley and S.H. Carr, ibid. 14, 111 (1976)
5.   K. Nishinari, D. Chatain and C. Lacabanne, The Dielectrics Society 1980 Meeting, Aussois, March 25-27 1980
6.   P. Berticat, Private Communication, Lyon (1980)
7.   K. Nishinari, H. Horiuchi and E. Fukada, Submitted to Rep. Prog. Polym. Phys. Japan
8.   K. Nishinari and E. Fudada, Submitted to J. Polym. Sci. Polym. Phys. Ed.
9.   M. Kimura, M. Usuda and T. Kadoya, Sen-i Gakkaishi, 30, T221 (1974)
10.   G.P. Mikhailov, A.I. Artyukhov and V.A. Shevelev, Vysokomol. Soyed. A11, 553 (1969)
11.   M. Norimoto, Wood Research, N° 59/60, 106 (1976)
12.   A. Sawatari, in "Charge Storage, Charge Transport and Electro-statics with their Applications" Ed. Y. Wada, M.M. Perlman and H. Kokado (Kodansha-Elsevier, 1979) p. 347
13.   R.G. Zhbankov, J. Polym. Sci. C16, 4629 (1969)
14.   J. Guillet, G. Seytre, D. Chatain, C. Lacabanne and J.C. Monpagens, J. Polym. Sci. Polym. Phys. Ed. 15, 541 (1977)
15.   J. Guillet, C. Seytre, J. May, D. Chatain and C. Lacabanne, ibid., 14 , 211 (1976)
16.   S. Sasaki and E. Fukada, ibid., 14, 565 (1976)
17.   A. Hiltner, J.M. Anderson and E. Baer, J. Macromol. Sci. B8, 431 (1973)
18.   H.K. Provorsek, R.H. Butler and H.K. Reimschuessel, J. Polym. Sci. A2, 9, 867 (1971)

FLUID DYNAMIC PROBLEMS OF POSTDILUTIONAL HEMOFILTRATION

WITH HIGH-FLUX MEMBRANES

A. Pozzi,[+] P. Luchini, and E. Drioli

Istituto di Principi di Ingegneria Chimica
Facoltà di Ingegneria, Università di Napoli
Italy

Hemofiltration with high-flux polymeric membranes is a rapidly developing technique which might compete with classical hemodialysis as an artificial kidney system. Clinical results have been in general quite satisfactory.[1]

A better control of the concentration of the various chemical species in the blood stream can be obtained with this process, particularly for intermediate-molecular-weight species (2000 - 5000 mw) whose poisoning effects have been identified.

One of the major problems in hemofiltration consists in the concentration polarization phenomena which take place upstream of the pressurized UF membrane, and which originate significant ultrafiltrate flux decay. Moreover, particularly when a protein gel layer is formed on the pressurized membrane face, significant variations of membrane rejections for the various chemical species present in the blood have been observed.[2]

A detailed fluid dynamic study of the ultrafiltration modulus used in a postdilutional hemofiltration process is necessary for a better understanding of the various phenomena due to concentration polarization which occur in the system. This analysis might also contribute to the evaluation of the influence of osmotic effects, generally neglected, on the overall performance of hemofiltration modulus.

[+]Istituto di Matematica, Facoltà di Ingegneria, Università di Napoli, Italy

In this paper the optimal geometry of flat-sheet membranes operating in a postdilutional hemofilter has been analyzed.

THE BASIC EQUATIONS AND THEIR SOLUTION

The basic equations governing the problem, in dimensionless incompressible form, are:

$$\text{div } \underline{V} = 0 \tag{1}$$

$$\underline{V}_t + V \cdot \text{grad } \underline{V} + \text{grad } p = (1/\text{Re})\Delta_2 V \tag{2}$$

$$c_t + \underline{V} \cdot \text{grad } c = (1/\text{ReSc})\Delta_2 c \tag{3}$$

where $\underline{V}$ is the velocity, c the concentration, and Re and Sc the Reynolds and the Schmidt numbers respectively.

The boundary conditions state that the component of the velocity along the membrane is vanishing, and that on the membrane, the cross-velocity V is related to c_y by $\text{ReScVc} = c_y$; moreover $V = \pm kp$ in the absence of a gel and $c = c_g$ when a gel is present; k is the dimensionless membrane constant and c_g is the gel concentration.

To solve the three-dimensional Stokes-Navier equations (1), (2), an expansion in series of the parameter ε (ratio of the half-distance d between the membranes to the square root of the area A of the membranes) after a stretching of the coordinate normal to the membrane and of the time is introduced; moreover a new pressure P is defined as $p\text{Re}\varepsilon^2/2$. To determine the basic quantities, the Fourier transform with respect to time is applied to the related equations. After some manipulations one obtains that the flow field parallel to the membranes can be expressed as a product of a function of the normal coordinate and of the frequency times a function of the other two coordinates x_1 and x_2.

The pressure P, if the filtrate volume flow is small with respect to the main volume flow, is given by the Laplace equation $Px_1x_1 + Px_2x_2 = 0$.

Once the velocity field is known, equation (3) can be reduced to the two-dimensional form, solved in Ref. 3 by introducing the new variable $\xi = \int (P_{x_1}^2 + P_{x_2}^2)^{-\frac{1}{2}} d\ell$ ($\xi = 0$ at the inlet); this integral must be evaluated along a streamline, $\psi = \text{const}$, where ψ is the stream function. The change of variables from x_1 and x_2 to ξ and ψ, neglecting the diffusion in the plane parallel to the membrane, leads to an equation independent of ψ: therefore the two-dimensional analysis can be used, and the curves $\xi = \text{const}$ are the curves $c = \text{const}$.

Two quantities of particular interest are the gel pressure p_g (i.e., the pressure at which gel formation begins) and the limiting velocity V_∞ (i.e., the filtrate velocity when the pressure drop through the membrane is infinite); in dimensional form these quantities can be written as

$$p_g = \xi_{max}^{-1/3} \cdot p_{gt} \tag{4}$$

$$V_\infty = V_{\infty t} \iint_S (2/3\xi^{1/3}) \, dS \tag{5}$$

where P_{gt} and $V_{\infty t}$ are the corresponding values in the two-dimensional case (see Ref. 3), given by

$$P_{gt} = (K_p/k)(3QD^2/4d^2A)^{1/3}$$

$$V_{\infty t} = K_V(3QD^2/4d^2A)^{1/3}$$

where K_p and K_V are functions of c_g (defined in Ref. 3), Q is the volume flow, D is the diffusion coefficient, and S is the surface of the membrane.

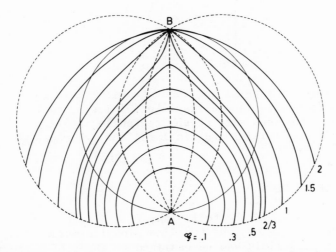

Fig. 1 Streamlines (---) and concentration lines (——) for an infinite plate ultrafilter with the flow inlet at A and the flow outlet at B. The same velocity and concentration field is valid in a circular device. The values of ξ marked on the diagram are referred to the circle; for other regions, the different areas must be taken into account.

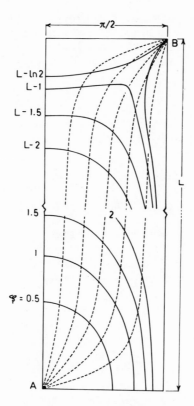

Fig. 2   Streamlines (---) and concentration lines (——) in a
         rectangular-plate ultrafilter; inlet A, outlet B. The
         dimensionless size of the rectangle is $\pi/2 \times L$ ($L \geq 4$);
         only the initial and final regions are reported;
         $\xi' = \xi \cdot L$.

DISCUSSION OF RESULTS

     A first result is the following: $V_\infty \leq V_{\infty t}$, $Pg \leq Pgt$; therefore,
from this point of view, the two-dimensional membrane is the best
one. Such a flow field can be realized by means of a rectangular
modulus with a uniform velocity at the inlet and outlet along two
opposite sides.

     Two other geometries have been considered, with the blood flow
inlet and outlet concentrated in a very small region of the membrane:
rectangular and circular (the first corresponds, e.g., to modules
RP6 and RP610 by Rhône-Poulanc). Figures 1 and 2 show the curves
at constant concentration and the streamlines. In the rectangular
case, the entrance and exit regions have been analyzed separately
(on the assumption that the ratio between the two sides is not less
than 2.5).

The limiting velocity in these two cases is nearly equal to the two-dimensional one ($V_\infty \simeq 0.99V_\infty t$); in the circular case, $p_g = 0.69 p_{gt}$, while in the rectangular case $p_g = 0$ due to the presence of corners, though for $p_g \gg p$ the area covered by the gel is very small. If the corners of the rectangle are rounded, $p_g > 0$ (e.g., a radius of curvature of 0.1 times the width yields $p_g \simeq 0.82 p_{gt}$ for the geometry considered). It must be noted that the comparison with the two-dimensional case is done for equal area and volume flow.

REFERENCES

1. H. Schneider and E. Streicher, Clinical observations of the polyamide hollow-fiber hemofilter in hemofiltration system, J. Dialysis, I, 737 (1977).

2. E. Drioli, V. Calderaro, B. Memoli, O. Albanese, and V. Andreucci, Influenza della polarizzazione per concentrazione nella emofiltrazione postdiluzionale del plasma umano, Ing. Chim. Ital., 15, 100 (1979).

3. A. Pozzi, P. Luchini, and E. Drioli, An unsteady state analysis of the hemofiltration process, J. Membrane Science, submitted for publication; presented at the 2nd Symp. on Synthetic Membranes in Science and Industry, Tubingen, September 1979.

SOME RESULTS ABOUT A NEW MONO-CUSP AORTIC VALVE PROSTHESIS

R. Brouwer, A. Cardon, Cl. Hiel (1), and W. Welch (2)

Free University of Brussels (V.U.B.)
Faculty of Applied Sciences (1), Faculty of Medicine (2)
Pleinlaan, 2 - 1050 Brussels (Belgium)

INTRODUCTION

Since the first implantation of an artificial heart valve by Hufnagel in 1952, many research workers proposed different types of such artificial valves. Up to this time none of the proposed models satisfied all the criteria necessary for an ideal valve[1] :

- biocompatibility
- no material fatigue
- limited pressure gradient across the valve, in order to restrict valve power loss
- backflow less than 5 % of forward flow
- limited turbulent flow in order to prevent hemolysis
- sufficient washout to avoid thrombogenesis
- no sharp boundaries and smooth closure.

In order to fulfil better these different conditions a new model of a mono-cusp aortic valve is proposed.

DESIGN AND SHAPE OF THE VALVE

The valve has a spherical form (see figure 1) and is made of flexible rubber material. The sphere is cut along a segment with radius about 80 % of the radius of the sphere, and along this section a circumferential reinforcement ring is fixed.
The valves under investigation at present have a radius of 23 mm. The valve is attached only over a part of the end segment on the aorta ring with the convex part in the left ventricular chamber (see figure 2). The design of the valve is such that the part of the sphere which is fixed to the aorta annulus has a more impor-

tant rigidity than the free edge. This increased rigidity is obtained by an increasing thickness of the sphere from the free
edge to the fixed part.

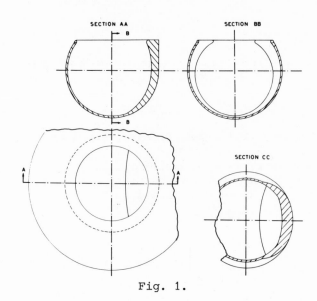

Fig. 1.

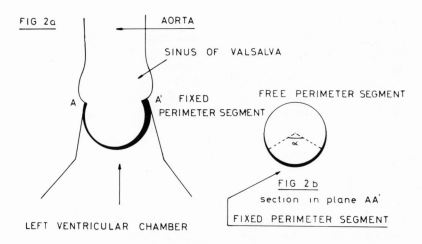

Fig. 2.(a). Position of the valve in the aortic root.
     2.(b). View of the valve from the aorta ; the valve is
            fixed along the angle α.

Figure 3 shows a photograph of the mono-cusp valve, placed in the pulse duplicator.

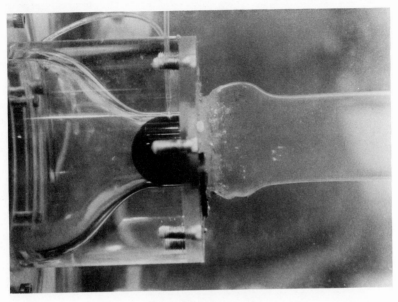

Fig. 3.

PHILOSOPHY OF OPERATION

During systole, as a result of the proposed geometrical shape, the most flexible part of the valve is moving to the more rigid fixed part and located into that part (see figure 4).

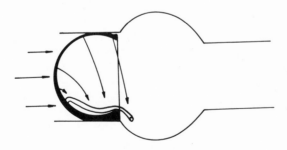

Fig. 4.

When the pressure in the aorta becomes more important than the
ventricular pressure, the flexible part moves back and the sphere
retakes his original form. The increasing interior pressure in
the sphere, that realises a little inflation of it, ensures the
final closure of the valve.

EXPERIMENTS

a) The test rig
   The first part of this research program concerns some in vitro
measurements. The most in vitro tests of artificial heart valves
were performed under experimental conditions without some clear
clinical significance. The reason of this gap is essentially an
incomplete simulation of the biological configuration.
The device used in our experiments, furnished by the University of
Sheffield[2,3], produces a physiological pressure curve in the left
ventricular simulator and in the aorta.
In the Sheffield-simulator the afterload is calculated in order to
correspond to the conditions measured in humans. A flexible ana-
tomical shaped ascending aorta, made of silicone rubber RTV 615A
(General Electric) has dimensions based on a model study of aortic
roots of pathological human cases[4]. The test section is enclosed
in a rigid fluid filled chamber allowing a wide range of aortic
root compliances. Testing each valve under conditions of constant
preload, afterload and contractility ensures that changes in measured parameters result only from the valve under test and that
the response of the model ventricle will be similar to that obtained in the beating heart[2,3].

b) Design parameters
   Having established the primary shape of the mono-cusp valve
three significant design parameters were chosen as a first stap to
the optimising process, i.e. : rigidity of the valve ; fixed perimeter α ; fixed height h.
In order to compare the different configurations of the valve, the
tests were carried out in similar experimental conditions. The
frequency of the positive displacement pump was 70 per minute, the
stroke volume 71 mℓ. The aortic root compliance was maintained at
$0.15 . 10^{-3}$ cm^{5}dynes^{-1}, and water was used as the test medium.
The performance parameters at the present investigation were the
pressure gradient across the valve and the amount of regurgitation. Measurements were digitised using a HP Digitizer and processed on a HP 92840 A mini-computer. It was found that the present optimum was obtained by :

        - the most rigid type of valve that was available
        - a fixed perimeter angle α of 180°
        - a fixed height of 10 mm.

This type of valve gave a mean systolic pressure difference of

7 mm Hg and a total amount of regurgitation in % of forward flow
of about 7 %. The ventricular and aortic wave form with this valve inserted in the test rig are shown in figure 5.

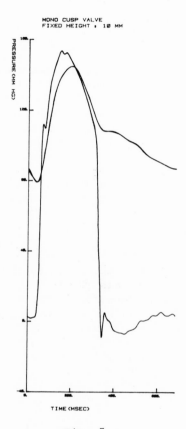

Fig. 5.

The optimum will now be modified in order to ameliorate valve performance. The further research program will include measurements
of :
    - the mean systolic pressure drop over the valves in function
      of the mean systolic flow
    - the peak pressure drop, ventricular and aortic pressure
    - the amount of regurgitation.

Also a further kinematic study of the opening and closing mechanism will be done, in addition to a visualisation study of the
flow pattern around the valve. The effect of extra-systoles is
investigated. The performance characteristics of the proposed
mono-cusp valve are compared with those of a caged ball valve,

a tilting disc valve and a xenograft.

REFERENCES

1. W.M. SWANSON & R.E. CLARK, Testing of prosthetic heart valves,
   ASME (1976).
2. T.R.P. MARTIN, R. PALMER and M.M. BLACK, A new apparatus for
   the in citro study of aortic valve mechanics, Engineering
   in Medecine, IMech. (1978), vol.7, N.4, pp. 229-230.
3. T.R.P. MARTIN and M.M. BLACK, Problems of in vitro testing of
   valve replacements, Proc. European Soc. Artificial Organs III,
   (1976).
4. C. KRAMER, H.J. GERHARDT, H. SCHWERIN and E.W. MÜLLER, Künst-
   liche herzklappen, modelluntersuchungen, Proc. Jahrestagung
   der Deutschen Gesellschaft für Kreislaufforschung, Wien,
   (1976).
5. P. YOGANATHAN and C. HARRISON, Pressure drops across prosthetic
   aortic heart valves under steady and pulsatile flow - in
   vitro measurements, J. Biomechanics, vol.12, (1979), pp. 153-
   163.

ACKNOWLEDGEMENTS

    This research is partially sponsored by the Research Council
of the Free University of Brussels (V.U.B.) and the Belgian Foundation of Medical Scientific Research (F.G.W.O.). We are very
grateful to Mr. J. VERZELEN, technical engineer for his help on
the subject of electronics and to Mr. A. VRIJDAG (photographs)
and technician Mr. G. VAN DEN NEST. We also thank Mrs. M. BOURLAU
for the incredible fast typing of this paper.

ASPECTS OF MEASURING TECHNIQUES IN HAEMORHEOLOGY WITHIN THE
LOW SHEAR RANGE

C.D.Meier

Contraves AG, Zurich, Switzerland

(Abstract)

The great variety of rheological measurements carried out in
the course of the last years with whole blood by rotational rheo
meters show the difficulties sometimes existing in establishing
arelationship between each measurement. It is common knowledge
that blood viscosity is determined by: 1) Haematocrit,
2) Erythrocytes aggregation, 3) Erythrocytes deformability.

When measurements are judged consideration should however
be given to measuring-technical factors that might strongly affect
results. Little attention is generally given to the factor "time"
prior to and in the course of the measuring process, erythrocytes
aggregation and erythrocytes' sedimentation speed being directly
influenced. Also the elastic properties present in whole blood
and even in its plasma must be considered when measuring
conditions are established. Further aspects are revealed in
discussing the Sigma phenomenon, shearing speed and surface
influences appearing on cylindrical-rotational and oscillation
rheometers when measuring non-Newtonian substances. The individual
factors are examined by a Low Shear 30-sinus on normal human blood
at a continuous and sine-wave shear stress, the effect of those
factors being explained on the basis of measuring results.

CONSTITUTIVE EQUATIONS FOR LARGE HUMAN ARTERIES

Stoyan Stoychev

Bulgarian Academy of Sciences
Institute of Mechanics and Biomechanics
1113 Sofia, ul. Akad. G.Bonchev, Bl.8

INTRODUCTION

The mechanical properties of large blood vessels have been intensively studied in two-dimensional experiments and different mathematical constitutive equations to describe their behaviour have been proposed [1-8] in the past years. In the papers[1,2,4] the strain-energy function (elastic potential) W is approximated by a polynomial in strain tensor components or deformation invariants. The unknown material constants are determined as to fitt the experimental data. Fung[8] approximated W by an exponent, Kasyanov and Knets[5] - by two exponents, in the papers[6,7] W consisted of an exponent and polynomial. To our knowledge, experimental data from human arteries to determine the material constants have been used only by Kasyanovand Knets[8].

The purpose of this paper is to present an analytical expression for the elastic potential W describing the mechanical properties of various large human arteries and to discuss the dependence of the material constants on the localization and age.

MATERIAL AND METHODS

The experiments reported herein were performed on segments of four areas: carotid, iliaca communis, iliaca externa and femoralis, which will be denoted by C, IG, IE and F, respectively. The material was taken

during autopsy in the Department of Pathologycal Anatomy
at Medical Academy. The age of the individuals was from
13 to 75 years.

The segments were mounted in the experimental
apparatus which has been described in detail[9]. The
apparatus enables measuring of the internal pressure,
internal volume, length, axial force and wall thickness
at constant length. The segments were inflated with
physiological solution between zero and 200 mm Hg in
steps of 20 mm Hg. The axial force and internal pressure
were measured after the relaxation of the material which
took place after about 100 sec.

Values of the experimental wall stress components
were computed from the equilibrium equations for the
thin-walled tube loaded by an internal pressure P and
axial force F.

$$S_1 = \frac{p.R}{2h} + \frac{F}{2\pi R} \; ; \; S_2 = \frac{p.R}{h} \tag{1}$$

where: R is the mid-wall radius, h is the wall thickness, the subscripts 1 and 2 correspond to the
axial and circumferential directions, respectively.

For the purpose of this work 83 arteries separated
in two age groups - between 14 and 40 years and 41-75
years - were investigated. In what follows these age
groups will be referred as the first and the second

AN ELASTIC POTENTIAL FOR LARGE HUMAN ARTERIES

We consider the blood vessel as a thin cylindrical
tube made of an orthotropic, incompressible and homogeneous matwrial. It is shown[10] that for an incompressible material the elastic potential (strain-energy
function per unit volume of the undeformed material)
could be written in the following form :

$$W = W (E_1, E_2) \tag{2}$$

where: E_1 and E_2 are the components of Green-Lagrange
tensor.

Kirchhoff's stresses $\tilde{S}$ are obtained by means of
the formulae

$$\tilde{S}_1 = \frac{\partial W}{\partial E_1}, \quad \tilde{S}_2 = \frac{\partial W}{\partial E_2} \tag{3}$$

It is established, in a prevous author's paper[11], that the physiological axial deformation E_O does not vary with the changes of the internal pressure P. If the axial deformation $E_1 \neq E_O$, the axial force F is a decreasing or increasing function of E_2. This means that the composit structure of blood vessels is optimised in such a way that the axial stress has a minimum at the physiological length of the vessel. In accordance with this experimental result we suppose the physiological axial deformation E_O to be a constant in the analytical form of the elastic potential W.

We assume W to be a polynomial-exponential function of the strain components

$$W = A E_1^4 + B(E_1 - E_O)^2 E_2^2 + C\left\{ \exp\left[D\left(\frac{E_1}{2} + E_2 \right) \right] - \frac{D}{2}E_1 - DE_2 - 1 \right\} \quad (4)$$

where the condition $W(0,0) = 0$ is satisfied.

According to (3) the wall stresses are

$$\tilde{S}_1 = 4AE_1^3 + 2B(E_1 - E_O) E_2^2 + \frac{CD}{2}\left\{ \exp\left[D\left(\frac{E_1}{2} + E_2 \right) \right] - 1 \right\}$$

$$\tilde{S}_2 = 2B(E_1 - E_O)^2 E_2 + CD\left\{ \exp\left[D\left(\frac{E_1}{2} + E_2 \right) \right] - 1 \right\} \quad (5)$$

with physical components

$$S_1 = \lambda_1^2 \tilde{S}_1; \qquad S_2 = \lambda_2^2 \tilde{S}_2 \quad (6)$$

It is seen that the five material constants A, B, C, D, E_O are involved in the first formula (5). This enables the constants to be determined from the experimental stresses in the axial direction by least squares analysis and their values to be used for a calculation of the stress $\tilde{S}_2$ using the second formula (5). The agreement between the experimental and calculated stresses was evaluated by means of the determination coefficient K defined as [12]

$$K = \frac{\sum_{i=1}^{N} \left[S_{2T}(i) - S_{2E}(i) \right]^2}{\sum_{i=1}^{N} \left[\bar{S}_{2E} - S_{2E}(i) \right]^2} \quad (7)$$

where: S_{2T} is the calculated stress by the second formula
       (5),
       S_{2E} is the experimental stress (1),
       $\bar{S}_{2E}$ is the mean of all N measured values S_2.
When $S_{2E}(i) = S_{2T}(i)$ for all i, $K = 0$.

The material constants determined in this way were
separated in eight groups - four localizations and two
ages - and student's criterion was used for establishing
statistically significant differences between the mean
values of the coefficients.

RESULTS

Each artery was inflated in four lengths. So 332
experiments were carried out and 747 experimental stress-
strain relations were calculated, 498 of which were used
to determine the material constants and 249 to evaluate
the determination coefficient K.

The mean values and the standard deviation at
$p < 0.05$ of the constants A, B, C, D, E_0 are given in
TABLE 1. The number of the specimens in each group is
shown in the last row. The distribution of A and B was
assumed to be LOG-normal, of C and E_0 - near to normal
and of D - normal.

The results of the statistical analysis could be
summarized briefly in the following points: 1). The
mean values of the constants in all localizations are
different for the both age groups - thus, the orthotropic mechanical properties of the arteries investigated
change significantly with the age; 2). The mean values
of A, C, D in the first age group do not differ for the
localizations below the bifurcation of the aorta IC, IE
and F; 3). The physiological axial deformation E_0 is
a constant for all blastic arteries C, IC and IE in
each age group.

The values of the determination coefficient K calculated from (7) are: in 65% of the specimens $K < 0.1$;
in 23% - $K < 0.2$; in 10% - $K < 0.3$ and only in 2% $K < 0.4$.
This means that the analytical expressions for the constitutive equations (5) fitt well the experimental results and the equations (5) describe well the main
features in the mechanical behaviour of the human arteries and their dependence on the localization and age.

TABLE 1. The mean values and SD of the constants A, B, C, D, E₀ for the constitutive equations (5). First age group is from 13 to 40 years, second age group is from 41 to 75 years. C - carotid artery, IC - iliaca communis, IE - iliaca externa, F - femoralis. N is a number of specimens

COEFFICIENT	AGE GROUP	C	IC	IE	F
$A \cdot 10^5 \; \dfrac{N}{m^2}$	I	$6.72^{+6.61}_{-3.31}$	$4.75^{+4.08}_{-2.19}$	$5.11^{+5.14}_{-2.55}$	$11.1^{+15.8}_{-6.4}$
	II	$22.5^{+20.9}_{-13.8}$	$34.4^{+35.1}_{-21.3}$	$18.4^{+41.6}_{-12.7}$	$37.2^{+92.1}_{-26.4}$
$B \cdot 10^5 \; \dfrac{N}{m^2}$	I	$3.96^{+1.72}_{-1.19}$	$13.7^{+14.9}_{-7.1}$	$7.33^{+6.16}_{-3.34}$	$19.1^{+20.1}_{-9.7}$
	II	$34.3^{+26.6}_{-14.9}$	$40.9^{+65.5}_{-25.2}$	$44.4^{+46.7}_{-22.7}$	$49.18^{+47.8}_{-24.2}$
$C \; \dfrac{N}{m^2}$	I	546^{+125}_{-102}	247^{+61}_{-49}	386^{+95}_{-74}	273^{+27}_{-24}
	II	181^{+42}_{-33}	336^{+80}_{-65}	300^{+72}_{-37}	248^{+24}_{-22}
D	I	6.17 ± 0.77	10.9 ± 2.57	9.68 ± 2.54	11.6 ± 1.71
	II	12.07 ± 2.28	13.2 ± 2.28	20.9 ± 5.31	18.6 ± 2.79
E_o	I	$0.196^{+0.086}_{-0.058}$	$0.184^{+0.041}_{-0.033}$	$0.252^{+0.081}_{-0.061}$	$0.314^{+0.133}_{-0.092}$
	II	$0.064^{+0.031}_{-0.021}$	$0.037^{+0.032}_{-0.017}$	$0.063^{+0.076}_{-0.034}$	$0.099^{+0.051}_{-0.033}$
N	I	11	10	11	11
	II	11	9	9	11

DISCUSSION

In this paper an analytical form of the elastic potential W with five material constants is proposed. The function W contains the experimental fact that the axial stress has an extremum at the physiological length of the blood vessel. The procedure for determination of the material constants from the experimental data in axial direction is controlled by the determination coefficient. In the paper mean value $\pm$ S.D. of the constants is presented and their variation with the localization and age is analysed. The results could be used in solving boundary-value problems in mechanics of circulation.

If we put $E_1 = E_0$ in (4), W simplifies

$$W(E_1=E_0,E_2) = AE_0^4 + C\left\{\exp\left[D(\frac{E_1}{2}+E_2)\right] - \frac{DE_0}{2} - DE_2 - 1\right\} \quad (8)$$

where E_0 is a known constant for each artery.

If the first term in (8) is neglected, the form (8) coincides with the proposed by Fung.[8]

REFERENCES

1. Tickner E.G., A. H. Sacks, A Theory for the Static Elastic Behavior of the Blood Vessels. - Biorheology, 4, 151-168, 1967.
2. Patel D.J., R.N.Vaishnav. The Rheology of Large Blood Vessels. - In: "Cardiovascular Fluid Dynamics", ed. by D.H.Bergel, Pergamon Press, London, 1972, 1-64.
3. Bergel D.H. The Properties of Blood Vessels. - In: "Biomechanics, its foundation and objectives". New Jersey, Prentice Hall Inc., Engewood Cliffs, 1972.
4. Cox R.H. Anisotropic Properties of the Canine Carotid Artery in vitro. - J.Biomechanics, 8, 1975, 293-300.
5. Kasyanov B.A., I.V.Knets. Strain Energy Function for Large Human Arteries. - Mech.Polimerov, 1974, 122-128.
6. Brankov G., A.Rachev, St.Stoychev. On the Nonlinear Elastic Behaviour of the Biological Membranes. - In: Proc. of the Second National Congress on Mechanics, Sofia, 1976, v.2.
7. Tong P., Y.C.Fung. The Stress-Strain Relationship for the Skin. - J.Biomechanics, 9, 1976, 649-657.

8. Fung Y.C. Biomechanics, - In: "Theoretical and
      Applied Mechanics", W.T.Koiter ed. North-Holland
      Publ.Company, 1976.
9. Stoychev St., L.Draganov, A.Rachev. A Device for
      Measuring the Mechanical Properties of the Blood
      Vessels in vitro. - Biomechanika, Sofia, 3, 1976,
      75-79.
10. Rachev A. Nonlinear Mechanical Behaviour of Elastic
      Orthotropic Membranes. Application in Biomecha-
      nics. - Biomechanics (Sofia), 1979, vol.9, 3-16.
11. Stoychev St. Stress-Strain Relationship for Human
      Arteries. - Proc.of the Third Int.Congress of
      Biorheology, 1978, La Jolla, California, p.159.
12. Roscoe J.T. "Fundamental Research Statistics for
      the Behavioral Sciences". Holt, Rinehart & Winston,
      New York, 1969.

VELOCITY PROFILE MEASUREMENTS BY LASER-DOPPLER VELOCIMETRY (LDV) IN
PLANE CAPILLARIES.   COMPARISON WITH THEORETICAL PROFILES FROM A TWO-
FLUID MODEL

J. Dufaux, D. Quemada, and P. Mills

Laboratoire de Biorhéologie
Université Paris VII
Paris, France

INTRODUCTION

Since the YEH and CUMMINS' work [1] the velocity measurement of
suspension flow with Laser Doppler Velocimeter has been widely developped. There are no important problems when the suspension concentration is low and when the tube dimensions are greater than
those of the volume where the light beams interfere.

Different authors [2][3][4] have done measurements in capillary tube
by reducing the spot. Several difficulties have appeared such as
the broadening of the spectrum of the scattered signal. This broadening is due to the rediffusion of the scattered light and, when the
speed is low to an inadequate separation of the signal from the
continuous level (pedestal).

We have constructed a system which avoid these disturbances
at least in part.

THE EXPERIMENT

The Capillary Viscometer
    In order to construct a capillary viscometer, we used the fol-
lowing equipment. The piston of a 10 cc syringe is moved forward by
the motor of a perfusion pump (Precidor, model 5003) expelling
fluid at constant flow rate between two rectangular glass plates
kept apart at a distance of 2a from one another by the means of mylar ribbons of a specific thickness. The flow rate can vary from
10^{-5} cm^3 s^{-1} to 10^{-2} cm^3 s^{-1} (corresponding to mean shear rate from
9.10^{-2} s^{-1} to 90 s^{-1}). The pressure measurement is done with a differential pressure transducer (Validyne 45, $0 < p \leqslant 20$ cm of water)

Laser Doppler Velocimetry (LDV)

     For a particle moving with velocity V through the interference
spot the well known Doppler effect leads to a frequency shift between the frequency of the hight scattered from the moving particle
and the incident hight frequency. The frequency shift f_D is given
by:

$$f_D = \frac{2n \, \sin\theta/2}{\lambda} \, V$$

where λ = light wave length
      n = refraction index
      θ = angle between the two incident light beams

     The mean dimension d of the interference volume is given by:

$$d = \frac{4}{\pi} \, \lambda \, \frac{f}{a}$$

where f = focal length of the focalisation lens
      a = light beam diameter.

     From the modification of these two parameters results the pos-
sibility to decrease the scattering volume. The use of a Bragg cell
allows to shift the Doppler signal frequency and to separate it
from the continuous level.

     Figure 1 shows one of the system that we have use. A spatial

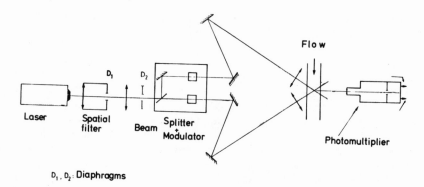

D_1, D_2: Diaphragms

Figure 1 : The modified Laser Doppler Velocimeter

filter gives a parallel light beam of 0,5 cm in diameter. The focal
length of two lens is equal to 4 cm. Then, the mean dimension of
the measurement volume is:

$$d = \frac{4}{\pi} \ 0.63 \ \frac{4}{0,5} \approx \ 6,5\mu$$

The use of a Bragg cell allows us to measure velocity down to
0,1 mm /s.

EXPERIMENTAL RESULTS

We have use different sort of suspensions. In this report we
give the results obtained with whole human blood. Figure 2a and
2b shows us the results obtained on a capillary plane tube of 350μ
in thickness.

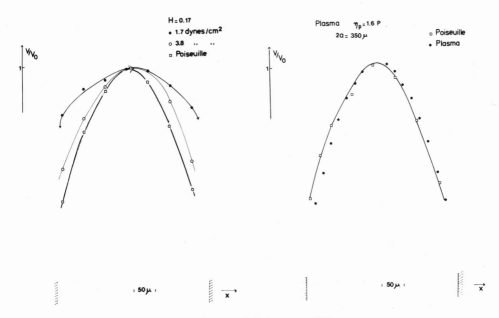

Figure 2 : Velocity profiles

In the case of a newtonian fluid such as plasma the Poiseuille
Haake profile is obviously found. We can observe a good agreement
between the measured flow rate and the integrated flow rate. We
have also noticed that the measurements are not very accurate near
the walls.

COMPARISON WITH A THEORETICAL MODEL

The model assumed that the fluid flowing through the capillary tube is divided into two phases a particle free marginal layer,
the viscosity of which is those of the plasma η_p, a central core,
the concentration of which is H. Its viscosity is described by a
relation given by one of the authors [5] :

$$\eta_r = \frac{1}{(1 - \frac{kH}{2})^2}$$

where:

$$k = \frac{k_0 + k_\infty \dot{\gamma}^{\frac{1}{2}} / \dot{\gamma}_c^{\frac{1}{2}}}{1 + \dot{\gamma}^{\frac{1}{2}} / \dot{\gamma}_c^{\frac{1}{2}}}$$

with $\dot{\gamma}$ = shear rate
k_0 , k_∞ , $\dot{\gamma}_c$ structural parameters which describe the rheological
state of the suspension.

It is possible to obtain the local velocity in the fluid by
integrating the following relations:

$$\dot{\gamma}(x) = - \frac{dv}{dx} = + \frac{\sigma_w}{\eta_p}$$

and

$$\dot{\gamma}^{\frac{1}{2}} = \frac{\sigma_w^{\frac{1}{2}}}{\eta_p^{\frac{1}{2}}} \left(1 - \frac{1}{2} (k_0 + k_\infty \alpha_c \dot{\gamma}^{\frac{1}{2}} / 1 + \alpha_c \dot{\gamma}^{\frac{1}{2}}) H \right)$$

where $\alpha_c = \frac{1}{\dot{\gamma}_c^{\frac{1}{2}}}$

σ_w= shear stress
in the core. The results are shown on figure 3.

The value of the marginal layer thickness is obtained with a method given in Professor Quemada's communication (see this issue).

The value characteristics parameters of the suspension are:

H = .48
Thickness of the plane capillary 370 μm
Thickness of the plasmatic layer   7 μm
k_0 = 4.15
k_∞ = 1.85
α_c = .225

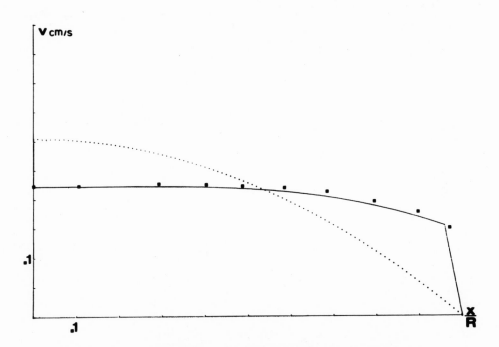

Figure 3 : Comparison of experimental and theoretical
results. The parameters k_0 , k_∞ , $\dot{\gamma}_C$ are
derived from the viscosity diagrams $\eta = f(\dot{\gamma})$
obtained at fixed hematocrit, using a
Couette viscometer [6].

CONCLUSION

The modification of a classical LDV equipment allows to
obtain velocity profils in plane capillary tubes up to 400µ in
thickness, for an hematocrit up to 0,50.

Using a viscosity law, a theoretical model can explain the
experimental results.

References

[1]   YEH Y. and CUMMINS H.Z., Applied Physics Letters, 4,176(1964)
[2]   RIVA C., ROSS B. and BENEDEK G.B., Investigative Ophtalmology
      11, pp.936-944 (1972).
[3]   EINAV S. et al., Biorheology 12, 203 (1975 a).
[4]   TANAKA T. and BENEDEK G.B., Applied Optics, vol 14 n°1,Janua-
      ry (1975).

[5]   QUEMADA D., Rheol.Acta, 16, 82:94 (1977); 17, 632:642 (1978);
      17, 643:653 (1978).
[6]   DUFAUX J., QUEMADA D., MILLS P., Submitted to Journal de Physi-
      que Appliquée.(1979).

AN OPTICAL METHOD FOR STUDYING RED BLOOD CELLS ORIENTATION AND

AGGREGATION IN A COUETTE FLOW OF BLOOD SUSPENSION

P. Mills, D. Quemada, and J. Dufaux

Laboratoire de Biorhéologie
Université Paris VII
Paris, France

INTRODUCTION

Red blood cell aggregation is certainly the most important factor in the non newtonian behaviour of normal human blood. Aggregated red blood cells build up rouleaux; this aggregation is reversible: at rest, the rouleaux form a three dimensional network structure; when a finite stress is applied to the suspension the structure breaks. Increasing the stress, the rouleaux are gradually disrupted and are finally reduced up to individual cells. With decrease in stress individual cells build up again rouleaux. In a steady state a dynamical equilibrium exists between the size of rouleaux and the stress applied. When the applied stress is high enough to break the rouleaux, cells can be oriented and deformed in the flow. There was studied an experimental method which allows us to rely the luminous flux backscattered by rouleaux and individual cells to their mean size and to their orientation.

A review on this subject is given in [1]. So we only mention the experimental results of H. SCHMID-SCHONBEIN and al. about the transmitted luminous flux through a blood sample in a cone plane viscometer [2] and in France DOGNON and HEALY's [3], and author's [1,4] works.

TRANSMITTED AND BACKSCATTERED FLUX - THEORY

We briefly expose theoretical findings about the interaction between a luminous beam and a suspension of aggregable particles. These ones can be deformed and oriented in the flow. Details of calculus are given in our paper [1].

Consider an homogeneous suspension between two planes which are separated by the distance ℓ ; if ϕ_0 is the incident luminous flux, ϕ_r is the reflected luminous flux, and μ is particle absorption by volume unit.

Let be f the isotropic scattering factor and s the specular reflecting factor. The ratio s/f depends from the particle shape and particle orientation.

The total reflected flux $[\phi_r]_{tot}$, and the specularly reflected flux ϕ_{rs} can be expressed as:

$$r_{tot} = \frac{\phi_r \text{ tot}}{\phi_0} = \frac{\dfrac{f + 2s}{2\sqrt{\beta}} \text{ sh } \ell\sqrt{\beta}}{\dfrac{2\mu + f + 2s}{2\sqrt{\beta}} \text{ sh } \ell\sqrt{\beta} + \text{ch } \ell\sqrt{\beta}} \tag{1}$$

$$r_s = \frac{\phi rs}{\phi_0} = \frac{\dfrac{s}{\sqrt{\delta}} \text{ sh } \ell\sqrt{\beta}}{\dfrac{2 + \dfrac{f}{2} + 2s}{2\sqrt{\delta}} \text{ sh } \ell\sqrt{\delta} + \text{ch } \ell\sqrt{\beta}} \tag{2}$$

where $\delta = (\mu + \dfrac{f}{2})(\mu + \dfrac{f}{2} + 2s)$ and $\beta = \mu + \dfrac{f}{2} + s.$

When the optical path values $\ell\sqrt{\beta}$ and $\ell\sqrt{\beta}$ are large enough for the assumption that the hyperbolic cosine and sine in equations (1) and (2) are equal; we can obtain simple expressions for r_{tot} and r_s. When $r_s = 0$ (no oriented particle), then $t_{tot} = r$ for particles in the disaggregated state and $r_{tot} = r'$ for the aggregates. In this case, $f = f_0 \sigma$ and $f' = f_0 \sigma'$, where σ is the total area of disaggregated particles and σ' the total area of aggregated particles

$$\sigma = \frac{4\mu}{f_0} \frac{r}{(1-r)^2} \quad ; \quad \sigma' = \frac{4\mu}{f_0} \frac{r'}{(1-r')^2} \tag{3}$$

from which:

$$\frac{\sigma'}{\sigma} = \frac{r'}{r} (\frac{1-r}{1-r})^2 \tag{4}$$

If we consider RBC like disks, with the radius a and the thickness b ($\alpha = a/b = 2.63$ for aggregated human RBC [5]) and rouleaux like stacking of disks, hence:

$$\frac{\sigma'}{\sigma} = \frac{\dfrac{\alpha}{m} + 1}{\alpha + 1}$$

where m is the mean number of particles per rouleau. We get the finally equation:

$$(\frac{1-r}{1-r})^2 \frac{r'}{r} = \frac{1}{3.63} (\frac{2.63}{m} + 1)$$ (5)

For non aggregating red blood cells, the ratio n_S of oriented particles to the total number N_O of particles per volume unit can be written:

$$n_S \neq k \frac{\Delta r}{r}$$

where Δr is the reflectivity difference between the random orientation state and the oriented state and where k is constant.

EXPERIMENTAL APPARATUS

The apparatus is schematicaly shown on Fig.1 . Blood sample placed between a transparent rotating cylinder and a coaxial cylinder (stator of a Couette type viscometer) was illuminated by a He Ne Laser (λ = 6328 Å) .

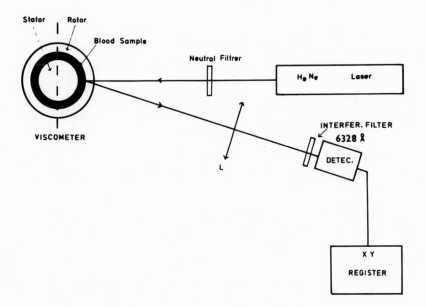

Figure 1 : Experimental apparatus

The blood sample filling the gap (ΔR = 0,5mm) was subjected to controlled shear rates in the range between $10^2 - 10^{-2}$ s^{-1} . Luminous flux backscattered with a θ angle incident beam was focused by an optical lens L. θ angle was choosen in such a way that no specular reflection was detected. The signal from the photometric detector located at the focus of L, was recorded by a X Y recorder.

As was checked by mean of a series of calibred variable density grey filters the detector worked in a linear regime. Interferen-
cial filter, located in front of the detector, filtered scattered
light in a narrow band centered at λ = 6328 Å .

EXPERIMENTAL RESULTS ($\dot{\gamma}$ = 0)

We can determine with the help of this apparatus the time kinetic of the rouleaux formation, starting from a state where RBC
are wholly dissociated.

Whole dissociation is obtained by applying sufficient shear
rates stress to blood suspension (by example, for normal blood
shear stress of about 1 dyne cm^{-2}, corresponding to a shear rate of
about $60s^{-1}$).

The curve in Fig.2 shows the time variation of the reflectivity for a normal blood sample. One can distinguish between three regions:

a) ($\dot{\gamma}$ = 128 s^{-1})
RBC are dissociated an oriented . A part of incident luminous
flux specularly reflected is not detected.

b) ($\dot{\gamma}$ = 0)
After having abruptly stopped the outer cylinder, particles
are yet dissociated but quickly get random orientations backscattered flux becomes a maximum.

c) ($\dot{\gamma}$ = 0)
Backscattered flux is decreasing when RBC are stacking to
form rouleaux.

The characteristic time τ_A for aggregation is about 5 sec (for
normal blood). For blood with dextrans with high molecular weight
this time is shorter [6] .

TRANSIENT RESPONSES TO INCREASING AND DECREASING SHEAR RATE STEPS

Orientation and Random Orientation Times.
When a suspension of three times washed RBCs in saline is submitted to a decreasing step of shear rate and then to an increasing
step, a typical response as shown in Fig.3 A is obtained. Orientation time τ_0 and random orientation time τ_{RO} are of the same order
of magnitude [$\tau_{Do} \sim 0,1s$ and $\tau_{Ro} \sim 0,2s$] . These transient times
depend on particle shape and on its internal viscosity [7] for a given suspension medium viscosity.

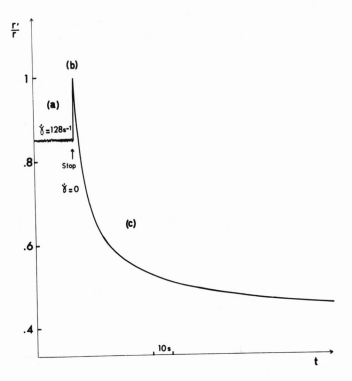

Figure 2 : Reflectivity variation with time. Normal blood.

Disaggregation and Aggregation Times.
        Applying a high shear rate ($\dot{\gamma}$ = 60s^{-1}) to a normal human
blood sample originally at rest, and then abruptly stopping the
flow typical response of Fig. 3B is obtained. The two peaks (i) and
(ii) correspond to alignment in flow and random orientation of the
dispersed cells fractions. Disaggregation time τ_{DA} is about 1s
(τ_{DA} < τ_A).

Comparison with Viscometric Data.
        Time variation of viscometer signals for the same suspension
flows as here before are very difficult to interpret. As a matter

of fact transient times depend not only on the physical characteristics of the sample but also on the apparatus response. Nevertheless photometric and viscometric results are in fair agreement when
the apparatus response can be discounted and when there is no viscoelastic effects.

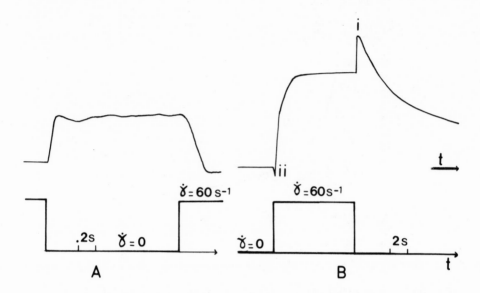

Figure 3: A) RBCs saline .   B) Normal human blood .
T = 21°C, C = .44

CONCLUSION

To conclude, new information concerning orientation and aggregation kinetics can be obtained by this method. Some microscopic
informations like rouleaux size and aggregate structure changes [6]
cannot be easily obtained by viscometry. On the other hand viscometry is more accurate in low shear rate range ($\dot{\gamma}$ < 1s^{-1}).

References

1 MILLS P., et al. Submitted to Revue de Physique Appliquée
   Paris.
2 KLOSE H.T. and al. Pflügers Arch.333,126-139(1972).
3 HEALY J.C., These, Univ.Paris VI, PARIS (1973)
4 MILLS P. et al."Hemorheology and diseases" 17-19 Nov 1979,
   Nancy, Douin Paris (In Press).
5 CHIEN S. et al.Biorheology vol 8,35-57 (1971).
6 MILLS P. et al.3rd Int.Conf.on Physico-Chemical Hydrodynamics
   Madrid March 30-April 2 (1980).
7 LERCHE D. et al.8thInt.Cong.on Rheology 1-6 Sept.(1980).

THE INFLUENCE OF THE INTERNAL VISCOSITY OF WASHED RED BLOOD CELLS ON THEIR RHEOLOGICAL BEHAVIOUR

D.Lerche, P.Mills[+], R.Glaser, J.Dufaux[+], D.Quemada[+]

Humboldt Universität zu Berlin, Germany
[+] Université Paris VII, France

(Abstract)

The rheological characterization of washed RBC suspensions has been performed both using a low shear viscosimeter and a optical light back scattering system. Changing the volume of the cells and in such a way the hemoglobin concentration in the cells and/or the structure of the cell membrane by treatments with the antibiotic Amphotericin B and different pH, osmotic pressure or ionic strengths of the medium it was found that the rheological behaviour (relative viscosity in dependence on the shear rate, $.02 \text{ s}^{-1} - 128 \text{ s}^{-1}$; intrinsic viscosities and the relaxation time corresponding to the orientation and desorientation) seems to be mainly connected with the change of the internal hemoglobin concentration.

THE QUANTITATIVE DESCRIPTION OF THE VISCOELASTIC
PROPERTIES OF HUMAN BLOOD

Pavel Říha

Institute of Hydrodynamics
Czechoslovak Academy of Sciences
166 12 Prague 6, Czechoslovakia

INTRODUCTION

Recently, various authors[1-6] have shown that human blood is
a viscoelastic material. The viscoelastic properties were conveniently measured by oscillatory tests[1-5] or by stability tests
in circular Couette flow[6].

In contrast to the large number of experimental investigations
a similar attention has not been paid to the theoretical description of rheological behaviour. Probably, only the Oldroyd constitutive model[6] and the generalized Maxwell model[7] have been
tested how good are in their capability of predicting the shear
viscosity and the complex viscosity of blood.

In view of this it then becomes of interest to investigate
the usefulness of further rheological models in predicting the
viscoelastic properties of blood. For this purposes, we shall
use experimental data of steady and oscillatory shearing flow
measurements for a fresh blood sample collected from a healthy
adult donor.

EXPERIMENTAL

The measurements of rheological properties of human blood
were carried out with the Contraves LS 30 viscometer and with

the Weissenberg Rheogoniometer /Model R 18/. The first concentric cylinder viscometer was used for determining steady
shearing properties. The oscillatory flow measurements were
taken with the Weissenberg Rheogoniometer with cone-and-plate
measuring section. The diameter of the cone-and-plate system
was 5 cm and the cone angle 2°. The frequency of oscillation
was varied from 0.1 to 10 Hz and the amplitude of oscillation
at the perimeter was 2000 μm. The viscous component η' and
elastic component η'' of the complex viscosity were calculated
according to the procedure of Walters [8]. All studies were performed at a temperature of 37 °C.

A fresh blood sample was drawn from normal human subject, with citrate solution as the anticoagulant. After centrifugation, red blood cells /RBCs/ were resuspended in the
plasma to yield hematocrit 45 % .

RESULTS

The dependence of the shear viscosity η and the compo-
nents η' and η'' of the complex viscosity on the shear rate $\dot{\gamma}$ and
the frequency of oscillations ω is shown in Fig. 1 . These findings are similar to that reported by Chien et al. [4] and Thurston[2,5,7].

The models selected for comparison with the experimental data in this study are the three-constant Oldroyd model[9],
the Spriggs model[9], the Meister model[9] and the Carreau
models[10]. The Oldroyd and Spriggs models represent rheological models of the differential type. The other models represent
rheological models of the integral type.

The material constants involved in rheological models are
determined by following procedures presented in[9,10]. The obtained values are given in Table 1. Since the data for $\eta'(\omega)$
are not convincing if $\eta'(\omega)$ is lower than $\eta(\dot{\gamma})$ or not, it is
assumed that $\eta'(\omega)$ is equal to the shear dependent viscosity
$\eta(\dot{\gamma})$ at $\omega = \dot{\gamma}$.

Comparisons of model predictions with the experimental
data for shear rate viscosity η are given in Figures 2 and 3.
Figures 4 and 5 show the model results compared with the data

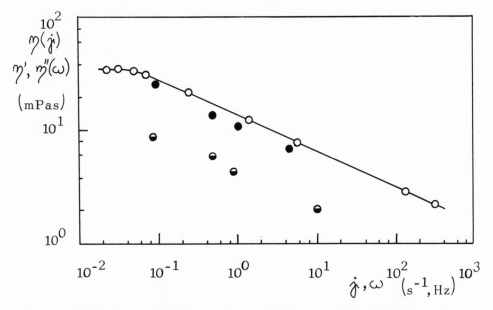

Fig. 1 Shear rate dependence of the viscosity η (o)   ,
and frequency dependence of the viscous component
η' (●) and the elastic component η'' (◐)   of the
complex viscosity.

Table 1. Model constants of human blood. Hematocrit 45 %,
temperature 37 °C.

Oldroyd	λ_1 = 2.29 sec	λ_2 = 0.21 sec	η_0 = 39 mPas	
Spriggs	α = 1.46	λ = 7 sec	c = 1	η_0 = 39
Meister	α = 1.3	λ = 6	c = 1	η_0 = 39
Carreau A	α = 1.46	λ = 11	t_1 = 22.9 sec	
	S = 0.16	η_0 = 39		
Carreau B	α = 1.46	λ = 11	t_1 = 11	
	S $\neq$ 1	η_0 = 39		

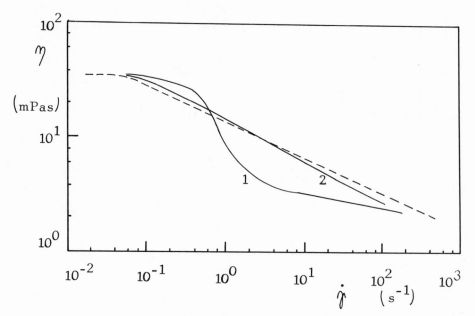

Fig. 2 Comparison of model predictions with the viscosity η
data. ----- Data. 1-Oldroyd model, 2-Spriggs model.

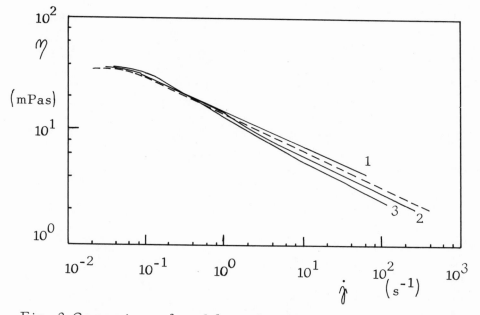

Fig. 3 Comparison of model predictions with the viscosity η
data. ----- Data. 1-Meister model, 2-Carreau model A,
3-Carreau model B.

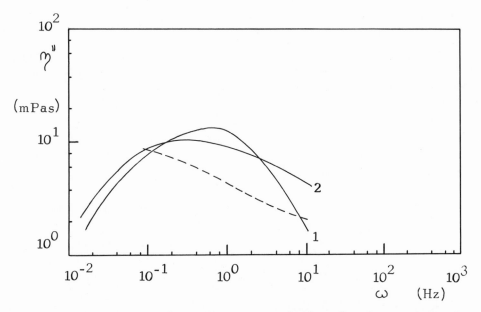

Fig. 4 Comparison of predictions with the elastic component
η'' data. ----- Data. 1-Oldroyd model, 2-Spriggs model.

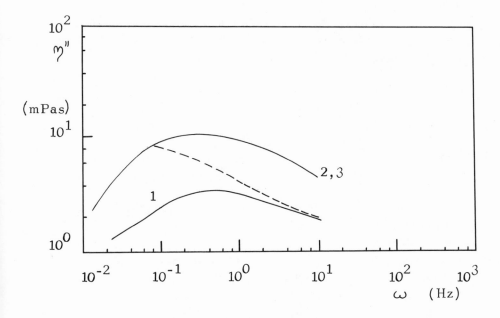

Fig. 5 Comparison of model predictions with the elastic
component η'' data. ----- Data. 1-Meister model,
2-Carreau model A, 3-Carreau model B.

for the frequency dependent elastic component η'' of the complex viscosity.

It is seen in Figs. 2 and 3 that, except the three-constant Oldroyd model, the selected models predict the viscosity function $\eta(\dot{\gamma})$ and thus the viscous component η' of the complex viscosity, reasonably well. On the other hand, the theoretical predictions of the elastic component η'' in Figs. 4 and 5 differ considerably from the measured values.

REFERENCES

1. A. Lessner, J. Zahavi, A. Silberberg, E. H. Frei, and F. Dreyfus, The viscoelastic properties of whole blood, in: "Theoretical and Clinical Hemorheology", H. H. Hartert, and A. L. Copley, eds., Springer-Verlag, Berlin-Heidelberg-New York (1971).
2. G. B. Thurston, Frequency and shear rate dependence of viscoelasticity of human blood, Biorheology 10:375 (1973).
3. A. L. Copley, R. G. King, S. Chien, S. Usami, R. Skalak, and C. R. Huang, Microscopic observations of viscoelasticity of human blood in steady and oscillatory shear, Biorheology 12:257 (1975).
4. S. Chien, R. G. King, R. Skalak, S. Usami, and A. L. Copley, Viscoelastic properties of human blood and red cell suspensions, Biorheology 12:341 (1975).
5. G. B. Thurston, Effects of hematocrit on blood viscoelasticity and in establishing normal values, Biorheology 15:239 (1978).
6. S. Deutsch, and W. M. Phillips, The use of the Taylor-Couette stability problem to validate a constitutive equation for blood, Biorheology 14:253 (1977).
7. G. B. Thurston, Rheological parameters for the viscosity, viscoelasticity and thixotropy of blood, Biorheology 16:149 (1979).
8. K. Walters, Rheometry, Chapman and Hall, London (1975).
9. Ch. D. Han, Rheology in Polymer Processing, Academic Press, New York (1976).
10. P. J. Carreau, Rheological equations from molecular network theories, Trans. Soc. Rheol. 16:99 (1972).

TEMPERATURE DEPENDENCY OF THIXOTROPIC BEHAVIOR OF WHOLE HUMAN BLOOD

C.R. Huang, J.A. Su, D. Kristol, J.D.Cohn[+]

[+] New Jersey Institute of Technology, USA
St.Barnabas Medical Center, Livingston, N.J., USA

(Abstract)

Whole human blood was recently established as a thixotropic
fluid whose apparent viscosity depends on the shear rate as well as
the duration of the shear. The rheograms which characterize a thixo
tropic fluid, the hysteresis loop and the torque-decay curve, were
observed in whole human blood samples. These rheograms can be
represented by a rheological equation developed by Huang. The thixo
tropic parameters of the equation can be used to calculate the
apparent viscosity of a whole human blood sample under different
flow conditions.In addition, they supply certain unique information
which can't be obtained by conventional viscosity measurement of
blood. This information includes the relative amount of erythrocytes
in rouleaux formation, the rate constant of breakdown of rouleaux
into individual red cells induced by a shear stress, the initial
stress to start the blood flow, the Newtonian and non-Newtonian
contribution of viscosity, etc. In this communication, we report
the temperature dependency of these thixotropic parameters of blood
samples taken from apparently healthy human subject. The temperature
ranged from 22.8 to 41°C. Rheograms were measured of each whole
blood sample via a Weissenberg Rheogoniometer modified with an
electronic continuous drive system. Thixotropic parameters of the
sample were calculated by a digital computer via a non-linear
regression analysis program. Our results show the following original
findings which will be of interest to biorheologists as well as
physiologists. When these thixotropic parameters were plotted

against the reciprocal absolute temperature, it was found that the
yield stress, the Newtonian contribution of viscosity, the relative
amount of rouleaux, the apparent viscosity and the non-Newtonian
contribution of viscosity are all at a minimum value at 37°C. The
rate constant of rouleaux breakdown to individual red cells has a
linear relationship with a negative slope, from which, the activ
ation energy of the rouleaux breakdown reaction can be evaluated.
It is also found that the mechanicism of rouleaux breakdown
reaction is independent of temperature.

MISCELLANEOUS

A RHEOLOGICAL STUDY OF A SUPERPLASTIC SHEET METAL FORMING PROCESS

M. Bidhendi and T.C. Hsu

Research Student and Professor of
Applied Mechanics, respectively

Department of Mechanical Engineering
University of Aston, Gosta Green
Birmingham B4 7ET, Great Britain

INTRODUCTION

Superplastic alloys, first known over half a century ago,
[Refs. (1).(2)] have been considered by engineers and metallurgists,
with less than perfect justification, to be materials of excep-
tionally high ductility. Actually, superplasticity is not a
matter of ductility - which is defined by the strain at fracture
under quasi-static conditions - but is due to the dependence of
flow stress on the strain rate, in other words, it is a rheological phenomenon. The basis of superplasticity is neither in the
late development of the localized neck in the material when
stretched, nor in the large strain at the fracture, but lies in
the fact that the neck does not aggravate itself after it is
formed, the strain-rate effect on the flow stress preventing the
formation of the highly localized necking in the tension test
specimens of ordinary ductile metals.

In the last decade, much research has been done on superplasticity, but mostly on the superplastic behaviour of materials
under uni-axial tension. The application of superplastic sheet
materials to difficult forming processes has given rise to an
interest in superplastic sheet metal forming, and some research
papers have appeared on the performance of such materials under
particular forming operations, with empirical results which,
naturally, hardly reveal the rheological properties of the materials. In the research project partly reported in this paper, a
zinc-aluminium superplastic alloy (Zn 77.5%,Al 22.0%,Cu 0.5%) in
sheets of 2.54, 1.91 and 1.27 mm thickness, is studied in the
bulge test. The bulge test, widely used for determining the
ductility of sheet materials, consists of clamping the test

material over a round hole and deforming the sheet with hydraulic
or pneumatic pressure on one side of it till it forms a bulge. The
ductility of the test material is represented by the height of the
bulge when fracture occurs. Some papers have also been published
on the bulge test of superplastic materials, nearly always with
the simplifying assumption that the bulge is spherical.

Experimental Technique

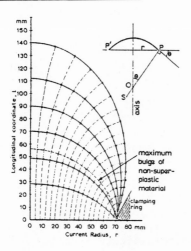

Figure 1. Half Sections of
the Bulge at Various Stages

The test used in the present
project is carried out at 250 °C
in an oven with transparent wall
and door, but it differs from
the ordinary bulge test in two
ways. Firstly, in the bulge test
for non-superplastic materials
the bulge rarely reaches a
height equal to the radius of
the hole (see Fig.1), but in our
tests the superplastic sheets
are formed into the shape of a
Rugby football before it breaks.
Secondly, concentric circles are
printed on the test material by
photo-resist and the positions
of these circles, as well as the
shapes of the bulge at various
stages of the forming process
are carefully measured. In
Fig.1 the solid lines represent
the meridional sections of the
bulge on one side of the axis of
symmetry, and the dotted lines
are the loci of the intersections of the printed circles with the
meridional sections. Some reflection will show that once the
meridional sections and the positions of these circles are known,
all the strains, strain-distributions and strain rates can be
determined[Ref.(3)].Samples of such results will be shown later
in this paper.

Actually, Fig.1 does not represent all the experimental data
that can be obtained in this test. The bulge grows to a height of
about 190 mm before it breaks at its pole, but only the data up to
a height of 140 mm are shown in Fig.1. Below 140 mm, the forming
process is controlled by supplying or shutting the supply of the
pre-heated nitrogen and the bulge ceases to grow once the gas
supply is stopped. However, at some height between 140 and 150 mm
the bulge grows by itself even after the gas supply is turned off,
showing that deformation continues in decreasing internal pressure.
This instable regime, which requires cinematographic experimental
techniques to study, is not investigated in this project, but a
qualitative analysis is given here for this peculiar behaviour.

Suppose the bulge is idealized as a spherical shell of radius a, so that, by elementary Strength of Materials, the stress σ is

$$\sigma = \frac{pa}{2t} \qquad (1)$$

where p is the internal pressure and t is the thickness of the shell. Considering that the volume of the shell material (v) remains constant at $(4\pi a^2 t)$ and that during expansion the pressure obeys the gas law

$$(\text{pressure}) \times (\text{volume}) = p.\frac{4\pi}{3} a^3 = \text{constant } k \qquad (2)$$

it can be readily seen in Eq.(1) that

$$\sigma = \frac{3k}{2v} \qquad (3)$$

which is constant. In other words, according to Eq.(3) a bulge can grow without additional gas supply, with thinning shell and decreasing pressure, only if there is no strain-hardening. Actually, the bulge is not a sphere, but in the theory of membrane stresses, discussed in the next section, it can be seen that the arguments for a constant or nearly constant flow stress is valid. Hence, the strain rate is the main factor governing the flow in this case.

Theoretical Background

In an axisymmetrical shell, the principal directions for the curvatures and the stresses are the circumferential and the meridional tangential directions. Thus, in the inset of Fig.1, PP' is the meridional section of a polar cap of the shell, the meridional slope of the surface at P is $(\tan \theta)$ and the principal directions are along the tangent through P as shown (meridional tangential) and in a plane perpendicular to the paper passed through the normal OP (circumferential). The well-known equation for the membrane stresses in thin shells then takes the form of

$$\frac{\sigma_s}{\rho_s} + \frac{\sigma_\theta}{\rho_\theta} = \frac{p}{t} \qquad (4)$$

where σ is for stress, subscript s is for the meridional tangential, subscript θ is for the circumferential and ρ is the radius of curvature. By considering the equilibrium of the polar cap shown in Fig.1 (inset), it can be shown that

$$\sigma_s = \frac{pr}{2t\sin\theta} = \frac{p}{2t} (\overline{OP}) \qquad (5)$$

By Geometry, $\qquad \overline{OP} = \rho_\theta \qquad (6)$

hence, combining Eqs. (6) (5) and (4) we get

$$\frac{\sigma_\theta}{\sigma_s} = 2 - \frac{\rho_\theta}{\rho_s} \qquad (7)$$

The left-hand side of Eq.(7) is the ratio of the two stresses which determine the mode of deformation and the change of shape of the shell, and on the right-hand side the ratio ρ_θ/ρ_s is a

characteristic of the local geometry related to a concept called
"prolateness". "Prolate" means egg-shaped and prolateness measures
the deviation from the perfect sphere. Thus, ρ_θ/ρ_s is equal to one
for a sphere, positive and less than one for an egg or egg-
shaped surface, and greater than one for an oblate spheroid, like
the shape of Dutch cheese.

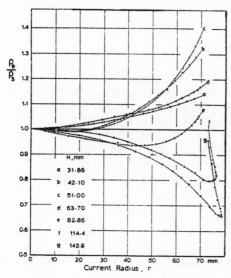

Figure 2. Deviations of the
Actual Bulge from the Sphere

The membrane stress formula,
Eq.(4), shows that the stresses
are related to the curvatures.
The curvatures, however, cannot
be directly measured on the
bulge. In this project they are
calculated from the measured
slope (θ in Fig.), The ratio
ρ_θ/ρ_s which, by Eq.(7),determines
the mode of deformation, is thus
determined experimentally and
plotted against the current
radius in Fig.2 where H is the
height of the bulge. If the
bulge were a sphere, the curves
in Fig.2 would all be the
horizontal line through $\rho_\theta/\rho_s = 1$
and it is obvious that in the
bulge test the bulge is not at
all like a sphere, hence the
results of any analysis involving
an assumption of the spherical
shape are inevitably of limited

Deviations of the Actual
Bulge from the Sphere.   Fig.2

value. In fact, in most cases
the bulge deviates from the
sphere in a complicated way.Thus,
in the shell represented in curve (e) the bulge is a sphere at the
pole (r = o), becomes egg-shaped at high latitudes, returns to a
sphere again and becomes oblate (flattened) near the rim.

Some Results

From the part of our results which may interest rheologists,
only two sets are presented here:the effect of strain rate on the
flow stress (Fig.3) and the strain paths at different parts of
the bulge (Fig.4). Each curve in Fig.3 represents a test at constant pressure and the stress determined is that at the pole of
the bulge. To determine these stresses and strain rates, each
test is run twice, first at constant pressure for the measurement
of the rate of growing height of the bulge and then another test
is run which is stopped at various stages for the detailed
measurements of displacements and slopes at those stages.There-
fore,the change in strain rate in each curve is due to the natural

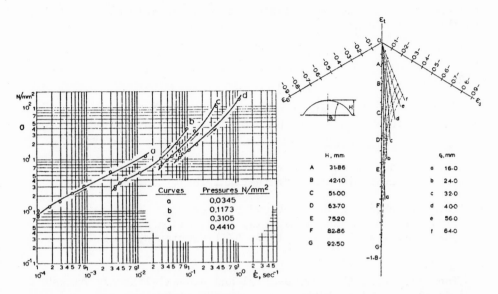

Figure 3. The Effects of Strain
Rates on the Flow Stress (Constant
Pressures)

Figure 4. Strain Paths
and Curves for Strain
Distributions

growth of the bulge. In Fig.3 the arrows indicate that during a
test the strain rate at the pole increases with the height of the
bulge. It can be shown by further analysis of the results in Fig
Fig.3 (not included in this paper) that the flow stress is a
function of both the strain rate and the strain.

In Fig.4 a triangular coordinate is used to represent the
simultaneous changes in the through-thickness strain (ε_t), the
circumferential strain (ε_θ) and the meridional tangential strain
(ε_s). The coordinate system is similar to that used by metallurgists for the properties of three-component alloys, and
accommodates the equation for the incompressibility of the metal,
[Ref.(4)] namely,

$$\varepsilon_t + \varepsilon_\theta + \varepsilon_s = 0 \qquad\qquad (8)$$

Some reflection will show that a vertical strain path pointing
downwards represents a balanced bi-axial stretch (ε_t negative)
and that a downward strain path leaning to the right is for a
thinning process in which the shell is locally stretched less in
the circumferential direction than in the meridional tangential
direction. The dotted lines in Fig.4 show the strain distributions because each of them shows the strains in the bulge at a
particular stage of the test. The maximum strain (ε_t) obtainable
in the bulge test is much greater than the value (-1.8) shown in

Fig.4 and is -3.44 when $\underline{H}$ is 142 mm and considerably exceeds-4.0 when the bulge breaks.

REFERENCES

(1) W.Rosenhain, I.L. Haughton and K.E.Bingham, "Zinc Alloys with Aluminium and Copper", J.Inst.Metals,1920, 23, 261.

(2) C.E.Pearson,"The Viscous Properties of Extruded Entectic Alloys of Lead-tin and Bismuth-tin", Ibid, 1934, 54, 111.

(3) T.C.Hsu, H.M.Shang, T.C.Lee and S.Y.Lee, "Flow Stresses in Sheet Material Formed into Nearly Spherical Shapes" , Trans.A.S.M.E.,1975, 97(1) Series H,57-65.

(4) T.C.Hsu "The Characteristics of Co-axial and Non-axial Strain Paths", J.Strain Analysis, 1966, 7(3), 216-222.

THE RHEOLOGICAL ASSESSMENT OF PROPELLANTS

F. S. Baker*, R. E. Carter*, and R. C. Warren**

*Propellants, Explosives and Rocket Motor Establishment,
  Waltham Abbey, Essex, England.
**Weapons Systems Research Laboratories,
  Salisbury, South Australia

INTRODUCTION

Propellants are made by incorporating fillers and volatile solvents with a paste of nitroglycerine and nitrocellulose, the former acting as a plasticiser. The extruded dough is cut to length and dried.

To monitor and understand the changes in the rheological properties of the materials during the incorporation stage, and the rheological behaviour of the dough on extrusion, a laboratory scale processing facility containing a torque rheometer and a capillary extrusion rheometer has been built.

EXPERIMENTAL

The torque rheometer was a fully instrumented jacketed mixer of capacity 100 grammes, modified for use with explosives. During incorporation, the mixer temperature could be controlled to better than $\pm 1^{o}C$ within the range $0 - 80^{o}C$.

The capillary extrusion rheometer was a fully instrumented, electrically driven ram extruder, of 100 grammes capacity, again modified for use with explosives. The operating temperature range was $0 - 80^{o}C$, with control accuracy to better than $\pm 0.1^{o}C$. This rheometer was a constant rate device with pressure sensors fitted both on the piston and at the die entrance. Die swell was determined using a line-scan camera.

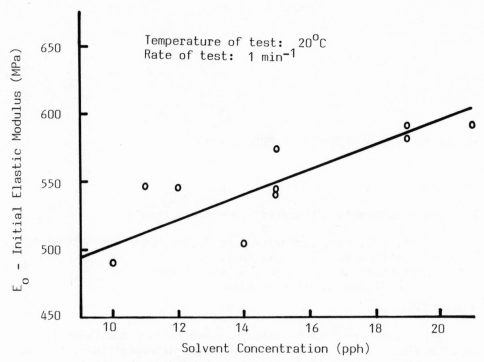

Figure 1:   The Variation in Initial Modulus with Solvent
            Concentration

RESULTS AND DISCUSSION

     Nitrocellulose has a fibrous nature and possesses a high
degree of crystallinity (1,2). During processing, much of this
fibrous nature is destroyed resulting in an appreciable loss of
the polymer's crystallinity, thereby affecting the rheological
properties of the nitroglycerine/nitrocellulose matrix, its interaction with fillers, and the physical properties of the finished
propellant. The breakdown of the nitrocellulose structure is
controlled by the concentration of the solvent, by the solvent
strength (3), by the temperature of the mix, and by the mixing
schedule to which it has been subjected.

     The variation of initial modulus of a finished, dried, filled
propellant with solvent concentration used in its preparation is
given in Figure 1. A similar trend was observed for the mean
tensile strength. The solvent was acetone, and the mixing time
and temperature were 90 minutes and $28^{o}C$ respectively. Figure 2
indicates that, for this system, the work unit (4), ie the total
work input per unit mass during a known period, is related to the
solvent concentration in a logarithmic-linear fashion.

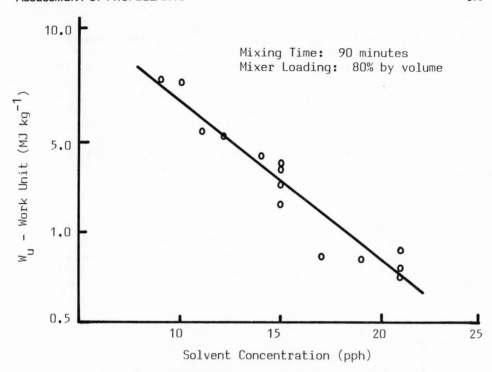

Figure 2:   The Dependence of Work Unit on Solvent
            Concentration

Figures 2 and 3 show that the rheological behaviour of the
dough is strongly dependent upon the solvent concentrations. The
flow curves demonstrate that these materials behave in a pseudoplastic manner. They do not obey however, a simple power law.

Examination of log apparent shear stress against log apparent
shear rate curves revealed that the line was convex to the log
shear rate axis at low shear rates, and concave to this axis at
high shear rates. All of the propellant systems so far examined
have shown this behaviour, and a typical example is shown in
Figure 4.   These curves indicated the existence of an initial
shear stress below which the material would not flow. A regression
analysis of the data showed that, at low shear rates, the relationship given in Equation (1) was obeyed.

$$\sigma - \sigma_o = k\dot{\gamma}^n \quad\quad\quad \text{................. (1)}$$

where σ is the shear stress, σ_o is the yield stress, k is the
consistency index, n is the flow index, and $\dot{\gamma}$ is the shear rate.

A "master" curve, Figure 5, was produced by using dies of
constant diameter but different lengths, and correcting the data

by the method described by Bagley (5). It is evident from this
plot that the longer the die, the lower is the shear rate at which
the flow behaviour deviates from that described by Equation (1).
A temperature rise of 30°C above the cylinder temperature was
measured at the exit of a die 50 mm long by 2 mm diameter at a
shear rate of 1000 s^{-1}. Thus it was deduced that the deviation
from the behaviour described by Equation (1) was due to shear
heating, and that this shear heating effect must be taken into
account in order to produce a true "master" curve. An attempt to
evaluate this shear heating effect following the method of Cox et
alia (6) is currently being made.

The full significance of the initial shear stress, σ_0, has
yet to be fully evaluated. Its value determines the extent to
which the extruded dough will deform and slump, and in some cases
a high value of σ_0 is desirable. Its magnitude also will influence
the efficiency of mixing since material in areas of the mixer

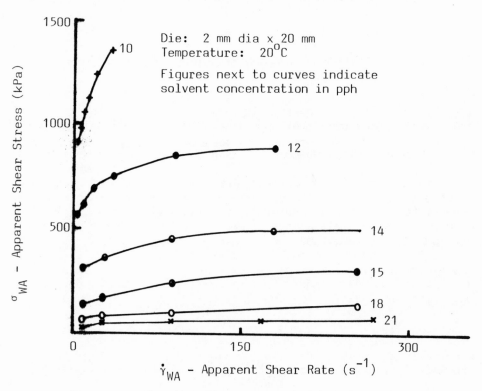

Figure 3:   The Effect of Solvent Concentration on the
            Flow Curves

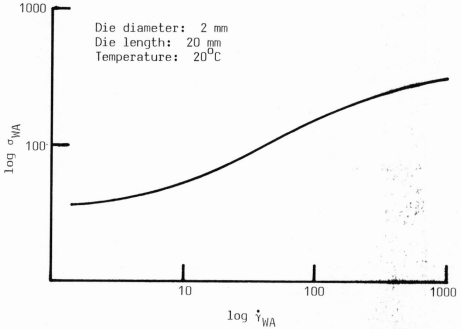

Figure 4:   Flow Curve showing Deviations from the Power Law

which is subjected to a shear stress of less than σ_0 will not
deform. In this case, either a machine providing a stress field
throughout the mix greater than the yield stress, or a mix with a
low value of σ_0 must be used.

CONCLUSION

     A rheological examination of nitrocellulose/nitroglycerine
propellant systems has revealed that their behaviour is complex
in that, like Bingham materials, they exhibit limiting shear
stresses below which no flow will occur. Above this limiting
shear stress, the materials flow in pseudoplastic manner, and
this type or complex flow behaviour has been described as
"plastic" flow (7). At higher shear rates, shear heating significantly affects the flow curves.

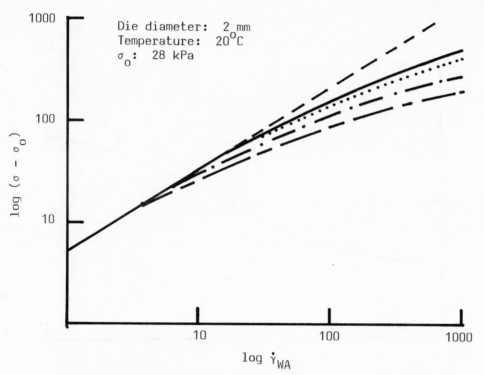

Figure 5: Flow Curve showing Deviations due to Shear Heating
           – – – –, theoretical curve; –·–·–, 50 mm die length;
           ———, 10 mm die length; — — —, 100 mm die length
           ·······, 20 mm die length;

REFERENCES

1   Miles, F.D., "Cellulose Nitrate", (Oliver and Boyd, 1955).

2   Trommel, J., Communication No 13, KNSF (1955).

3   DeBruin, G., Witte, C.H.D., Trommel, J., Communication
     No 14, KNSF (1957).

4   Van Buskirk, P.R., Turetzky, S.B., Gunberg, P.P., Rubber
     Chemistry and Technology 48 577-591 (1975).

5   Bagley, E.B., J App. Phys. 28 626 (1957).

6   Cox, H.W., Macosko, C.W., Am. Inst. Chem. Eng. J., 20 785
     (1974).

7   Lenk, R.S., "Polymer Rheology", 3 (Applied Science, London,
     1978).

ON THE PLASTIC DEFORMATION OF SUPERCONDUCTORS

P. Feltham

Brunel University, Uxbridge, England

(Abstract)

In the transition from the normal (n) to the superconducting (s) state the creep rate of metals flowing under constant stress increases while, if deformation occurs at a constant strain-rate, a small drop in the flow stress, $\Delta\sigma_{ns}$, is observed. These "softening" effects have been the subject of an extensive literature in the last few years. The material shows however that, so far, no satisfactory explanation of the observations and of the basic origin of the phenomena exists. A model is here outlined in which inelastic scattering of Cooper pairs by the stress fields at crossing points of dislocations enhances the probability of intersection: the principal rate-determining process. The effect of pair-scattering on the strain rate is considered equivalent to a small increase in the temperature of the metal.

RHEOLOGICAL PROPERTIES OF TREATED SUB-BENTONITE

E.Z.Basta, M.A.Maksoud, A.T.A.Aziz

Cairo University, Egypt

(Abstract)

The possibilities of the application of atreated sub-bentonite in drilling were emphasized; its yield was improved by alkali-activation. The alkali-activated sub-bentonite clay could resist the action of salt water especially when being activated at an elevated temperature of 100°C and one atmospheric pressure; i.e., using steam under normal pressure. This activated form resisted the effect of high temperature and pressure similar to the conditions prevailing at variable depths of wells. Smaller cake thickness (2 mm) and loss filtrate were produced with that clay slurry activated by applying the steaming method.

FLOW BEHAVIOUR OF LUBRICANTS IN SERVICE CONDITIONS

Kurt Kirschke

Bundesanstalt für Materialprüfung
Fachgruppe 5.2 "Rheologie und Tribologie"
1000 Berlin 45

1. INTRODUCTION

In application lubricants are exposed to high rates of shear. As a result of their high content of additives of different kind, including high polymers, they can show a non-Newtonian rheological behaviour in hydrodynamic and elastohydrodynamic (EHD) lubrication. In viscometry with standard viscometers under low rates of shear, this anomalous flow behaviour does not appear. These lubricants can, therefore, become effective in bearing and gear application with another viscosity than is measured in viscometry. In addition to this, other flow effects can occur. The rheology of the lubricants certainly determines its behaviour in the hydrodynamic and elastohydrodynamic regimes. For this reason suitable rheological properties are a principle demand.

2. VISCOMETRY OF LUBRICANTS

For investigations of the flow behaviour of lubricants up to high rates of shear a device was developed, which enables the viscometry of fluids by pressurized flow in capillaries of different length and diameters. With this device Newtonian oils and oils with certain high polymer additives were measured. The blended oils have a marked non-Newtonian flow behaviour already at ranges of shear rates which are still accessible to measurement. In the case of oils with polymer additives of lower molecular weight, as they are used in practice, an analogous behaviour can be expected in a range of higher rates of shear. This was shown by the experiments of R.C. Rosenberg in a simulated engine bearing, where the temporary viscosity decrease of lubricant blends containing commonly used viscosity index improvers and of commercial multigraded engine oils

has been determined [1]. Results indicate that low shear rate viscosity determination is not sufficient to characterize the behaviour
of these lubricants in high-speed bearings. Relevant proposals for
standardization of high shear viscometers were made a fairly long
time ago and a new possibility for a graphic representation of the
flow behaviour of non-Newtonian lubricants were specified [2].

3. HYDRODYNAMIC LUBRICATION WITH NON-NEWTONIAN LUBRICANTS

The anomalous flow behaviour of certain lubricants appearing
in a reduction of the viscosity with increasing rate of shear can
be a disadvantage if these rheological properties are left out of
account, because as a result of the lower viscosity the hydrodynamic
load capacity of the bearing is lower than expected. This effect,
therefore, has to be considered, and even advantages are possible
if the characteristics of such fluids are taken into account. In
plain bearings operating at greatly varying sliding velocity, the
application can be advantageous. Disadvantages can be expected in
cases of changing loads. As a result of the dependence between
viscosity and shear rate, these lubricants change their viscosity
with the number of revolutions per minute respectively with the
sliding velocity in a desirable manner, and with the load in an un-
favourable manner. Because lubricating oils of the type described
show a variable viscosity only within a large but limited region
of shear rate which is dependent on the kind, the molecular weight
and the molecular weight distribution of the contained high polymer,
the suitable selection of such an oil requires an approximate determination of the medium shear rate, which occurs at the bearing.
In the following, an experimental result of the behaviour of a
lubricant with special high polymers in a plain bearing is reported.
Fig. 1 shows a favourable behaviour of an oil with a special high
polymer in comparison with a plain oil, at changing speed. The influence of increasing self-heating with rising speed was eliminated
by appropriate cooling. The reason for this phenomenon is the dependence of the viscosity of a non-Newtonian oil on the rate of
shear and the fact that the shear rate in a bearing is high and
variable with the operating conditions. Besides the described effect,
other rheological effects can occur.

There are high polymer solutions showing a marked normal
stress effect under shear [3, 4]. A number of solutions of different
polyisobuthylenes were investigated by shear between rotating parallel plates. The non-rotating lower plate of the apparatus was
held self-adjusting. Springs were used to apply the load. In the
case of the investigated non-Newtonian oils the friction torque
and the thickness of the layer after an initial increase of the
torque and decrease of the thickness of the layer is independent
of the time of operation. In the fluid layer between the plates a
pressure is induced which holds the balance to an external axial

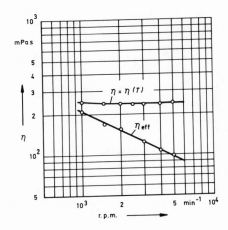

Fig. 1: Effective viscosity of a Newtonian and a non-Newtonian
lubricant (lower curve) in a plain bearing at changing
speed.

load of the plates. This was investigated in a long-term test in
comparison to the behaviour of Newtonian oils (Fig. 2). The results
of the experiments show the possibility of hydrodynamic lubrication
without complying with the precondition of a wedge-shaped gap when
using fluids with appropriate normal stress effect. Incidentally,
another property of such fluids can be the resistance to centrifu-
ging from a rotating shaft of bearing and the action as a sealing
mechanism in radial face seals.

4. EHD – LUBRICATION AND RHEOLOGICAL PROPERTIES OF LUBRICANTS

In hydrodynamic lubrication, which is characterized by surfaces that are conforming, the flow behaviour of the lubricant is
dependent on the temperature, and in case of non-Newtonian lubricants on the temperature and shear rate. Other parameters can be
neglected. Mating gear teeth, cam and followers and rolling element bearings with nonconforming surfaces are working in elastohydrodynamic (EHD)-lubrication. In this kind of lubrication Hertzian pressures are much higher than in hydrodynamic lubrication and
the thickness of the oil film in the contact region is very small.
Therefore, at least the elastic deformation of the surfaces and the
influence of pressure on the viscosity of the oil have to be considered. At typical Hertzian pressures the fluid viscosity can increase several orders of magnitude. At present, there are many
cases with insufficient agreement between analytically predicted
and experimentally measured data for film thickness and traction
(friction). I think, the reason is the non-observance of rheological effects in EHD-lubrication such as shear and time dependent
effects on viscosity. On the other side the stronger consideration
of rheological effects is the only possibility to influence resp.

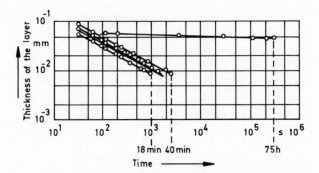

Fig. 2: Comparative investigations of Newtonian oils and a non-
         Newtonian oil. Mineraloil + 10% Polyisobutylene.
         Load 15 N, r.p.m. 486. Diameter of disks 30 mm.

improve EHD-lubrication. That could be managed by developing lubricants with appropriate rheological properties. Other parameters
such as constructional and operating conditions can be changed
scarcely or are of little influence. The main rheological property
of a lubricant is the viscosity as a function of temperature,
pressure, and shear rate. Beyond that we have to pay attention to
that a liquid element passes through the contact area in a time
of about 10^{-5}s. Because the shear relaxation time for mineral oils
with polymeric additives is about the same order of magnitude as
the transit time, the shear viscosity can also depend on time. In
addition hereto, normal stress effects can occur under certain cir-
cumstances which will increase the load carrying capacity.
      A number of authors have studied the relationship between
observed and theoretical film thicknesses in EHD-contacts. The predictions are based on a Newtonian theory. They found good agreement

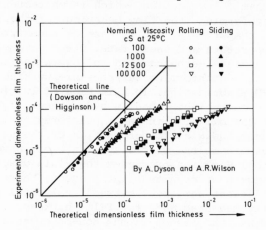

Fig. 3: Comparison of experimental film thickness given by sili-
         cone fluids and predictions based on a Newtonian theory.

for Newtonian lubricants and more or less deviations for non-Newtonian lubricants when using the low-shear viscosity values of a commercial standard viscometer. In Fig. 3 the experimental film thickness given by non-Newtonian silicone fluids are compared with theoretical values (Newtonian theory) [5]. There is a considerable influence of the molecular weight and small differences between rolling and sliding. Relevant deviations have been observed for other high polymer fluids, solutions containing high polymers, and water in oil emulsions. The investigation of the relationship between the viscometric properties of motor oils and performance in European engines indicated comparable results. The tested oils showed a poor correlation with their low shear-rate viscosities but a good correlation with their viscosities measured at shear rates of $10^5 s^{-1}$ to $10^6 s^{-1}$.

The build-up of a lubricant film can be enhanced by normal stress effects. Under high pressure, or in a region of high shear rates, the normal stress could become large. If obtainable, normal stresses could be important as load-bearing mechanisms in high speed machine elements. Results provide strong evidence for elastic behaviour of the lubricant film, demonstrating a transition from a predominantly viscous to a predominantly elastic response with increasing pressure at constant temperature and, separately, with decreasing temperature at constant pressure.

For determining lubricant flow behaviour, extensive work with various types of viscometers has been done. The main limitation of all of these approaches is that the lubricant is not subjected simultaneously to all of the conditions which it can experience in real application. Detailed knowledge of the properties which the lubricant exhibits in shear and compression over wide ranges of time scale, pressure and temperature is essential for a clearer understanding of EHD-lubrication.

REFERENCES

1   R.C. Rosenberg
     Influence of Polymer Additives on Journal Bearing Performance
     SAE Prepr. 750692 for Meeting June 3-5, 1975

2   K. Kirschke
     Arbeiten zur Normung eines Kapillarviskosimeters für hohe
     Schergefälle
     Rheologica Acta Vol. 7 (1968) Nr. 4, p. 354-360

3   K. Kirschke
     Experimentelle Ergebnisse zum Normalspannungseffekt
     Proc. IVth International Congress on Rheology, Aug. 1963,
     Brown University, Providence, Rhode Island, USA

4. J.F. Hutton
    Recent Advances in Lubricant Rheology
    Proc. VIIth International Congress on Rheology, August 1967,
    Chalmers University of Technology, Gothenburg, Schweden

5. A. Doysen and A.R. Wilson
    Film Thickness in EHD-Lubrication by Silicone Fluids
    Proc. Inst.Mech. Engrs., vol. 180, pt. 3 K, 1965 - 66,
    p. 97 - 112.

MECHANICS OF FAILURE IN TISSUE SPECIMENS OF THE APPLE CORTEX

M.G.Sharma, N.N.Mohsenin

Pennsylvania State University, University Park
Pennsylvania, U.S.A.

(Abstract)

During harvesting and transportation, fruits and vegetables
are generally subjected to mechanical forces that may bring about
damage in the form of bruises and browing. In order to design
fruit handling and harvest systems that do not produce any damage
in fruits, a knowledge of a threshold value of loading, a fruit may
sustain before bruises or browing is very much needed. The threshold
value of load for a given fruit can be evaluated if a valid
criterion for the tissue failure of the fruit under all possible
types of loading that may occur in practice is available.

The paper represents a first attempt for the development of a
general criterion of failure for the tissue specimens of an apple
cortex based upon the mechanistic principles. The paper describes
the experimental devices and specimen grips specially developed
for studying the behavior under various combination of biaxial
and triaxial stress states.

As part of this investigation several types of failure
experiments were conducted on samples of materials subjected to
uniaxial and combined loads. The experimental program included
such tests as uniaxial tension on "dogbone" type of specimens,
uniaxial compression, pure torsion and torsion combined with
compression on solid cylindrical samples. The program also
included pure internal pressure and external pressure combined
with axial compression tests on hollow cylindrical samples. In

each of the above tests specimens were subjected to monotonically
increasing loads up to a desired level and then unloaded and any
permanent (plastic) strains remaining in the specimen were measured.
The above tests were conducted for several magnitudes of loads
under the ultimate loads at which the specimens failed due to
excessive deformation or rupture. Principal stresses and strains
corresponding to various levels of loads below and at the ultimate
loads were evaluated. From the data loci of stress states correspond
ing to a desired magnitude of plastic strains were obtained. In
addition, failure data was plotted on the principal stress space
and examined to see if it corresponds to any one of the well known
criterion based upon the maximum stress, maximum shear, maximum
distortion energy, Mohr-coulomb and Nadai's theories. The results
indicate that failure data reasonably fits well to the criterion
based upon Nadai's theory. The implication of this result and the
isoplastic strain surfaces in the stress space in predicting the
threshold value of loads at which bruising or browning occurs,
in the whole fruit has been discussed.

RHEOLOGICAL CHARACTERIZATION OF

TIME-DEPENDENT FOODSTUFFS

D. DeKee, G. Turcotte and R.K. Code

Department of Chemistry and Chemical Engineering
Royal Military College of Canada
Kingston, Canada K7L 2W3

Department of Chemical Engineering
Queen's University
Kingston, Canada K7L 3N6

INTRODUCTION

Improved rheological characterization of foodstuffs, particularly time-dependent effects, is required for process and equipment design. This is a challenging task. Foodstuffs, because of their complex structure and microscale heterogeneity, exhibit a variety of non-Newtonian effects such as pseudoplasticity, yield stress and thixotropy. In addition, rheological measurements may be a function of the previous shear and thermal history of the sample.

In this paper, the descriptive abilities of two models, the Herschel-Bulkley[1] and a model proposed previously by the authors[2], are tested on equilibrium viscosity data for mayonnaise, tomato juice, yogurt and honey. The second model is extended to time-dependent behavior using a second order kinetic model to describe the rate of structural change with shear.

RHEOLOGICAL MODELS

The two models are:
i)   the three parameter Herschel-Bulkley model[1]

$$\eta(\dot{\gamma}) = \tau_o \dot{\gamma}^{-1} + m\dot{\gamma}^{n-1}$$

and

ii) a model proposed by the authors[2]

$$\eta(\dot{\gamma}) = \tau_o\dot{\gamma}^{-1} + \sum_{P=1}^{M} \eta_P e^{-t_P\dot{\gamma}} + \sum_{P=M+1}^{\infty} \eta_P e^{-t_P\dot{\gamma}} \tag{2}$$

Equation (2) contains three parameters if one considers M=1 and neglects the rest term, which may be interpreted as a limiting viscosity (η_∞) at high rates of shear. This model may be extended to time dependent behavior by the introduction of a structural parameter which is assumed to obey second order kinetics.

$$\frac{d\eta}{dt} = -\frac{\eta}{\dot{\gamma}} \frac{[\dot{\gamma}^2 t_1 \eta_1 e^{-t_1\dot{\gamma}} + \tau_o]}{[\eta_1\dot{\gamma}e^{-t_1\dot{\gamma}} + \tau_o]} \frac{d\dot{\gamma}}{dt} - \frac{k_1\dot{\gamma}[\eta-\eta_e]^2}{[\tau_o + \eta_1\dot{\gamma}e^{-t_1\dot{\gamma}}]} \tag{3}$$

The rate constant k_1 is a function of the rate of shear.

DISCUSSION

The test results are summarized in Figure 1. The viscosities are equilibrium values corresponding to long shearing times. The measurements were made on a Ferranti-Shirley cone-and-plate viscometer. Details on the parameter estimation and their numerical determination are available elsewhere[2].

The Herschel-Bulkley model is quite successful in describing the data. Honey is a Newtonian fluid. This is a limiting case of the models considered here. The data on yogurt are also very well represented by the three parameter exponential model. The data on tomato juice exhibit a 'discontinuity' at shear rates in the vicinity of 1000 s^{-1}. The curve fitting procedure in the shear rate range below 10^3 s^{-1} results in unacceptable oscillations for the exponential model with M=1 and η_∞=0. This problem does not arise at very high shear rates. The deviations at the lower shear rates are reduced by extending the exponential model to M=2 and again neglecting the rest term. For clarity, this improved fit is not shown for the tomato juice data in the low shear rate range. A similar oscillation problem occurred in the case of mayonnaise for M=1 and η_∞=0. The solid line through the data represents the exponential model with M=2 and η_∞=0.

Figure 2 shows photomicrographs of tomato juice before and after the observed drop in viscosity. From the photographs it appears that the plant cells are broken down at shear rates higher than 1800 s^{-1} but are only deformed at shear rates below 1000 s^{-1}.

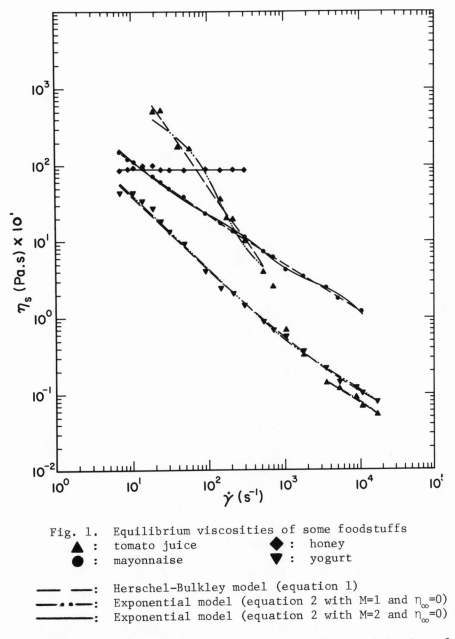

Fig. 1. Equilibrium viscosities of some foodstuffs
  ▲ : tomato juice   ◆ : honey
  ● : mayonnaise   ▼ : yogurt

—— —— : Herschel-Bulkley model (equation 1)
——•—•— : Exponential model (equation 2 with M=1 and η_∞=0)
———— : Exponential model (equation 2 with M=2 and η_∞=0)

Additional data were collected on the transient behavior of
the products in Figure 1. Viscosity decay rates, at constant
rate of shear, were measured over a range of shear rates. Hysteresis loops of τ vs. $\dot{\gamma}$ covering a range of cycle times to reach

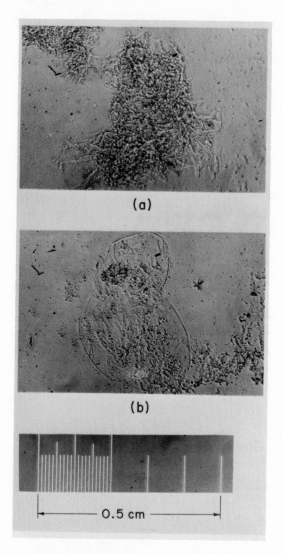

Fig. 2. Photomicrograph of tomato juice after
         shearing the sample to its equilibrium
         viscosity at a shear rate of 549 s^{-1}(a)
         and at a shear rate of $1,812s^{-1}$(b)

a maximum rate of shear were also examined. The description of
these results by the extension of the exponential model to time-
dependent behavior will be presented at the conference.

TIME DEPENDENT FOODSTUFFS

613

Table 1. Model Parameters

Product	Exponential Model			Herschel-Bulkley Model		
	η_1	t_1	τ_o	m	n	τ_o
	$(Pa \cdot s) \times 10^2$	(s)	$(Pa) \times 10^2$	$(Pa \cdot s^n) \times 10^2$	–	$(Pa) \times 10^2$
Yogurt	1.15	4.52×10^{-5}	4.17×10^3	18.7	0.644	3.95×10^3
Tomato Juice $\dot{\gamma} < 550 s^{-1}$	4.13×10^3	2.03×10^{-2}	2.59×10^4	4.96×10^5	-0.469	145
$\dot{\gamma} > 3600 s^{-1}$	0.791	4.00×10^{-5}	2.77×10^3	196	0.400	145
Honey	917.			917.		
Mayonnaise	130.	6.99×10^{-3}	1.05×10^4	2.94×10^3	0.404	4.60×10^3
	η_2 $(Pa \cdot s)$	t_2 (s)				
Mayonnaise	77.1	6.99×10^{-4}				

CONCLUSION

In terms of only three parameters, the Herschel-Bulkley model describes the data very well. However, at high shear rates, this model erroneously predicts a viscosity which decreases to zero. Also, because of the particular units, the parameters cannot be related to a molecular model. The exponential model on the other hand contains parameters which may be related to the molecular structure of time-dependent materials[3]. The η_p have units of viscosity and the t_p have units of time. For a case where two terms of the series (M=2) is judged necessary, one can achieve a parameter reduction by letting $t_p = 10 \ t_{p+1}$. This procedure yields excellent results for the mayonnaise data of Figure 1 over a shear rate range of three decades. The exponential form of the second model is an advantage when carrying out mathematical operations and the model may be readily extended to time-dependent materials.

NOMENCLATURE

k_1	rate constant
m,n	constants in the Herschel-Bulkley model
t	time
t_p	time constants in the exponential model
$\dot{\gamma}$	shear rate
η	non-Newtonian viscosity
η_p	constants of dimensions of viscosity
η	equilibrium viscosity
η_∞	limiting viscosity at high rates of shear
τ	shear stress
τ_o	yield stress

REFERENCES

1. C. Tiu, and D.V. Boger, J. Text. Studies 5:329 (1974).
2. D. DeKee, and G. Turcotte, submitted to Chem.Eng. Commun.
3. D. DeKee, and P.J. Carreau, J. Rheol., in press.

RHEOLOGICAL PROPERTIES OF NON-AQUEOUS SUSPENSIONS

OF TITANIUM DIOXIDES STABILIZED WITH LECITHIN

Umeya Kaoru and Kanno Takashi

Faculty of Engineering
Tohoku University
Aramaki-Aoba, Sendai, Miyagi 980, JAPAN

INTRODUCTION

It is well known that the flocculated suspensions show thixotropic and rheopectic flow as well as shear-thinning and shear-thickening flow. These types of time-dependent, non-Newtonian behavior are usually attributed to the structural breakdown or rebuildup within a suspension under shear. The objective of this study is to discuss the mechanism of the shear-thinning behavior observed for non-aqueous suspensions of titanium dioxides stabilized with different amounts of lecithin.

EXPERIMENTAL

Rheological measurements were carried out with a rotational viscometer for the 20vol% suspensions of TiO_2 in liquid paraffin, CCl_4 and the mixture of them. The suspensions with different stabilities were prepared by adding different amounts of lecithin. Sedimentation tests as well as the adsorption measurements of lecithin were also made as a means of evaluating the degree of dispersion and the dispersion stability of the particles in suspension.

RESULTS AND DISCUSSION

Figure] shows the effect of the amount of lecithin on the shear-rate dependences of relative viscosity for the 20vol% suspensions of TiO_2 in liquid paraffin. At a fixed shear rate, the relative viscosity of suspension

615

decreases with increasing the amount of lecithin. This may be attributed to an increase in the degree of dispersion under shear due to the steric stabilization between the suspended particles resulted from the adsoption of lecithin. On the other hand, the relative viscosities of each suspension decrease with increasing shear rate. This may be attributed to an increase in the degree of dispersion under shear due to the hydrodynamic forces, which should be distinguished from the changes in the degree of dispersion due to the colloidal forces between the particles in suspension.

The limiting values of the relative viscosity at high shear rate (2nd Newtonian viscosity) may be expected to be independent of colloidal force between the particles in each suspension. As the shear rate decreases, the relative viscosities of each suspension come to depend on both shear rate (hydrodynamic force) and the amount of lecithin (colloidal force), however, these values show a tendency to reach a constant value of $]0^4$ (]st Newtonian viscosity) in relatively low shear rates. When the shear rate is decreased furthermore, the relative viscosities of each suspension are found to increase again. In these regions, the shear stresses of each suspension are almost independent of shear rate (plastic flow). Consequently, all flow curves observed can be referred to as "the extended Ostwald flow". It is noted that, with increasing the amount of lecithin, the extended Ostwald flow behavior appears in lower shear rates.

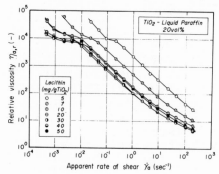

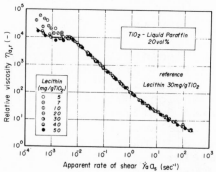

Fig.   Effect of the amount of lecithin on the shear-rate dependences of relative viscosity for the 20 vol% suspensions of TiO_2 in liquid paraffin.

Fig. 2 Master curve for the 20vol% suspensions of TiO_2 in liquid paraffin at a reference of 30mg/gTiO$_2$ of lecithin.

Figure 2 shows a composite master curve obtained by shifting the flow curves in Figure 1 horizontally to be superposed each other on the reference curve of the suspension with 30mg/gTiO$_2$ of lecithin. Except for the plastic flow region in low shear rates, the dependences of relative viscosity on shear rate (Ostwald flow curves) for the suspensions with different amounts of lecithin can be reduced to that of a single master curve by finding appropriate shift factors.

Similar result was also obtained for the 20vol% suspensions of TiO$_2$ in CCl$_4$. The shift factors obtained for both suspensions in different dispersion mediums, assuming as reference curves those with 30mg/gTiO$_2$ of lecithin, are shown in Figure 3 as a function of the amount of lecithin added. In this figure, the characteristic shear rates at which the changes in relative viscosity from a limit of low shear rates ($\eta_{r\,0}$) to that of high shear rates ($\eta_{r\,\infty}$) attain a half value, are also plotted. The shift factors may indicate the changes in colloidal forces between the suspended particles relative to those of respective reference suspensions associated with increasing the amount of lecithin. On the other hand, the characteristic shear rates which locate the Ostwald flow behavior may be correlated with the changes not only in colloidal forces but also in hydrodynamic forces resulted from the difference in the viscosities of dispersion mediums.

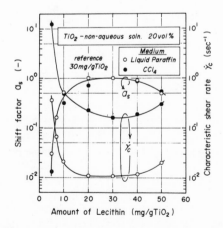

Fig. 3 Effect of the amount of lecithin on the shift factors and the characteristic shear rates for the 20 vol% suspensions of TiO$_2$ in different dispersion mediums.

Figure 4 shows the effect of the composition of dispersion medium on the shear-rate dependences of relative viscosity for the 20vol% suspensions of TiO_2 in the mixture of liquid paraffin and CCl_4 stabilized with 30 mg/gTiO$_2$ of lecithin. At a fixed shear rate, the relative viscosity of suspensions decreases with decreasing a fraction of CCl_4 in medium. This may be attributed to an increase in the degree of dispersion due to the hydrodynamic force resulted from an increase in the viscosity of dispersion medium, even at a constant shear rate. By comparing the shear-rate dependences of relative viscosity for the suspensions in different dispersion mediums each other, it can be recognized that a similar Ostwald Flow curve is located in different shear rate regions depending on the viscosity of dispersion medium.

Figure 5 shows a composite master curve obtained by shifting the flow curves in Figure 4 horizontally to be superposed each other on the reference curve of the suspension in liquid paraffin. Except for the plastic flow region, Ostwald flow curves for the suspensions in different dispersion mediums can be reduced to that of a single master curve.

The shift factor as well as the characteristic shear rate are plotted in Figure 6 against the viscosity of constituent dispersion mediums. The result obtained from the suspensions stabilized with 10mg/gTiO$_2$ of

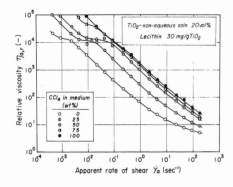
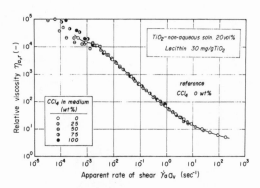

Fig. 4 Effect of the composition of dispersion medium on the shear-rate dependences of relative viscosity for the 20vol% suspensions of TiO$_2$ stabilized with 30 mg/gTiO$_2$ of lecithin.

Fig. 5 Master curve for the 20vol% suspensions of TiO$_2$ stabilized with 30mg/gTiO$_2$ of lecithin at a reference medium of liquid paraffin.

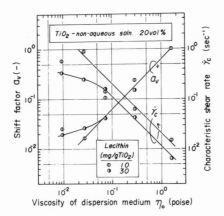

Fig. 6 Effect of the viscosity of constituent dispersion medium on the shift factors and the characteristic shear rates for the 20vol% suspensions of TiO_2 stabilized with different amounts of lecithin.

lecithin is also plotted. The characteristic shear rates are recognized to be inversely proportional to the viscosity of constituent dispersion medium. This means that an increase in the viscosity of dispersion medium is equivalent to that in the shear rate with regard to the structural breakdown due to the hydrodynamic forces.

     In Figure 7, volume fraction of particles in sediment obtained with centrifugal sedimentation at 1400G is plotted against the amount of lecithin for the suspensions in both liquid paraffin and CCl_4. For the suspensions in CCl_4, the result obtained with gravitational sedimentation is also shown in this figure.

     In any cases, the volume fraction of particles in sediment increases, at first, with increasing the amount of lecithin added, and attains saturation, respectively, at the amount of lecithin over 20-30mg/$gTiO_2$. Comparing the result with that of the shift factor in Figure 3, the effect of the amount of lecithin on the characteristic shear rate can be confirmed to be correlated closely to the changes in the colloidal forces between the suspended particles. Both of them may be interpreted by the decrease in the colloidal forces between the particles associated with the steric stabilization resulted from the increase in the amount of lecithin adsorbed on the particles. The amount of lecithin adsorbed on particles is shown in Figure 8 as a function of the amount of lecithin added to the suspension.

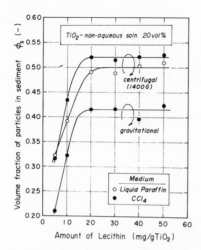

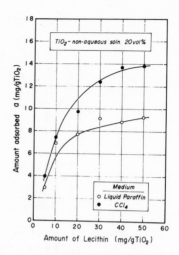

Fig. 7 Effect of the amount
of lecithin on the volume
fraction of particles in
sediment for the 20vol%
suspensions of TiO₂ in different dispersion mediums.

Fig. 8 The amount of lecithin
adsorbed on TiO₂ versus the
amount of lecithin added to
the suspensions in different
dispersion mediums.

The saturated value of the volume fraction of
particles in sediment for the suspensions in CCl₄ is
larger than that for the suspensions in liquid paraffin.
This may suggest that the steric stabilization of the
former is more excellent than the latter, which can be
interpreted by the differences in the saturated values
of the amount of lecithin adsorbed, as shown in Figure 7.
Small deviations from a linear line in the low viscosity
region in Figure 6 can be interpreted by the difference
in the colloidal forces associated with the difference
in dispersion mediums.

From the comparison of the result of gravitational
sedimentation with that of centrifugal, it can be confirmed that the flocculated structure can be formed at
rest even if the amount of lecithin adsorbed is approached saturation, however, it may possibly be disrupted under a centrifugal forces to form a closely
packed sediment. The shear-rate dependences of the
relative viscosity can be concluded to be resulted from
the shear-induced breakdown of the flocculated structure
in suspension which has been formed by colloidal forces
between the particles under a rest period.

CHARACTERIZATION OF CHITOSAN FILM

Carlos A. Kienzle-Sterzer, Dolores Rodriguez Sanchez
and ChoKyun Rha

Food Materials Science and Fabrication Laboratory,
Dept. of Nutrition and Food Science, Massachusetts
Institute of Technology

INTRODUCTION

Chitosan [(1-4)-2 amino-2-deoxy-β-D glucan] is deacetylated
chitin usually in various degrees of deacetylation and depolymer-
ization. The removal of some or all of the acetyl groups from the
chitin liberates amine groups, thus imparting the polycationic
nature to the polymer (Muzzarelli, 1977). The cationic and amine
containing nature make the chitosan a novel polymer belonging to the
group the potential of which has not yet been exploited industrially.

The film-forming qualities of chitosan and the different methods of film preparation have been reported (Rigby, 1936; Muzzarelli,
1973). The film-forming quality is expected to depend on the structure of the bulk chitosan from which it is cast, as reported by
Averbach (1977). The mechanical properties in terms of breaking
stress and tensile elongation have been studied (Averbach, 1977),
however, the relationship between the mechanical properties and
structure in chitosan film has not been established. In this study,
reduced stresses, the changes in cross-linking density and the deviation from the classical rubber elasticity of the chitosan film
under strain at maximum swelling were examined.

EXPERIMENTAL

Sample Preparation

Chitosan (FLONAC-N, Velsicol, Hampton, Va.) was dissolved in
acetic or propionic acid (0.1665 M) to make 0.8, 1.0, 1.2 and
1.4% w/w solutions. The solutions were allowed to rest over night

and filtered. Films were cast by drying a chitosan solution for
approximately 24 hours. The films were then neutralized with NaOH
(1 N) for 20 minutes, washed several times until the pH of the wash
water remained constant at near neutral, cut and kept in distilled
deionized water.

Determination of Swollen Volume

Chitosan films were totally immersed in distilled deionized
water at a constant temperature of 24°, 30° or 37°C. The weights
of the films were determined at 24 hour intervals until they reached
a constant value. The samples were then dried and weighed to determine the volumes of dry films.

Stress-Strain Measurements

Tensile measurements of swollen chitosan films were made under
immersion in deionized water at 24°C at a deformation rate of
1 mm/min. The samples were stretched 10-40% and after 5, 10 and 15
minutes of relaxation the lengths of the samples which sustained
zero force were determined using a Instron Universal Testing Machine
(Model 1122, Instron, Inc., Canton, Ma.)

RESULTS AND DISCUSSION

The equilibrium volume fraction, v_2, of chitosan in the swollen
state calculated assuming isotropic swelling increases with increasing chitosan concentration, and also with increase in temperature
(Figure 1). The stress-strain relationship of the swollen chitosan

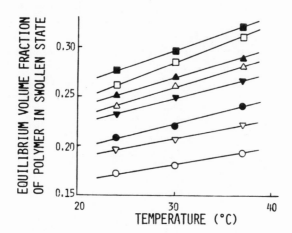

Fig. 1.   Equilibrium volume fraction of chitosan films versus tem-
          perature. $\mathbf{O}, \nabla, \Delta, \square$, in acetic acid; $\bullet, \blacktriangledown, \blacktriangle, \blacksquare$, in propionic
          acid for 0.8, 1.0, 1.2, 1.4% w/w respectively.

films was interpreted using the semi-empirical equation of Mooney-Rivlin (Mooney, 1948; Rivlin, 1948).

$$[f^*] \equiv fv_2^{1/3}[A(\lambda-\lambda^{-2})]^{-1} = 2C_1 + 2C_2\lambda^{-1}$$

where: $[f^*]$ = "reduced stress", f = equilibrium elastic force, v_2 = equilibrium volume fraction of the polymer in the swollen state, A = the cross-sectional area of the network in the unrestricted state, λ = L/L_O (L_O = initial swollen unstrained length, L = final strained length), C_1, C_2 = constants independent of λ.

However, because of the high degree of swelling and ionogenic characteristics of chitosan films which induce changes in volume during strain, it is difficult to obtain an equilibrium elastic force solely due to viscoelastic effects. In order to obtain the equilibrium elastic force as accurately as possible, the chitosan films were allowed to relax for 5, 10 and 15 minutes, and the changes in volume with deformation, V_λ, were calculated with the method used by Hasa and Ilavsky (1975). The logarithm of the time versus the change in volume with deformation, V_λ, are shown in Figure 2. The times at which the changes in volume initiate, decrease with the decrease in the concentration of chitosan, therefore with increase in swelling possibility due to the decrease in the friction coefficient between chains (Treolar, 1975).

The elastic force at the time when V_λ was equal to one was selected to calculate the reduced stress. The elastic force obtained was considered mainly due to the viscoelastic relaxation and did not have an appreciable effect due to the changes in swelling under

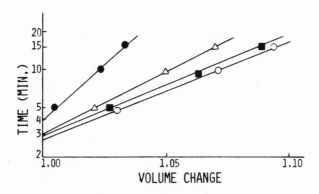

Figure 2. Log of the time of relaxation versus the change in volume of chitosan network under strain. ●,■, 1% w/w of chitosan in acetic acid at 1.1 and 1.3 strain respectively; Δ, O, 0.8 w/w of chitosan in acetic acid at 1.2 and 1.4 strain.

the strain. The reduced stresses [f*] versus the reciprocal of
strain (λ^{-1}) are given in Fig. 3, where the least-square analysis was
used to get the best fitted line. All the samples had negative
slope ($2C_2$) and therefore deviated from the Gaussian network behavior. This can be explained by the increase in inter- and intramolecular interactions which in this case may be due to ionic or
hydrogen bonds between amines and hydroxyl groups caused by closer
alignment or stretching of the chains. Such rearrangement would
impair the validity of the Gaussian single chain statistics, leading
to a negative strain-dependent correction.

Values of $2C_1$ and $2C_2$ calculated are presented in Table 1. The
magnitude of negative values for $2C_2$ are greater for increasing
values of v_2. The deviation from Neo-Hookean behavior is represented as $\Psi = C_2/(C_1 + C_2)$, (Ferry and Kan, 1978), in column 5 in
Table 1. It increases in magnitude with increases in the chitosan
concentration and thus can be explained by an increase in the fraction of network strands in the trapped entanglement network (Ferry
and Kan, 1978).

The shear modulus in the statistical theory of rubber-like

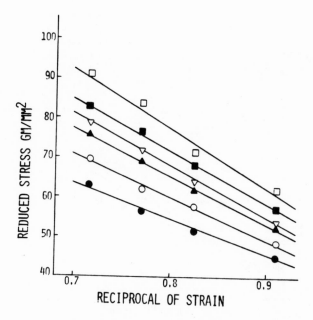

Fig. 3.   Reduced stress ([f*]) versus the reciprocal of strain
(λ^{-1}) for chitosan films. ■ ▲,●, 1.40, 1.20, 0.80 % w/w
of chitosan respectively in propionic acid; ☐,▽,○, 1.40,
1.00, 0.80% w/w of chitosan respectively in acetic acid.

Table 1.   Non-Gaussian Behavior of Chitosan Films
           Network parameters of chitosan films
           under uniaxial extension.

Films Cast From	Conc. (% w/w)	v_2	$2C_1$ (gm/mm^2)	$2C_2$ (gm/mm^2)	$2C_1+2C_2$ (gm/mm^2)	Ψ	$\dfrac{\delta_{1D}}{\delta_{sD}}$
Propionic	0.80	0.208	121	-82	39	-2.10	6.21
Acid	1.00	0.232	147	-106	41	-2.58	7.17
Solution	1.20	0.251	158	-116	42	-2.76	7.52
	1.40	0.276	175	-129	46	-2.80	7.61
Acetic	0.80	0.174	144	-104	40	-2.60	7.20
Acid	1.00	0.196	168	-124	44	-2.82	7.64
Solution	1.20	0.242	188	-142	46	-3.09	8.18
	1.40	0.262	201	-154	47	-3.28	8.55

elasticity for large deformations can be represented by $2C_1$, and for small strains by $2C_1 + 2C_2$ (Treolar, 1975). Although, $2C_1+2C_2$ overestimates the limit value of the modulus as the elongation decreases toward unity (Fasina, 1977). Assuming this effect negligible the modulus can be defined as:

$$G = C(\delta kT)\ (\frac{v}{v_o})^{2/3} \qquad \text{(Mark and Sullivan, 1977)}$$

where G = the modulus of elasticity, δ = the crosslinking density, k = Boltzmann's constant, T = absolute temperature, v = volume of the unswollen network, v_o = volume at network formation, C = the structure factor.

Then under the assumption of v equal to v_o and taken C equal to 0.5 for nonaffine deformation of tetrafunctional network and C equal to unity at low strain where the deformation should be affine (Flory, 1977), the relation between the crosslinking density at large deformation and at small deformation is given by:

$$\frac{\delta_{1d}}{\delta_{sd}} = 2(\frac{2C_1}{2C_1+ 2C_2})$$

where δ_{1d} = crosslinking density at large deformation, δ_{sd} = crosslinking density at small deformations.

Values of δ_{1d}/δ_{sd} for the network studied (column 7, Table 1) show that the crosslinking density increases with deformation and more with the increase in chitosan concentration therefore with the increase in v_2, confirming again that the network responsible for chitosan film is non-Gaussian.

The differences observed between the values obtained for films cast from both acetic and propionic acids are due to differences in the volume fraction of polymer in the system during crosslinking because of the different solubility of chitosan in these acids and the differences in the volatile losses of the solvent during the crosslinking process.

CONCLUSIONS

The swollen chitosan films cast from dilute acetic and propionic acid solutions do not exhibit ideal rubber-like elasticity under 10 to 40% uniaxial extension. The non-Gaussian behavior is likely due to the increase in crosslinking density upon elongation. The increase in the crosslinking density may be attributed to the increase in electrostatic or hydrogen bonds between the chains which become more aligned during the deformation.

REFERENCES

Averbach, B.L.A., 1977, Film-forming capability of chitosan, in: "Proc. Ist. Int. Conf. Chitin/Chitosan, R.A.A. Muzzarelli and E.R. Pariser, ed., MIT Sea Grant Report MITSG 78-7.

Dusek, K. and Prins, W., 1969, Structure and elasticity of non-crystalline polymer networks, Adv. Polymer Sci., 6:1.

Fasina, A.B., 1977, Elasticity of rubbery networks in the region of zero strain, Rubber Chem. Tech., 50:780.

Ferry, J.D. and Kan, H.C., 1978, Interpretation of deviations from neo-Hookean elasticity by a two-network model with cross-links and trapped entanglements, Rubber Chem. Tech., 51:731.

Flory, P.J., 1977, Theory of elasticity of polymer network. The effect of local constraints on junctions, J. Chem. Phys., 66:5720.

Hasa, J. and Ilavsky, M., 1975, Deformational, swelling and potentio-metric behavior of ionized poly(methacrylic acid) gels. II. Experimental results,J. Polym. Sci., Polym. Phys. Ed.,13:263.

Janacek, J. and Vojta, V., 1975, Negative C2 values of inhomogeneous polymer networks measured in elongation, J. Polym. Sci, Polym. Symp., 53:291.

Mark, J.E. and Sullivan, J.L., 1977, Model networks of end-linked polydimethyl-siloxane chains. I. Comparisons between experimental and theoretical values of elastic modulus and the equilibrium degree of swelling, J. Chem. Phys.,66:1006.

Mooney, M., 1948, Thermodynamics of a strained elastomer. I. General analysis, J. Appl. Phys., 19:434.

Muzzarelli, R.A.A., 1973, "Natural Chelating Polymer", Pergamon Press, Oxford.

Muzzarelli, R.A.A., 1977, "Chitin", Pregamon Press, Oxford.

Rigby, G.W., 1936, Process for the preparation of films and filaments and products thereof, U.S. Patent No. 2,030,880.

Rivlin, R.S., 1948, Large elastic deformation of isotropic materials. IV. Further developments of the general theory, Phil. Trans. Roy. Soc. (London), A241:379.

Treolar, L.R.G., 1969, Volume changes and mechanical anisotropy of
        strained rubbers, Polymer, 10:279.
Treolar, L.R.G., 1975, "The Physics of Rubber Elasticity, 3rd Ed.",
        Clarendon Press, Oxford.

ACKNOWLEDGEMENTS:

    This study was partially supported by M.I.T. Sea Grant No.
88133 "Synthesis of Chitosan Structure Matrix for Food" from NOAA,
Department of Commerce, Office of Sea Grant, NA79AA-D-00-101. The
authors wish to thank Mr. Dean Horn and Prof. J. Connor of the M.I.T.
Sea Grant Office for their support, and Mr. Peter Perceval and Mr.
R. Pariser for providing Flonac-N.

RHEOLOGICAL PROPERTIES OF HIGHLY DILUTE VISCOELASTIC AQUEOUS

DETERGENT SOLUTIONS

Signe Gravsholt

The Royal Danish School of Pharmacy
2, Universitetsparken
DK 2100 Copenhagen Ø, Denmark

INTRODUCTION

Highly dilute aqueous solutions containing 0.2 - 0.3 mmol dm^{-3} (80 - 120 p.p.m.) of some cationic detergents show visible elastic recoil at 25^{0}C[1] although the viscosity of the solutions is as low as 0.9 - 1.0 mPa s[2]. Some of the rheological properties of aqueous solutions of one of the detergents (cetyltrimethylammonium salicylate (CTA-Sal)) have been studied[3,4]. At concentrations between 0.3 and 1.0 mmol dm^{-3} CTA-Sal the solutions behave like ordinary Newtonian solutions at suitably low shear rates, whereas at higher shear rates the solutions show rheopectic behaviour as illustrated in fig. 1[4].

In the present work some rheological properties of aqueous solutions containing 3 or 5 mmol dm^{-3} CTA-Sal have been studied at shear rates from 0.0006 to 4.6 s^{-1}.

EXPERIMENTAL

Materials

N-cetyl-N,N,N-trimethylammonium bromide (CTAB): Merck zur Analyse; sodium salicylate: Merck zur Analyse. Preparation and purification of CTA-Sal have been described previously[1] 40.02

EXPERIMENTAL

Materials

N-cetyl-N,N,N-trimethylammonium bromide (CTAB): Merck zur Analyse; sodium salicylate: Merck zur Analyse. Preparation and pu-

rification of CTA-Sal have been described previously[1,4]. In some
of the experiments equimolar mixtures of aqueous solutions of CTAB
and sodium salicylate were used giving essentially the same results
as the pure CTA-Sal solutions.

Measurements

A Couette type viscometer (Contraves 30 + LS 100[5]) was used
for the rheological experiments and all measurements were made at
25.0 C.

RESULTS AND DISCUSSION

In fig. 1. the shear stress (τ) versus time (t) and shear rate
($\dot{\gamma}$) is plotted for 0.6 mmol dm^{-3} (253 p.p.m.) CTA-Sal[3]. The solution show Newtonian behaviour at $\dot{\gamma}$ < 0.99 s^{-1} and rheopectic behaviour at higher shear rates.

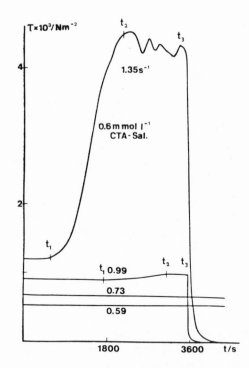

Fig. 1. Shear stress (τ) versus time (t) and shear rate (given as
numbers on the curves) for an aqueous solution containing
0.6 mmol dm^3 (253 p.p.m.) CTA-Sal at 25.0°C[3]. At the time
t_3 the rotation is stopped abruptly and the latter parts
of the two upper curves show the shear stress relaxation
after cessation of flow. Experimental details have been
described previously[4].

Measurements on 3 and 5 mmol dm^{-3} CTAB + sodium salicylate or on pure CTA-Sal solutions show that the solutions are pseudoplastic at low shear rates, pseudoplastic and thixotropic at intermediate shear rates and might even exhibit first thixotropic and then rheopectic behaviour at a suitable constant shear rate. In table 1. are given the shear rate regions where the different types of rheological behaviour have been detected.

Table 1.   Shear Rate Regions where the Different Types of Rheological behaviour have been detected at 25.0°C. for 3 and 5 mmol dm^{-3} CTAB + sodium salicylate.

	3 mmol dm^{-3} $\dot{\gamma}/s^{-1}$	5 mmol dm^{-3} $\dot{\gamma}/s^{-1}$
Pseudoplasticity	0.01 - 0.16	0.0006 - 0.01
Pseudoplasticity + thixotropy	0.21 - 2.48	0.0135 - 0.085
Tixotropy + rheopexy	3.37 - 4.59	

In the pseudoplastic region the relation between the shear stress (τ) and the shear rate (γ̇) can be expressed by the equation

$$\tau = K \dot{\gamma}^{k} \tag{1}$$

where k has the value 0.5 for 5 mmol dm^{-3} and 0.9 for 3 mmol dm^{-3}.

In fig. 2. is shown the shear stress (τ) versus time (t) for 5 mmol dm^{3} CTA-Sal at a constant shear rate (γ = 0.046 s^{1}). During the first 60 s the shear stress increases to a maximum value and then gradually decreases reaching a constant value after some 2000 s. Then the rotation is stopped abruptly and the shear stress decreases to zero during the next 2000 s. The part of the curve showing the stress relaxation at constant rate of shear can for 3 mmol dm^{3} be expressed by the equation proposed by Stoltz et al.[6]

$$\tau = b \gamma \left[1 - (1 - c/a) e^{-t/a}\right] \tag{2}$$

where a is the stress relaxation time at constant rate of shear. In table 2. are given the values of a and c/a together with the apparent maximum viscosity (η_m) and the apparent steady state viscosity (η_s) for 3 mmol dm^{3} CTAB + sodium salicylate at 25.0°C. It can be seen that both the apparent maximum viscosity (η_m), the apparent steady state viscosity (η_s) and the relaxation time (a) at constant shear rate decrease with increasing shear rate, whereas c/a is nearly constant in the shear rate region used.

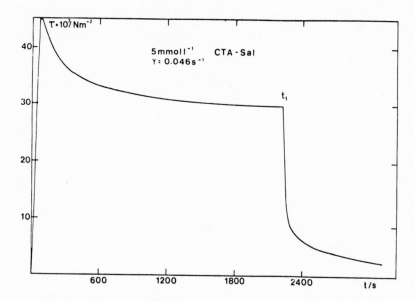

Fig. 2.   Shear stress (τ) versus time (t) at constant shear rate
          for 5 mmol dm^{-3} CTA-Sal at 25.0°C. After 2220 s the rota-
          is stopped abruptly.

Table 2. The Apparent Maximum Viscosity (η_m), the Apparent Steady
         State Viscosity (η_s), the Stress Relaxation Time at Con-
         stant Shear rate (ã) and the Value of c/a as a Function
         of Shear Rate ($\dot{\gamma}$) for 3 mmol dm^{-3} CTAB + sodium salicylate
         at 25.0°C.

$\dot{\gamma}/s^{-1}$	η_m/mPa s	η_s/mPa s	a/s	c/a
0.212	184.0	111.0	136	1.33
0.728	74	49	48	1.23
	73	49	47	1.27
1.35	55	35	20	1.23
	53	34	20	1.41
0.116		25	36	1.80[6]

In the last line in table 2. are quoted the results found by Stoltz et al.[6] from their measurements on blood at 37°C. It is seen that their results are comparable with the present measurements on 3 mmol dm^{-3} CTAB + sodium salicylate at 25.0°C.

The stress relaxation curves at constant shear rates for 5 mmol dm^{-3} CTA-Sal (fig. 2.) can not be expressed by equation (2) and it seems that at least two relaxation times is required to describe the shape of the relaxation curves.

CONCLUSION

The rheological properties of highly dilute viscoelastic aqueous solutions containing cetyltrimethylammonium salicylate (CTA-Sal) are very complex and thus difficult to explain. Depending on the concentration and/or the shear rate used the solutions may exhibit pseudoplastic, pseudoplastic + thixotropic, thixotropic + rheopectic, rheopectic or Newtonian behaviour. When the solutions exhibit pseudoplastic behaviour the shear stress versus shear rate relation can be expressed using equation (1). A solution containing 3 mmol dm^{-3} CTAB + sodium salicylate has at 25.0°C rheological properties comparable to those of blood at 37°C.

ACKNOWLEDGEMENTS

This work was supported by Danish Natural Science Research Council and was made at Fysisk-Kemisk Institut, The Technical University of Denmark.

REFERENCES

1. S. Gravsholt, Viscoelasticity in Highly Dilute Aqueous Solutions of Pure Cationic Detergents, J. Colloid Interface Sci. 57:575 (1976).
2. S. Gravsholt, Physico-Chemical Properties of Highly Dilute Aqueous Detergent Solutions showing Viscoelastic Behaviour, in "Physical Chemistry of Surface Active Substances, Proc. VII Intern. Congr. Surface Active Substances, Moscow 1976", 2:906 (1978).
3. S. Gravsholt, Rheopectic Behaviour of Highly Dilute Viscoelastic Aqueous Detergent Solutions, Naturwissenschaften 66:263 (1979).
4. S. Gravsholt, Physico-Chemical Properties of Viscoelastic Aqueous Detergent Solutions, in "Polymer Colloids II", R. Fitch, ed., Plenum, New York (1980).
5. F. R. Spinelli and Ch. D. Meier, Measurement of blood viscosity (1), Biorheology, 11:301 (1974).
6. J. F. Stoltz, S. Gaillard, M. Lucius et M. Guillot, Recherche d'un modèle viscoélastique applicable au comportement rhéologique du sang, J. mécan., 18:593 (1979).

REINFORCED HYDROPHILIC MATERIALS FOR SURGICAL IMPLANTS

C. Migliaresi, J. Kolařík[+], M. Štol[+] and L. Nicolais

University of Naples, Italy

[+]Academy of Sciences, Prague, Czechoslovakia

INTRODUCTION

The possibility of using polymeric materials for constructing artificial organs has been object of many scientific publications and patents[1]. However, polymers which are suitable for particular applications often display undesidered side effects once implanted in the human body. In fact the specifications which apply to polymeric materials used as substitutes for internal organs are very severe[2].

In the last years, a growing interest in a new class of biomaterials, the hydrogels, has been shown. These polymers are swollen extensively in water (30-70%) and display excellent biocompatibility also for long term applications[3]. In fact, hydrogels appear to absorb proteins and to adhere with cells more gently than low water content foreign interfaces[4]; moreover, the water molecules included in the polymer seem to be associated within the three-dimensional network to form a quasi-organized structure similar to the one formed in the proteins[5]. This enables rapid ingrowth of cells and capillaries and permits modelling of all tissue characteristics. In contrast with this high biocompatibility, the hydrogels generally display very poor mechanical properties reducing very much the possibility of their applications as a material for artificial organs.

In the present work the possibility of using the concepts of

composite materials[6], for designing artificial tendons, is investigated.

MATERIALS AND METHODS

The hydrogel used in this study is the poly-2-hydroxyethylme-
tacrylate (PHEMA), developed by Wichterle and Lim[7] and widely applied
for contact lenses. The commercial 2-hydroxyethylmethacrylate (HEMA)
monomer was redistilled before use. The radical polymerization was
accomplished at T=65°C for 2 hours using 0.1% by weight of 2-azo-
iso-butyronitrile (AIB) as initiator and 0.1% by weight of ethylene—
dymethacrylate (EDMA) as crosslinking agent. Sets of PHEMA containing respectively 10, 20, 30 and 40 percent by weight of diacetine
were prepared. The mixture was poured between two glass plates for
preparing flat specimens with a thickness ranging from 0.25 to 2.5
mm. The fibers used were commercial PET fibers (a bundle of 36 fibers,
overall titre 167 dtex) with false twist. To eliminate the plastic
deformation the fibers were opportunely treated[8]. The artificial
tendons were obtained by crosslinking the HEMA/diacetine mixture at
40% diacetine content in a silicon rubber tube, 2 mm inside diameter,
in which the desidered amount of uniaxially oriented fibers was placed.
Once the reaction was completed, the samples were dipped in distilled water for two months at room temperature to eliminate any
unreacted material and to eliminate diacetine in the case of HEMA/
diacetine mixtures.

Permeability measurements were performed by using a classical
permeability cell apparatus[9].

The creep of fibers and tendons was determined with an apparatus previously described[8].

Tensile properties were determined by using an Instron universal testing machine mod. 1112, at a strain rate of 0.2 min^{-1}. The
straining of tendons was recorded by means of an extensometer, because no way could be found to eliminate the sample slippage in the
jaws of the apparatus. During the tensile tests the tendons were
frequently wetted by water in order to prevent the drying of the
sample. In the other cases the samples were immersed in distilled
water, kept at the constant temperature of 37°C.

RESULTS AND DISCUSSION

The asympthotic values of the water flow J, expressed as JxL, where L is the membrane thickness, measured on membranes prepared from HEMA/diacetine at different relative content, are reported in Figure 1 as function of the applied pressure. It can be observed that the effect of diacetine is to modify the permeability at each value of the applied pressure ΔP.

The permeability coefficient k, reported in Table I, has been calculated as[10,11]:

$$k = JL\eta /\Delta P \qquad (1)$$

where η is the water viscosity at 37°C. With J measured in ml/cm^{2}sec, L in cm, ΔP in dynes/cm^2 and η in poises, k results expressed in cm^2.

Using the analysis proposed by Paul[12], from the permeability results, one can conclude[13] that a large part of the water transport is of viscous type. Moreover the presence of viscous flow in the water transport mechanism can also be inferred from the fact that

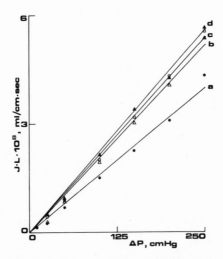

Fig. 1. Normalized water flow versus applied pressure for membranes prepared from HEMA/diacetine mixtures at initial diacetine content of: 10% (a), 20% (b), 30% (c), 40% (d).

the initial values of permeation flow are much higher than the asympthotic values which are generally reached after several days of operation. This long time necessary for the compaction phenomenon could
be due to the slow decrease of pore radii for the effect of the applied pressure. These permeability measurements allow us to have an
idea on the degree of porosity of the material[13]. In particular the
system PHEMA/diacetine 40% shows a high dimensional stability once
swollen in water and a good degree of microporosity which is important for the biocompatibility of the implants[4].

   Typical stress-strain curves of PHEMA for different initial
diacetine contents are shown in Figure 2. The presence of diacetine
reduces the elastic modulus E (i.e. the initial slope of the stress
strain curve), and the strength σ_b, while increases the elongation
at break ε_b. The moduli E are decreasing for PHEMA with increasing
percent of diacetine added.

   With reference to the rubber elasticity theory[14] and the Flory-
Huggins equation[14], has been possible to obtain, from previous re -
sults, the values of M_c, i.e. the molecular weight between two cross-
links, and of χ, the Flory-Huggins interaction parameter between
polymer and swelling water. The results from the theoretical investigation[13], are reported in Table 1 together with the calculated

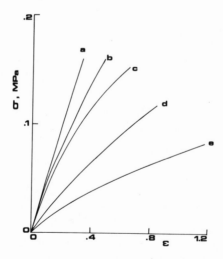

Fig. 2. Stress-strain curves for pure PHEMA(a) and PHEMA with ini-
        tial diacetine content of: 10% (b), 20% (c), 30% (d) ,40%(e).

Table 1.   Physical parameters of PHEMA at different initial
            diacetine content

Vol.% of diacetine	$k \times 10^{16}$ cm^2	v_2	M_c Kg/Kmole	χ calc.	χ pred [15].
0	/	0.543	15.24	0.811	0.809
10	0.83	0.544	16.45	0.813	0.810
20	1.07	0.542	18.69	0.811	0.808
30	1.11	0.534	28.39	0.804	0.801
40	1.15	0.528	67.18	0.798	0.795

values of v_2, i.e. the volume fraction of polymer in the swollen
gel. They point out that the crosslinking degree decreases with increasing diacetine content and that the calculated χ values well
compare with those predicted by Hasa and Janacek[15].

The tensile properties of the PET fibers used in this work are
shown in Figure 3 together with those of our SMT, of a rat tail ten-

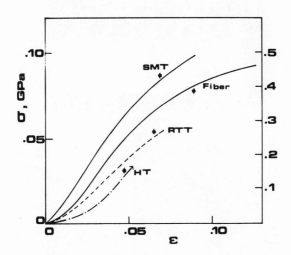

Fig. 3. Stress-strain curves for horse tendon (HT), rat tail ten-
        don (RTT), synthetic model tendon (SMT) and PET fiber
        used

don[16] and of a horse tendon[17]. If we compare the modulus and strength
of the PET fibers and collagen[16] (E=1 GPa, σ_b=60 MPa), it is clear
that the required modulus of synthetic tendons is reached with almost
20% vol. of fibers, while in the case of strength the respective a-
mount is only 10%. Hence, 15-20% vol. of fibers was used in the preparation of synthetic tendons; such fraction appeared to be optimal
also with respect to a uniform filling of the tendon with fibers (see
Figure 4). A SMT with a fiber fraction of 19% is discussed in this
paper. The stress-strain dependence of SMT (Figure 3) is much less
S-shaped at its beginning than that observed with fibers, so that
stress increases linearly with strain almost from the very beginning.
The shape of the time dependence of creep of model tendon is similar
to the dependence determined for the parent fibers (Figure 5), i.e.
the effect of the gel matrix is almost negligible. In the range be
tween 1 and 100 min., creep proceeds at the almost constant rate
d logD(t)/d logt=0.025, so that compliance does not approach an equilibrium value, but overall changes are relatively small, as has al
ready been pointed out for fibers alone. It can be said, moreover,
that a long term creep followed by an interval during which the tendon recovers can be considered as much more frequent. Such a regime
was imitated in a simplified manner: after one minute of creep at
σ_c= 12.5 MPa, the stress was removed for one minute, and the cycle

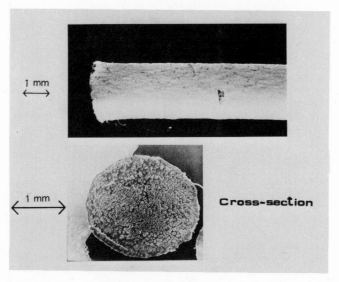

Fig. 4. General vue of prepared synthetic model tendon (SMT)

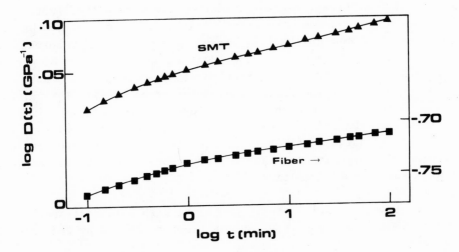

Fig. 5. The time dependence of compliance for fibers and synthetic
         model tendon (SMT) at the stress of 12.5 MPa

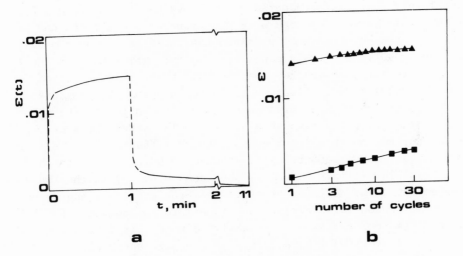

a   b

Fig. 6. a) Time dependence of strain of SMT during one-minute
            creep and the following recovery
         b) Dependence of strain after one-minute creep (upper
            curve) and strain after the following one-minute recovery
            (lower curve)

was repeated thirty times without interruption. Figure 6a shows that
one minute was sufficient only for the recovery of 90-95% of the arising deformation. If, however, recovery took up a time ten times longer than that of creep, recovery was complete within the limits of
experimental error, i.e. the sample length before and after the experiment was the same. In cyclical creep the nonrecovered deformation
was gradually increasing with the logarithm of the number of cycles
(Figure 6b). In contrast, deformation attained after one minute of
creep became established after ten cycles at an equilibrium value.
It can be said therefore that with repeated loading the model tendon lose some 10% of stiffness in the first ten cycles and that further cyclic loading does not affect it distinctly. In agreement with
the preceding experiments, the recovery of the samples was complete
after a sufficiently long time (10 h).

The tensile modulus of the composite with unidirectionally oriented continous fibers is given for the fiber direction by the rule of
mixing[6]:

$$E = E_m v_m + E_f v_f \qquad (2)$$

where E_m, E_f are the moduli and v_m, v_f are the volume fractions of
the matrix and fibers respectively. As the modulus of the matrix is
completely negligible compared with the value for the fiber, E_f/v_f
(8.46 GPa) can be compared with the properties of fibers and tendon.
Hence it can be said that in composites the fibers have a modulus
higher by 30%. The likely cause lies in that all fibers in the gel
matrix are strained in the same way, and there is no shift of the
deformation curves of fibers with different geometric texture. The
strength of the tendon corresponds to the strength of fibers contained therein with reasonable accuracy. The strain at break, $_b$, is
lower for tendon than for fibers, also as a result of the elimination
of changes in the geometrical texture of fibers present in the incompressible matrix. The values obtained, 9.6%, is in the range of
values found for collagen fibers. Since the stiffness and strength
of model tendon is directly proportional to the content of PET fibers, these properties may be easily adjusted, if the volume frac-
ction of texturized fibers varies between 0.1 and 0.25.

REFERENCES

1.   "Polymers in Medicine and Surgery", R. L. Kronenthal, Z. Oser,
      and E. Martin eds., Plenum Press, New York (1975)

2. D. J. Lyman, Biomedical Polymers-An Introduction, in "Polymers in Medicine and Surgery", R.L. Kronenthal, Z. Oser, and E. Martin, ed., Plenum Press, New York (1975)

3. L. Sprincl, J. Vacik, J. Biomed. Mater. Res., 7:123 (1973)

4. B. D. Ratner, and A. S. Hoffman, Synthetic Hydrogels for Biomedical Applications, in "ACS Symposium Series",N°31 , J. D. Andrade, ed., (1976)

5. S. D. Bruck, J. Biomed. Mater. Res., 7:387 (1973)

6. L. Nicolais, Polym. Eng. Sci., 15:137 (1975)

7. O. Wichterle, and D. Lim, Nature, 165:117 (1960)

8. J. Kolarik, C. Migliaresi, M. Stol, and L. Nicolais, "Mechanical Properties of Model Synthetic Tendons", submitted to J. Biomed. Mater. Res.

9. M. F. Refojo, J. Appl. Polym. Sci., 9:3427 (1965)

10. M. L. White, J. Phys. Chem., 64:1563 (1960)

11. S. Madras, R. L. McIntosh, and S. G. Mason, Can. J. Res., 27B: 764 (1949)

12. D. R. Paul, J. Polym. Sci., Polym. Phys. Ed., 12:1221 (1974)

13. C. Migliaresi, L. Nicodemo, L. Nicolais, and P. Passerini, "Physical Characterization of Foamed Poly-2-hydroxyethylmethacrylate Gels", submitted to J. Biom. Mater. Res.

14. L. R. G. Treloar,"Physics of Rubber Elasticity", Clarendon Press, Oxford (1958)

15. J. Hasa, and J. Janacek, J. Polym. Sci., Part C, 16:317 (1967)

16. S. Torp, E. Baer, and B. Friedman, Colston Papers, 26:223 (1975)

17. N. C. Herrick, H.B. Kingsbury, and D.Y.S. Lou, J. Biomed. Mater. Res., 12:877 (1978)

UPON VISCOELASTIC BEHAVIOUR OF COMPOSITIONS OF THE TYPE OF

POLYMER CONCRETE

J. Hristova

Bulgarian Academy of Sciences, Sofia, Bulgaria

(Abstract)

In the present study the viscoelastic behaviour of polymer concrete prepared on the basis of polyester resin was analyzed.

The polyester resin at low stress and deformation shows linearity in its viscoelastic behaviour in conditions of creeping and relaxation. In this case there is a possibility to use for the description of the behaviour the integral equations of Boltzman-Volterra with singular nuclei of the type :

$$K(t - \tau) = \frac{e^{-\beta t}}{t} \sum_{n=1}^{\infty} \frac{[A \Gamma(\alpha)]^n t^{\alpha n}}{\Gamma(\alpha n)} \tag{1}$$

$$T(t - \tau) = A e^{-\beta t} t^{\alpha - 1} \tag{2}$$

where α, β and A are parameters. The values of this parameters, determined in conditions of compression by experimental data at creeping and at relaxation were identical.

Because the viscoelastic behaviour of a polymer concrete was determined by the behaviour of the polymeric component there is a likeness between the behaviour of the discussed compositions and polyester resin. The experimental data showed that also the compositions exhibit a linear field of viscoelastic behaviour, in

which integral equations of Boltzmann-Volterra with nuclei of the
discussed type are applicable. The comparison of the parameters of
the nuclei determined for compositions with different quantities
of aggregates in conditions of compression showed that the values
of α, and β of the various compositions are identical.

THE RHEOLOGY OF POLYMERS WITH

LIQUID CRYSTALLINE ORDER

Donald G. Baird

Departments of Chemical Engineering and Engineering
Science and Mechanics
Virginia Polytechnic Institute and State University
Blacksburg, Virginia 24061

INTRODUCTION

Liquid crystals are substances that possess mechanical properties resembling those of fluids yet are structured enough to
diffract X-rays and transmit polarized light. The liquid
crystalline state is believed to be an intermediate phase between
the isotropic fluid state and the crystalline solid state. For
this reason these fluids are also referred to as mesophases. Because of the unique optical properties they are also termed
anisotropic fluids. Low molecular weight compounds which form
liquid crystals have been studied for over a century. However, it
has been only in the last thirty years that liquid crystalline
order has been found to exist in polymeric systems.

Interest in polymeric systems with liquid crystalline order is
due to the use of these systems in the manufacture of ultra-high
strength and modulus materials [1,2] and their function in biological
systems. Fibers processed from liquid crystalline solutions of
poly-p-phenyleneterephthalamide (PPT) in sulfuric acid have been
reported to have moduli as high as 13×10^{11} dyn/cm^2 and tensile
strengths as high as 3×10^{10} dyn/cm^2. For comparison, steel fibers
have moduli of the order of 20×10^{11} dyn/cm^2 and tensile strengths
of the order 3×10^{10} dyn/cm^2. Many features of the living state
such as specificity, asymmetry, dynamic transformation, rhythmicity,
control, and communication in molecular domains and evolution can
be attributed to the liquid crystalline state.[3]

Liquid crystalline order arises in polymeric systems through
three mechanisms. These include intermolecular interaction of

rodlike chains, intramolecular interaction in vinyl type polymers
with mesogenic side groups, and domain formation in blockcopolymers
with incompatible blocks. Our interest in this paper will be in those
systems in which liquid crystalline order forms through intermolecular arrangements of rodlike molecules.

Just as for low molecular weight compounds, three types of mesomorphic arrangements have been identified for polymer systems.
These are shown schematically in Figure 1. In the nematic mesophase
the centers of gravity of the molecules are arranged randomly as in
an isotropic fluid but there are local regions in which molecules all
have the same orientation. In the cholesteric mesophase, again
there is no long range order in the centers of gravity but the
molecules are confined to planes in which the orientation of the
molecules is the same. The orientation varies periodically through
the liquid sample. In the smectic mesophase the molecules lie
within planes of well defined spacing. Rodlike polymeric systems
form both nematic and cholesteric mesophases and will be the
systems of interest in this review.

Liquid crystalline order has been observed in polymer solutions
in which case we refer to the system as lyotropic. It has also
been identified in undiluted polymer or in polymer melts in which
case we refer to the system as thermotropic. We will consider both
lyotropic and thermotropic liquid crystals here.

A study of the processing of liquid crystalline solutions and
melts reveals some interesting facts. Probably the most important
observation is that fibers and injection molded specimens are highly
oriented. Molecular orientation in the solid state comes directly
from flow induced orientation since no further drawing of the
specimens occurs. Because of the direct correlation of flow
history to structure-property relations, an understanding of the
rheological properties of these systems is essential.

Rheological Properties of Lyotropic Systems

Interest in processing these systems arose not only because
of the exceptionable physical properties found in the processed
materials but because of the ease of processing. As the concentration is increased the viscosity of these solutions increases
until reaching a critical concentration C*. At this concentration,
solutions become anisotropic and the viscosity decreases with increasing concentration. This behavior is illustrated in Figure 2
and is characteristic of cholesteric and nematic mesophases.[5,6]
Other material functions such as the primary normal stress difference (N_1), the dynamic moduli (G' and G"), and the relaxation
time (λ_1) exhibit the same behavior as shown in Figures 2 and 3.

Measurements of the linear viscoelastic properties by Baird[7]

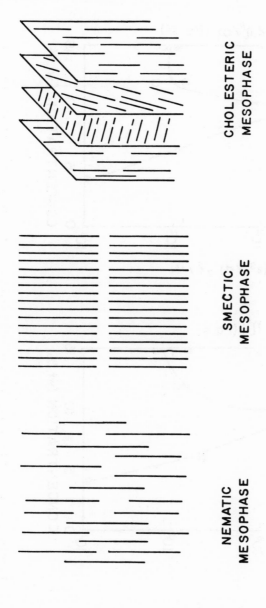

Figure 1. Schematic drawing of three mesomorphic arrangements.

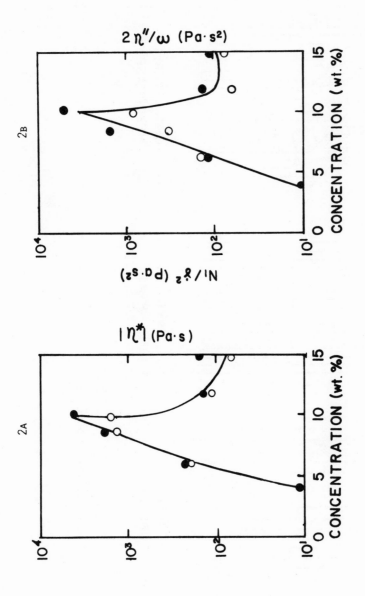

Figure 2. (a) Viscosity versus concentration for solutions of PPT in 100%
                 H_2SO_4. ○ Steady Shear ● Dynamic

           (b) Primary normal stress difference and 2 G'/w^2 versus concentration
                 for solutions of PPT in 100% H_2SO_4: ○ $N_1/\dot{\gamma}^2$ ● $2G'/w^2$

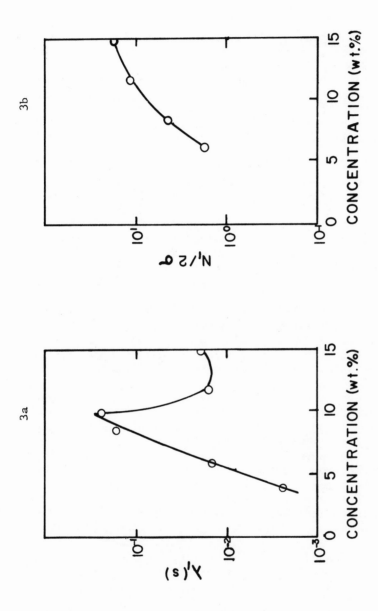

Figure 3.   (a)   Relaxation time versus concentration for solutions of PPT in 100% H_2SO_4

(b) $N_1/2\sigma$ (Weissenberg Number) versus concentration for the same solutions.

Izuka,[8] and Aoki and co-workers[9] reveal that lyotropic systems
can dissipate and store energy during a cyclic deformation. The
shape of the storage and loss moduli versus frequency demonstrate
clearly that the liquid crystalline state is fluid in nature.

The shear rate dependence of viscosity (η) has been reported
by several researchers.[5,6,10] With the onset of liquid crystalline
order η becomes less dependent on shear rate ($\dot{\gamma}$). If the molecular
weight distribution is narrow enough the viscosity of the liquid
crystalline state can be nearly constant. At very low $\dot{\gamma}$, Papkov
and co-workers reported yield stresses.[10] However, Baird[11] and
White and co-workers[9] reported that yield stresses disappeared with
temperatures slightly above room temperature.

The primary normal stress difference has been measured for
both isotropic and anisotropic solutions of poly-p-phenyleneterephthalamide in sulfuric acid.[10,11] At high shear rates N_1
becomes independent of $\dot{\gamma}$ in the anisotropic state.

Correlations between the linear viscoelastic and steady shear
properties at low $\dot{\gamma}$ values is not good for anisotropic solutions.
However, for isotropic solutions of the same polymer the correlation
is quite good.

Although geometric effects have been reported in the rheometry
of low molecular weight liquid crystals, this does not seem to be
the case for polymer systems. Baird[7] found for anisotropic solutions
of PPT that η versus $\dot{\gamma}$ data was identical to within experimental
error using the cone-and-plate and plate-plate geometries.

Only one study has been reported using a capillary rheometer.[12]
Entrance pressure losses (ΔPent) were found to follow the characteristic viscosity behavior. Entrance pressure losses increased
with concentration until C* and then decrease with increasing concentration. Although ΔPent increased with $\dot{\gamma}_w$, die swell became
constant and ranged from 1.0 to 1.3 for the anisotropic solutions.
The large entrance pressures are due to the orientation of the
molecules in the entrance of the capillary. The lack of die swell
can be interpreted to indicate that very little recovery at the die
exit occurs.

In summary, it appears from the few rheological studies that
have been carried out that lyotropic systems are viscoelastic:
i.e. they both store and dissipate energy during deformation. However, it is believed that some of the elastic properties may be due
to regions of isotropic fluid dispersed in anisotropic regions. We
can expect lyotropic systems to have less elasticity the higher
the polymer concentration and the more narrow the molecular weight
distribution.

Thermotropic Liquid Crystals

In the last several years a number of highly aromatic polyester copolymers have been reported to form liquid crystalline melts.[13] There have been only two reports in the literature of studies of rheological properties of these unique systems.[14,15,16] Kuhfuss and Jackson [14,15] studied the viscous behavior of copolymers of polyethyleneterephthalate (PET) and p-hydroxybenzoic acid (PHB) of various molar ratios of PHB. With compositions of 40% and more of PHB, the melts became turbid and the viscosity dropped as much as two orders of magnitude at some shear rates. The viscosity was shear dependent for $\dot{\gamma}$ less than 100 sec.$^{-1}$ but was constant for $\dot{\gamma}$ above 100 sec.$^{-1}$

Jerman and Baird[16] studied similar polymer compostions using a capillary rheometer. They measured in addition to viscosity , ΔPent and die swell. The ratio of the extrudate diameter to the capillary diameter (De/Do) was found to be less than one (i.e. the jet contracted) independent of the capillary length for a fixed diameter, and to increase with $\dot{\gamma}_w$ from about .90 to 0.96. Representative viscosity data is given in Figure 4, while die swell data is presented in Figure 5. Even though the extrudate contracted entrance pressure losses were reported to be about 60 times the wall shear stress (τ_w) values. The ratio of ΔPent/τ_w was found to be independent of $\dot{\gamma}_w$ as shown in Figure 6.

In summary we know very little about the rheological properties of thermotropic polymeric liquid crystals. Capillary data suggest that the fluids may not be viscoelastic. The shear dependent viscosity may be due to structural changes or orientation of the molecules during flow. More measurements involving transient behavior, steady state normal stresses, and linear viscoelastic properties are needed.

Rheological Equations of State

Rheological equations of state have been developed by several researchers to describe the dynamical behavior of liquid crystalline systems.[17-22] At the time these theories were developed, no consideration was given to the possibility that macromolecular fluids could exhibit liquid crystalline order. However, because there are numerous similarities in the behavior of polymeric and low molecular weight organic liquid crystalline systems and the constitutive equations were developed following proper laws of continuum mechanics, it is believed that the theories reviewed here may contain some of the essential features needed to describe the flow behavior of polymeric liquid crystalline systems. We review here the theories of Ericksen[17,19] and Hand.[18] The theories reviewed here primarily pertain to the nematic and cholesteric mesophases.

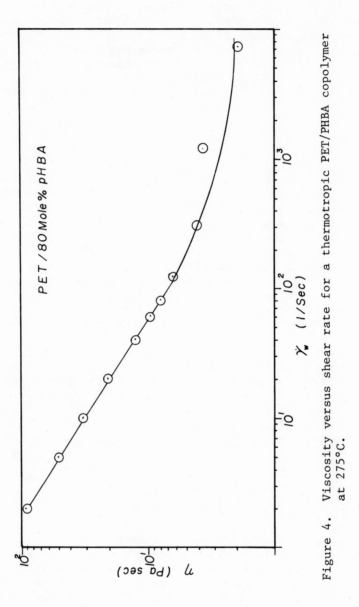

Figure 4. Viscosity versus shear rate for a thermotropic PET/PHBA copolymer at 275°C.

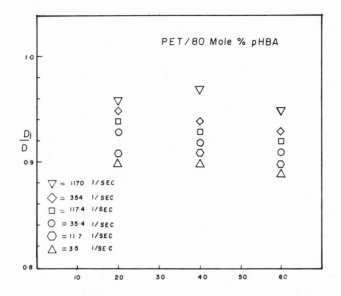

Figure 5. Ratio of extrudate diameter to capillary diameter versus L/D for different shear rates for copolymer of PET/80 mole % PHB at 275°C.

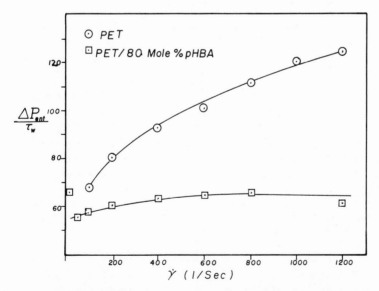

Figure 6. Ratio of $\Delta Pent/\tau$ versus shear rate for PET and a thermotropic copolymer of PET^W/80 mole % PHBA at 275°C.

Liquid crystals are viewed to consist of tiny packets of rodlike molecules whose orientation is given by the vector $\underset{\sim}{d}$. Ericksen believed that stresses generated during flow were due to both the rate of deformation and the orientation of the molecules. He expressed the stress tensor as follows:

$$\tau_{ij} = f_{ij} \; (\dot{\gamma}_{mn}, d_k) \tag{1}$$

where $\dot{\gamma}_{mn}$ is the rate of deformation tensor. The orientation vector $(\underset{\sim}{d})$ was coupled to the flow by the following relation:

$$\frac{D}{Dt} \, d_i = g_i \; (d_k, \gamma_{mn}) \tag{2}$$

where $\frac{D}{Dt}$ is the material time derivative. Following an approach similar to that used by Ericksen to develop the second order fluid theory, Ericksen expanded these functions in powers of $\dot{\gamma}_{mn}$ and d_k. The following constitutive equation was obtained after linearizing the expression in terms of $\dot{\gamma}_{mn}$:

$$\tau = -p\underset{\approx}{\delta} + 2\mu\underset{\approx}{\dot{\gamma}} + (\mu_1 + \mu_2 \, \underset{\sim}{d} \cdot \underset{\approx}{\dot{\gamma}} \cdot \underset{\sim}{d})\underset{\sim}{d}\underset{\sim}{d} +$$
$$+ 2\mu_3 \; [(\underset{\approx}{\dot{\gamma}} \cdot \underset{\sim}{d})\underset{\sim}{d} + ((\underset{\approx}{\dot{\gamma}} \cdot \underset{\sim}{d})\underset{\sim}{d})^t] \tag{3}$$

where the μ_i's are material constants. The orientation vector was coupled to the flow by:

$$\frac{D}{Dt} \, \underset{\sim}{d} = \lambda [\underset{\approx}{\dot{\gamma}} \cdot \underset{\sim}{d} - (\underset{\sim}{d} \cdot \underset{\approx}{\gamma} \cdot \underset{\sim}{d})\underset{\sim}{d}] \tag{4}$$

where $\frac{D}{Dt}$ is the Jauman derivative and λ is a constant. In simple shear flow (i.e. $v_1 = \dot{\gamma}X_2$, $v_2 = v_3 = 0$) this theory predicts both primary (N_1) and Secondary (N_2) normal stress differences. However, N_1 and N_2 are linear in shear rate and η is independent of $\dot{\gamma}$. In the limit as $\dot{\gamma} \to 0$, yield stresses are predicted. As $\dot{\gamma} \to \infty$ and the fluid becomes totally oriented, $N_1 = N_2 = 0$ and η remains constant Some of the experimental data presented in the last section qualitatively agree with these predictions.

Hand[18] attempted to generalize the concept of structured fluids by introducing a tensor $\underset{\approx}{A}$ which described the microscopic structure of the fluid. His constitutive equation reduces to the same form as Ericksen's under the same assumptions as used by Ericksen. However, if he assumed that the fluid was initially isotropic and became anisotropic during flow his theory predicted that N_1 depended on $\overset{\bullet}{\gamma}{}^2$ at low $\overset{\bullet}{\gamma}$ and became constant at high $\overset{\bullet}{\gamma}$. The model predicted viscosity with high (η_∞) and low (η_0) shear rate limits.

Ericksen[22] reformulated his theory for nematic liquid crystalline systems because they did not reduce to the static theory of Oseen and Frank, and it was not possible to prescribe the boundary conditions for preferred directions. The major changes consisted of formulating general conservation laws (i.e. conservation of mass, momentum, energy, and angular momentum) and making sure that his constitutive equation satisfied the entropy inequality due to Mueller. The new balance laws included the momentum and energy transport due to the angular motion of the rodlike packets. For simple shear flow this new constitutive equation predicts that the stress tensor is no longer symmetric. Furthermore, it does not predict yeild stresses as his first theory. However, similar predictions for N_1, N_2 and η are given in the limit of high $\overset{\bullet}{\gamma}$.

In conclusion, all fluid models for polymeric liquid crystals neglect the possibility that these systems can be viscoelastic. They are primarily valid for cases in which the Deborah number is zero. Some data seem to indicate that this may not be too serious of an assumption.

Conclusions

In summary, we know very little about the rheological properties of polymeric liquid crystals. Before models for these fluids can be constructed, more data involving the nonlinear and transient response of these fluids is needed.

References

1. S. L. Kwolek, U. S. Patent No. 3,671, 542, 1972.

2. H. Blades, U. S. Patent No. 3,767, 756, 1973.

3. R. K. Mishra, Mol. Cryst. Liq. Cryst. 29:201 (1975).

4. J. H. Wendorf, Scattering in Liquid Crystalline Polymer Systems in "Liquid Crystalline Order in Polymers" A Blumstein, ed., Academic Press, New York (1978).

5. J. Hermans, Jr., J. Colloid Sci., 17:638 (1962).

6. D. G. Baird, Rheology of Polymers with Liquid Crystalline Order, in "Liquid Crystalline Order in Polymers," A. Blumstein, ed., Academic Press, New York (1978).

7. D. G. Baird, Rheological Properties of Liquid Crystalline Solutions of Poly-p-Phenyleneterephthalamide in Sulfuric Acid, J. Rheol., In Press.

8. E. Izuka, Mol. Cryst. Liq. Cryst., 25:287 (1974).

9. H. Aoki , Synthesis, Characterization, Rheological and Fiber
    Formation Studies of p-linked Aromatic Polyamides,
    Polym. Sci. and Engr. Report No. 108, University of
    Tennessee (1977).

10. S. P. Papkov, V. G. Kulichiphin, and V. D. Kalymykova,
    Rheological Properties of Poly-p-Benzamide Solutions,
    J. Polym. Sci: Polym. Phys., 12:1753 (1974).

11. D. G. Baird, Viscometry of Anisotropic Solutions of Poly-p-
    Phenyleneterephthalamide in Sulfuric Acid, J. Appl.
    Polym. Sci., 22:2701 (1978).

12. D. G. Baird, Capillary Rheometry of Anisotropic Solutions of
    Poly-p-Phenyleneterephthalamide in Sulfuric Acid, In Prep.

13. J. Preston, Synthesis and Properties of Rodlike Condensation
    Polymers, in "Liquid Crystalline Order in Polymers" A.
    Blumstein, ed., Academic Press, New York (1978).

14. W. J. Jackson, Jr. and H. F. Kuhfuss, Liquid Crystalline Polymers.
    I. Preparation and Properties of p-Hydroxy Acid Copolyesters,
    J. Polym. Sci., Polym Chem. Ed., 14:2043 (1976).

15. F. E. McFarlane, V. A. Nicely, and T. G. Davis, Liquid Crystal
    Polymers. II. Preparation and Properties of Polyesters
    Exhibiting Liquid Crystalline Melts, Cont. Top. Polym.
    Sci., 2:109 (1977).

16. R. E. Jerman and D. G. Baird, Capillary Rheometry of Thermotropic
    Liquid Crystalline Polyesters, In Prep.

17. J. L. Ericksen, Transversely Anisotropic Fluids, Köll. Zeit,
    173:117 (1960).

18. G. L. Hand, A Theory of Anisotropic Fluids, J. Fluid Mech. 13:
    33 (1962).

19. J. L. Ericksen, Conservation Laws for Liquid Crystals, J. Rheol.,
    5:23 (1961).

20. F. M. Leslie, Some Constitutive Equations for Anisotropic Fluids,
    Quart. J. Mech. App. Math., 19:357 (1966).

21. A. C. Eringen and E. S. Shuhubi, Int. J. Eng. Sci. 2:189 (1964):
    2:205 (1964).

22. J. L. Ericksen, Continuum Theory of Liquid Crystals of Nematic
    Type, Mol. Cryst. Liq. Cryst., 7:153 (1969).

A THIXOTROPIC MODEL FOR CEMENT PASTES

Romano Lapasin, Vittorio Longo, Sandra Rajgelj[*]

Istituto di Chimica Applicata e Industriale
(*) Istituto di Scienza delle Costruzioni
Università di Trieste, Italia

INTRODUCTION

Several researches[1,9] have indicated the thixotropic behaviour
of cement pastes. When subjected to shear, cement pastes undergo
a structural breakdown process. Pastes can be defined as partially
thixotropic systems since, even if a long rest period follows the
shear action, they have no appreciable tendency to build up their
structure.

Generally the shear and time-dependent behaviour of cement
pastes has been described according to empirical approaches, except
for Tattersall's approach[8] in which an interpretation of the thixotropic breakdown is suggested. With the aim of finding one or more
parameters suitable to characterize the rheological behaviour and
to be connected with the composition parameters of pastes, the empirical approaches are generally based on experimental tests such
as stress transients at constant shear rate or hysteresis cycles.
In both cases it is difficult to represent the whole thixotropic
behaviour, and with hysteresis cycles even to give a significant
rheological characterization.

The object of the present work is to define a rheological
model capable of describing all possible steady and unsteady flow
conditions of the material. The Cheng approach[10] seems to meet
these requirements: if some hypotheses are satisfied, the rheological behaviour of the material in any kinematic condition can
be described in terms of scalar constitutive equations consisting
of an equation of state

$$\tau = f(\dot{\gamma}, \lambda)$$

and a rate equation

$$\frac{d\lambda}{dt} = g(\dot{\gamma}, \lambda)$$

in which the function f and g are subjected to certain restrictions and λ is an arbitrary structural parameter.
The state equation is generally assumed to be linear in λ and is expressed, in the more general form, by

$$\tau = f_0(\dot{\gamma}) + \lambda f_1(\dot{\gamma})$$

For a partially thixotropic material, the rate equation contains only the breakdown term:

$$\frac{d\lambda}{dt} = -k(\dot{\gamma})\lambda^n$$

where $k(\dot{\gamma})$ is the breakdown rate constant, and n is the order of the breakdown process. It follows that, at any shear rate, the equilibrium condition corresponds to

$$\lambda_e = 0$$

In order to define the model, a set of shear stress-time transients is determined at constant shear rate, until a steady condition is attained. For each shear rate, a new sample is empoyed. If the initial conditions of the samples are equivalent, it is then possible to investigate the breakdown kinetics at different shear rates and the shear dependent behaviour of the material in two limiting conditions In fact, the initial and final shear stress values of each transient can be assumed to correspond to maximum and equilibrium structural conditions, respectively.
For t = 0   : $\lambda = 1$

$$\tau_{max} = f_0(\dot{\gamma}) + f_1(\dot{\gamma})$$

For $t \to \infty$: $\lambda \to 0$

$$\tau_e = f_0(\dot{\gamma})$$

It follows that, for each transient, the structural parameter λ at the time t can be obtained from the corresponding shear stress τ, according to

$$\lambda = \frac{\tau - \tau_e}{\tau_{max} - \tau_e}$$

The order n of the breakdown reaction can be determined from the

shear transients, indipendently of the state equation assumed. In
fact

$$\frac{d\lambda}{\lambda^n} = -k(\dot{\gamma})dt$$

$$\lambda = \exp(-k(\dot{\gamma})t) \qquad\qquad \text{for } n = 1$$

$$\lambda = ((n-1)k(\dot{\gamma})t+1)^{1/(1-n)} \qquad\qquad \text{for } n \neq 1$$

Therefore it is possible to determine separately the parameters
of the state equation and of the rate equation, from data relative
to maximum and equilibrium shear stresses and to stress transients,
respectively.

EXPERIMENTAL

The cement pastes were prepared from tap water (hardness 20°F,
pH 7.5) and two types of Portland cements (PTL 325 and 425). The
w/c ratios were .40 and .45 for PTL 425 and .40 for PTL 325.
Each mix was prepared first by hand with a spatula and then mixed
in a vanomixer, at two speeds, according to the following stirring
time sequence: 30s at v_1, 30s at v_2, 90s at rest, 60s at v_2 ($v_1 =$
$= 36.65$ rad/s, $v_2 = 45.03$ rad/s).

The rheological measurements were performed with the rotating
coaxial cylinder viscometer Rotovisco Haake, using the measuring
heads Mk 50 and 500 and the device MVIII (cup diameter 42 mm, bob
diameter 30.4 mm, bob height 60 mm). The measurable shear stress
range is $15 \div 510$ Pa and the shear rate range is $4 \div 216$ s^{-1}.

RESULTS AND DISCUSSION

For the definition of f_0 and f_1, the equation suggested by
vom Berg[1] was taken into consideration

$$\tau = \tau_f + A \sinh^{-1}\frac{\dot{\gamma}}{C}$$

because it was used satisfactorily to fit the flow behaviour of
different series of cement pastes and the parameters τ_f, A and C
were related to two main composition parameters, namely, cement volume concentration C_V and specific surface of the cement S_{vB}.

If a common C value can be determined for the maximum and equilibrium flow curves with an acceptable degree of fitting , a simple
state equation linear in λ can be established. Accordingly, the maximum and equilibrium stress data were fitted simultaneously in order
to determine the corresponding values of τ_f and A_0 of the two curves
and the C value. The least square fitting was performed by employ-

Table 1.   Parameters of state equation

PTL	w/c	$\tau_{f,0}$ (Pa)	$\tau_{f,1}$ (Pa)	A_0 (Pa)	A_1 (Pa)	C (s^{-1})
325	.40	6.94	8.29	11.4	15.1	24.3
425	.40	8.43	8.43	30.1	80.2	26.6
425	.45	7.68	7.68	17.9	38.4	17.1

ing the simplex algorithm according to the Nelder-Mead formulation. The resulting state equation is

$$\tau = \tau_{f,0} + (\tau_{f,1} - \tau_{f,0})\lambda + (A_0 + (A_1 - A_0)\lambda)\sinh^{-1}\frac{\dot{\gamma}}{C}$$

where

$$f_0(\dot{\gamma}) = \tau_{f,0} + A_0 \sinh^{-1}\frac{\dot{\gamma}}{C}$$

$$f_1(\dot{\gamma}) = (\tau_{f,1} - \tau_{f,0}) + (A_1 - A_0)\sinh^{-1}\frac{\dot{\gamma}}{C}$$

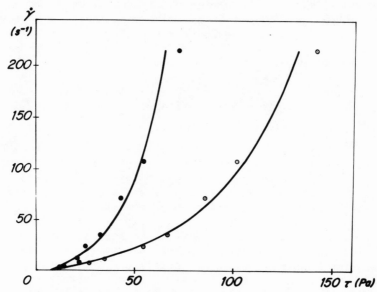

Fig. 1.   Equilibrium and maximum curves for PTL cement 425 at w/c ratio .45.

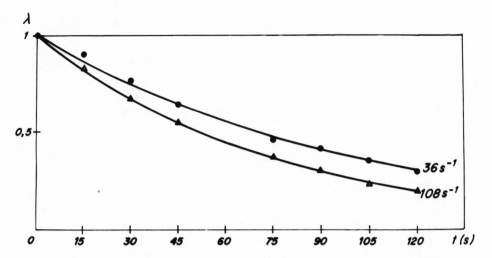

Fig. 2.   Transients for PTL cement 425 at w/c ratio .40 at different shear rates.

The values of $\tau_{f,0}$, $\tau_{f,1}$, A_0, A_1 and C are reported in Tab. 1.
The fitting of maximum and equilibrium curves is satisfactory, as
is shown in Fig. 1.
As regards the rate equation, a preliminary test showed that the
order n does not vary appreciably with varying shear rate for each
system. By simultaneously fitting all the transients data the n
value was determined together with all the k values relative to the
different shear rates.
An example of the fitting of transient data is reported in Fig.2.
The dependence of k on $\dot\gamma$ can be expressed by:

$$k = \frac{\dot\gamma}{a+b\dot\gamma}$$

Table 2.   Parameters of rate equation

PTL	w/c	n	a (s^{-1})	b
325	.40	1.36	678	26
425	.40	1.00	1356	62.4
425	.45	1.09	2634	27

Table 2 reports the values of n, a and b for the three systems examined.

     At this point, having defined the model and indicated the pro-
cedure for the calculation of the parameters of the state and rate
equations, it is interesting to analyze the correlation between
rheological and compositive parameters of the pastes.

CONCLUSIONS

     A rheological model was defined for the description of the
rheological behaviour of cement pastes, according to Cheng's approach. Because of the partially thixotropic nature of the system,
the values of the parameters of the state and rate equations can
be determined separately.
The state equation was defined on the basis of the equation suggested by vom Berg for the flow curve of cement pastes. The equations
were tested on cement pastes with different compositive characteristics. The fitting was satisfactory and allowed the rheological
parameters to be established.

REFERENCES

1. W. vom Berg, private communication (1979).
2. J.P. Bombled, Rhéologie des mortiers et des bétons frais: étude
   de la pâte interstitielle de ciment, Rev. Mater. Construct. 688:
   137 (1974).
3. M. Collepardi, Il comportamento reologico delle paste cementi-
   zie, Il Cemento 3: 99 (1965).
4. R. Lapasin, V. Longo and S. Rajgelj, Thixotropic Behaviour of
   Cement Pastes, Cem. Concr. Res. 9: 309 (1979).
5. C. Legrand, Contribution à l'étude de la rhéologie du béton frais,
   Matériaux et constructions 29: 275 (1972).
6. A.A. Nessim and R.L. Wajda, The Rheology of Cement Pastes and
   Fresh Mortars, Mag. Concr. Res. 17 (51): 59 (1965).
7. E.M. Petrie, Effect of Surfactant on the Viscosity of Portland
   Cement - Water Dispersions, Ind. Eng. Chem. 15 (4): 242 (1976).
8. G.H. Tattersall , The Rheology of Portland Cement Pastes, Brit.
   J. Appl. Phys. 6: 165 (1955).
9. O.J. Uzomaka, A Concrete Rheometer and its Application to a Rheological Study of Concrete Mixes, Rheol. Acta 13: 520 (1974).
10. D.C.-H.Cheng and F. Evans, Phenomenological Characterization of
    the Rheological Behaviour of Inelastic Reversible Thixotropic
    and Antithixotropic Fluids, Brit. J. Appl. Phys. 16: 1599 (1965).

ON THE RHEOLOGICAL BEHAVIOUR OF CEMENT SUSPENSIONS TESTED

IN A COUETTE RHEOMETER

Wolfgang van Berg

Institut for Building Research

Technical University Aachen, Germany, F.R.

1. INTRODUCTION

The rheological properties of cement suspensions are with regard to
the workability of both mortars and concretes of particular significance for building technology.
Cement suspensions show a very complex rheological behaviour. They
have a yield stress, they show shear thinning and sometimes shear
thickening, and their rheological properties are time dependent both
in respect to chemical reactions (i.e. to the age of the suspension)
and to the duration of mechanical loading. Flow curves of cement suspensions were measured with a specially built rheometer. Details of
the rheometer and the testing procedure are given in[1].
Usually the measured flow curves showed shear thinning at low and
shear thickening at higher shear rates. (s. Fig. 1)
The degree of shear thickening depends on the concentration of solids
and the specific surface of the cement. The smaller the specific
surface and the higher the concentration of solids the more distinct
is the shear thickening.
The hysteresis between the up curve and the down curve gives an
information about the time dependent rheological properties. If at
each shear rate the up curve is in the range of higher shear stresses
than the down curve, this is a result of time dependent shear thinning; the other case indicates time dependent shear thickening.
In the tests mentioned above in general time dependent shear thinning was observed. In some cases, however, in the upper range of the
flow curves there occured an overlapping of up and down curve, that
means that there ist time dependent shear thinning at low and time
dependent shear thickening at high shear rates (s. Fig. 1, curve on
the left).
Further tests, the results of which will be published in detail later

on, show that the degree of time dependent shear thickening depends on the change of shear rate with time (Ḋ) in the experiment.
If Ḋ, which is constant at each test, is decreased in further
experiments with a fresh sample, time dependent shear thickening
grows sharper, as it is shown by some examples of measured flow
curve in Fig. 2.

2. INVESTIGATION OF THE INFLUENCE OF SHEAR RATE UPON SHEAR STRESS-TIME BEHAVIOUR

2.1 Constant shear rate test

In the Institute of Buildung Research at the Technical University
of Aachen tests were carried out in order to examine the causes
of the observed rheological behaviour of cement suspensions, in
which shear stress time curves at constant shear rates were
measured. Shear reates were measured in steps of about 50 s^{-1}
between 50 s^{-1} and 1800 s^{-1}. The tested cement suspension had a
solid concentration of $C_v = 0,39$ and the cement used had a
specific surface of about 290 m^2/kg (Blaine). Some of the test
results are presented in Fig. 3.
Two different types of curve were found. At low shear rates (with
these special test conditions up to a shear rate of about
D = 1065 s^{-1}) a steady decrease of shear stress with time was
measured (time dependent shear thinning); in contrast to that at
shear rates higher than D = 1065 s^{-1} there was observed a re-
increase of shear stress after a decrease at the beginning of the
experiment; this phenomen can be called time dependent shear
thickening. The shear stress minimum was observed to be reached
the earlier the higher the shear rate was.

2.2 Evaluation of shear stress time curves

For the mathematical description of the shear stress τ in dependence upon the test time t the following equation was chosen.

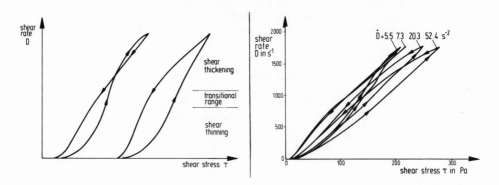

Fig.1: Schematic flow curves   Fig.2: Flow curves at different
       of cement suspension   changes of shear rate with
                                          time (Ḋ)

$$\tau = PO + P1 \times \exp(-t/T1) + P2 \times \exp(-t/T2) + P3 \times (1-\exp(-t/T3)) \quad (1)$$

From equation (1) results for the shear stress τ

at the time $t = 0$ $\quad \tau(t = 0) = PO + P1 + P2$ $\qquad\qquad\qquad\qquad$ (2)

and for $\quad t \to \infty$ $\quad \tau(t \to \infty) = PO + P3$ $\qquad\qquad\qquad\qquad$ (3)

For the shear stress time curves with steady decreasing shear stress P3 und T3 become sero.
So in tests with $D = 1065 \ s^{-1}$ the part P3 $\times$ (1-exp(-t/T3)) of equation (1) disappears.
Figure 4 schematically shows the portions of structural breakdown for conditions with steady shear stress decrease and figure 5 shows the combination of portions of equation (1) to the shear stress time curve when shear thickening is present.
The portions P1 $\times$ exp (-t/T1) and P2 $\times$ exp (-t/T2) of equation (1) describe the decrease of shear stress which results, according to the literature, from changes in the internal structure of the ce-ment suspension due to the mechanical loading and is therefore called structural breakdown.
In widening of the results of Tattersall[2] who supposed a simple exponential relationship to describe the breakdown curve, two ex-ponential functions were necessary here to describe the measured decrease of shear stress. A similar suggestion was made by Nessim and Wajda[3].
From the evaluation results that the portion P1 $\times$ exp (-t/T1),which will be called primary breakdown in the following part of the text, happens very quickly and has already finished maximum 5 s after the beginning of loading. In contrast to that the "secondary breakdown" P2 $\times$ exp (-t/T2) is much more slowly.
PO + P3 means the remaining stress when equilibrium conditions are reached.

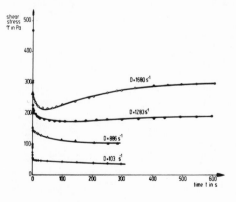

$PO = 20.29 + 9.35 \times 10^{-2} \times D$

$P1 = 20.29 + 9.35 \times 10^{-2} \times D + D/(1.84 \times 10^{-2} \times D + 4.41)$

$P2 = D/(1.40 \times 10^{-2} \times D + 5.02)$

$P3 = -213 + 0.200 \times D \qquad$ FOR $D > 1065$

$P3 = 0 \qquad\qquad\qquad$ FOR $D \leq 1065$

$T1 = (5.06 \times 10^{-3} \times D + 1)/(1.46 \times 10^{-2} \times D)$

$T2 = (6.77 \times 10^{-4} \times D + 1)/(3.40 \times 10^{-5} \times D)$

$T3 = (D - 1065)/(4.11 \times 10^{-3} \times D + 4.16) \qquad$ FOR $D > 1065$

$T3 = 0 \qquad\qquad\qquad\qquad$ FOR $D \leq 1065$

Fig.3: Shear stress time curves   Table 1: Dependence of equation
       at different shear rates   parameters upon shear
                                             rate D

The equation parameters PO to P3 and T1 to T3 were determined
by stepwise fit of measuring points by the method of least
squares by means of a computer program. The procedure cannot be
described here in detail, but it will be published in future[4].
The correspondence between the observed and the calculated courses
of the shear stress time curves was very good for most of the tests
made. The calculated course of the curves are marked as lines for
the examples in Figure 3.

2.3 Dependence of equation parameters upon shear rate

To describe shear stress time curves as a function of shear rate
the paramters of equation (1) which were calculated by least
squares fit from the single tests, as it has been described above,
were drawn in dependence upon shear rate and by least squares fit
were approached by mathematical equations.
Although a considerable scattering of data was obtained, the dependence of the equation parameters upon shear rate could be
described by the equations presented in Table 1.
To verify the quality of the in this way calculated and the observed
shear stress time curves, there was carried out a regression between
the calculated shear stresses cal τ and the observed shear stresses
obs τ. The regression shows a good correspondence between calculation
and measurement, the coefficient of determination is 98 % for
n = 1638 measuring points.

3. DISCUSSION

From the dependence of equation parameters PO to P3 and T1 to T3
upon the shear rate, some interesting informations can be deducted
concerning the flow behaviour of the cement suspension tested.
 The results of Tattersall[2] concerning the exponential decrease
of shear stress which is caused by the breakdown of the internal
structure of the cement suspensions must be widened. In the first
few seconds after the beginning of loading a strong decrease of

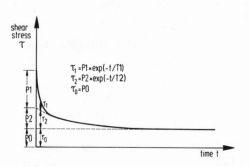
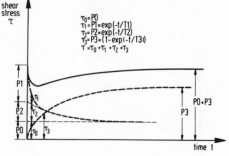

Fig.4: shear stress time curve   Fig.5: shear stress time curve
       without shear thickening   with shear thickening

shear stress is observed, which is here called primary breakdown.
This primary breakdown has to be combined with a secondary breakdown, which is much more slowly.
Whether or in which degree the primary breakdown is influenced by
an apparatus conditioned stress overshoot could not be examined
yet. (compare Lockeyer and Walters[5])

The time constants T1 and T2 of the structural breakdown depend
in a high degree on shear rate. The dependences can be described
by hyperobolic functions (see Table 1).

The paramter P2 indicates the decrease of shear stress due to the
secondary breakdown. According to Lapasin, Longo und Rajgelj[6] the
dependence of P2 upon D can be described by a hyperbolic function.
This could also be verified for these tests.

Beginning with a certain shear rate (D = 1065^{-1} in these tests) the
shear stress, after a decrease in the beginning of the test,
increases until it reaches a state of equilibrium. This behaviour
can be described mathematically by superimposing the breakdown with
a time dependent exponential increase of shear stress.
An indication of the possible causes of the observed behaviour
results from tests which have been carried out with water in the
same measuring equipment, i.e. with serrated cylinder walls in the
rheometer described in[1].
It was found out that probably due to the serrating of the cylinder
walls there was a turbulent flow behaviour, that starts at shear
rate of 1065· s^{-1} (s.Fig.6). In contrast to that laminar flow was
observed up to D = 1800 s^{-1} when smooth cylinders were used. Laminar
flow in the whole shear rate range was also observed with oil the
viscosity of which was about 0,14 Pa·s.
According to Roscoe[7] in suspensions turbulent flow can occur at
shear rates which are lower than that shear rate where turbulent
flow of the suspending medium starts. Roscoe attributes this
phenomen to the fact that in suspensions with internal structure
a maximum shear rate in single ranges of the flow field arises,

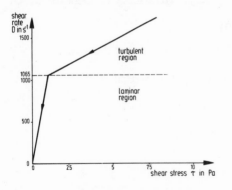

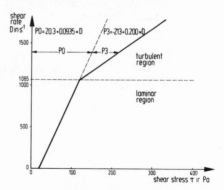

Fig.6: Flow curve of water, measured in Couette rheometer
with serrated cylinders

Fig.7: Equilibrium flow curve of
cement paste, calculated
from equation (1)

which is higher than the average shear rate used for calculation.
In consequence of these considerations one can suppose that in the
case of the tests made the time dependent shear thickening behaviour
is a result of turbulent flow and not a case of rheopexy in the
original meaning of the word, i.e. an increase of shear stress due
to changing of internal structure in consequence of shear stress.
Turbulence occured for the first time at the same shear rate for
cement suspensions and water.
The degree of turbulence at a constant shear rate seems to increase
with time of loading until it reaches a state of equilibrium.This is
suggested by the exponential increase of shear stress with time.
In the state of equilibrium τ (t = ∞) = PO + P3 Binghamien flow
behaviour results from the mathematical evaluation for the laminar
region (s.Fig. 7). The increase of shear stress due to turbulence
is proportional for equilibrium conditions up to (D - 1065).

REFERENCES

1. W.   vom Berg, Influence of specific surface and concentration
        of solids upon the flow behaviour of cement pastes.
        Mag. Concr. Res. 31: 211 (1979).

2. G.H. Tattersall, The Rheology of Portland Cement Pastes,
        Brit. J. Appl. Phys. 6:165 (1955).

3. A.A. Nessim and R.L. Wajda, The Rheology of Cement Pastes and
        Fresh Mortars,
        Mag. Concr. Res. 17:59 (1965).

4. W.   vom Berg, Zum Fließverhalten von Zementsuspensionen,
        Diss. RWTH Aachen (to be published).

5. M.A. Lockeyer and K. Walters, Stress overshoot: Real and
        apparent,
        Rheol. acta 15: 179 (1976).

6. R.   Lapasin, V. Longo und S. Rajgelj; Thixotropic behaviour of
        Cement Pastes,
        Cem. Concr. Res. 9:309 (1979).

7. R.   Roscoe, Suspensions, in "Flow properties of Disperse
        Systems, J.J. Hermann ed.
        North-Holland Publishing Company, Amsterdam (1953).

CORRESPONDENCE BETWEEN TRANSIENT AND DYNAMIC

VISCOELASTIC FUNCTIONS IN THE CASE OF ASPHALTS

B. de la Taille and G. Boyer

Centre de Recherches ESSO S.A.F.
B.P. 6, rue du Tronquet
76130 Mont-Saint-Aignan, FRANCE

1. INTRODUCTION

   For assessing the mechanical properties of asphalts[*], a variety
of empirical tests are currently used : penetration, softening point,
ductility, Fraass breaking point, etc... Although these tests provide
useful comparative information, their interest is very limited since
the ranges of temperature and loading time where they yield results
are very small compared to service conditions. Further, due to the
complexity of the loading patterns in such tests, it is difficult
to analyze their rheological significance. Therefore, starting from
these tests, the field behaviour of asphalts is difficult to analyze
or to predict. Consequently, there has been a considerable interest
for several years in determining asphalt rheological properties
through more fundamental parameters. The mechanical behaviour of
asphalt is complex and can be generally described adequately through
the viscoelasticity pattern. The range of loading times to which
asphalts may be subjected is very wide, depending on the type of
application envisaged.

   In road applications; loading times can be either very long
(parking and traffic light areas, thermal stresses) or very short
(fast lorries rolling on a pavement). On the other hand, asphalt for
roofing applications can also be subjected to stresses or strains
over a variety of loading times : short at the laying stage (unroll-
ing the felts), long in service conditions (dilatation/shrinkage of
cement concrete roofs, creep on slopes). Thus, to obtain asphalt

---

[*]Asphalt : terminology employed here for bituminous binder, and not
  rock asphalt nor paving mix.

rheological properties in field conditions of interest, experimental
design should be set up in such a way that a wide range of loading
times is covered.

As long loading time results are accessible through transient
experiments and short loading time results through dynamic experiments, it is desirable to combine these two testing modes in order
to extend the measurement range of each one. This is made possible
through theory of linear viscoelasticity which shows that transient
and dynamic modes are equivalent in nature, and sets interrelations
allowing calculation of transient results from dynamic ones, and
reverse[1].

2. OBJECTIVE OF THE STUDY

The objective of the present study was to verify for asphalts
the validity of these interrelations between transient and dynamic
viscoelastic functions.

Two different penetration grade asphalts (called D1 and D2)
from the same crude origin were considered. They were prepared by
blowing a light Middle East crude vacuum resid (600 cSt at 100 °C).
Their penetration (25 °C/100 g/5 s) were respectively 158 for D1
and 45 for D2. They were chosen because of their rheological and
practical interest : temperature susceptibility of blown asphalts
is low, and they are expected to give good low temperature performance in the pavement.

These two asphalts were tested with two apparatus : the Sliding
Plate Rheometer which has been especially developed for testing
asphalts[3], and the Balance Rheometer, developed by Kepes mainly for
polymer study, whose theoretical bases have been discussed in many
papers[4,5,6]. The test achieved with the Sliding Plate Rheometer is
a shear creep test with a transient loading pattern, and results
are expressed as time dependent compliances $J(t)$ (t : loading time).
The Balance Rheometer test is a dynamic shear test and provides
complex shear moduli $G^*(\omega)$ (ω : angular frequencies). The variations
of these viscoelastic functions $J(t)$ and $G^*(\omega)$ versus time or angular
speed were established. Ranges of loading time and angular frequencies were chosen in such a way that there was an overlap region.
The agreement between the two types of results was then checked by
using the afore-mentioned interrelations.

3. EXPERIMENTAL

3.1. Sample Conditioning

As the Sliding Plate Rheometer is a very simple apparatus, the
test procedure is quite fast and it is possible to run several tests
during one single day. But this is not the case for the Balance

Rheometer which needs a whole day for one test (by "one test", we mean the procedure to obtain results at one temperature for different values of loading time or angular frequency). For this reason, procedures of sample conditioning adopted for each apparatus were different.

The main difference consists in the time the sample is allowed to cool down at ambient temperature after pouring in the measurement system. For the Balance Rheometer, the sample is poured the day preceding the test and is kept overnight at ambient temperature while this time is only 45 minutes for the Sliding Plate Rheometer.

For the Sliding Plate Rheometer, another procedure (called procedure 2) was also used for asphalt D1 tested at 20 °C : the specimen in the measurement system was cooled at 0 °C for 15 minutes before setting it in the apparatus itself, in order to avoid any deformation of the specimen.

3.2. Experimental Design

The two asphalts were tested with both apparatus at two temperatures 0 and 20 °C. Each test was triplicate in randomized order.

For the Sliding Plate Rheometer, complex shear compliances were calculated from the creep curve at 4 loading times : 1, 10, 100 and 1000 seconds.

For the Balance Rheometer, complex shear moduli were measured for 10 angular frequencies from 1.96×10^{-2} up to $7.85 \times 10^{1} s^{-1}$, at 2 levels of shear strain amplitude. Unfortunately, the instrument capacity imposes to apply a low shear strain amplitude for measuring high moduli and a high shear strain amplitude for measuring low moduli. Consequently, it was not possible to use the same levels of strain for all tests.

4. EXPERIMENTAL RESULTS

4.1. Transient Results

Means of results for both asphalts at the two temperatures are plotted in Figure 1. A statistical analysis had been carried out previously to determine the repeatibility of the test, which was found to be 23 %.

An interesting information should be pointed out : when the mold with specimen was cooled at 0 °C before testing at 20 °C (asphalt D1, procedure 2), lower compliances were measured, that is the product is stiffer. By means of a statistical analysis of variance, it was proved that the difference between the two procedure results is significant.

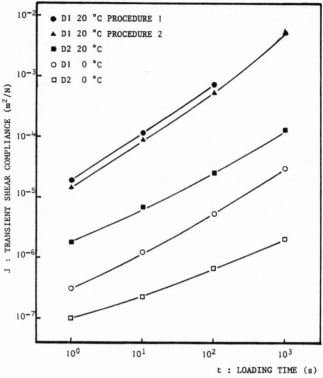

Fig. 1   Transient Results

4.2. Dynamic Results

It appears that the viscoelastic functions are slightly dependent upon shear strain. Figure 2 shows the variations of the absolute value $|G*|$ of the complex modulus $G*$ versus frequency : $|G*|$ is slightly lowered when shear strain is increased. At the present time we wonder   whether this dependence has to be attributed to a non-linear behaviour of blown asphalts or to the measurements with the Balance Rheometer (at too high shear strains, wall slip effects can yield a possible bias). This point would need further investigations in the future.

In Figure 3 are depicted the variations of the loss angle δ of the complex shear modulus versus angular frequency. As expected, at low temperature/high frequency, the behaviour is more elastic (low value of δ) while it is more viscous at high temperature/low frequency. The nature of asphalt also affects this behaviour : D2 (which suffered more severe air blowing) is more elastic than D1.

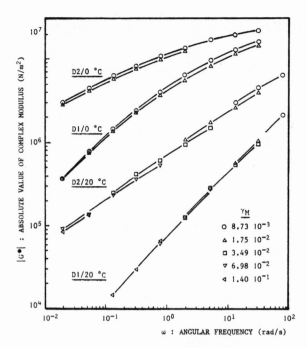

Fig. 2   Dynamic results - Dependence upon shear strain

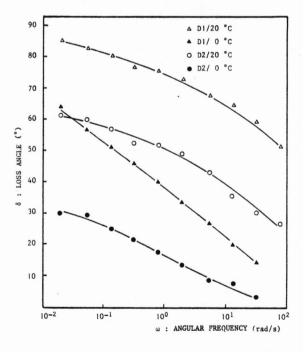

Fig. 3   Dynamic results - Loss angle

4.3. Correspondence between transient and dynamic results

The Ferry-Ninomiya approxation method[1] was used in this study. It consists in assessing the value of the transient compliance J(t) from the components J' and J" of the complex compliance J*.

$$J(t) = J'(\omega) + 0.4 \ J'' \ (0.4 \ \omega) - 0.014 \ J'' \ (10 \ \omega) \qquad t = \frac{1}{\omega}$$

$$\text{with } J' = \frac{G'}{|G*|^2} \qquad J'' = \frac{G''}{|G*|^2}$$

(G' and G" are the components of G*)

A comparison was carried out between transient compliances calculated according to this method and measured with the Sliding Plate Rheometer (procedure 1), in the frequency/loading time overlap range. As illustrated by Figure 4, there is a good overall correspondence between transient and dynamic measurements.

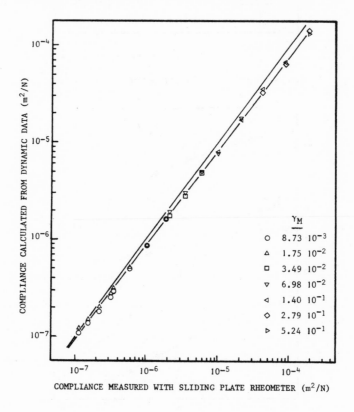

Fig. 4   Comparison between transient and dynamic results

But it can be seen that calculated values are always slightly smaller than measured ones. This could be explained by the difference existing between sample conditioning procedures for each apparatus : the longer cooling period adopted for the Balance Rheometer could lead to a higher structural hardening of the asphalt than what occurs with the Sliding Plate Rheometer procedure. This point needs some further investigation by adopting the same procedure for both apparatus.

5 CONCLUSION

The correspondence between transient viscoelastic functions directly measured with the Sliding Plate Rheometer and calculated from dynamic data of the Balance Rheometer is satisfactory within the overlap loading time region of the two apparatus. When the remaining problems are solved, that is assessing the influence of the sample conditioning procedure on the results and clearing up the origin of the effect of shear strain on dynamic results, it will be possible to widen the loading time range of the Sliding Plate Rheometer by combining the two sets of results. However this extension will be in fact limited by the maximum level of modulus measurable with the Balance Rheometer, which is rather low ($\simeq 2 \times 10^7$ N/m^2).

Of course, an alternate means to extend this loading time range is to apply the time-temperature superposition principle[1] to draw master curves, the validity of which has been recognized by several authors in the case of asphalts.

REFERENCES

1. J.D. Ferry, "The viscoelastic properties of polymers", Wiley, New-York, London (1961).
2. W.J. Gan, "Measurement and prediction of asphalt stiffnesses, and their use in developing specifications to control low temperature pavement transverse cracking", Presentation to ASTM symposium, Chicago (July 1, 1976).
3. J. Fenijn, R.C. Krooshof, "The Sliding Plate Rheometer : a simple instrument for measuring the viscoelastic behavior of bitumen and related substances is absolute units", Presentation to Canadian Technical Asphalt Association Winnipeg (November 16, 1970).
4. T.E.R. Jones, K. Walters, "A theory for the Balance Rheometer", Brit. J. Appl. Phys., 2 : 815 (1969).
5. M. Yamamoto, "Theoretical analysis of new rheometers", Jap. J. of Appl. Phys., 8 : 1252 (1969).
6. K. Walters, "Rheometrical flow systems – Part I : flow between concentric spheres rotating about different axes", J. Fluid Mech., 40 : 191 (1970).

VISCOELASTIC ANALYSIS OF A THREE-LAYER PAVING SYSTEM

SUBJECTED TO MOVING LOADS

G. Battiato,* G. Ronca,** and C. Verga*

 * Assoreni, ENI Group, S. Donato Milanese
** Politecnico di Milano, Italy

INTRODUCTION

The structural design of asphalt pavements is generally achieved by elastic methods based on the multilayer elastic theory developed by Burmister. The application of these methods requires some approximations, the asphalt surface layers being characterized by a typical viscoelastic behavior.

As previously shown by the authors (1), the assumption of a moving load system allows for correct evaluation of the critical deformations inside the structure which are responsible for fatigue behavior and rutting phenomena in asphalt pavements. Simplified approaches based on the analysis of static loads involve substantial underestimation of the deformations (1).

A complete viscoelastic method for the calculation of permanent deformations in asphalt layers has been developed by the authors (2).

In the present paper the viscoelastic effects associated with the passage of a single moving load on a three-layer paving system are considered. The first layer is representative of all the asphalt layers, whose mechanical characteristics are described by a simple analytical expression fitting experimental data rather closely over a wide time interval at all temperature of interest:

$$J(t) = J_1 t^\alpha$$

The sub-base layer and the subgrade semi-infinite layer can be considered elastic. The layers are supposed to be rigidly bounded and incompressible.

MATHEMATICAL APPROACH

The viscoelastic problem is solved by a combination of Hankel transforms over the horizontal space variable and Fourier transforms over the time variable. By this procedure the mathematical problem is reduced to an equivalent elastic one. Transformations to the frequency spectrum can be performed analytically for all deformation components, although different techniques have to be used for different cases.

The practical feasibility of our approach rests upon this fortunate circumstance, thanks to which the amount of numerical work can be reduced to an acceptable size.

The two lower elastic layers (sub-base and subgrade) are considered to be equivalent to a single layer with an effective modulus whose value depends through an approximate formula on the mechanical and geometric parameters of the problem.

Extension of our work to a complete characterization of a three-layered system and prediction of the important effect of the subgrade Poisson ratio are being developed.

RESULTS

The tensile longitudinal strain at the base of the asphalt layer is the critical design parameter generally considered for the prediction of the fatigue life of a road structure. In Figure 1 the longitudinal deformation has been reported as a function of distance from the load center along the direction of the load motion. The calculation has been performed with the following load and structure parameters:

Load system: circular load corresponding to a single wheel axle of 6.5 tons, with a load radius of 11.37 cm

Velocity: 30 km/h

Thickness of the layers: 1st layer (asphalt concrete): 10 cm
                         2nd layer: 20 cm

Mechanical characteristics of the layers at 20°C:

Asphalt concrete: $\alpha = 0.38$; $J_1 = 0.93 \times 10^{-4} cm^2/kg_f$

Sub-base: $E_2 = 2050\ kg_f/cm^2$
Subgrade: $E_3 = 1000\ kg_f/cm^2$

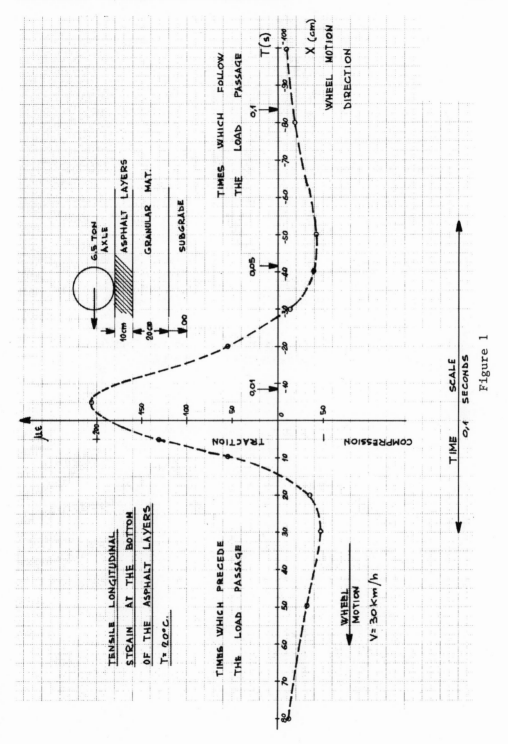

Figure 1

The lack of symmetry between deformations calculated at
x > 0 and x < 0, i.e., between times which follow and precede the
load passage, is a typical viscoelastic effect.

The above results are very close to experimental findings in
deformation tests performed at the Nardò Test Road (3).

REFERENCES

1. G. Battiato, G. Ronca, and C. Verga. The viscoelastic analysis
    of a two-layer paving system subjected to repeated moving loads:
    the contribution of residual deformations. Proceedings of the
    VIIth International Congress on Rheology, Gothenburg, August
    23-27, 1976.

2. G. Battiato, G. Ronca, and C. Verga. Design of flexible pave-
    ments by the viscoelastic "VESTRA" method. The Arabian Journal
    for Science and Engineering, Volume 4, Number 1, January 1979
    (University of Petroleum and Minerals, Dhahran, Saudi Arabia,
    John Wiley & Sons).

3. Italian Road Federation - Measurement of the "aggressiveness"
    of goods traffic on road pavements, Autostrade, No. 10, 1979.

PREDICTION OF THE BEHAVIOR OF CONCRETE BY MEANS OF

A RHEOLOGICAL LAW

J.C. Robinet - H. Di Benedetto

Ecole Nationale des Travaux Publics de l'Etat

Rue Maurice Audin - 69120 - Vaulx-en-Velin - France

INTRODUCTION

The description of concrete behavior with an <u>incremental</u> <u>rheological</u> law is relatively recent ; in fact the first attempt of modelisation, using the hypo-elastic laws, dates back to ten years ago. These models have been completed, taking into consideration the viscous part which is necessary to the description of the creep and relaxation phenoma. These very simple models define the laws not taking into account the <u>direction of the incremental</u> <u>stress vector</u>. The experimental verifications, showing the "directional dependence" of the response, have lead the authors to propose a new model for the description of concrete behavior. This paper proposes a non-linear incremental law with an infinity of domains of expression(or a continuous domain) and to which a validity domain is associated. This new model has also been carried out in order to narrow down the observed divergence between reality and calculation with hypoelastic laws describing the cyclic phenomena. In fact, the irreversibilities of the material are amplified through the successive modifications of the stress path.

A comparison between numerical integration of the proposed rheological model and experimental results over cyclical and monotonous path are presented in the last section.

RHEOLOGICAL LAW IN INCREMENTAL CHARACTERS

Characterisation of a rheological law rests on four fundamental principles[1], [2], (principle of objectivity, principle of determinism, axiom of neighborhood, thermodynamic principles) which are dictated by the physical reality of the phenomenon they describe.

Deterministic principles applied to incremental form allow us to write G ($\delta\underset{\sim}{\sigma}$, $\delta\underset{\sim}{\varepsilon}$, δt) = 0^* where δt is the time increment

This description does not take into account the thermal, chemical and electical effects.

It is classical, in rheology, to dissociate two large classes of media :
- the rate independent materials Truesdell and Noll[3] for which only the elasto-plastic part is considered F ($\delta\underset{\sim}{\sigma}$, $\delta\underset{\sim}{\varepsilon}$) = 0
- the rate-of-stress dependent materials G ($\delta\underset{\sim}{\sigma}$, $\delta\underset{\sim}{\varepsilon}$, δt) = 0

A first order linearisation of G could be considered over a semi straight support of $\delta\underset{\sim}{\sigma}$[4] $\delta\underset{\sim}{\varepsilon} = \underset{\sim}{M} \cdot \delta\underset{\sim}{\sigma} + \underset{\sim}{v}\, \delta t$ (1)
where $\underset{\sim}{M} \cdot \delta\underset{\sim}{\sigma}$ represents the rate independent part and $\underset{\sim}{v}\,\delta t$ the viscous part.

In the following sections of this paper we only consider the rate independent part of the law : $\delta\underset{\sim}{\varepsilon} = \underset{\sim}{M} \cdot \delta\underset{\sim}{\sigma}$.

Proposed law

The law that we propose rests on four hypoteses which are imposed by the fundamental principles and justified by the physical behavior of the material. Only one more hypothesis is necessary to describe the viscous part[6].

H.1 We consider that the non viscous matrix $\underset{\sim}{M}$ is orthotropic when expressed in principal axes of the Cauchy-Euler strain tensor**($\underset{\sim}{\varepsilon}_r$)

$$M_r = \begin{vmatrix} \dfrac{1}{E_1} & \dfrac{\nu_2^1}{E_2} & \dfrac{\nu_3^1}{E_3} & & & \\[2mm] \dfrac{\nu_1^2}{E_1} & \dfrac{1}{E_2} & \dfrac{\nu_3^2}{E_3} & & 0 & \\[2mm] \dfrac{\nu_1^3}{E_1} & \dfrac{\nu_2^3}{E_2} & \dfrac{1}{E_3} & & & \\[2mm] & & & 2G & & \\[1mm] & 0 & & & 2H & \\[1mm] & & & & & 2I \end{vmatrix}$$

In this set of axes the non viscous part of the law takes the following form :

$$\delta\underset{\sim}{\varepsilon}_{Ar} = M_r\, \delta\underset{\sim}{\sigma}_r$$

---

* $\delta\underset{\sim}{\sigma}$ ($\delta\underset{\sim}{\varepsilon}$) Cauchy stress (Almansi strain) increment tensor expressed in co-rotational axes.

**If F is the linear tangent transformation

$\underset{\sim}{\varepsilon}_1 = \underset{\sim}{F}^{-1} \cdot F^{-1} = 1 - 2\,\underset{\sim}{\varepsilon}_A$; $\underset{\sim}{\varepsilon}_A$ Almansi strain tensor

H.2   we would assume that the stress paths simulate the memory of
      the material if they lead to the same state of stress and the
      same specific volume.

H.3   this hypothesis allows us to determine the functions E_i and v_i^j,
      which are necessary for the description of M_r. In this study
      we have only kept a linear variation with θ_i (figure 1). Figu-
      res 1 and 2 permit to define the chosen variation.

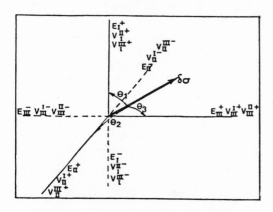

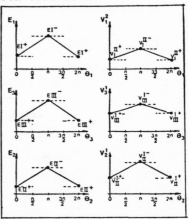

Fig. 1. Definition of particular
values of the functions E and v
in the axes of orthotropy.

Fig. 2. Variation of E and
v vers θ_i.

H.4   The preceding three hypotheses do not permit to define the
      fonctions G, H and I which determine the distorsion of the
      material. As pointed out by Boehler[5] following the experimen-
      tal studies performed on rocks, the inverse of the Young
      modulus follows an ellipsoïdal law, once the coordinate axes
      of the expression of the law rotate. These experimental re-
      sults constitute the fourth hypothesis. After calculation[6]
      we can get the G, H and I fuctions.

      The reader is in a position to establish through these four
hypotheses that the proposed law interprets a continuous evolution of the response of the material as a function of the stress
tensor "orientation".

VALIDITY DOMAIN OF THE RHEOLOGICAL LAW

      The following developpement allows us to determine   incremental
limits for the law. Considering the continuum, the first limit expresses the begining of plasticity ($\delta\sigma = 0$). In order to obtain a
non trivial solution, the incremental law which writes $\delta\sigma = \underset{\sim}{C}$
$\delta\varepsilon = 0$, implies that   det $(\underset{\sim}{c}) = 0$ (2). The equality (2) defines
the plasticity criterion of the incremental rheological law.

      A second validity condition of the law is the appearance of
discontinuity in the material. The theoretical interpretation of

this phenomenon (Kinematical and stress compatibility conditions[7], [8], [9], [10]) defines the instability criterion of the model

$$\det (n\underset{\sim}{C}n) = 0$$

Remark : These creteria are not global ones but are only applied incrementally.

Simulation of the memory

We determine the elements of the rate independent part $\underset{\sim}{M}$ from triaxial tests. In this type of test it is generally admitted that the principal directions of the stress tensor are coincident with those of the strain tensor if concrete is initialy isotropic. It is supposed that the anisotropy introduced when making the concrete, is negligible. The principal coordinate of stress are then the axes of orthotropy according to hypothesis H.1.

For any state of the concrete the value E_I^{+-}, V_J^{I+-} defined in three directions of orthotropy, figure 1, are obtained by tests consisting in loading or unloading in every one of these three directions. The expressions E_i and V_j^i corresponding to the values of the matrix $\underset{\sim}{M}$, for a general stress case depend on the values of E_J^{+-} V_J^{I+-} for the considered state (cf. H.3). The definition of these six values (E_J^{+-}, V_J^{+-}) imposes the conducting of six tests from a sample having the characteristics of the concrete. This is a very difficult if not impossible experimental task. That is why we introduce the memory of the media. The desired values are not obtained anymore from the real considered state of the concrete but from simple tests on a sample which simulates the state of the concrete. Simulation is done with a path we called "elementary path". It is in this part that the second hypothesis manifests its importance.

The two parameters we first choose to simulate the memory are the state of stress and the specific volume. The numerical integration shown later in the text shows that this simulation resembles, in a very satisfying manner, the behavior of concrete.

Application of the rheological law

The law whose parameters have been found over the elementary path permits to "Summarize" the history of the material. It is then possible, according to the second hypothesis, to define the values of the rate independent matrix in the three directions of orthotropy of the material. The functions of interpolation defined by the 3 rd and 4 th hypotheses, determine the rate independent matrix for an arbitrary direction of stress. Then the rheological law can be integrated for any loading history. We present in this paragraph some numerical integrations carried out on the concrete tested by Newman[11].

Simulation of a deviatoric triaxial test

Figure 3, shows the results of the numerical integration over a deviatoric triaxial path of revolution. The dashed curves represent the instability domain (probability of crack) of the rheological law.

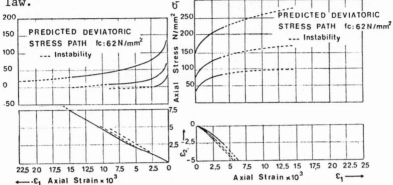

Fig. 3.

Comparisons of the direction increment vectors of theoretical and experimental deformation at rupture over the triaxial path of revolution (figure 4)

Validity domain of the rheological law

For a material considered initially as homogeneous we have searched for the possibility of the appearance of fissure at incremental level over different stress-paths. Results of this search are shown, figures 4 and 5.

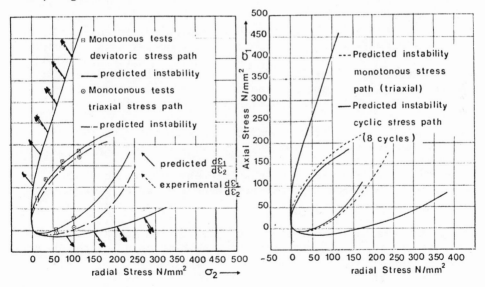

Fig. 4.   Fig. 5.

CONCLUSION

We have presented in this paper, a non linear rheological incremental law with a continuous domain of expression for concrete.
The hypotheses developped result in a law which yields an output depending on the "direction" of the stress path, on the past
history of the material, as well as on the begining of plastic
flow and the appearance of cracks.
Worked numerical examples illustrate a good agreement between
observed and calculated behaviour, and prove the potentialities of
the proposed model for practical computations with concrete.

REFERENCE

1 Mandel Cours de Mécanique des Milieux Continus - Tome 1
        Gauthier-Villars - Paris (1966)

2 C. Truesdell   Introduction à la Mécanique rationnelle des
                  milieux continus - Masson - Paris (1974)

3 C. Truesdell   The non-linear field theories - vol. III.3 -
   and W. Noll   Handbook of physics - Springer - (1965)

4 J.C. Robinet   Présentation d'une loi rhéologique non linéaire
   H. Di Benedetto   en écriture incrémentale pour les sols, et com-
                     paraison avec les différents modèles existants
                     Cahier du Groupe Français de Rhéologie (1980)

5 Boehler   Contribution à l'étude de l'équilibre limite des sols
             anisotropes - Thèse de doctorat de spécialité -
             Grenoble (1968)

6 J.C. Robinet H. Di Benedetto   Loi rhéologique en écriture incré-
   Comtet - P. Mathieu   mentale pour les sols - Journée
                                  Rhéologie - E.N.T.P.E. (1979)

7 S.W. Rudnicki   Conditions for the localization of deformation
   J.R. Rice   in pressure sensitive dilatant materials -
                   J. Mech. Phys. Solids, 23 - (1975)

8 R. Hill   Bifurcation phenomena in the plae tension test
   J.W. Hutchinson   J. Mech. Phys. Solids, 23 - (1975)

9 J. Mandel   Conditions de stabilité et postulat de Drucker
               In rheology and soil Mechanics symp. - Grenoble
               (1964)

10 I. Vardoulakis   Bifurcation and localization modes of the tria-
                    xial test on dry sand samples - Journée de Rhéo-
                    logie - Vaulx-en-Velin (1979)

11 M.D. Kotsovos   Behavior of concrete under multiaxial stress ACI
   S.B. Newman   Journal Proceedings - vol. 74 - (1977)

LATE PAPERS

Editors' Note

The excruciating slowness of the Italian mail services (suffice
it to say that a letter posted in India on December 3, 1979 reached
us on February 22, 1980), the occurrence of a postal strike in
Canada at a crucial time, and some communication problem with the
French Society of Rheology contributed to the accumulation of
a backlog of very late papers submitted for presentation at the
Congress after we had prepared the indexes to this book. The titles
of all these papers and the authors' addresses are given on the
next three pages. Furthermore, whatever additional information we
could obtain (sometimes the entire paper, sometimes an abstract)
is reported in the following pages. We wish to thank Plenum Press
for arranging to have this material, incomplete though it may be,
included in the Proceedings.

TITLES

- J.P.LALLEMAND, A.LAGARDE, Laboratoire de Mécanique des Solides,
  40, Avenue du Recteur Pineau, 86022 Poitiers Cedex, France.
  "Methods of Studying the Rheooptical Dynamic Characteristics
  of Birefringent Materials".
- J.POUYET, J.L.LATAILLADE, A.PARMAR, Laboratoire de Mécanique
  Physique, 351 Cours de la Libération, 33405 Talence Cedex, France,
  and Department of Engineering Science, University of Oxford,
  Parks Road, Oxford OX1 RPJ, England. "Investigation of Visco-
  elastic and Viscoplastic Behavior of Solid Polymers at High
  Strain Rates".
- H.GIESEKUS, M.KWADE, R.SCHARF, Abteilung Chemietechnik, Lehrstuhl
  fur Stromungsmechanik, Universität Dortmund, Postfach 500500,
  4600 Dortmund 50, West Germany. "The Influence of Drag Reducing
  Additives on Coherent Structures in a Turbulent Mixing Layer".
- H.GIESEKUS, M.KWADE, R.SCHARF, Address as above. "Space-Time
  Correlation Measurements in a Turbulent Mixing Layer of Newtonian
  and Drag-Reducing Fluids using two Laser-Doppler Anemometers".
- F.R.EIRICH, Polytechnic Institute of New York, 333 Jay Street,
  Brooklyn, N.Y.11201, U.S.A. "Polymer Adsorption and Flow in
  Capillaries".
- D.S.PEARSON, Bell Laboratories, 600 Mountain Avenue, Murray Hill,
  N.J. 07974, U.S.A. "Non-Linear Dynamics of Polymer Solutions".
- J.K.GILLHAM, Department of Chemical Engineering, Princeton
  University, Princeton, N.J. 08540, U.S.A. "A Generalized Time-
  Temperature Phase Diagram for the Thermosetting Process".
- M.MORTON, M.F.TSE, Institute of Polymer Science, The University
  of Akron, Akron, Ohio 44325, U.S.A. "The Rheology of ABA Triblock
  Copolymers".
- S.S.STERNSTEIN, Department of Materials Engineering, Rensselaer
  Polytechnic Institute, Troy, N.Y.12181, U.S.A. "Viscoelastic
  Characterization of Glassy Polymers and Composites".
- N.W.TSCHOEGL, Department of Chemical Engineering, California
  Institute of Technology, Pasadena, California 91125, U.S.A.
  "Energy Storage and Dissipation in a Linear Viscoelastic Material"
- J.M.THUEN, R.J.HINRICKS, Narmco Materials, Costa Mesa, California,
  U.S.A. "Rheological Characterization of the Curing Dynamics of
  Composite Materials".
- D.R.OLIVER, M.SHAHIDULLAH, Department of Chemical Engineering,
  The University of Birmingham, P.O.Box 363, Edgbaston, Birmingham
  B15 2TT, England. "The Stretching Flow of Multigrade Motor Oil

in Planar Flow Nozzles".
- J.B.SKULA, Indian Institute of Technology, Kanpur 208016, Kanpur,
  India. "Peristaltic Transport of Biofluids".
- Q.T. NGUYEN, Laboratoire de Chimie-Physique Macromoléculaire,
  1, Rue Grandville, 54 Nancy, France. "Behavior of Macromolecular
  Solutions through Ultrafiltration Membranes".
- L.B.B.M.JANSSEN, D.J.VAN ZUILICHEM, Laboratorium voor Fysische
  Technologie, Prins Bernardlaan 2, Delft, The Netherlands.
  "Rheology of Reacting Biopolymers in Extrusion".
- A.K.DEY, N.C.F.SHAH, K.S.S.IYER, S.PRASAD, M.E.R.A.D.O., and C.M.E.,
  Poona, India, "Optimal Design of Pipe Line for Transportating
  Hot Non-Newtonian Fluids with Heat Transfer to the Surrounding".
- V.S.AU YEUNG, C.W.MACOSKO, Polymer Materials Program, Department
  of Chemical Engineering, Princeton University, Princeton, N.J.
  08544, U.S.A., and Department of Chemical Engineering, University
  of Minnesota, Minneapolis, MN 55455, U.S.A. "Extensional
  Rheometry of Several Blow Molding Polyethylenes".
- C.V.MACOSKO, M.A.OCANSEY, H.H.WINTER, Department of Chemical
  Engineering, University of Minnesota, Minneapolis, MN 55455,
  U.S.A., and Department of Chemical Engineering, University of
  Massachusetts, Amherst, MA 01003, U.S.A. "Steady Planar Extension
  with Lubricated Dies".
- M.GOTTLIEB, C.W.MACOSKO, Department of Chemical Engineering,
  Ben Gurion University, Israel, and Department of Chemical
  Engineering, University of Minnesota, Minneapolis, MN 55455,
  U.S.A., "Experimental Testing of Molecular Theories of Network
  Elasticity".
- J.J.ULBRECHT, V.R.RANADE, Department of Chemical Engineering,
  State University of New York at Buffalo; Buffalo, N.Y. 14214,
  U.S.A. "Flow of Viscoelastic Polymer Solutions in Curved Tubes:
  Velocity Profile Measurement by Laser-Doppler Anemometry".
- L.CHOPLIN, P.J.CARREAU, Chemical Engineering Department, Ecole
  Polytechnique of Montreal, C.P.6079, Succ."A", Montreal, Que.
  Canada H3C 3A7."Excess Pressure Losses at the Entrance of a Slit".
- H.ESSABBAH, B.AUVERT, C.LACOMBE, Unité de Biorhéologie, Faculté
  de Médicine Pitié-Salpêtrière, Paris, France. "Comparative
  Hemorheology, in Transient Flow, of Patients with Polycythaemia
  and a Control Group".
- I.ANADERE, H.CHMIEL, W.LASCHNER, Fraunhofer-Institut fur
  Grenzflächen-und Bioverfahrenstechnik, Eierstrasse 46, 7000
  Stuttgart 1, and Abteilung für orthopädische Rheumatologie,
  Klinik für Orthopädie und Unfallchirurgie Dr.Baumann e V.,

Alexanderstrasse 5 - 7a, 7000 Stuttgart 1, West Germany,
"Viscoelastic Parameters of Pathological Synovial Fluids".
- V.B.GUPTA, Department of Textile Technology, Indian Institute
  of Technology, New Delhi, 110028, India. "The Viscoelastic
  Properties of Heat-Set Polyethylene Terephthalate and their
  Correlation with Structure".
- M.FRIAS, University of Lisbon, Portugal. "Thermodynamically
  Structured Flows".
- J.ZARKA, Laboratoire de Mécanique des Solides, Ecole Polytechnique,
  91128 Palaiseau Cedex, France. "A Simple Analysis of the Assembling
  of Rheological Models".
- J.BETTEN, Institut für Werkstoffkunde, RWTH Aachen, Augustinerbach
  4, 51 Aachen, West Germany. "Integrity Basis for a Second-Order
  and a Forth-Order Tensor".
- G.CHAUVETEAU, M.MOAN, Institut Français du Petrole, B.P.n.311,
  92506 Rueil Malmaison, France, and Laboratoire d'Hydrodynamique
  Moléculaire, Faculté des Sciences, 29283 Brest Cedex, France.
  "Dilatant Laminar Flow of Dilute Polymer Solutions in non-Uniform
  Velocity Gradient Fields".
- C.K.RHA, J.A.MENJIVAR, E.R.LANG, Food Materials Science and
  Fabrication Laboratory, Department of Nutrition and Food Science,
  Massachusetts Institute of Technology, Cambridge, Mass.02139,
  U.S.A.   "Rheology of Bronchial Secretion: Structural Model".
- R. VERA, T.FORT, Instituto de Investigaciones en Materiales,
  Apdo Postal 70-360, Ciudad Universitaria, México, D.F.Mexico.
  "Effects of High Bonding Pressure Cycles on the Strength of
  Adhesive Joints".

INVESTIGATION OF VISCOELASTIC AND VISCOPLASTIC BEHAVIOUR OF SOLID POLYMER AT HIGH STRAIN RATES

J. Pouyet, J.L. Lataillade, A. Parmar[+]

Laboratoire de Mécanique Physique, Université de Bordeaux, France
[+] Dept.of Engineering Science, University of Oxford, England

(Abstract)

The problem of viscoelastic and viscoplastic solid subjected to a stress step input of a duration between 0.1 and 1 millisecond at high strain rates and for large strains is studied in one dimensional situations for compressional and shear states:

$$10^2 \text{ s}^{-1} < \dot{\varepsilon} < 5 \; 10^3 \text{ s}^{-1}, \; \varepsilon_{max} \sim 0.5 - 10^2 \text{ s}^{-1} \lesssim \dot{\gamma} \lesssim 10^3 \text{ s}^{-1}, \; \gamma_{max} \sim 1$$

The experimental procedure is the Hopkinson pressure bar technique in either a compressional or a torsional configuration: a short specimen – a thin disc in the former case and a tube with a small wall thickness in the latter one – is firmly sandwiched between two elastic rods, instrumented with strain gages for obtaining either ($\dot{\varepsilon}$(t), σ(t) or ($\dot{\gamma}$(t), τ(t)) in the gage, which is supposed to remain on a quasi-static state of stress. A computer approach is developed to get a formulation of the mechanical relationship in both viscoelastic and viscoplastic ranges. Some results are given for a polypropylene homo and copolymer and a polyethylene tested with the compressional apparatus and a polyethylene tested with the torsional one. A Graphical method in the ($\dot{\varepsilon}$, t, ε, σ) space is shown to solve the problem of an analog modelisation when the rate of deformation does not remain steady thoughout the sollicitation. A non-linear model with two relaxation

times is proposed as example for the polypropylene homopolymer
tested at room temperature.

ENERGY STORAGE AND DISSIPATION IN A LINEAR VISCOELASTIC MATERIAL

N. W. Tschoegl

Division of Chemistry and Chemical Engineering
California Institute of Technology
Pasadena, California 91125, USA

INTRODUCTION

The purpose of this paper is to present general equations for the energy stored and the energy dissipated in the deformation of a linear viscoelastic material. The equations can be particularized to any Laplace transformable strain or stress excitation. Some consequences of the application of the equations to the harmonic and to the triangular strain excitation (hysteresis) are discussed.

ENERGY STORAGE AND DISSIPATION

The rate at which energy is absorbed at time t per unit volume of a material during a deformation is equal to the rate at which work is performed on it. This rate is the product of the instantaneous stress, $\sigma(t)$, and the rate of strain, $\dot{\epsilon}(t)$, so that

$$\dot{W}(t) = \sigma(t)\dot{\epsilon}(t). \tag{1}$$

The total work of deformation or, in other words, the mechanical energy absorbed per unit volume of material in the deformation up to time t then becomes

$$W(t) = \int_0^t \dot{W}(u)\,du = \int_0^t \sigma(u)\dot{\epsilon}(u)\,du. \tag{2}$$

This energy is the sum of the energy stored, $W_s(t)$, and the energy dissipated, $W_d(t)$, i.e. $W(t) = W_s(t) + W_d(t)$.

In a *viscoelastic* material inertial effects are considered negligible. In this case the stored energy is potential energy only. If the behavior is *linear*, expressions for $W_s(t)$ and $W_d(t)$ may be derived by representing the energy storing and energy dissipating mechanisms by the usual spring and dashpot models. By definition, energy is dissipated in the dashpots and the dashpots alone. Also, *all* of the energy required to deform the dashpots is dissipated. Summing over all dashpots yields the rate at which energy is dissipated as

$$\dot{W}_d(t) = \sum_n \eta_n \dot{\varepsilon}_{dn}^2(t) = \sum_n \phi_n \sigma_{dn}^2(t) \tag{3}$$

where η_n is the coefficient of viscosity of the nth dashpot, ϕ_n is its coefficient of fluidity ($\phi_n = 1/\eta_n$), and $\dot{\varepsilon}_{dn}(t)$ and $\sigma_{dn}(t)$ are the rate of strain across the dashpot, and the stress through it, at time t. Energy is stored in the springs, and the springs alone. Furthermore, *all* the energy required to deform the springs is stored. Summing over all springs gives

$$W_s(t) = (1/2) \sum_n G_n \varepsilon_{sn}^2(t) = (1/2) \sum_n J_n \sigma_{sn}^2(t). \tag{4}$$

for the energy stored. In Eq. (4) G_n is the modulus of the nth spring, J_n is its compliance, and $\sigma_{sn}(t)$ and $\varepsilon_{sn}(t)$ are the stress through, and the strain across the nth spring at time t. Eqs. (3) and (4) are the basic relations for determining energy storage and dissipation during a deformation. Because of space limitation, only strain excitations will be considered here. The equations for stress excitations are easily obtained through an analogous derivation.

Strain Excitations

The response to a strain excitation is modelled most conveniently with a parallel combination of several Maxwell units, each consisting of a spring and a dashpot in series. Letting $\bar{\varepsilon}(s)$ be the Laplace transform of the imposed strain, the transforms of the strain across the spring and across the dashpot of the Maxwell unit become

$$\bar{\varepsilon}_{sn}(s) = \tau_n s \bar{\varepsilon}(s)/(1+\tau_n s) \quad \text{and} \quad \bar{\varepsilon}_{dn}(s) = \bar{\varepsilon}(s)/(1+\tau_n s) \tag{5}$$

where $\tau_n = \eta_n/G_n$ is the relaxation time. The same strain acts across all Maxwell units because they are in parallel. For an isolated spring (which must be present if the material does not show steady-state flow) the relaxation time may be considered to be infinite. Thus, the energy stored in response to a strain excitation up to time t is

$$W_s(t) = (1/2)\{G_e\}\epsilon^2(t) + (1/2)\sum_n G_n \epsilon_{sn}^2(t) \qquad (6)$$

where G_e is the equilibrium modulus associated with the isolated spring and the braces signify that G_e may be zero. The rate at which energy is dissipated is given by Eq. $(3)_1$. For a continuous distribution of relaxation times we have

$$W_s(t) = (\tfrac{1}{2})\{G_e\}\epsilon^2(t) + (\tfrac{1}{2})\int_{-\infty}^{\infty} H(\tau)\epsilon_s^2(t;\tau)d\ln\tau \qquad (7)$$

and

$$\dot{W}_d(t) = \int_{-\infty}^{\infty} \tau H(\tau)\dot{\epsilon}_d^2(t;\tau)d\ln\tau \qquad (8)$$

where $H(\tau)$ is the relaxation spectrum, and $\epsilon_s(t;\tau)$ and $\dot{\epsilon}_d(t;\tau)$ are obtained from

$$\epsilon_s(t;\tau) = L^{-1}\frac{\tau s \bar{\epsilon}(s)}{1 + \tau s} \qquad \text{and} \qquad \epsilon_d(t;\tau) = L^{-1}\frac{s\bar{\epsilon}(s)}{1 + \tau s} \qquad (9)$$

in which L^{-1} denotes the inverse Laplace transform.

APPLICATIONS

Because of space limitation only two applications can be considered and they must be discussed in a rather cursory manner. Nevertheless, they will serve to illustrate the method and to emphasize some noteworthy aspects.

Harmonic Excitation

The stress in response to a sinusoidal strain excitation in the steady state is

$$\sigma(t) = \epsilon_o G'(\omega)\sin\omega t + \epsilon_o G''(\omega)\cos\omega t \qquad (10)$$

where ϵ_o is the strain amplitude and $G'(\omega)$ and $G''(\omega)$ are the storage and loss moduli, respectively, ω being the radian frequency. Eq. (2) furnishes

$$W(t) = (\epsilon_o^2/4)[G''(\omega)(2\omega t + \sin 2\omega t) + G'(\omega)(1 - \cos 2\omega t)] \qquad (11)$$

for the energy absorbed up to time t, letting $W(t)=0$ for $t=0$. One might be tempted to consider the first term on the right to represent the energy stored, and the second term to represent the energy dissipated. However, this viewpoint would neglect the lack of phase coherence between the energy storing and between the energy dissipating mechanisms, each of which is characterized by its own relaxation time. The correct procedure which accounts for the lack of phase coherence is outlined below.

Substitution of the transform of $\varepsilon(t)=\varepsilon_o \sin\omega t$ into Eq. $(9)_1$ gives, after decomposition into partial fractions,

$$\bar{\varepsilon}_s(s;\tau) = \frac{\varepsilon_o \omega(\tau s + \omega^2\tau^2)}{(s^2 + \omega^2)(1 + \omega^2\tau^2)} - \frac{\varepsilon_o \omega(\tau s + \omega^2\tau^2)}{(1 + \tau s)(1 + \omega^2\tau^2)} \quad (12)$$

The two terms on the right represent the strains associated with the steady-state response and the transient response, respectively. Rejecting the latter, retransforming and substituting the result into Eq. (7) leads to

$$W_s(t) = (\varepsilon_o^2/4) \left\{ G'(\omega) - \left[G'(\omega) - \frac{dG'(\omega)}{d\ln\omega} \right] \cos2\omega t \right.$$

$$\left. + \left[G''(\omega) - \frac{dG''(\omega)}{d\ln\omega} \right] \sin2\omega t \right\} \quad (13)$$

after making use of the relations

$$G'(\omega) - \frac{dG'(\omega)}{d\ln\omega} = \{G_e\} - \int_{-\infty}^{\infty} H(\tau) \frac{\omega^2\tau^2(1 - \omega^2\tau^2)}{(1 + \omega^2\tau^2)^2} d\ln\tau \quad (14)$$

and

$$G''(\omega) - \frac{dG''(\omega)}{d\ln\omega} = \int_{-\infty}^{\infty} H(\tau) \frac{2\omega^3\tau^3}{(1 + \omega^2\tau^2)^2} d\ln\tau. \quad (15)$$

The rate at which energy is dissipated is obtained by an analogous derivation making use of the relations

$$\frac{dG'(\omega)}{d\ln\omega} = \int_{-\infty}^{\infty} H(\tau) \frac{2\omega^2\tau^2}{(1 + \omega^2\tau^2)^2} d\ln\tau. \quad (16)$$

and

$$\frac{dG''(\omega)}{d\ln\omega} = \int_{-\infty}^{\infty} H(\tau) \frac{\omega\tau(1 - \omega^2\tau^2)}{(1 + \omega^2\tau^2)^2} d\ln\tau. \quad (17)$$

Integration subject to the initial condition $W_d(0)=W(0)-W_s(0)$ leads to

$$W_d(t) = (\varepsilon_o^2/4) \left[2\omega t G''(\omega) + \frac{dG''(\omega)}{d\ln\omega}\sin2\omega t - \frac{dG'(\omega)}{d\ln\omega}\cos2\omega t \right]. \quad (18)$$

for the energy dissipated. It should be noted that Eqs.(13) and

(18) add to reproduce Eq. (11).

Triangular Excitation (Hysteresis)

The energy absorbed in a cyclic deformation is given by the hysteresis loop

$$W(\text{loop}) \;=\; \oint \sigma(u)d\varepsilon(u). \tag{19}$$

Equation (19) follows directly from Eq. (2). *In the steady state* W(loop) represents the energy dissipated during the complete cycle because the energy stored during the deformation half-cycle is completely released during the recovery half-cycle and the stress and strain at the end of the cycle assume the same values they had at the beginning of it. A well-known example of the hysteresis loop in the steady-state is the Lissajous ellipse obtained by displaying the stress given by Eq. (10) against the strain, $\varepsilon(t) = \varepsilon_0 \sin\omega t$.

Another hysteresis experiment consists of the imposition of a triangular strain excitation such as is shown in the right half of Fig. 1.

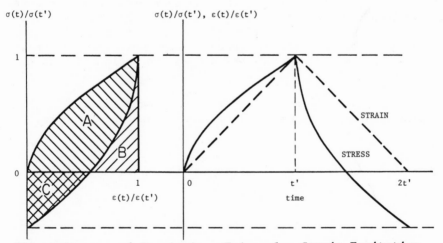

Fig. 1 Stress and Strain in a Triangular Strain Excitation

A constant rate of strain, $\dot{\varepsilon}_0$, is applied up to time t'whereupon the direction of the deformation is reversed. It reaches zero at time 2t'. This excitation is *not* a steady-state excitation. Not all of the energy stored during the deformation half-cycle is returned to the driving system during the recovery. The total energy absorbed during the deformation is represented by the area under the stress-strain curve in Fig. 1 for t=0 to t=t'. Thus, $W(t')_{\text{def}}$ = A+B where A and B are the hatched areas. The energy absorbed over the loop is given by

$$W(\text{loop}) = \dot{\varepsilon}_0^2 \int_{-\infty}^{\infty} \tau^2 H(\tau) [2t'/\tau - (1-\exp-t'/\tau)(3-\exp-t'/\tau)] d\ln\tau \qquad (20)$$

But the area of the loop is A+C where C is the crosshatched area in Fig. 1. Substituting the transform of the triangular strain excitation,

$$\bar{\varepsilon}(s) = \varepsilon_0 (1 - \exp-t's)/s^2 \qquad (21)$$

into Eq. $(9)_2$ and retransforming leads to $\dot{\varepsilon}_d(t;\tau) = \dot{\varepsilon}_0 (1-\exp-t/\tau)$ for $0 \le t' \le \bar{t}$. Substitution into Eq. (8) and integration furnishes the energy dissipated during the deformation half-cycle (i.e. up to time t') as

$$W_d(t')_{\text{def}} = \dot{\varepsilon}_0^2 \int_{-\infty}^{\infty} \tau^2 H(\tau) [t'/\tau - (1-\exp-t'/\tau)(3-\exp-t'/\tau)/2] d\ln\tau. \qquad (22)$$

Hence, $2W_d(t')_{\text{def}} = W(\text{loop})$ and, therefore, $W_d(t')_{\text{def}} = (A+C)/2$. The energy stored during the deformation half-cycle follows by difference as $W_s(t')_{\text{def}} = B+(A-C)/2$. Therefore, from the response to a triangular excitation the energy dissipated can be obtained from the hysteresis loop only if the area C is known. If the recovery is stopped when $\sigma=0$, the energy represented by the area of the loop contains an unknown amount of stored energy. It should be noted that repetition of the excitation until a steady-state is reached results in a closed loop whose area then represents the energy dissipated over a complete cycle, just as in the harmonic case.

A more detailed account of energy storage and dissipation in a linear viscoelastic material will be presented elsewhere.

THE STRETCHING FLOW OF MULTIGRADE MOTOR

OIL IN PLANAR FLOW NOZZLES

D.R. OLIVER AND M. SHAHIDULLAH

Department of Chemical Engineering
University of Birmingham

ABSTRACT

Jet thrust nozzles are described which produce planar
extensional flow, with superimposed shear. The extensional strain
rate is not constant in the nozzle, but increases towards the exit.

Thrust measurements on multigrade-type oils show that the
axial stresses developed in the nozzles are slightly higher than
those obtained in uniaxial extensional flow at the same nozzle
exit velocity, which is not unreasonable.

The significance of these stresses in lubrication processes
is pointed out; recently derived equations are used to show how
the stresses cause a pair of squeeze film surfaces to approach
each other more slowly, under constant load. The effect is most
pronounced in the early stages of the squeezing process.

Introduction

Present interest in squeeze film flow derives largely from the
problem of lubrication of car engines; in particular the cases of
gear teeth, cams, piston rings and little end bearings. Modern
motor oils contain polymeric additives which make the viscosity
less temperature dependent, but it is important to enquire whether
elastico-viscous behaviour in the oils is beneficial to the
lubrication process. Recent experimental work by Grimm [1] and by
Tichy and Winer [2] suggest considerable load enhancement effects
for aqueous elastico-viscous liquids at high squeeze rates; some
theoretical work suggests load enhancement [2,3] but the majority of
earlier work does not (see review in ref. 1).

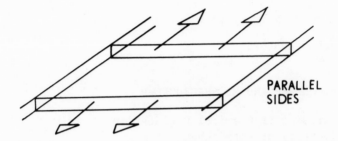

Fig.1. Planar Extensional Flow

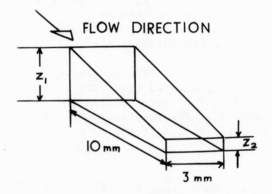

Fig.2. Planar Extensional Flow Nozzle

In most lubrication problems, the oil film lies between strips
of bearing surface rather than between disc-like surfaces; the oil
near the centreline thus undergoes "planar" extension, which is
illustrated in figure 1. This may be simulated in nozzle flow by
a wedge-shaped convergence (figure 2), two sides of which are
parallel. The jet thrust method may then be used to investigate
the axial stress present in a flowing liquid, as described
previously [4,5]. Essentially, the jet thrusts of Newtonian and
non-Newtonian liquids are compared at fixed Reynolds numbers, the
thrust difference per unit cross-sectional area giving the axial
stress in the non-Newtonian liquid.

Planar Flow Nozzles

These were made from brass and had the following dimensions
(figure 2). The length

NOZZLE	z_1 (mm)	z_2 (mm)
A	2.0	0.490
B	3.22	0.456
C	3.53	0.560

of all nozzles in the flow direction was 10 mm and across the flow
direction 3 mm. The outside dimensions were similar to those of
previously-used nozzles [4].

Owing to the wedge-shaped design of nozzles, the extensional
strain rate of the flowing liquid is not constant but increases
in the flow direction. Making the simple assumption of plug flow
through the nozzle, with little velocity profile development, it
is possible to show the variation of extensional strain rate with
time (figure 3). Also shown on this graph is the time dependence
of extensional strain rate for liquid being extruded between a
strip and a wall, when the strip approaches the wall with a
constant velocity of 25 cm. s^{-1}. The extensional strain rate
varies rapidly, as in other cases which are being examined at
present, which shows that the strain rate profile in the nozzles
is of some practical significance.

Oils Used

The oils were simulated 10W/30 motor oils, with polymer
additives, but no detergent-inhibitor present. The oil referred to
in figure 4 contained 3.9 per cent of dispersed polyalkylmethacrylate
additive; it had a shear viscosity of 107 cp. at 24°C and 17.0 cp
at 84°C, at a shear rate of 10^4 sec^{-1}. It is slightly shear-
thinning in character, with a flow behaviour index of 0.87 (24°C)
or 0.91 (84°C).

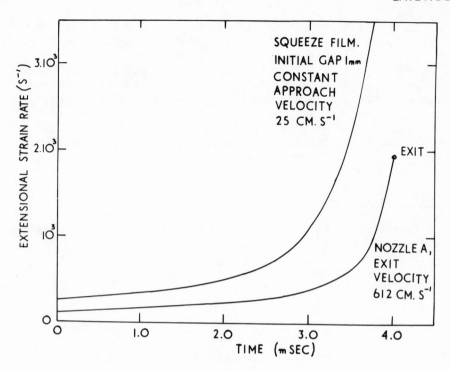

Fig.3.   Variation of Strain Rate with Time

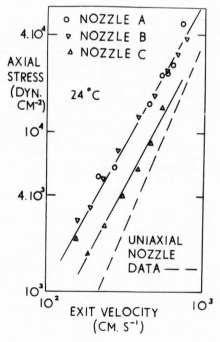

Fig.4. Variation of Axial Stress with Exit Velocity

Results

Figure 4 illustrates the relationship between axial stress in the fluid at the nozzle exit and the exit velocity. Also shown on the graph is the line obtained for the same oil tested in the uniaxial nozzles described previously [4]. The results for only one oil are shown in the interests of brevity. By using the equations quoted in reference 3, it is possible to use the data of figure 4 in order to demonstrate load enhancement in squeeze film flow. This is done by relating the squeezing force to the normal approach velocity for a purely-viscous oil (Scott equation) and comparing the results with those of an oil with similar viscous properties but possessing inertia and also showing elastic stresses. A step-wise procedure may be used to give a gap-time relationship for a 1 cm x 1 mm strip being driven towards a flat surface by a force of 10N, as shown in figure 5.

Discussion of Results

When axial stress is plotted against fluid exit velocity (figure 4), the data obtained using planar extensional flow nozzles lie above those obtained using uniaxial nozzles. This is reasonable, since for a Newtonian liquid the extensional viscosity is four times the shear viscosity in planar flow and three times the shear viscosity in uniaxial flow. A possible reason for the lower stresses developed in nozzle C is that the shear rates in the exit plane are lower than for the other nozzles; the combined effects of shearing and stretching the fluid are thus reduced. The results, though reasonable, illustrate the complexity of real extensional flow problems, significant variables being
(i)   the extension ratio (or velocity ratio),
(ii) the extensional strain rate,
(iii) the time-dependence of extensional strain rate,
(iv) the nature of the flow (uniaxial, planar or biaxial extension)
(v)   the superimposed shear and its interaction with extension.

Fortunately, [4] high speed stretching flows of polymer solutions seem to be dominated by item (i) and this makes it possible to plot the results in the way shown in figure 4. Planar extensional flow nozzles produce rather higher stress levels than uniaxial nozzles; the latter data may therefore be used as a conservation estimate of stress level in squeeze film lubrication problems.

Figure 5 shows that in the early stages of a constant-load squeezing process, the "purely viscous" case gives rapid motion, though the rate of approach of the surfaces falls as the plate spacing is reduced. If the inertia of the moving parts had been included, the plates would have moved closer together at this stage. The early stages of the case involving elastic and fluid

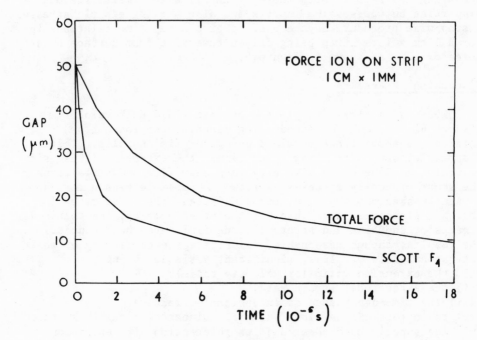

Fig. 5. Gap-time relationship for squeeze
film of oil under constant force conditions

inertia forces gives much slower motion, with the viscous force making little contribution to load bearing at this stage and the elastic component dominant. Later, the approach rate of the surfaces falls, but the spacing is larger at all times than for the "purely-viscous" case. It follows that the elastic stresses should particularly benefit the lubrication process for short squeeze times; the effect is enhanced if the strip is more heavily loaded or if the inertia of the metal parts is included.

The main conclusions are listed in the Summary.

References

1. R.J. Grimm, "Squeezing Flows of Polymeric Liquids", A.I.Ch.E.J. 24: 427 (1978).

2. J.A. Tichy and W.O. Winer, "The Influence of Fluid Viscoelasticity on a Squeeze Film Bearing" Trans. A.S.M.E. 100:56 (1978).

3. D.R. Oliver, "Influence of Fluid Inertia, Viscosity and Extra Stress on the Load Bearing Capacity of a Squeeze Film of Oil" App. Sci.Res. 35:217 (1979).

4. D.R. Oliver and R.C. Ashton, "The Flow of Polymer Thickened. Motor Oils in Convergent Jet Thrust Nozzles" J.Non-N.Fl.Mech. 2:367 (1977).

5. D.R. Oliver and R.C. Ashton, "The Flow of Polymer Thickened Oils in Convergent Jet Thrust Nozzles at a Temperature of 84°C". J.Non.NFl.Mech. 4: 345 (1979)

RHEOLOGY OF REACTING BIOPOLYMERS DURING EXTRUSION

L.P.B.M. Janssen

D.J. van Zuilichem

Lab. for Physical Technology
Delft University of Technology
Delft, the Netherlands

Dept. for Process Engineering
Agricultural University
Wageningen, the Netherlands

INTRODUCTION

Extrusion is used more and more to process starch and protein rich materials. In starch processing (e.g. maize, patato and rice) the function of the extruder is to rupture the particles and to gelatinize their contents to a certain extend for animal feed or to produce complete gelatinization for human food applications, such as snacks. When texturising protein based materials (e.g. defatted soymeal, cotton seed and sesam seed) the extruder function is to produce lamellar protein strands. These "crosslink" reactions give, apart from the usual shear and temperature effects on the viscosity, also rise to an extra "history based" viscosity effect.

Since the extrusion of biopolymers is often associated with pressure cooking cavioles in the material and temperature effects occure due to a flashing process as soon as the pressure is lowered at the die exit. In order to overcome the associated viscosity influence and the entrance and exit effects a slit viscosimeter is used at the exit of the extruder[1,2].

VISCOSITY FUNCTION

It is well known that within normal operating ranges starches and protein rich materials are shear thinning, which justifies the use of a power law equation for the viscosity. For changing temperature effects this equation may be corrected by multiplication of the power-law and the temperature dependence[3], thus giving for the apparent viscosity

$$\eta_a = K|\dot{\gamma}|^{n-1} \exp(-\beta T)$$

The viscosity will also be influenced by the processing history
of the material as it passes the extruder.
If we assume that the "cross linking" process may roughly be
described as a first order reaction and that the viscosity is
proportional to the thus formed concentration it is easy to show
that the history dependence of the viscosity can be expressed as

$$\eta = \eta_o \{1 - \exp \int_o^t - k(\dot{\gamma}, T) \, Dt\}$$

where k is the reaction constant, η_o a reference viscosity and
Dt a convective derivative accounting for the fact that the
coordinate system is attached to a material element as it moves
through the extruder. It has to be realized, that interactions
between the molecules generally occur through breaking and building
of hydrogen and other physico-chemical bonds. This cross linking
effect can be thought of as dependent on two mechanisms: a temperature effect determines the frequency of breaking and formation of
bonds and a shear effect determines whether the ends of a bond
that breaks meets a new end or will be attached to its old counter-
part again. We may assume that this last effect will not be a
limiting factor as soon as the actual shear stress is higher than
a critical value and the shear stress levels are high enough
within the extruder, thus the process may be described by an
Arrhenius model, giving for the viscosity at the exit of the
extruder:

$$\eta_a = \eta_o |\dot{\gamma}|^{n-1} \exp(-\beta T) \{1 - \exp \int_o^\tau \exp(- \frac{\Delta E}{RT(t)}) \, Dt\}$$

ΔE is the activation energy and R the gas constant.
Under the restrictive assumptions that the constants η_o, n, β and
ΔE are temperature independent it shows that at least four different
measurements have to be done in order to characterize the material
properly.

THE POWER NUMBER

It appears that the proposed procedure for the viscosity
function can be applied adequately to materials with sufficient
homogeneity as there is soy and very fine maize grits. Normal
maize grits however contain sufficient cell walls and other
impurities in order to change the rheological behaviour in such a
way that no reliable conclusions could be drawn.

Once the apparent viscosity is known it is possible to
establish a relationship between the power number Π and the
Reynolds number Re:

$$\Pi = \frac{P_o}{\eta_a N^2 D^3} \cdot \frac{D}{L} \cdot \frac{\delta}{1}$$

where P_o is the torque applied to the screw, D the screw diameter, L the channel length, δ the flight gap width and 1 the thickness of the flight.

This relationship is presented for soy in figure 1 and for maize in figure 2. It can be seen that for soy within reasonable limits the defined power number is independent of moisture content and Reynolds number. For normal maize grits however the power number is a function of moisture content and at low Reynolds numbers it also may depend on Reynolds.

DISCUSSION

From engineering point of view a semi quantitative analysis of the extrusion of biopolymers is given. The phenomena involved are remarkably complex and a broad general discussion of the model necessitates the adoption of some approximative assumptions. Although the measurements justify an optimistic point of view, more experimental verification is definitely needed.

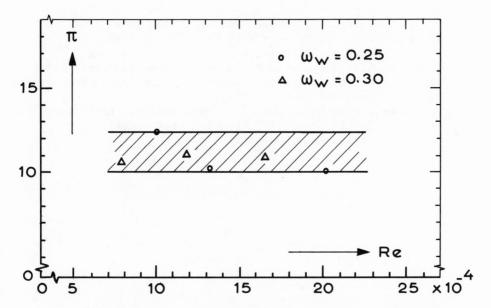

Fig. 1. Power number as function of Reynolds at different water contents (ω_w) for soy.

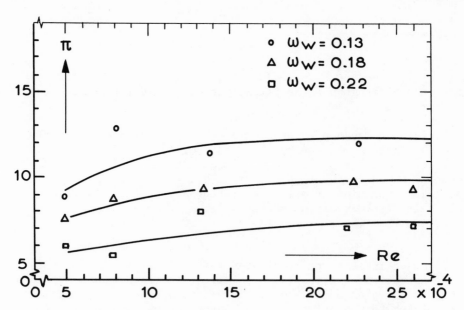

Fig. 2. Power number as function of Reynolds at different
water contents (ω_w) for maize grits.

REFERENCES

1. D. J. van Zuilichem, G. Buisman and W. Stolp, Shear behaviour
     of extruded maize, Proc. IUFOST Conf., Madrid (Sept. 1974).
2. D. J. van Zuilichem, S. Bruin, L. P. B. M. Janssen and
     W. Stolp, Single screw extrusion of starch- and proteinrich
     materials, Proc. Second International Symposium on Engineering
     and Food, Helsinki (August 1979).
3. A. B. Metzner, Flow behaviour of thermoplastics, in: "Processing
     of Thermoplastic Materials" (ed. E.C. Bernhardt),
     Van Nostrand-Reinhold, New York (1979).

OPTIMAL DESIGN OF PIPE LINE FOR TRANSPORTING HOT NON-NEWTONIAN

FLUIDS WITH HEAT TRANSFER TO THE SURROUNDING

A.K.Dey, N.C.F.Shah, K.S.S.Iyer[+], S.Prasad[+]

[+] Merado, Poona, India
    C.M.E., Poona, India

(Abstract)

The optimal design of pipe lines for transporting hot viscous fluids is of great importance in process industries as well as petroleum refineries for the distribution network of their products. For example while transporting Phenol extract through its pipe line to furnace for manufacturing Carbon Black, one has to take care to see that an optimat temperature is maintained for continual flow. The design of pipe line should not only aim towards economy but also ensure the proper flow temperature of fluid at all time. The transportation process occurs under unstead state condition. Such a situation demands an analytical approach to design a pipe line for transporting hot non-newtonian fluid minimising the heat transfer to surroundings. Literature available in the non-newtonian heat transfer studies are more concentrated -nder steady state conditions with constant wall temperature or constant heat flux.

In this paper an attempt is made to study the heat transfer under unsteady state conditions analytically with the help of temperature profiles to be ontained from experimental investigation.

The model pipe line has been designed to obtain the tempera ture and also the pressure profiles. The rheological data are collected from capillary tube viscometer for each fluid under study. They are incorporated in the mathematical model to obtain

a generalized solution under unstead condition.   This approach
will be significant to obtain economical design for commercial
pipe lines.

EXTENSIONAL RHEOMETRY OF SEVERAL BLOW MOLDING POLYETHYLENES

V. S. Au-Yeung and C. W. Macosko*

Department of Chemical Engineering
Princeton University
Princeton, NJ 08544

In free parison blow molding the development of parison shape appears to be the most critical step in proper polymer design.[1] At least three parison shape factors can be related to final bottle properties: wall thickness swell which controls bottle weight, parison diameter swell which controls the amount of flash or trim and parison sagging which leads to nonuniform wall thickness. The challenge to the rheologist is to relate these factors to basic rheological measurements which can be done on small samples. Most previous studies have attempted to correlate steady shear or linear viscoelastic properties to parison behavior.[1,2] In this study we examine extensional measurements on three similar high density blow molding polyethylenes.

MATERIALS

Three similar high density polyethylenes which showed different behavior when blow molded using free parison extrusion were selected for this study. All three samples have similar molecular characterization: $M_w=10.8\pm1.3\times10^4$ and $M_n=17.3\pm4\times10^3$. Other characterizations and some of the blown bottle properties are shown in Table 1.

_____

*Permanent address: Dept. of Chemical Engineering & Materials Science, University of Minnesota, Mpls.,MN 55455.

717

Table 1

Sample	Density g/cc	Melt Index (190°C, 44 psi)	Shear Viscosity (100 s⁻¹, 190°C) Poise	Bottle Weight g	Trim g
1	0.954	0.45	13,000	80	4.32
2	0.954	0.25	13,000	74	4.11
3	0.954	0.18	13,000	71	4.07

APPARATUS

Rheometrics Extension Rheometer, RER, was used in this study. The test frame is based on the recent design of Münstedt.[3] It employs a long cylindrical static bath, approximately 6×70 cm, with a separate heating fluid circulated through a surrounding jacket. A second surrounding jacket is evacuated and silvered to reduce connection and radiation losses. Temperature at each end of the bath can be maintained to within ±0.2°C.

Cylindrical samples can be successfully prepared by extruding from a melt index apparatus into a silicone oil bath or by using a vacuum compression mold. Samples were annealed for ~15 min in silicone oil at near test temperature, ends cut square, etched with concentrated H_2SO_4 and cleaned with acetone. Cut samples were bonded to small T-shaped aluminum clips with a high temperature epoxy (Uhu, Plus). The inhomogeneity of such samples at high extension is comparable to that reported by Münstedt.[4] Other sample shapes, such as long rectangles and rings, also gave comparable results but ruptured at lower strains. We have recently investigated bonding of samples by melt penetration into a screen. This appears to be a useful method and is much faster than epoxy bonding.

The performance of the Rheometrics Extensional Rheometer was verified by comparison of data on low density polyethylene (IUPAC-A) and polystyrene. Fig. 1 shows the comparison of the constant rate measurements on IUPAC-A with Münstedt's data and with Laun's data which were performed using a rotating clamp instrument based on Meissner's design.[5] With a combined analog and digital feedback control system the instrument can be used to perform constant stress experiments. The tensile creep data on IUPAC-A resulting from an applied constant stress of $3.2×10^4$ Pa are shown in Fig. 2. The experimental data compare favorably with those measured by Münstedt's creep apparatus.[3,4] Measurements of polystyrene at constant rates of 0.1 and 0.003 sec⁻¹ have also been carried out and are in excellent agreement with Münstedt's data.

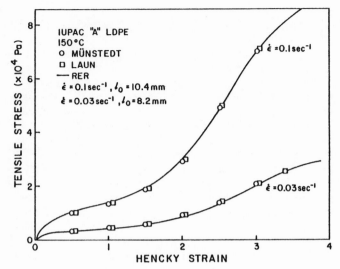

Fig. 1. Comparison of constant stretching rate results.

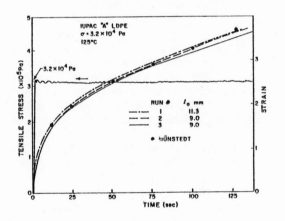

Fig. 2. Comparison of constant stress results.

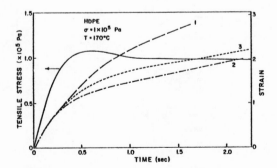

Fig. 3. Constant stress experiments on three HDPE samples at high stress.

EXTENSIONAL RHEOMETRY OF HDPE

The experimental creep curves of the three blow molding resins under constant stress of $\sigma=1\times10^5$ Pa and at 170°C are plotted in Fig. 3. Large differences in the creep behavior can be observed and under constant stress it is possible for the three resins to achieve a steady extensional flow from which a steady-state elongational viscosity, $\mu_s=\sigma_o/\dot{\varepsilon}_o$, can be obtained. At such a high stress level the experimental time scale is only about two seconds before the samples reach the maximum extensional ratio (i.e. rupture, typically at Hencky strain of about 2-2.5 for HDPE). As indicated in Fig. 3 the feedback control system in the Extensional Rheometer takes about one second before it can maintain the specified stress level. Thus care should be taken in interpreting such high stress experiments on these resins. However, the constancy of the stress level at longer time scales can be seen in Figs. 2 and 4.

A creep test followed by a recoil experiment can be performed in order to determine the elastic strain recovery behavior of these three similar resins as a function of recovery time. Their recovery curves at short recovery times (first thirty seconds) after a maximum Hencky of 2.35 at constant stress of $\sigma=2\times10^4$ Pa are plotted in Fig. 4. The strain recovery measurement is done by dropping the feedback reference to a low stress value, 8.2×10^2 Pa in the case of Fig. 4. We are developing a control strategy to give stress-free recovery. The equilibrium values in Fig. 4 were obtained by releasing the sample and measuring its final length. Fig. 4 shows that sample 1, which has the highest bottle weight, also has the highest recoverable strain at both short and long times. In comparing results, however, it is important to note that due to the low viscosity of sample 1 the rate of deformation is also much higher at the same stress level.

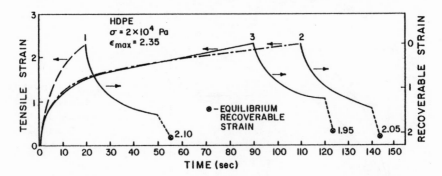

Fig. 4.   Recoverable strain as a function of recovery time for HDPZ after creep experiments. Transient recovery at 8.2×10^3 Pa. Equilibrium recovery under no load.

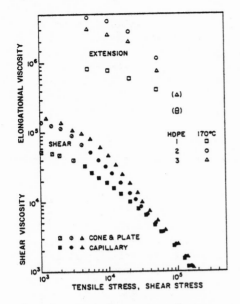

Fig. 5. Steady-state elonga-
         tional and shear vis-
         cosity as a function
         of stress.

The steady state elonga-
tional viscosity μ_s of the three
resins measured under constant
stress is shown in Fig. 5. Their
corresponding shear viscosity at
similar stress levels measured
with cone and plate in the Rheometrics Mechanical Spectrometer
and with a capillary rheometer
is shown in the same figure for
comparison. The tensile stress
dependence of the elongational
viscosity for high density polyethylenes over the limited range
of stress is much stronger than
that of polystyrene[3] and low
density polyethylene.[6,7] In
Fig. 5 the elongational viscosity
decreases more than tenfold over
a decade increase in stress.

The time dependent elon-
gational viscosity of the three
resins in a constant stretching

rate experiment has also been measured. Fig. 6 shows one of the
results (sample 2 at 170°C). In constant stretching rate experi-

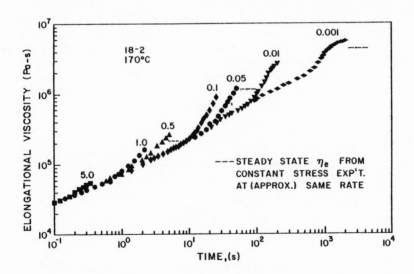

Fig. 6. Time-dependent elongational viscosity at different
         stretching rates for sample 2 at 170°C.

ment, no steady-state elongational flow can be achieved owing to the fact that high density polyethylene melt ruptures at small extensional ratio.   Steady-state elongational flow can be achieved at a smaller extensional ratio in constant stress experiments. The corresponding steady-state elongational viscosities associated with those constant rates in Fig. 6 are indicated in the same figure.   The fact that the steady-stage elongational viscosity is smaller than the final time-dependent elongational viscosity before rupture may mean that the time-dependent elongational viscosity reaches a maximum value before arriving at a steady state.

DISCUSSION

      The parison swell which takes place after the polymer melt has emerged from the extrusion die determines the polymer distribution of this cylindrical tube, and hence greatly affects the service performance of the blow-molded container.   If the parison swells extensively, the added flash or excess material will increase the percentage of scrap polymer (trim) and produce an overweight container.

      Elastic recoil after constant-stress experiments gives a strong correlation to bottle swell.   A similar correlation has been established in elastic recoil measurements after steady shear flow.[8]   Since the typical cycle time for bottle blow-molding is of the order of 5 sec, the short time elastic recovery is much more important than the equilibrium recovery for correlation to the dynamics of the parison formation.

      The elongational viscosity for sample 1 is much lower than for samples 2 and 3 at low stress levels.   The same is true to a lesser extent for the shear viscosity.   This suggests that during the time interval between dropping and inflation of the parison, gravity, which produces a relatively low stress, will cause the parison of sample 1 to sag more than the other two samples.   This in turn causes highest trim for sample 1.

REFERENCES
1.   J. S. Schaul, M. J. Hannon and K. F. Wissbrun, Trans. Soc. Rheol. 19:351 (1975).
2.   J. C. Miller, Trans. Soc. Rheol. 19:341 (1975).
3.   H. Münstedt, J. Rheol. 23:421 (1979).
4.   H. Münstedt, Rheol. Acta 14:1077 (1975).
5.   J. Meissner, Rheol. Acta 8:78 (1969).
6.   H. M. Laun and H. Münstedt, Rheol. Acta 17:415 (1978).
7.   H. Münstedt and H. M. Laun, Rheol. Acta 18:492 (1979).
8.   V. S. Au-Yeung and B. Maxwell, paper given at the 50th Meeting of the Society of Rheology at Boston, Oct. 1979.

STEADY PLANAR EXTENSION WITH LUBRICATED DIES

C.W. Macosko, M.A. Ocansey
Chemical Eng. & Materials Sci.
University of Minnesota
Minneapolis, MN 55455

H. H. Winter
Chemical Engineering
Univ. of Massachusetts
Amherst, MA 01003

Measurement of extensional flow behavior is still relatively undeveloped in comparison to shear rheometry. The measurement of very viscous materials in uniaxial extension has made great progress, however, relatively few results are available in planar or biaxial extension.[1] Recently, we have suggested that lubricated stagnation flow dies can be used for generating steady extensional flow.[2] Here we report our initial experimental results with a lubricated planar stagnation die using a polystyrene melt and silicone oil as lubricant.

THEORY

The kinematics of steady planar extensional flow are

$$(\underset{\sim}{v}) \;=\; (\dot{\varepsilon}x, \; -\dot{\varepsilon}y, \; 0). \tag{1}$$

The rate of extension is assumed to be constant throughout the fluid. Then material elements move along stream surfaces

$$xy \;=\; constant. \tag{2}$$

The extension of a material element when moving from x_0 to x is

$$\varepsilon \;=\; \ell n(x/x_0). \tag{3}$$

The volumetric flow rate is proportional to the extension rate,

$$Q \;=\; h\ell \, \dot{\varepsilon}x_0 \tag{4}$$

where ℓ is the stream width and h the thickness of the streams at x_0 as indicated in Fig. 1.

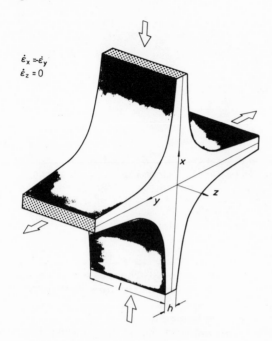

Two viscosity coefficients can be defined to describe the material behavior in steady planar extension

$$\eta_{p_1} = \frac{T_{xx} - T_{yy}}{\dot{\varepsilon}} \quad (5)$$

and

$$\eta_{p_2} = \frac{T_{xx} - T_{zz}}{\dot{\varepsilon}} \quad (6)$$

For a Newtonian fluid with viscosity η_0 these coefficients become

$$\eta_{p_1} = 4\eta_0 \quad (7)$$

and

$$\eta_{p_2} = 2\eta_0 \quad (8)$$

Fig. 1. Planar stagnation flow.

The stress on the stream surface has the components[2]

$$T_{nn} = \frac{y^2}{x^2 + y^2} (T_{xx} - T_{yy}) + T_{yy} \quad (9)$$

$$T_{nt} = \frac{xy}{x^2 + y^2} (T_{xx} - T_{yy}) \quad (10)$$

where the indices t and n indicate the tangent and the normal to the stream surface.

In an effort to achieve this ideal planar stagnation flow, the test fluid is pumped through a die with lubricated walls. We assume that the lubricant layer can be made thin. Then the die wall can be made to follow a stream surface. The lubricant must satisfy the following conditions:

(1) Constant lubricant flow rate (conservation of mass).
(2) No slip at the die wall.
(3) Velocity match at the fluid-lubricant interface, v_0.

(4)   Continuity of stress across the interface (ignore interfacial tension).

For a thin layer, δ, curvature can be neglected and the velocity distribution is that of a lubrication film

$$v_t(x_n) = v_o\left[1 - \frac{x_n}{\delta}\right] + \frac{\partial T_{nn}}{\partial x_t}\frac{x_n\delta}{2\mu}\left[\frac{x_n}{\delta} - 1\right]. \qquad (11)$$

In order to have planar extension throughout the test fluid, the velocity at the fluid-lubricant interface, v_o, must match the velocity of planar extension along the stream surface $v_t = \underset{\sim}{v}\cdot\underset{\sim}{t} = \dot{\varepsilon}(x^2 + y^2)^{\frac{1}{2}}$. The stresses at the interface must match those of eqs. (9) and (10). We are carrying out a solution of this boundary value problem. However, for our initial analysis of the experimental data we assume that the velocity field in the lubricant can satisfy all the conditions given above and that the velocity and stress fields are fully developed in the melt. These assumptions are tested to some extent in the following experiments.

EXPERIMENTAL

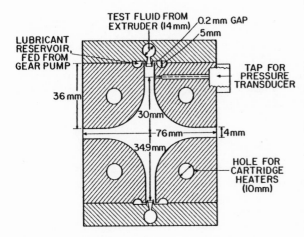

Fig. 2. Cross-section of lubricated planar stagnation die, length 100 mm.

Fig. 2 shows a cross-section of the planar stagnation die used for this study. The minimum total strain for a material element passing through the die is 3. There are two end plates (not shown). The end plate nearer to the extruder exit divides the melt stream equally and directs it into the two circular channels. The other end plate distributes the lubricant into the four semicircular channels which subsequently flows into the die as the sheath component.

Fig. 3 shows the set up for the lubrication experiment. A 1" diam. Killion extruder with a maximum output of 10 kg/hr delivered a continuous flow of a polystyrene melt to the die. The polystyrene, a general purpose injection molding grade from

Monsanto (Lustrex 101, $M_w \simeq 4 \times 10^5$ and $M_w/M_n \simeq 2$ by GPC), has a zero
shear rate viscosity of 1.8×10 Pa·s at 200°C. A Zenith gear pump
was used to meter the flow of the lubricant. Hydrocarbon lubricants were found to be unstable at the temperature used. Silicone
oils worked quite well. Both Dow Corning and General Electric
fluids of about 1 Pa·s were used. A set of temperature con-
trollers maintained the die and oil line temperature. A pressure
transducer, Dynisco Model PT420A, was mounted in the tap indicated
in Fig. 2 and connected to a digital readout and a chart recorder.
The pressure hole was filled with silicone oil during experiments.

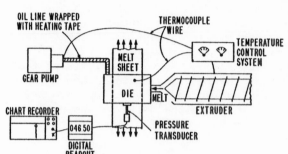

Fig. 3. Arrangement of
lubricated stagnation die
with extruder for feeding
polymer melt and gear
pump for the silicone oil
lubricant.

The pressure transducer should read $p_o = -T_{nn}$. At the
pressure tap location $x \gg y$ so if the stress follows eq. (9) and
there is no additional pressure due to pumping the lubricant, then
$T_{nn} = T_{yy}$. At the exit $x \ll y$ and $T_{nn} = T_{xx}$. We expect no pressure
jump at the exit so $T_{xx} = 0$ and thus

$$p_o = T_{xx} - T_{yy} \tag{12}$$

Flow birefrigence was also used to measure the stresses.
Janeschitz-Kriegl and coworkers have shown that the stress optical
law holds for polystyrene melts over a very wide range of deformation rate.[3,4] The optical system used consisted of a tungsten
light source, a lens, a slit, a polarizer, a wedge-shaped glass
window 6 mm wide across each exit plane of the die and a Babinet
compensator. As indicated in Fig. 1, the birefringence Δn is
measured in the x,z plane. For planar extension the principal
stresses will be aligned with the coordinate direction, and the
birefringence is

$$\Delta n = C(T_{xx} - T_{zz}) \tag{13}$$

where C = the stress optical coefficient, $4.9 \times 10^{-9} m^2/N$ for polystyrene.[4]

PROCEDURE AND RESULTS

   After the die and extruder had reached the test temperature,
the gear pump was started, flooding the die with the lubricant
and ensuring that all the walls were covered. Melt was then
pumped into the die under the lowest extruder flowrate. When melt
was first pumped into an empty die it was not possible to achieve
lubrication. Also, it was generally not possible to reestablish
lubrication after stopping the experiment.

   When lubricated stagnation flow had been established the
extruder flowrate was increased. Timed samples were cut from the
die and weighed to determine the flowrate of the melt. This was
used to calculate the extension rates by eq. (4). The pressure at
the entrance was measured.

   Another set of experiments were performed without pumping the
lubricant. A piece saved from the previous experiment was coated
with lubricant and inserted into the die. The die and the extruder barrel were then heated up to the test temperature. The
extruder was then started up at a known flowrate and the response
of the pressure as a function of time was recorded. To determine
the effect of the lubricant on the pressure, a set of unlubricated
experiments were performed. In this case, the die was thoroughly
cleaned to ensure that there was no lubricant anywhere on the
surface.

   Fig. 4 shows typical pressure readings as a function of time.
Without lubricant the pressure rises to a high value after the
extruder is started up. The steady state value of 9.6×10^4 Pa compared well to 13.8×10^4 Pa calculated by approximating the die shape
with two wedges and then using Hamel's exact solution for a Newtonian fluid.[5] When lubrication was achieved following the procedure of pumping the lubricant the pressure dropped nearly a
factor of 7 indicating slip at the wall. When this value was
analyzed according to eqs. (12) and (5), we found that the pressure reading was considerably higher than expected by eq. (7).
The source of the error is suspected to be due to the hydrostatic
pressure in the oil line. We are presently studying this problem
with simultaneous pressure measurements at several points in the
die and in the oil line. However, when lubrication was achieved
following the procedure of precoating the walls of the die, we
found that the results agree well with four times the shear value
as shown in Fig. 5.

   Using the method of flow birefringence, Δn was measured as a
function of flowrate. Since the birefringence is measured in the
y-direction, through line stagnation, the influence of shear from
the walls should be minimal. Under unlubricated conditions, the

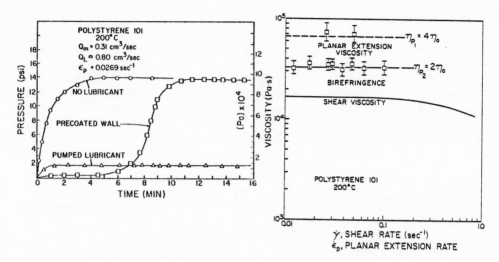

Fig. 4. Typical inlet pressure   Fig. 5. Shear viscosity
         traces for three con-   data compared to
         ditions.   extensional vis-
                                                 cosity results.

values were then analyzed according to eqs. (13) and (6) and found
to be in good agreement with twice the shear viscosity as shown in
Fig. 5.

CONCLUSION

     The results obtained from the experiments so far indicate
that lubrication can be achieved and used to generate steady extensional flow in the planar stagnation die. Work is in progress
to determine the effect of the oil line pressure and flowrate on
the total pressure measured.

     The flow birefringence approach also gives good results and
the use of a new die with glass end windows will give us the
opportunity to study the whole flow field.

     This work was supported by grants from the Mobil Oil Corp.
and the National Science Foundation. Silicone oil was donated by
Dow Corning Corp.

REFERENCES

1.   J. M. Deally, J. Non-Newt. Fluid Mech. 4:9 (1978).
2.   H. H. Winter, C. W. Macosko and K. E. Bennett, Rheol. Acta
     18:323 (1979).
3.   H. Janeshitz-Kriegl, Adv. in Polymer Sci. 6:170 (1969).
4.   H. M. Laun, M. H. Wagner and H. Janeschitz-Kriegl, Rheol.
     Acta, in press.
5.   S. Middleman, "Fundamentals of Polymer Processing," McGraw
     Hill, New York (1977).

EXPERIMENTAL TESTING OF MOLECULAR THEORIES OF NETWORK ELASTICITY

M. Gottlieb, C.W. Macosko[+]

[+] Ben Gurion University, Israel
University of Minnesota, U.S.A.

(Abstract)

The classical theory of rubber elasticity uses the phantom network model in which a network chain possesses no properties other than exerting force on the junctions to which it is connected. Experiments have shown that this model does not describe correctly the stress-strain relationship in extension. Also the small strain modulus value predicted by it is lower than the experimental value. Recently models by Flory and by Edwards have been developed to account for these discrepancies. In the present work well characterized model network were prepared by random crosslinking of linear polydimethylsiloxane molecules. Under uniaxial extension, compression, and shear stress-strain data were obtained as a function of the network structure. Results indicate that restrictions on junction fluctuations (Flory) cannot solely account for the experimental modulus but describe qualitatively the stress-strain behavior in compression and extension. Edwards' model describes correctly the behavior in extension but not compression.

FLOW OF VISCOELASTIC POLYMER SOLUTIONS IN CURVED TUBES: VELOCITY

PROFILE MEASUREMENT BY LASER DOPPLER ANEMOMETRY

J.J. Ulbrecht, V.R. Ranade

State University of New York at Buffalo, U.S.A.

(Abstract)

Among the various manifestations of viscoelasticity in polymer solutions under laminar flow conditions, the interference with the inertia-driven flows is one of the more specular ones. Is there a link between the reduction of the friction factor and the secondary flow? An attempt to answer this question will be presented in this contribution. The three velocity components (axial, radial, and vertical) were measured in two toroids having the radius of curvature 5 and 10 inches, respectively. The diameter of the tube was one inch in both cases. The velocities were measured using the DISA 55X Modular Laser Doppler Anemometer with a 15 mW He-Ne Laser. Two components were measured by rotating the beam splitter and a flat mirror was used to measure the third component. The liquids used in this work were ethylene-glycol to represent the viscous Newtonian behaviour and a 0.2% aqueous solution of polyacrylamide as a non-Newtonian viscoelastic liquid. The first experimental data available show clearly that, in the range of Dean numbers between 40 and 120, the secondary flow pattern in the PAA solutions differs considerably from that in the Newtonian glycol, although the primary profiles in the two liquids are the same. The most conspicuous change is the distortion of the symmetry of the secondary flow pattern with a strong enhancement of the vertical component in the region close to the outer wall of the toroid.

EXCESS PRESSURE LOSSES AT THE ENTRANCE OF A SLIT

L. Choplin, P.J. Carreau

Chemical Engineering Department, Ecole Polytechnique
of Montreal, Canada

(Abstract)

Excess pressure losses at the entrance of a slit are reported
for the flow of viscous fluids under large ranges of Reynolds and
Weissenberg numbers. Polyacrylamide (Dow Separan AP-30) solutions
in water and in corn syrup, and non elastic fluids such as Carbopol
solutions and pure corn syrup were used. The excess pressure losses
were obtained from pressure measurements with ten flush-mounted
transducers, hence eliminating hole errors and differentiating
between entrance and exit losses. The contribution of the viscous
dissipation to the excess pressure losses was calculated from a
macroscopic energy balance, correcting for the kinetic energy
losses and the normal stresses arising in the fully developed flow
region. It is shown that the normal stress term is negative and
that the correction factor should be written correctly as

$$n_{ent} = n_{vis} - n_{\psi 1} = \Delta P_{ent}/2\tau_w$$

where n_{vis} is the correction term due to the total viscous
dissipation at the entrance and $n_{\psi 1}$ is the correction due to the
normal stress term in the fully developed flow region.

The entrance losses are important only when the fluid exhibits
both elastic and shear-thinning properties. In absence of elastic
or shear-thinning properties, the entrance losses are smaller or
close to values obtained with Newtonian fluids. For the solutions
of polyacrylamide in corn syrup (approximately second order fluids)

in creeping flow, $n_{\psi 1}$ is of the order of n_{vis} , resulting in very
low excess pressure losses. These results support the numerical
analysis of J.R.Black, M.M.Denn and G.C.Hsiao (in "Theoretical
Rheology" edited by J.F.Hutton, J.R.A.Pearson and K.Walters,
Applied Science, 1975). For shear-thinning viscoelastic fluids,
the large entrance losses are attributed to stress overshoots
under high rate of deformation at the wall of the slit near the
entrance.

COMPARATIVE HEMORHEOLOGY, IN TRANSIENT FLOW,

OF PATIENTS WITH POLYCYTHEMIA AND A CONTROL GROUP

H. Essabbah, B. Auvert and C. Lacombe

Unité de Biorhéologie du département de Biophysique
91 boulevard de l'Hôpital
75013 Paris France

INTRODUCTION

The rheological behaviour of blood has been studied for some-time using primarily in vitro techniques. The stationary viscosity of normal human blood has been measured many times (see for example Copley[1]). The clinical interest of such studies has been shown after examination of blood from pathological cases (see for example Dintenfass[2]). In vitro transient flow studies allow the measurement of some rheological parameters which are ignored in stationary studies. These parameters are of interest mainly because physiologically blood is in unsteady motion. Up to now only few in vitro studies have been performed using transient flow techniques. Indeed there exists only a few viscometers that allow these kind of measurements.

In the first part of this paper we describe improvements made to the viscometer used by our group. In the second part we study, in transient flow, the rheological behaviour of blood from patient with polycythemia. This study is of importance since for these patients the only preventive treatment against complications due to an increase in blood viscosity remains the venesection.

METHODS

For our studies we employed a Couette type viscometer (Healy and Joly[3]), the principle of which is shown in Fig. 1. This device was connected to a microcomputer in order to obtain an automatic rheological apparatus. Prior to this connection each experimental protocol had to be prepared manually with a signal generator which commanded the motor. This procedure was tedious and time consuming.

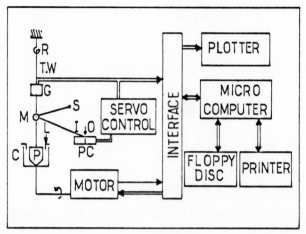

Fig. 1.   Schematic diagram of the automatic Couette type viscometer
          with coaxial cylinders. The outer cylinder (C) is driven
          by the motor. Its movement is transmitted by the liquid
          (L) to the inner cylinder (P) suspended by a torsion wire
          (TW) to a spring (R). The wire hangs also a galvanometer
          (G) and a mirror (M). A spot (S) is reflected by the
          mirror and an image (I) is obtained in the middle (O) of
          photoelectric cells (PC). A servo control device sends
          into the galvanometer (G) a current (IS : input signal)
          which maintains the image (I) in O during the rotation of
          the outer cylinder (C).

In addition each rheogram was recorded and analysed manually,
statistical analysis required the re-entering of the data into an
off-line computer via the interface (Fig. 1). The interface includes two digital-to-analog-converters for the command of the
motor plus the plotter device, and a data acquisition system which
was connected to the input signal. This system includes a variable
gain amplifier, Bessel's filters and a analog-to-digital-converter.
The software system is mainly composed of four major subroutines in
order to control the motor, to read the input signal and to plot
and analyse rheograms. This last subroutine requires, in general,
less than three minutes. After analysis of a rheogram, rheological
parameters are stored in floppy disc files. There are numerous
advantages of this type of automatic system :
     1) The motor is servo-controlled by the computer which can
generate any type of signal.
     2) The filtered signals, as well as the original signal, are
memorized by the system and computer programs are used to analyse
rheograms.
     3) Files of results are kept for further investigations and
statistical analysis.

Finally this integrated instrument can be used as an entirely automatic viscometer. It allows us to envisage systematic studies into the rheological influence of either biological, physical or chemical factors.

Blood was obtained by venepuncture and anticoagulated with EDTA. The experiments were performed the same day at 25°C. The measurements were first taken at the hematocrit of the patient and then once again after the hematocrit was adjusted to 45% in order to do comparative studies. The adjustment of the hematocrit was done by diluting a part of the sample with autologous plasma obtained by centrifugation. The rheological determinations are performed under transient flow conditions by measuring the dynamic response of blood to rectangular or triangular steps of shear rate. This method is particularly sensitive to exhibit the non Newtonian properties of blood : shear-thinning, viscoelasticity and thixotropy. It has been developed by Joly et al.[4,5].

For the rectangular steps the used shear rates are 0.05 s^{-1}, 1 s^{-1} and 20 s^{-1}. The recorded rheograms give the shear stress variation as a function of time. We shall call them respectively rheogram type I, type II, type V (Fig. 2). For the triangular steps the rates of variation of the shear rates as a function of time are 0.019 s^{-2} and 0.044 s^{-2}, the corresponding maximum values attained by the shear rates being respectively 0.1 s^{-1} and 1 s^{-1}.

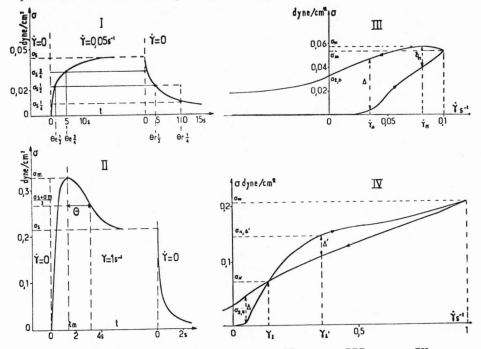

Fig. 2.   Rheograms type I, type II, type III, type IV.

The hysteresis loops so obtained give the shear stress variation as a function of shear rate and will be called type III and type IV respectively (Fig. 2). Every measurement is performed 5 minutes after homogenization by agitation[3] ; the duration of any rheogram recording must be shorter than 1 minute.

For each rheogram empirical rheological parameters were defined elsewhere[4,5] (Fig. 2,3). For instance we used the empirical index of thixotropy :

$$\xi = \frac{\sigma_m - \sigma_s}{\dot{\gamma} \, \theta \, \sigma_s}$$

and the empirical coefficients $\sigma_{2,0}/\sigma_m$ and Δ/σ_m characterizing viscoelasticity. The rheograms type I and type IV mark particularly the viscoelastic behaviour ; the type I gives information about the size of the rouleaux. The rheograms type II and type V are more characteristic of the thixotropic and shear-thinning behaviour.

RESULTS

We have studied 24 polycythemic patients without taking into consideration the etiology of their disease. They were selected according to their elevated hematocrit which was greater than 50% as well as their hemoglobin level greater than 16 g/100 ml.

The stationary viscosities determined at the following shear rates 0.05 s^{-1}, 1 s^{-1} and 20 s^{-1}, for all samples of polycythemic blood adjusted to 45% of hematocrit, were comparable to those values obtained for the control group, under the same experimental conditions. On the other hand, these stationary viscosities were compared to those at the original hematocrit (greater than 45%). In all cases they were found to be less. At a same value of hematocrit, polycythemic and normal blood have similar stationary viscosities. Plasma viscosity measured at a shear rate of 20 s^{-1} is normal for all polycythemic patients. Thus a difference between polycythemic and hemoconcentrated normal blood, from a rheological view-point, cannot be distinguished by measurements taken in stationary flow.

However the comparison of the thixotropic parameters for poly-cythemic and for control blood adjusted to hematocrit of 45% allowed us to distinguish three types of polycythemics (Fig. 3) : we found 6 cases where the thixotropic coefficient was normal, 11 cases where it increased, and 7 cases where it decreased. At the original hematocrit of the pathologic blood (greater than 45%) we found a similar type of classification : 7 cases where the thixotropic co-efficient was normal, 4 cases where it showed an increase and 13 cases where the coefficient decreased. It can be seen that the ratio of patients with an increased thixotropic coefficient is lower in the second classification (17%) than in the first (46%).

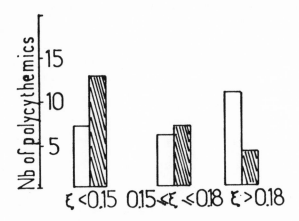

Fig. 3. Polycythemic classification taking into account the
         thixotropic coefficient

☐ hematocrit = 45 %   ▨ hematocrit > 45%

Consequently hemodilution induces a decrease of the stationary
viscosity and an increase of the thixotropic coefficient ξ interpreted as a better desaggregation of the rouleaux. It should be
noted that for the control group, the thixotropic coefficient did
not vary appreciatively as a function of the hematocrit.

The viscoelastic behaviour of blood can be studied by the
method of triangular steps which permits the distinction at 45%
of hematocrit of three types of polycythemics : 11 cases with a
normal viscoelastic behaviour, 7 cases where it increased and 6
cases where it decreased. All those polycythemics with a diminished
viscoelasticity belong to the group whose thixotropic coefficient
was increased at 45% of hematocrit. But it is important to note
that the different classes obtained for the viscoelastic behaviour
and for the thixotropic coefficient are not interchangeable. On the
other hand, the rheological distribution of the polycythemics is
independent of the etiology of their disease.

CONCLUSION

This preliminary study constitutes the first application of
an automated viscometer on pathological blood. We have only examined
24 cases and further study is needed to confirm the system of
classification we have proposed. This technique of measuring in
transient flow shows that hemodilution promotes a better desaggregation of the rouleaux for some polycythemic patients. As for
other polycythemic patients, eventhough the stationary viscosity

decreases, the thixotropic coefficient remains below to the normal value, interpreted as a greater stability of the rouleaux.

These findings could perhaps be explained by hematological desorders such as modifications of plasmatic factors or charges in red blood cells properties.

It is to remark that the proposed classification of the polycythemic patients based on the rheological behaviour does not correspond to the classification generally recognized in hematology.

REFERENCES

1. A. L. Copley, C. R. Huang and R. G. King, Rheogoniometric studies of whole human blood at shear rates from 1000 to 0.0009 sec^{-1}. Part I. Experimental findings, Biorheology, 10:17 (1973).
2. L. Dintenfass, Theoretical aspects and chimical applications of blood viscosity equation containing a term for the integrated viscosity of the red cell, Blood cells, 3:367 (1977).
3. J. C. Healy and M. Joly, Rheological behaviour of blood in transient flow, Biorheology, 12:335 (1975).
4. M. Bureau, J. C. Healy, D. Bourgoin and M. Joly, Etude rhéologique en régime transitoire de quelques échantillons de sangs humains artificiellement modifiés, Rheol. Acta, 18:756 (1979).
5. M. Bureau, J.C. Healy, D. Bourgoin and M. Joly, Rheological hysteresis of blood at low shear rate, Biorheology, (1980) in press.

THE VISCOELASTIC PARAMETERS OF PATHOLOGICAL SYNOVIAL FLUIDS

I. Anadere[+], H. Chmiel[+], W. Laschner[++]

+ Fraunhofer-Institut für Grenzflächen- und Bioverfahrenstech-
nik, Eierstrasse 46, 7000 Stuttgart 1, FRG
++ Abteilung für orthopädische Rheumatologie
Klinik für Orthopädie und Unfallchirurgie Dr. Baumann e.V.
Alexanderstrasse 5-7a, 7000 Stuttgart 1, FRG

INTRODUCTION

The viscoelastic flow properties of synovial fluids (SF) have been subject of research in recent years (1,3,4,6,7). In pathological cases the viscoelasticity of SF are altered according to the kind and stage of disease existing in the joints. The viscoelasticity of SF results mainly from the interaction of the chains of a biopolymer, hyaluronic acid (HA) and its protein complex present in these fluids. The degree of polymerisation and the concentration of HA affects the viscous and the elastic components of the complex viscoelasticity. In this study these parameters and their relationship to HA content are investigated for pathological SF.

MATERIAL AND METHODS

The SF were prepared for measurement as described previously (1). The viscoelasticity was determined as a function of shear rate ($\dot{\gamma}$) with an oscillating capillary viscometer (5, 1) at 22° C and 2 Hz. The viscoelasticity values at $\dot{\gamma} = 10 \text{ s}^{-1}$ were used for the comparative evaluation of the different disease groups. The concentration of HA was determined with a modified carbazole reaction (2). The values were evaluated statistically with the Student t-test. The theory of linear viscoelasticity was applied which assumes a complex coefficient of viscoelasticity (η^*), composed of viscous (η') and elastic components (η''). (5).
This theory is valid only for low shear rates and frequencies.

$$\eta^* = \eta' - i\eta''$$

741

RESULTS

The various joint effusions investigated were divided into following groups

(a) patients without actual joint disease (meniscus defects) (n = 28)
(b) degenerative (osteoarthrosis) (n = 26)
(c) traumatic (n = 37)
(d) rheumatic arthritis (R.A.)   i) seronegative (n = 52)
                                        ii) seropositive (n = 33)
The flow curves of 3 different samples are shown in Fig. 1.

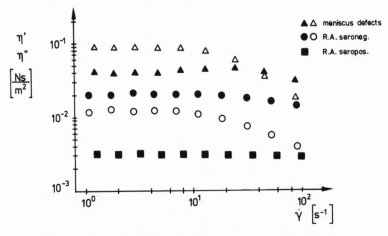

Fig. 1   Viscous (η') (filled signs) and elastic (η'') components of visco-
            elasticity versus shear rate for several joint diseases
            (F = 2 Hz, T = 22° C)

The viscous (η') and the elastic (η'') components of viscoelasticity plotted versus shear rate ($\dot{\gamma}$) show apparent differences. Both parameters are con-stant till to $\dot{\gamma}$= 10 s^{-1}. The further increase of shear rate is followed by a decrease in both components. Only the sample of seropositive R.A., which has almost newtonian flow properties remains constant. By this sample the elastic components cannot be detected with the measuring apparatus. On the other hand, the elastic components of the sample with meniscus defects are quite high, and at lower shear rates the values are higher than the values of the viscous component.

Table I The complex coefficient of viscoelasticity (η^*), its viscous (η') and elastic (η'') components and the concentration of HA for various joint diseases ($F = 2$ Hz, $\dot{\gamma} = 10$ s^{-1}, $T = 22°$ C)

Disease	n	η^* $\times 10^{-3}$ Ns/m^2		η' $\times 10^{-3}$ Ns/m^2		η'' $\times 10^{-3}$ Ns/m^2		C_{HA} mg/g	
		$\bar{x}$	$\pm$ s	$\bar{x}$	$\pm$ s	$\bar{x}$	$\pm$ s	$\bar{x}$	$\pm$ s
Meniscus def.	28	66.40	$\pm$ 12.0	39.12	$\pm$ 9.2	53.66	$\pm$ 18.6	1.09	$\pm$ 0.30
degenerat.	26	44.90	$\pm$ 9.2	30.16	$\pm$ 11	33.16	$\pm$ 12.2	0.97	$\pm$ 0.32
traumatic	37	20.35	$\pm$ 8.8	18.22	$\pm$ 10.5	11.95	$\pm$ 6.1	0.51	$\pm$ 0.28
R.A. a) seroneg.	52	16.30	$\pm$ 8.1	15.92	$\pm$ 6,3	11.90	$\pm$ 5.2	0.46	$\pm$ 0.21
b) seropos.	33	9.80	$\pm$ 3.4	9.0	$\pm$ 3.2	3.20	$\pm$ 2.1	0.34	$\pm$ 0.16

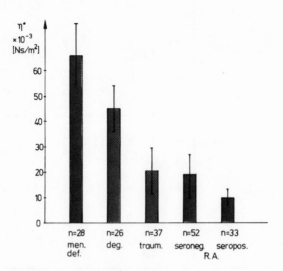

Fig. 2   Mean values of complex coefficient of viscoelasticity (η^*) for various joint diseases
$(F = 2\ \text{Hz}, \quad \dot{\gamma} = 10\ \text{s}^{-1},\ T = 22°\ \text{C})$

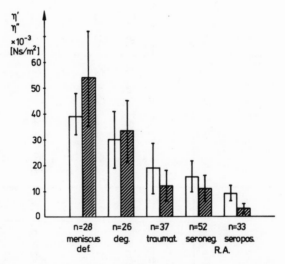

Fig. 3   Mean values of viscous (η') and elastic components (η'') components of viscoelasticity for various joint diseases
$(F = 2\ \text{Hz}, \quad \dot{\gamma} = 10\ \text{s}^{-1},\ T = 22°\ \text{C})$

To compare the statistical differences between the groups of diseases
investigated, the η , η' and η'' values at $\dot{\gamma} = 10\ s^{-1}$ were used. These
values and concentration of HA of the samples are listed in Table I.
Fig. 2 shows the η^* values of the various groups of joint diseases. The mean
values are at highest with patients with meniscus defects and at lowest with
patients with seropositive R.A. If the viscous and elastic parameters are
investigated separately (see Fig. 3), it becomes evident that the elastic
components of the synovial fluids with meniscus damages and degenerative
joint diseases lie higher than the viscous components. This effect is reversed
in the case of traumatic and rheumatic arthritis, for both seropositive and
seronegative cases. The changes in both parameters correlate with changes
of HA concentration. To answer the question what kind of relationship
exists between the viscoelastic parameters and the concentration of HA,
both the viscous and elastic components were investigated separately as
a function of HA concentration. These parameters are plotted on Figs. 4 and
5 as a function of HA concentration.

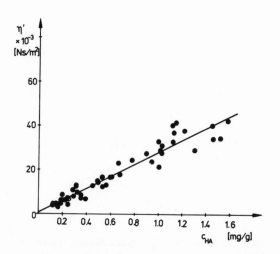

Fig. 4   Viscous component of viscoelasticity η' versus HA
              concentration
              $(F = 2\ Hz,\quad \dot{\gamma} = 10\ s^{-1},\ T = 22°\ C)$

To simplify the calculation only data from 50 random determinations were
used. The relationship between η' and HA concentration can be interpreted
with linear regression analysis of 1st degree. The resulting correlation
coefficient between the two parameters is r = 0,918 and η' yields to

$$\eta' = (0,23 + 27,36 \, C_{HA}) \times 10^{-3} \, Ns/m2$$

However the elastic component can be better fitted to an exponential curve using the exponential variant of the linear regression analysis. yields thus to

$$\eta'' = 0,7 \; \exp \; (3,46 \, C_{HA}) \times 10^{-3} \, Ns/m2$$

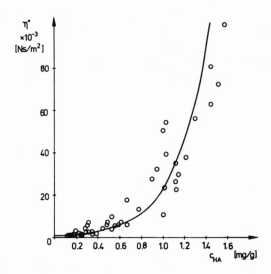

Fig. 5   Elastic component of viscoelasticity η'' versus HA concentration
($F = 2$ Hz, $\dot{Y} = 10 \, s^{-1}$, $T = 22°$ C)

DISCUSSION

It has been demonstrated that the viscoelasticity of SF results from the interaction of the HA protein complex chains. The viscous component in linear and the elastic component exponential dependent on the HA concentration. Joint diseases are accompanied by changes of the HA concentration and of the rheological parameters i.e. of the viscous and elastic components which are responsible for the energy dissipation and storage in the joints respectively. The non-inflammatory joint diseases exhibit higher elastic components which result in better storage of mechanical energy and the ability to act as a shock adsorber. In inflammatory diseases both parameters of viscoelasticity are diminished, which lead to severe consequences for the normal joint function. The changes in the elastic component are more pertinent, therefore the determination of this parameter proves to be a more sensible method to detect changes that occur in joint diseases.

Acknowledgements
The authors want to thank the Deutsche Forschungsgemeinschaft for financial help and Mrs. H. Schreier for technical assistance.

REFERENCES

1.   I. Anadere, H. Chmiel and W. Laschner
      "Viscoelasticity of "normal" and pathological synovial fluid"
      Biorheol. 16:179 (1979)
2.   T. Bitter and H.M. Muir
      "A modified uronic acid carbazole reaction"
      Anal. Biochem. 4:330 (1962)
3.   D.A. Gibbs, E.W. Merrill and K.A. Smith
      "Rheology of hyaluronic acid"
      Biopolymers 6:777 (1968)
4.   R.R. Myers, S. Negami and R.K. White
      "Dynamic mechanical properties of synovial fluid"
      Biorheol. 3:197 (1966)
5.   G.B. Thurston
      "Viscoelasticity of blood"
      Biophys. J. 12:1209 (1972)
6.   G.B. Thurston and H. Greiling
      "Viscoelastic properties of pathological synovial fluids for a wide
      range of oscillatory shear rates and frequencies"
      Rheol. Acta 17:433 (1978)
7.   H. Zeidler, S. Altman, B. John, R. Gaffga and W.-M. Kulicke
      "Rheologie pathologischer Gelenkflüssigkeiten
      Weitere Ergebnisse zur Viskoelastizität"
      Rheol. Acta 18:191 (1979)

THERMODYNAMICALLY STRUCTURED FLOWS

M. Frias

University of Lisbon, Portugal

(Abstract)

   Nonlinear transport has been analysed as a nonmarkovian functional process or, in more usual rheological terminology, under a functional dependence between stress and the time history of deformation gradient,

$$T(t) = \mathcal{F} (E (t - s)) \tag{1}$$

The functional response for a fluid being independent of the reference configuration for density preserving transforms, a simple fluid , i.e. a locally materialized and analised fluid in the sense of Coleman and Noll, is an isotropic fluid.

   A purely viscous fluid is one for which the stress at any material point and time is a function of the velocity gradient evaluated at the point and time, so

$$T(t) = F (\nabla v) \tag{2}$$

within an isotropic pressure, F being a symmetric tensor valued function of ∇v.

   The fact that a simple fluid is an isotropic fluid permits the reduction of the deformation gradient to its symmetric part, E, i.e. being

$$2 \ E \ = \ \nabla v + (\ \nabla v \ )^T \tag{3}$$

$(\nabla v)^T$ being the transpose of ∇v. So (2) is simplified to

$$T \ = \ F \ (E) \tag{4}$$

Under a Taylor type approximation, using Reiner's suggestions and taking into account that the strain matrix must satisfy its own Cayley-Hamilton equation, the previous equation assumes the simple form

$$T \ = \ a \ I + b \ E + c \ E^2 \tag{5}$$

So, the symmetry properties of the purely viscous simple fluid reduce the functional form to this simple quadratic tensorial function, a, b, c being functions of the three principal invariants of E.

Flow is an essentially irreversible transport process and its eventual mechanical simplicity concealsthe basic indeterminacy left to the Hamilton of the many body system that it physically represents. The respective statistical approach is made through time correlation functions. These ones display various symmetries, namely the time reversal one, that emerges at the condensed matter level on the form of the so called reciprocity relations, endowed with transport coefficients. Time correlation functions display also positivity properties, that emerge at the condensed matter level as strong ellipticity properties, conditioning nonlinear processes.

In the linear domain, time reversibility and the corresponding reciprocity relations assure the existence of a dissipation potential i.e. a simple structure for the purely viscous transport. Incompressible newtonian flow, power fluids, Ostwald-de Waele and Bingham fluids obey the tensorial equation (5) but, with the exception of the first, do not belong to the linear domain, i.e. they are both mechanically and thermodynamically nonlinear. The mathematical symmetry e.g. the common principal directions for T and E assure a higher hierarchy of reciprocity relations and so a thermodynamical admissibility.

DILATANT LAMINAR FLOW OF DILUTE POLYMER SOLUTIONS IN NON-UNIFORM VELOCITY GRADIENT FIELDS

G. Chauveteau, M. Moan°

Institut Français du Petrole, Rueil Malmaison, France

° Faculté des Sciences, Brest, France

(Abstract)

The molecular interpretation of dilatant behavior of dilute polymer solutions flowing through porous media is still a contro versial subject. A better understanding of this behavior has been obtained through an experimental study of various "complex" flows in which macromolecules are subjected to both shear and extensional stresses.

In all experimental conditions tested (in bi- and tridimensional flow through one or several successive constrictions and with various types of polymers, solvent viscosities and polymer concentrations), a strong extra pressure drop is observed beyond a critical deformation rate which: 1) is closely related to the relaxation time of the isolated macromolecule, 2) is nearly independant of polymer concentration in the concentration range tested, 3) doesn't depend on constriction number N. Moreover the viscous dissipation due to the polymer in converging flow is proportional to the concentration logarithm at a given N and increases with N up to an asymptotic value.

All these results suggest an elongation of macromolecules as soon as a critical stretch rate in converging flow region is obtained, inducing a high viscous dissipation. In geometries with successive constrictions, the macromolecule relaxation time is increased by stretching in the first constriction. Thus, a previously elongated molecule arrives in the second constriction, causing an increased pressure drop.

As a conclusion, dilatant behavior at high flow rates in porous media is explained by viscous dissipation on stretched molecules.

RHEOLOGY OF BRONCHIAL SECRETION: STRUCTURAL MODEL

ChoKyun Rha, Juan A. Menjivar and Elizabeth R. Lang

Food Materials Science and Fabrication Laboratory, Dept. of Nutrition and Food Science, Massachusetts Institute of Technology, Cambridge, Mass. 02139

RHEOLOGY AND FUNCTION OF BRONCHIAL MUCUS

The function of bronchial mucus depends on its rheological characteristics. Bronchial secretions, particularly those produced in disease and malfunction of the airways are often described in rheological terms, such as "the abnormally high viscosity of mucus" (Faillard and Schauer, 1972), "abnormally increased viscosity" (Denton, 1960), "the thick, tenacious material which fills and obstructs the bronchi" (Jakowska, 1968), "thick and tenacious and tend to form long strings or strands" (di Sant'Agnese and Davis, 1976), "lift ... 6 inches and would not break" (Braddock, 1969), and "increased mucus viscosity, producing an increase in both yield stress and flow resistance of the mucus and thus impairing mucociliary transport" (Sanchis et al., 1973). Clearly rheological properties are of predominant importance in bronchial mucus, yet bronchial mucus has not been the subject of complete rheological characterization. In addition, no consistent quantitative difference in a rheological parameter of bronchial mucus has been demonstrated when normal secretions are compared with those produced during pulmonary disease.

The mucociliary transport system consists of a cell layer containing both ciliated and mucus-secreting cells (Havez, 1977). The mucus is transported by the tips of the cilia over a layer of low viscosity interciliary fluid. The layer of mucus acts as a mechanical link between the energy supplied by ciliary beats and the translocation of particles (Florey, 1955; Silberberg et al., 1977) due to its characteristic viscoelastic properties.

The glycoproteins in mucus preferentially associate to form

753

a reversible gel-like phase of high density and limited hydration
ratio. The inherent viscoelastic nature of the gel-like associated
macromolecules is responsible for the function of bronchial mucus.
On the other hand an excessive consistency burdens the ciliary
transport impeding the function.

This study makes a theoretical examination of bronchial mucus
rheology based on the available biochemical information in order to
understand the rheological material functions and the rheological
mechanism of bronchial mucus.

Biochemical Properties of Mucus

Normal human mucus contains, by weight, about 1% salt and other
dialysable components, 1% free protein, 1% glycoproteins and 95% or
more water (Clamp, 1978). The glycoproteins in human bronchial mucus
vary in size, amino acid composition, sugar content, sulfation and
sialylation (Roberts, 1978). Only a fraction of this heterogeneous
class of molecules contain cysteine residues. The limited information on human bronchial glycoproteins available in the literature
indicate molecular weights ranging from approximately 500,000
(Lhermitte et al., 1976; Roussel et al., 1975) to a polydisperse
form of 1.3×10^6 to 3.0×10^6 (Davidson, 1978 and Roberts, 1974).
It is established that glycoprotein molecules can be fractionated
into two portions after proteolytic digestion, one containing all
the carbohydrate and most of the peptide backbone, and a small fraction containing peptides only (Lhermitte et al., 1976; Hatcher and
Jeanloz, 1974).

The bronchial mucus may be considered to be composed of glycoprotein units of about 400,000 MW. Thus glycoproteins contain approximately 500 to 700 residues in the peptide backbone, based on
a protein content of 15% (Roussel et al., 1975). Biochemical evidence suggests that this backbone is composed of repeating units of
50 to 100 amino acid residues (Bhushana-Rao and Mason, 1977;
Lhermitte et al., 1976; Pigman et al., 1970) each with a major portion to which the polysaccharide chains are attached at every third
or fourth residue and a smaller portion (up to one third) of polypeptide free of sugars. The polysaccharide-containing region will
be stiff while the polysaccharide-free region will be more flexible.

RHEOLOGY OF BRONCHIAL SECRETION

The summary of the rheological properties of bronchial secretions, tracheal mucus, and sputum reported in the literature (Table
1) indicates the fragmented and scattered nature of information currently available. Since bronchial secretion contains macromolecular
complexes, then shear sensitivity, time dependency and viscoelasticity are inherently expected, and indeed these have been shown
experimentally (Table 1). On the other hand, a comprehensive study

TABLE 1. RHEOLOGICAL PROPERTIES OF BRONCHIAL SECRETIONS

Sample	Rheological Parameter	Frequency, Shear Rate or Shear Stress	References
Bronchial Mucus Chronic Bronchitis	Intrinsic Viscosity = 810-880 ml/gm		White et al., 1959
Bronchial Mucus	Viscosity = 10-0.02 poise	Shear Rate = 0-18,000 1/sec	Charman and Reid, 1972
Bronchial Secretion	Viscosity = 70-0.5 poise	Shear Rate = 0.07-10,000 1/sec	Mitchell-Heggs, 1977
Bronchial Secretion	Charactersitic Time = 100 sec		Mitchell-Heggs, 1977
Sputum	Viscosity = 200-2 poise	Shear Rate = 0.02-10,000 1/sec	Mitchell-Heggs, 1977
Sputum	Dynamic Viscosity = 8000-10 poise	Frequency = 0.01-1.0 c/sec	Mitchell-Heggs, 1977
Sputum Asthma	Elastic Modulus = 0.34-2.0 dynes/cm^2	Frequency = 0.01-0.79 c/sec	Mitchell-Heggs, et. al., 1974
Chronic Bronchitis	Elastic Modulus = 0.60-2.09 dynes/cm^2	Frequency = 0.01-0.79 c/sec	Mitchell-Heggs, et al., 1974.
Bronchiectasis	Elastic Modulus = 0.18-0.66 dynes/cm^2	Frequency = 0.01-0.79 c/sec	Mitchell-Heggs, et al., 1974
Cystic Fibrosis	Elastic Modulus = 0.13-0.52 dynes/cm^2	Frequency = 0.01-0.79 c/sec	Mitchell-Heggs, et al., 1974
	Elastic Modulus = 0.54-6.85 dynes/cm^2	Frequency = 0.01-0.79 c/sec	Mitchell-Heggs, et al., 1974
Sputum	Viscosity = 23-2800 poise		Glover, 1954
Sputum	Creep	Shear Stress = 3.6 dynes/cm^2	Davis and Dippy, 1969
	Residual Elastic Component = 100 dynes/cm^2 Residual Viscous Component = 2x10^4 poise Retardation Time = 150 sec Elastic Component = 5.25 x dynes/cm^2 Viscous Component = 7.31 x 10^3 poise		
Sputum Chronic Bronchitis (reconst. 4% solids)	Shear Viscosity = 2-20 poise	Shear Rate = .05 sec^{-1}	Bornstein et al., 1978
Sputum Mucoid	Shear Viscosity = 1000 - 12000 poise	Shear Rate = .85 sec^{-1}	Chen and Dulfano, 1978
Mucopurulent	Shear viscosity = 1000 - 10000 poise	Shear Rate = .148 sec^{-1}	Chen and Dulfano, 1978
Purulent	Shear Viscosity = 2800 - 24000 poise	Shear Rate = .85 sec^{-1}	Chen and Dulfano, 1978
	Shear Viscosity = 2900 - 19000 poise	Shear Rate = .148 sec^{-1}	Chen and Dulfano, 1978

of rheological parameters which is required for the general understanding and the construction of an overall picture of flow behavior
of bronchial mucus is not available.

Structural Model

The structural model for mucus glycoproteins proposed by
Silberberg and coworkers (Meyer and Silberberg, 1978; Meyer, 1977)
satisfies most rheological considerations and current biochemical
information. The proposed model has bristly carbohydrate rich regions interconnected with the non-carbohydrate-coated third of the
peptide chain which is rolled into a ring stabilized by intramolecular disulfide bridges.

Theoretical Evaluation

A theoretical evaluation of the structure of bronchial secretions should consider the information contained in the biochemical
data as well as rheological functions. The biochemical evidence already discussed identifies features of a possible structural model
as follows:

1. The major axis of the 400,000 MW unit glycoprotein can be
represented by a fairly extended chain of 500 to 700 amino acid
residues. Assuming a fully extended peptide backbone of 600 residues,
the chain length is 2200Å for the unit glycoprotein, based on 3.6Å
per residue (Dickerson and Geis, 1969).

2. The minor axis of the glycoprotein can be considered to be
at maximum length, the sum of the width of the peptide plus two polysaccharide chains each consisting of 20 sugars (Roussel et al., 1975).
If the chain of 20 sugars is assumed to be 40Å long (Flory, 1953),
then the width of the glycoprotein is estimated at roughly 100 Å.

3. The approximately two thirds of the polypeptide chain to
which the sugar chains are attached, is dispersed between short bare
peptide segments which are flexible (Bettleheim and Block, 1968;
Hatcher and Jeanloz, 1974; Lhermitte et al., 1976). The bristle-
like structure of the carbohydrate rich segments causes extension of
the chain which will as a result be very stiff and posses limited
capability to entangle in solution.

Since a glycoprotein unit with molecular weight 2×10^{6} daltons
is generally accepted as the primary unit isolated from native mucus,
the subsequent rheological analysis is based on five glycoproteins
linked in a linear fashion. Based solely on the molecular dimensions of the main structural glycoprotein unit, a first approximation
to the rheological properties of native mucus has been made. The
calculated rheological properties for a 1% solution of the glycoprotein unit are shown in Table 2, where three alternatives are compared,

Table 2. Calculated Rheological Properties of Glycoprotein
(M.W. = 2×10^6) Based on Estimated Molecular Dimensions

	Fully Extended Rigid Rod	Semi-Rigid Elongated Macromolecule	Bead-Spring Chain
Viscosity (Poise) (1% glycoprotein)	19	—	—
Characteristic Time (Sec)	0.12	0.12	0.01
Elastic Modulus ($\frac{dyne}{cm^2}$)	1.1	0.47	0.02
Loss Modulus ($\frac{dyne}{cm^2}$)	11.9	5.6	1.4
Loss Tangent	11	12	95
Intrinsic Viscosity ($\frac{ml}{g}$)	23,000	—	4,000
Overlap Parameter	230	—	40

a rigid rod-like macromolecule (Kirkwood and Auer, 1951), a semi-rigid elongated molecule (Ferry, 1978) and a chainlike bead-spring macromolecule (Rouse, 1953; Zimm, 1956).

The calculations of viscosity and intrinsic viscosity [η] were made using Einstein-Simha's equation (Einstein, 1906, 1911; Simha, 1940) for the rigid model and Mark-Houwink-Sakaruda eq. (1) for the chainlike model using the K value for denatured proteins (Tanford, 1967) and a = 0.8 for maximum solvent-solute interaction.

$$[\eta] = KM^a \tag{1}$$

The rheological properties in Table 2 must be interpreted as the contribution of the size, asymmetry and flexibility of the glycoproteins, in the absence of any type of interaction except solvent-solute. The intrinsic viscosity values calculated for the most rigid and flexible glycoprotein units are 23,000 ml/g and 4000 ml/g respectively (Table 2). These values are 30 times and five times larger than those found for bronchial mucus for chronic bronchitis (White et al., 1959) (Table 1). For comparison it is interesting to note that sodium carboxymethylcellulose of equivalent molecular weight has intrinsic viscosity of approximately 66,000 ml/g.

The elastic and loss moduli G' and G" at 1 rad/sec, and the relaxation time (τ) of the 1% glycoprotein solution were calculated using the following equations (2,3, and 4) for a rod-like glycoprotein,

$$G' = (\frac{3cRT}{5M}) \, \omega^2\tau^2/(1 + \omega^2\tau^2) \tag{2}$$

$$G'' = \omega\eta_0 + (\frac{3cRT}{5M}) \, \omega\tau \, [1/(1 + \omega^2\tau^2) + 1/3] \tag{3}$$

$$\tau = \pi\eta_0 L^3/18kT \, \ln(L/d) \tag{4}$$

where c is the concentration of glycoprotein (~0.01 g/ml), M the molecular weight (2 x 10^6 daltons), ω the frequency, τ the relaxation time (sec), R the universal gas constant, T the absolute temperature, L the length of the glycoprotein (~11,000Å), d the diameter (~100Å) and k is Boltzmann constant. The equivalent equations were used for the semi-rigid and the chainlike bead-spring model (Ferry, 1978).

The elastic modulus at 1 rad/sec for the rigid and semi-rigid molecules are within the range of experimental data (Mitchell-Heggs, 1977), however the bead-spring model gave a value one order of magnitude lower than the lowest value reported in the literature (Table 1). There is virtually no data available for loss modulus of human bronchial secretion. However, the loss modulus calculated for a fully extended rigid rod is similar to, while those calculated for the other two chains are lower than, experimental data reported for canine tracheal mucus by Litt and Kahn (1974). Loss tangent for a fully extended rigid rod is nearly one order of magnitude less than that estimated for the bead-spring model indicating the higher elasticity inherent in more expanded rigid-chain model.

The characteristic times are 0.12 sec for both the fully extended rigid rod and semi-rigid elongated macromolecules and 0.01 sec for the bead-spring model. These are about three and four orders of magnitude, respectively, lower than that of bronchial secretion obtained by extrapolation of Mitchell-Heggs (1977) data. This difference could well be due to the concentration effect promoted by the large hydrodynamic volume and axial ratio of the glycoproteins therefore causing considerable constraint at 1% concentration in mucus. These calculations indicate that fully extended rigid conformation may be the most likely for glycoprotein and show that the glycoprotein unit by virtue of its geometry alone can exhibit viscosity and elastic modulus similar to that determined experimentally in bronchial secretion.

If indeed the glycoprotein units are so stiff and bulky and generate the viscoelastic properties due to geometry, and yet the function in the physiological system (King et al., 1974; Silberberg, 1977) mandates the cross-linking of these glycoprotein units, then what type of cross-linking would be able to hold these chains together? These extended and bulky glycoproteins would require cross-linking of considerable strength in order to maintain the chains in a network. If disulfide bonds are responsible for the network, as proposed, (Roberts, 1976; Allen, 1976; Bhushana-Rao and Masson; 1976), probably at least three or four disulfide bonds per junction would be required considering that even an unbranched linear polysaccharide chain such as alginate requires twenty or so calcium bonds per junction zone (Rees, 1969). Since there are at most only a few (Roussel et al., 1975; Lhermitte et al., 1976; Reid, 1978;

Cheng and Boat, 1976) sulfuhydryl groups in human bronchial mucus, the interactions primarily responsible for network structure are not likely to be the disulfide bonds alone, but rather more complex and flexible multitype secondary interactions which also involve a large intermolecular surface.

CONCLUSION

Theoretical analysis based on the biochemical information available indicates that the glycoprotein unit (M.W. 2×10^6) itself, solely by virtue of its dimensions, may exhibit rheological properties similar to those reported for bronchial secretions. The crosslinks leading to a network may involve a variety of weak intermolecular interactions, in a junction zone and not likely to be provided solely by disulfide bridges in human bronchial mucus.

REFERENCES

Allen, A., 1977, Structure and junction of gastric mucus, in: "Mucus in Health and Disease," M. Elstein and D.V. Parke, ed., Plenum Press, New York.

Bettelheim, F.A. and Block, A., 1968, Water vapor sorption of bovine and ovine submaxillary mucins, Biochim. Biophys. Acta, 165: 405.

Bhushana-Rao, K.S.P. and Masson, P.L., 1977, A tentative model for the structure of bovine oestrus cervical mucin, in: "Mucus in Health and Disease," M. Elstein and D.V. Parke, eds., Plenum Press, New York.

Bornstein, A.A., Chen, T-M. and Dulfano, M.J., 1978, Disulfide bonds and sputum viscoelasticity, Biorheol., 15:261.

Braddock, L., 1969, in: "Proceedings of the 5th International Cystic Fibrosis Conference," D. Lawson, ed., Cystic Fibrosis Research Trust, London, 384.

Charman, J. and Reid, L., 1972, Sputum viscosity in chronic bronchitis, bronchiectasis, asthma and cystic fibrosis, Biorheol., 9:185.

Chen, T.M. and Dulfano, M.J., 1978, Physical properties of sputum VIII. The effect of lyophilization and reconstitution on the viscosity of sputum, Biorheol. 15:269.

Chen, P.W. and Boat, T.F., 1976, Properties of a sulfated mucous glycoprotein from bronchial washings of a B-secretion of cystic fibrosis, Fed. Proc., 35:1444.

Clamp, J.R., Allen, A., Gibbons, R.A. and Roberts, G.P., 1978, Chemical aspects of respiratory mucus, Brit. Med. Bull., 34:39.

Davidson, E.A., 1978, cited in: "CIBA Found. Symp. 54," R. Porter, J. Rivers and M. O'Conner, eds., Elsevier, N.Y., 157.

Davis, S.S. and Dippy, J.E., 1969, The rheological properties of sputum, Biorheol., 6:11.

Denton, R., 1960, Bronchial obstruction in cystic fibrosis: rheo-
     logical factors, Pediatr., 26:611.
Dickerson, R.E. and Geis, I., 1969, "The Structure and Action of
     Proteins," Harper and Row, New York.
di Sant'Agnese, P.A. and Davis, P.B., 1976, Research in cystic
     fibrosis, N. Eng. J. Med., 295:481,534,597.
Einstein, A., 1906, Eine neue bestimmung der molekuldimensionen,
     Ann. Physik., 19:289.
Einstein, A., 1911, Berichtigung zu meiner arbeit: Eine neue Besti-
     mmung der molekuldimensionen, Ann. Physik., 34:591.
Faillard, H. and Schauer, R., 1972, Glycoproteins as lubricants,
     protective agents, carriers, structural proteins and as parti-
     cipants in other functions, in: "Glycoproteins," A. Gottschalk,
     ed., Elsevier, New York.
Ferry, J.D., 1978, Viscoelastic properties of dilute polymer solu-
     tions, Pure Appl. Chem., 50:299.
Flory, P.J., 1953, "Principles of Polymer Chemistry," Cornell
     University Press, Ithica, N.Y.
Florey, H., 1955, Mucin and the protection of the body, Proc. Roy.
     Soc. Lond. Ser. B., 143:147.
Glover, F.A., 1954, in: "Proc. 2nd. International Congress on
     Rheology", Butterworths, London.
Hatcher, V.B. and Jeanloz, R.W., 1974, Studies on cyansgen bromide
     fragments from monkey cervical glycoprotein, in: "Methodolo-
     gie de la Structure et du Metabolisme des Glycsuonjuges,"
     M.J. Montreiul, ed., Vol.1, Centre National de la Researche
     Scientific, Paris.
Hafez, E.S.E., 1977, Functional anatomy of mucus-secreting cells, in:
     "Mucus in Health and Disease," M. Elstein and D.V. Parke, eds.,
     Plenum Press, N.Y.
Jakorwska, S.,1968, "Cystic Fibrosis and Related Human and Animal
     Diseases," Gordon and Breach, N.Y.
King, M.,Gilboa, A., Meyer, F.A. and Silberberg, A., 1974, On the
     transport of mucus and its rheological stimulants in ciliated
     systems, Amer. Rev. Resp. Dis., 110:740.
Kirkwood, J.G. and Auer, P.L., 1951, The viscoelastic properties of
     solutions of rod-like macromolecules, J. Chem. Phys.,19:281.
Lhermitte, M., Lamblin, G., Lafitte, J.J., Rousseau, J., Degand, P.
     and Roussel, P., 1976, Properties of human neutral bronchial
     mucins after modification of the peptide or the carbohydrate
     moieties, Biochimie.,58:367.
Litt, M., Khan, M.A., Chakrin, L.W., Wardell, J.R. and Christian, P.,
     1974, The viscoelasticity of fractionated canine tracheal
     mucus, Biorheol., 11:111.
Meyer, F.A., 1977, Comparison of structural glycoproteins from mucus
     of different sources, Biochim. Biophys. Acta, 493:272.
Meyer, F.A. and Silberberg, A., 1978, Structure and function of
     mucus, CIBA Found. Symp., 54:203.
Mitchell-Heggs, P.F., Palfrey, A.J. and Reid, L., 1974, The elasti-
     city of sputum of low shear rates, Biorheol. 11:417.

Pigman, W., Payza, N., Moschera, J. and Weiss, M., 1970, Homologous
    substitutions in the primary structure of the core proteins
    of submaxillary and blood group glycoproteins, Fed. Proc.,
    29:599.
Rees, D.A., 1969, Structure conformation and mechanism in the forma-
    tion of polysaccharide gels and networks, in: "Advances in
    Carbohydrate Chemistry and Biochemistry, Vol. 24," Academic
    Press, N.Y.
Reid, L., 1978, General discussion IV, CIBA Found. Symp., 54:267.
Roberts, G.P., 1974, Isolation and characterization of glycoproteins
    from sputum, Eur. J. Biochem.,50:265.
Roberts, G.P., 1976, The role of disulfide bonds in maintaining the
    gel structure of bronchial mucus, Arch. Biochem. Biophys.,
    173:528.
Roberts, G.P., 1978, Chemical aspects of respiratory mucus, Brit.
    Med. Bull, 34:39.
Rouse, P.E., 1953, A theory of the linear viscoelastic properties
    of dilute solutions of coiling polymers, J. Chem, 21:1272.
Roussel, P., Lamblin, G., Degand, P., Walker-Nasire, E. and Jeanloz,
    R.W., 1975, Heterogeneity of the carbohydrate chains of
    sulfated bronchial glycoproteins isolated from a patient
    suffering from cystic fibrosis, J. Biol. Chem., 250:2114.
Sanchis, J., Dolovich, M., Rossman, C., Wilson, W. and Newhouse, M.,
    1973, Pulmonary mucociliary clearance in cystic fibrosis,
    New.Eng. J. Med., 288:651.
Silberberg, A., Meyer, F.A., Gilboa, A. and Gelman, R.A., 1977,
    Function and properties of epithelical mucus, in: "Mucus in
    Health and Disease," M. Elstein and D.V. Parke, eds.,
    Plenum Press, N.Y.
Simha, R., 1940, The influence of brownian movement on the viscosity
    of solutions, J. Phys. Chem., 44:25.
White, J.C., Elmes, P.C. and Whitley, W., 1959, Mucoprotein of bron-
    chial mucus gel, Nature, 183:1810.
Zimm, B.H., 1956, Dynamics of polymer molecules in dilute solution:
    viscoelasticity, flow bifrefringence and dielectric loss, J.
    Chem. Phys., 24:256.

ACKNOWLEDGEMENT

    This study was partially supported by NIH Biomedical Research
Support Grant No. 87804 from the Office of the Provost, M.I.T.

EFFECTS OF HIGH BONDING PRESSURE CYCLES ON THE STRENGTH OF ADHESIVE JOINTS

Ricardo Vera G. and Tomlinson Fort, Jr.

Instituto de Investigaciones en Materiales
Universidad Nacional Autónoma de México
Apdo. Postal 70-360, Cd. Universitaria, México 20,D.F.

INTRODUCTION.

The mechanical strength of an adhesive joint is related to both the degree of interfacial contact and the extent of shrinkage stresses arising from contraction of the adhesive with respect to the adherend during bond formation[1,2,3,4].

Several techniques have been developed to improve the degree of real contact between adhesive and adherend, among them is the use of pressure to force spreading of the adhesive[5,6]. However, the shrinkage stresses are considered to be an inherent characteristic of each system.

The effect of four different high bonding pressure cycles on joint strength for a thermoplastic adhesive bonded to a rigid adherend has been reported[4]. These bonding cycles were designed to study the influence of shrinkage stresses and the degree of interfacial contact on joint tensile strength. It was shown that under insovolume bonding conditions the shrinkage stresses can be eliminated concurrent with optimization of interfacial contact.

Increments in joint tensile strength up to 100 percent were obtained in glass-polystyrene butt joints under isovolume bonding conditions[4].

These studies have been extended to aluminum-epoxy and aluminum-polystyrene joints to test the advantages of isovolume bonding for both thermoplastic and thermosetting adhesives. This work also shows the effect of shrinkage stresses on adhesive layer thickness.

763

TABLE I

MEASURED PROPERTIES OF EPOXY RESIN

Property		
Volumetric contraction	4.3 %	
Density of cured resin, gr/cm^3	1.23	
Glass transition temperature,°C	130.0	
Tensile strength, Kg/cm^2	420.0	
Young Modulus, Kg/cm^2	1.1×10^4	

	Time, hrs.	
	25°C	60°C
Curing time	12.0	4.0
Gel effect	4.5	1.5
Minimun postcuring time	240.0	

EXPERIMENTAL

The adhesives used were high molecular weight amorphous poly_styrene (215,000), an eposy resin (AWlOG, araldite, Ciba Geigy and aluminum (type 6062-T6). The curing characteristics of the epoxy resin were determined by thermal analysis, Table 1. The aluminum was machined to the shape recommended by ASTM Procedure D1062-65 with a simple modification to control adhesive layer thickness. The bonding surface of the aluminum were polished up to CeO powder (Buehler procedure[7]) and treated chemically with chromic acid solution (ASTM D2651-67, Method A).

The adhesive joints were prepared in a specially designed mold following a procedure described elsewere[8].

Aluminum-polystyrene joints were prepared under two bonding conditions referred to as Cycles III and IV[4] to check the results obtained in the previous work[4]. The bonding cycles described below were designed for thermosetting adhesives and used to make the aluminum-epoxy joints.

Cycle V. To improve interfacial contact. After heating the samples to initiate the curing reaction (60°C), line AB in Figure 1a, a bonding pressure (360 Kg/cm^2) is applied, line BD. After a holding time (30 min.), line DE, the pressure is brought down (to 2 Kg/cm^2), line EF. Bonding takes place under this pressure while the temperature is allowed to drop naturally, line FC. After 12 hours the bonding pressure is released.

Cycle VI. To minimize shrinkage stresses and to improve interfacial contact. After heating the specimens to initiate the curing reaction, line AB in Figure 2b, a bonding pressure, 360 Kg/cm^2, is applied line BD. The pressure is reduced as a function of time so that the specific volume of the resin be the same during curing, line DC. After 12 hours, the joints are removed from the mold.

The dependence of pressure on time required for curing the epoxy resin under constant volume, Figure 2, was determined experimentally. The data shown in Figure 2 was obtained by trial and error in the following way. The curing contraction of the epoxy resin was determined at atmospheric pressure with the aid of the dilatometer of a thermo-mechanical analyzer, as shown in Figure 3. It was assumed that the decompression rate required for curing under isovolume conditions should be similar to the contraction rate of the resin at atmospheric pressure. A given amount of resin was placed a high pressure mold at 60°C and a initial pressure was applied. The pressure was then reduced at a rate equal to the contraction rate of the resin at atmospheric pressure. If the

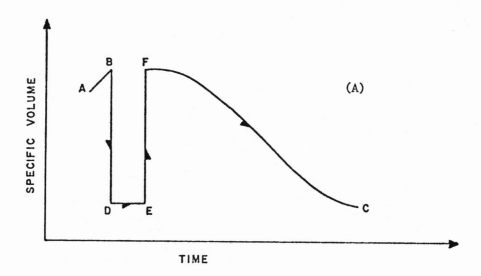

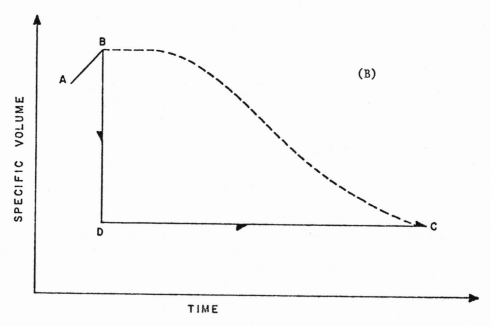

Figure 1.   High Bonding Pressure Cycles for Thermosetting Adhesives
            A) Shows Cycle V and   B) Shows Cycle VI

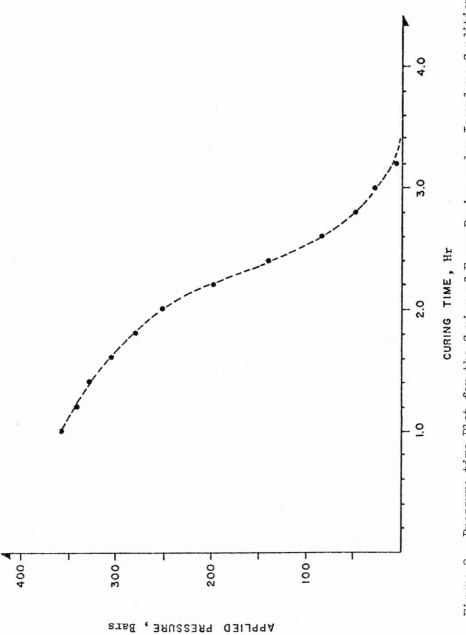

Figure 2.   Pressure-time Plot for the Curing of Epoxy Resing under Isovolume Conditions

applied initial pressure was different from zero after or before
the curing time limit, the experiment was repeated at another initial pressure. Pressure was measured directly with the aid of a
high pressure transducer.

The samples used as the control were prepared under a bonding
pressure of 2 Kg/cm^2.

The adhesive joints were tested in cleavage by following the
ASTM Procedure D1062-65 at a strain rate of 0.7 cm/min. All specimens were tested after a post-curing time of 10 days.

RESULTS.

A.   Aluminum-polystyrene. The cleavage strength of aluminum-
polystyrene joints are shown in Table 2. Isothermal compression-
decompression of the molten adhesive, Cycle III, increased interfacial contact and, thus, joint strength in 50 percent with respect to the control.

Bonding under isovolume conditions, Cycle IV, yielded a 130
percent increment in cleavage strength with respect to the control
and 56 percent with respect to Cycle III.

The mechanical behaviour of aluminum-polystyrene joints was
very similar to that observed in glass-polystyrene joints made under the same bonding cycles and tested in uniaxial tension; development of interfacial cracks produced by contraction of the polystyrene during bonding and similar increments in joint strengths.

B.   Aluminum-epoxy. The effect of the bonding Cycles V and VI
on the cleavage strength of aluminum epoxy joints was studied as
a function of adhesive layer thickness. The experimental results
are shown in Figure 4.

By forcing complete interfacial contact, Cycle V, the cleavage strength increased 50 percent with respect the control regardless adhesive layer thickness. The cleavage strength decreased
inversely with adhesive layer thickness for Cycle V and for the
control. On the other hand, the cleavage strength of joints made
under isovolume conditions, Cycle VI, is independent of adhesive
layer thickness. For thin adhesive layers (0.01 cm.) the difference in cleavage strength between Cycles V and VI is small while
for thicker layers (0.1 cm.) the difference is more than 100%.

The mechanical behaviour of joints made under Cycle V and
the control agrees well with that reported in other works[6,9,11].
However, the argument that the decrements in joint strength with
increasing layer thickness is due to the probability of increas-

TABLE 2

CLEAVAGE STRENGTH OF ALUMINUM-POLYSTYRENE JOINTS

	Control			Cycle III			Cycle IV	
Thickness, mm.	Cleavage Strength, Kg/cm.		Thickness mm.	Cleavage Strength, Kg/cm.		Thickness, mm.	Cleavage Strength, Kg/cm.	
0.219	5.0		0.198	6.0		0.236	15.0	
0.224	4.0		0.216	7.5		0.245	10.5	
0.230	7.5		0.220	10.5		0.250	13.0	
0.237	6.0		0.221	9.0		0.252	14.0	
		A v e r a g e						
0.227	5.62		0.217	8.4		0.245	13.1	

ing defects within the adhesive layer is incorrect. The results for Cycle V are a strong evidence against that argument. Non bonded sites and voids in the adhesive layer act as stress concentration risers that reduce joint strength[12,13]. But, in samples made under Cycle V non bonded sites and voids were completely minimized. Furthermore, the results for Cycle VI confirm that the effect of adhesive layer thickness is due to the shrinkage stresses, as explained in other works[9,12]; the shear stress caused by contraction of the adhesive during bond formation increases as adhesive layer thickness increases. As the shear stress increases joint tensile strength should decrease[8,14]. A similar behaviour is observed in joints tested in cleavage providing rigid adherends and semi-brittle adhesives[15], because under these conditions the load is perpendicular to the plane of the adhesive layer. The materials used in this work approach these conditions.

The degree of shrinkage stresses on cleavage strength was studied by making joints at different initial bonding pressures under the conditions of Cycle VI. As shown in Figure 5, cleavage strength increases initially with bonding pressure and then becomes independent of bonding pressure between 300 and 400 Kg/cm^2. The same effect was observed in glass-polystyrene joints made under iso-volume conditions.

CONCLUSIONS

The previous and the present works show that high pressures can be used to improve interfacial contact and to minimize shrinkage stresses in adhesive joints by means of proper bonding cycles. Increments in joint strength up to 50 percent can be achieved by optimizing interfacial contact. But increments up to 100 percent can be obtained when interfacial contact is optimized concurrent with minimizing shrinkage stresses.

The effect of adhesive layer thickness on joint strength is due to the shear stress generated by contraction of the adhesive during joint formation rather by the probability of finding defects within the adhesive layer. The experimental results confirm previous theoretical studies.

A simple method developed to determine the initial bonding pressure and the decompression rate required to cure a thermosetting adhesive under isovolume conditions also proved to be satisfactory.

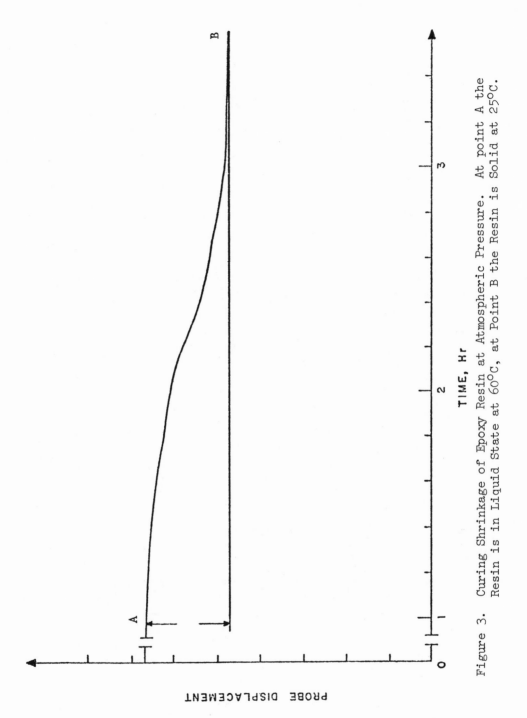

Figure 3. Curing Shrinkage of Epoxy Resin at Atmospheric Pressure. At point A the
Resin is in Liquid State at 60°C, at Point B the Resin is Solid at 25°C.

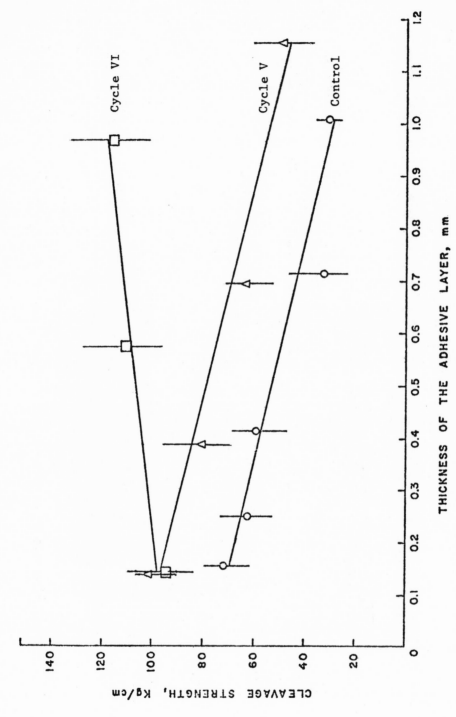

THICKNESS OF THE ADHESIVE LAYER, mm

Figure 4.   Cleavage Strength of Aluminum-Epoxy Joints as a Function of Adhesive
            Layer Thickness, for Cycles V, VI and the Control.

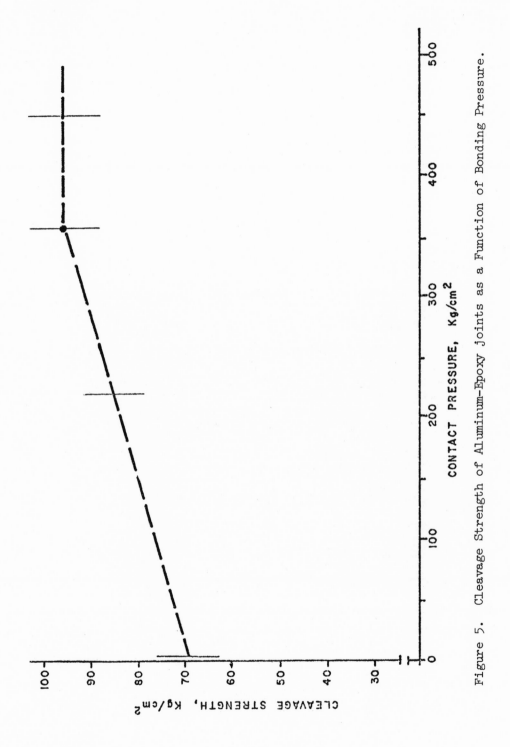

Figure 5. Cleavage Strength of Aluminum-Epoxy joints as a Function of Bonding Pressure.

REFERENCES

1. "Adhesion", D.D.Eley, ed., Oxford Univ. Press, London U.K.
    1961.

2. J.J. Bikerman, "The Science of Adhesive Joints", Academic
    Press, New York, N.Y., 1968.

3. A.N. Gent, J. Poly. Sci., A2, 9, 283 (1971).

4. R. Vera, E. Baer, T. Fort, Jr., J. Adhesion, 6, 357 (1974).

5. W.A. Zisman, Adv. Chem. Ser., 43, 1 (1964).

6. C.V. Cagle, "Adhesive Bonding Techniques and Applications",
    Mc Graw Hill, New York, N.Y., 1968.

7. Buehler Ltd., Metal Digest, 11, (2/3), 19 (1971).

8. Ricardo Vera G., "The Effects of High Bonding Pressures on
    Fracture of Adhesive Joints and Reinforced Polymers", Ph.D.
    Thesis, Case Western Reserve Univ., August 1975.

9. G. Kraus and J.E. Manson, J. Poly Sci., 6 (5), 625 (1950).

10. "Treatise on Adhesion and Adhesives", R.L. Patrick, ed.,
    Marcel Dekker, New York, N.Y., 1967.

11. W.A. Dukes and R.W. Bryant, J. Adhesion, 1, 48 (1968).

12. G.R. Irwin, "Structural Mechanics", J.N. Goodier and N.J.
    Hoft, eds., Pergamon Press, New York, N.Y., 1968.

13. J.L. Gordon, "Treatise of Adhesion and Adhesives".

14. E. Orowan, J. Franklin Inst., 290(6), 493 (1970)

15. D.H. Kaelble, "Treatise of Adhesion and Adhesives".

AUTHOR INDEX

775

SUBJECT INDEX

L971-3   720
          ———
          3/8/3